1/06

Environmental and Health Risk Assessment
and Management

ENVIRONMENTAL POLLUTION

VOLUME 9

Editors

Editorial Board

The titles published in this series are listed at the end of this volume.

Environmental and Health Risk Assessment and Management

Principles and Practices

by

Paolo F. Ricci

University of Queensland (NRCET)
Brisbane, Australia
and
University of San Francisco
San Francisco, California, USA

 Springer

Library of Congress Cataloging-in-Publication Data

ISBN-10 1-4020-3775-9 (HB)
ISBN-13 978-1-4020-3775-7 (HB)
ISBN-10 1-4020-3776-7 (e-book)
ISBN-13 978-1-4020-3776-4 (e-book)

Published by Springer,
P.O. Box 17, 3300 AA Dordrecht, The Netherlands.

www.springeronline.com

Printed on acid-free paper

CONTENTS

PREFACE

This textbook is about the law, economics, practical assessment, and the management of risky activities arising from routine, catastrophic environmental and occupational exposures to hazardous agents. The textbook begins where emission and exposure analysis end by providing estimates or predictions of deleterious exposures. Thus, we deal with determining the nature and form of relations between exposure and response, *damage functions*, and with the principles and methods used to determine the costs and benefits of risk management actions from the vantage point of single and multiple decision-makers. Today, national and international laws, conventions and protocols are increasingly concerned with reducing environmental and health risks through minimizing exposure to toxic substances, bacteria, viruses and other noxious agents. They do so through risk methods. The reason for the now worldwide use of risk assessment and management is that individuals and society must decide when, and at what cost, past and future hazardous conditions can either be avoided or minimized. In this process, society must account for the limited resources it can spend to remain sustainable.

Risk-based methods play a pivotal role in identifying and ranking alternative, sustainable choices, while accounting for uncertainty and variability. Specifically, most reductions in risks require a balancing of the costs and benefits associated with the action to reduce exposure to a hazard and thus risk. This balancing necessarily involves linking exposure and response through causation. This essential aspect of risk assessment and management, if done incorrectly, can be costly to society. Moreover, if an agency sets risk-based pollution standards on incorrect causation, such choice can be legally challenged. Fundamentally, in risk assessment and management, science and law intersect through legal and scientific causation to the point that the failure to provide a sound causal argument can make an otherwise beneficial law or regulation invalid.

A complex flow of heterogeneous scientific information characterizes risk assessment and management. That information is an input for setting environmental standards or guidelines mandated by health, safety and environmental statutes. Scientific information is also the *output* of scientific work of environmental, health and safety agencies that develop, gather, and analyze data and potential relations that contribute to establishing an acceptable environmental or health risk. Scientific causal arguments integrate that heterogeneous information into empirical causal structures. Risk-based causal arguments use physical laws and empirical relationships as building blocks. For example, a building block may consist of the biochemical reactions that occur between exposures to a virus in a cell. Another is the dose-response model.

Risks arise from routine events, such as from smoking or drinking contaminated water, as well as catastrophic events, such as the collapsing of a dam or a fire in a refinery. Those at risk are known individuals, such as when my child is scalded when entering the bathtub, or are statistical entities, such as the expected number of prompt deaths from the release of an infectious agent in a subway station, before that event occurs. Although the data and processes leading to risk of illness or death are different, risk assessors fundamentally use the same methods for the analysis of risks. Thus, regardless of whether we estimate cancer or asthma cases from indoor air pollution or we calculate the probability of failure of a pump with redundancies in a power plant, we rely on the methods discussed in this textbook.

Although we can reconstruct hazardous situations through risk assessment, after an adverse consequence has occurred, risk assessment and management are most useful when they are predictive. As private or public entities, we seek to predict future adverse events, their consequences and probable magnitude so that we can decide on how best to manage them. A preventive action is likely to create its own set of hazardous situations. The combination of risk assessment and management is an exciting, but often complicated, combination of art and science. In risk assessment, we use chemistry, physics, physiology, cell biology, mathematics, statistics, epidemiology and toxicology to develop a causal structure of the effect of exposure on adverse response. This structure is probabilistic and statistical. The assessment of risks combines analysis with judgment because it reflects technical and scientific choices, assumptions or defaults. In risk management, we link the results from risk assessment to decision-making principles and methods. We bring economic, social and perceptual factors to represent the seriousness of adverse outcomes and to select options that are optimal for the decision-maker. When dealing with benefits, costs measured in some monetary unit and health or environmental risks measured as failure rates, we also must account for uncertainty and variability. Risk management methods, from such disciplines as economics and decision sciences, formally describe the potential effect of private or public decisions intended to increase health and welfare of a country, region or area. This is exciting. The complications include:

- The combinatorial explosion of models, formulae, data sets and options to choose from

- Uncertain causal relations of relevance to risk management

- Many scientific questions are not yet answered by science with the accuracy that the stakeholders, including the decision-makers, need for acting with prudence

- Legal standards of proof and legal causation can be at odds with scientific standards of proof, causation and peer review

- Stakeholders can have different and conflicting values and objectives

- There can be undisclosed interests and motives behind using risk-cost-benefit analysis to justify public choices

This book is about these complications. Although some formulae involve elementary calculus, these formulae are inessential to the general understanding of the methods and processes discussed. A danger in writing this textbook is that its diverse topics are not developed at the level of detail that can satisfy specialists. In addition, others may seek a more rigorous treatment of statistical methods, while others still may want discussion of alternatives to cost-benefit analysis, such as bounded rationality. We can only suggest that dealing with heterogeneous data and diverse methods of analysis in a single and manageable source requires simplifications and places limits on form and content. We focus on applications and examples: analytical rigor and proofs concede to them. The form and content adopted consist of providing the essential elements of the process of risk assessment and management at a level of detail that is appropriate for third or fourth year students and first year graduate students.

TO THE READER

My intent is to provide a set of practical discussions and relevant tools for making risky decisions that require actions to minimize environmental or health risks. Risks are probabilistic quantities or hazard rates; it follows that using them is synonymous with uncertainty and variability. Moreover, there should be a defensible causal relation between exposure and the probability of adverse response. This involves statistical analyses with experimental and non-experimental data. I have simplified the statistical and probabilistic material to make it useful to environmental and health risk assessors *and* managers. Some readers may question our references to works published early in the history of the subject. I do this for three reasons. The first is that those early references deal most eloquently with the fundamental issues that risk assessors and managers deal with routinely. The second is that those early references are available in most libraries. The third is that advances in the state-of-the-art belong to textbooks that are either more theoretical or that deal with technical or legal topics in detail. Quotations are given either in italics or as separate paragraphs, depending on their length.

ACKNOWLEDGMENTS

Many theoretical and practical insights in risk assessment and management are due to L. A. "Tony" Cox, Jr. His works, some of which we have collaborated on over more than two decades, continues to shape my thinking; Chapter 10 includes his contributions on the Fundamental Causal Diagram, principles and methods. Over these past years, others have also enriched my understanding of risk assessment and management. Richard Wilson steered me towards this field at the Kennedy School of

Government, Harvard University. Others include Andrew Penman, at the NSW Health Department (Australia), Richard Zeckhauser at Harvard, and George Apostolakis, now at MIT, who contributed in different ways to my thinking about the public health, economic, social statistical and mathematical aspects of the disciplines. Alan Neal, at Leicester University, and Frank Wright, at Salford University, were instrumental in explaining European Union and UK law relevant to risk assessment, health and safety in graduate law school at Leicester University, UK. I thank Tom MacDonald, University of San Francisco, for corrections made while reviewing some of the chapters of this textbook. Tom Beer, CSIRO Australia, found calculation errors, corrected them and provided useful comments to earlier drafts. Alan Hubbard, at UC Berkeley, reviewed and clarified many of the statistical discussions. Finally, at least three graduate classes have read initial drafts of the entire textbook. Their comments have changed and restructured arguments, simplifying prose and examples.

DEDICATION

To Andrea, Alexander, and Luke.

CHAPTER 1.

LEGAL PRINCIPLES, UNCERTAINTY, AND VARIABILITY IN RISK ASSESSMENT AND MANAGEMENT

Central objective: to describe key legal principles of precaution, their relation to assessment and management, and a framework for dealing with uncertainty and variability

In this chapter we review:

- Uncertainty and variability as used in risk assessment and management
- Probabilistic reasoning for risk assessment and management, with Bayesian reasoning as a means to combine expert judgments with experiment
- The legal basis for managing risks in national and international jurisdictions, such as Australia, Europe and the United States
- The use of scientific evidence that relates to risks in legal causation, a required component in regulatory (administrative), environmental and health and safety laws

Today, risk assessment and management principles permeate environmental, health and safety statutes, secondary legislation and legal cases of many countries. These include the US, Australia and the European Union. Specifically, national and international laws that deal with environmental and health protection require risk assessment, management and cost-benefit analysis to rank and to guide the selection of optimal choices. The concerns of those who regulate hazardous conditions span a wide range of human and natural activities. These range from technological choices for reducing emissions, from cleaning-up contaminated sites to eradicating communicable disease, from providing scientific information when drafting laws and legal instruments (e.g., directives, guidelines, and standards) to informing the public. For instance, EPA's risk characterization policy (US EPA, 2000) states that:

"Each risk assessment prepared in support of decision making ... should include a risk characterization that ... should be ... clear, transparent, reasonable, and consistent ... (and) will depend upon the information available, the regulatory application of the risk information, and the resources (including time) available. In all cases, however, the assessment should identify and discuss all the major issues ... and provide commentary on any constraints limiting fuller exposition."

Understanding the methods of risk assessment and management, with a comparable understanding of how the law looks at scientific evidence and causation, is useful to risk assessors and managers. The methods of risk assessment consist of models that describe and predict how potential sequences of events, resulting from human or natural failures, can lead to exposure, while accounting for the magnitude and severity of the consequences. Exposure events that lead to risk range from the routine release of carcinogens from trucks operating daily to the catastrophic, such as the failure of a dam or a fire in a refinery. The methods of risk assessment and management can support arguments for and against making scientific predictions and assigning probabilities to those predictions. This may seem puzzling. Let us explore why it is not. The Third Assessment Report (Allen, Raper and Mitchell, 2001) provides climate change predictions for the period 1990 to 2100. The climatic changes due to human activities are asserted to result in temperature changes ranging from *1.4* to *5.8* ^0C. However, there is no estimation of the uncertainty in those predictions because (Allen, Raper and Mitchell, 2001). They state that:

"It was the unanimous view of the TAR lead authors that no method of assigning probabilities to a 100-year climate forecast is sufficiently widely accepted and documented in the refereed literature to pass the extensive IPCC review process."

Some of the reasons for not explicitly dealing with uncertainty include (Allen, Raper and Mitchell, 2001):

"The difficulty of assigning reliable probabilities to socioeconomic trends (and hence emissions) in the latter half of the 21st century, … , and the possibility of nonlinear response in the carbon cycle or ocean circulation to very high late 21st century greenhouse gas concentrations."

Yet, Wigley and Raper (2001), in the same issue of *Science*, do develop probabilistic estimates of these changes. Their *90%* probability interval is *1.7* to *4.9* ^0C. Their analyses use probability distributions for the parameters of physical processes governing the relationship between human activities and changes in temperature, unlike TAR. Both papers were published in *Science* and thus passed *scientific* peer review. The question is: What is an appropriate and adequate representation of uncertainty, variability and cause and effect such that risk management interventions can be made with *more* confidence than would be the case without such representation? This question has four parts. These are:

- Are there appropriate *representations* of uncertainty and variability; and, if so, what are they?
- How do the managers know that the chosen representations are adequate for their purposes?
- How are *empirical* cause and effect modeled?
- What are the *net benefits* to making decision using these representations?

This chapter lays the basis for answering these questions. The chapters that follow provide the answers. Finally, it might seem that the connection between probabilistic reasoning and legal reasoning is remote. As discussed in this chapter, that connection is immediate. The practical position we take is that risk assessment and management are based on probabilities that can be *attached* to the magnitude of the adverse consequences as well as to the net cost or benefits of reducing a specific hazard. We make this premise to set the stage for answering an important question posed by environmental managers confronting risky choices.

The principal elements of this chapter are connected as follows.

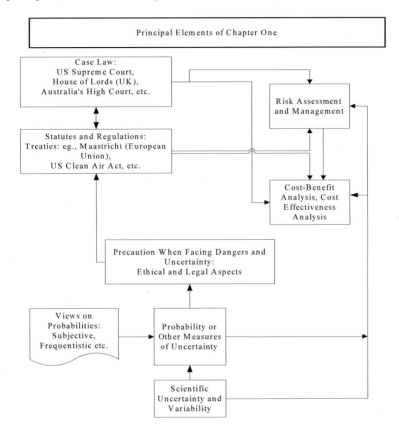

1.1. INTRODUCTION

Intuitively and practically, a *risk* is the probability of an adverse effect of a specific magnitude. For example, the risk of prompt death due to a specific choice can be stated as *1%* per year, above background. This means that there is a *0.01* probability of death per year of exposure to that activity.[1] Intuitively and mathematically, the probability of surviving that particular hazard is *0.99* per year. If the activity is beneficial, the value of the benefit can be expressed in monetary units (dollars, yen, or other unit of value) or some other unit, such as *utility*. Clearly, there will be several combinations of probability and magnitude. All of them, in the appropriate order, form the distribution of the magnitude of a particular adverse outcome. For example, for the same type of activity, there can be *0.10* probability of *two* prompt

---

[1] A probability, *pr*, is a number in the closed interval *[0 $\leq pr \leq$ 1]*.

deaths, *0.05* probability of *five* prompt deaths, and so on, until the probabilities of those (mutually exclusive events) sum to *1.00*. The reason for summing the probabilities is that these adverse outcomes (*two* prompt deaths, *five* prompt deaths and so on, for the same type of activity) are mutually exclusive. The probabilistic view of risks emphasizes distributions rather than deterministic numbers. Emphasis on using single numbers, the deterministic view of decision-making, can lead to gross over- or under-evaluations of the magnitude of costs, benefits and risks of a decision. Moreover, such emphasis can result in rankings of actions that are incorrect. When this happens, society is placed at a disadvantage relative to what it could achieve; some suffer unduly while others benefit without merit.

Environmental choices are often characterized by incomplete scientific knowledge. Practically, those choices involve scientific conjectures, incomplete, inconsistent and possibly incorrect data. This forecloses using single numbers or deterministic predictions with confidence. To wait for science to provide the certainty inherent to single numbers, if it could, may be neither practical nor sufficiently precautionary. Thus, the *principal* question that confronts risk assessors and managers is: What measures of uncertainty and what causal analysis can improve the management of potentially severe, irreversible or dreaded environmental outcomes?

The answer to this question can be difficult because it involves an adequate resolution of the difficulties inherent to modeling uncertain causation. Those difficulties have lead to many environmental law principles that set the boundaries of what should be done, but do not provide an operational construct to answer our question. Those principles, ranging from the *precautionary principle* to protecting human health from *a significant risk of material health impairment*, do not explain how to make management choices when incomplete, inconsistent and complex scientific evidence characterizes probable adverse outcomes. Rather, they pass the task to lower jurisdictions. To achieve the goals of principles of precaution, those who appeal to it as a justification for action must deal with scientific causal conjectures, partial knowledge and variable data. Managing hazards under the precautionary principle requires inductive, empirical methods of assessment. However, acting on a scientific conjecture can be socially unfair, costly, and detrimental when applied to complex environmental choices in which the stakes are serious or imply irreversible impacts on the environment.

1.1.1. Precautionary Acts as an Answer to the Principal Question

An early example of legal precaution is Justinian's statement, in 527 AD, that *the maxims of the law are to live honestly, to cause no harm unto others and to give everyone his due*. These maxims are a basis of the modern proportionality and due process principles. The proportionality principle is a heuristic balancing of social interests where an economic activity can be prohibited if the means to do so are appropriate and necessary. On this basis, the least costly alternative is the legally preferable policy choice, as was held in *Fromancaise SA* v. *F.O.R.M.A*, (1983), by European Court of Justice, and in the U.K. in *R.* v. *Minister for Agriculture, Fisheries and Food, ex parte Fedesa* (1991).

The essence of the precautionary principle is that the absence of scientific certainty about potentially serious or irreparable environmental damage does not prevent, by itself, a costly intervention to either mitigate or eliminate that hazard. This principle is both anticipatory and conservative. Specifically, the precautionary principle stands for the proposition that *when there are threats of serious or irreversible damage to the environment, scientific uncertainty should not prevent prudent actions to be taken to prevent potential damage.* An important view on this principle is that taking precautionary actions on behalf of society should be made without considering economic costs and benefits. Those advocating this view focus on either the magnitude or the severity of the consequences of actions not taken. Scientific uncertainty and costs should not prevent a precautionary act under egregious circumstances. Others believe in the opposite proposition. Their view is that precautionary decisions must be guided by balancing risks, costs and benefits that result in the ranking of choices guided by economic efficiency, broadly understood to be based on welfare economic principles.

Although there are different legal interpretations of the precautionary principle, it is the instrumental aspects of the principle that can create legal and practical difficulties, not the principle itself. Acting with prudence and anticipation goes beyond the legal norm *primum non nocere* (Justinian's *first cause no harm onto others*) by imposing a positive and anticipatory duty to act in the face of scientific uncertainty. That is, potentially catastrophic, cumulative or other types of hazard have led to different legal principles of precaution, such as *zero risk, significant risk* and others. In this chapter, we are concerned with the representation of uncertain facts and empirical causation, to make operational the precautionary principle. These concerns are part of the controversies that surround the implementation of the principle and its variants. An important aspect of the controversies arises from the difficulties of determining the magnitude, equitable distribution, severity and probability of the impacts of human or natural events on the environment and health.

1.2. THE PRECAUTIONARY PRINCIPLE

Legislation and legal oversight of administrative processes through judicial or other forms of checks and balances establish different statements of principles of precaution. Those principles are generally interpreted by agencies that develop secondary legislation, such as Directives or Regulations in the European Union. Administrative processes lead to the actual numerical standards or guidelines that we encounter in our daily lives. These can be issued as numbers to achieve the appropriate level of environmental protection, after scientific peer review and public comments, which are legally enforceable. The process can be reviewed at the administrative levels (through the *exhaustion of remedies*), often moving through the courts or even be remanded to the legislature for reconsideration. As an instance of such supervision, in American federal environmental law, some key statutes have a clause that requires Congressional re-authorization after some specified period of time, e.g., five years. In the European Union, a new Treaty can contain re-written Articles or have Articles deleted from the previous Treaty. These administrative and legal processes are lengthy and complex, with the potential for inaccurate, inefficient and costly outcomes. A principal reason for this potential, as the Comptroller General of the United States stated in 1979, is that:

"Major constraints plague the Environmental Protection Agency's ability to set standards and issue regulations. The most important factor is the inconclusive scientific evidence on which it must often base decisions. Numerous court suits result."

This statement and its warning are particularly cogent for actions to be taken under the principles of precaution discussed in this chapter because acting with precaution is anticipatory and is thus related to uncertain outcomes. At the international level, perhaps the best-known statement of the precautionary principle is found in (the legally *unenforceable*) Principle 15 of the United Nations Framework Conference on Environment and Development (1992). It states that:

"Where there are threats of serious or irreversible damage, lack of full scientific certainty shall not be used as a reason for postponing cost-effective measures to prevent environmental degradation."

This goal of prudently allocating scarce resources implies that an open invitation to act without understanding causation and balancing risks, costs and benefits can be as damaging as not acting. This is probably why Principle 15 does not state that *cost-benefit* analysis is inappropriate to rank environmental choices. The realization that public choices are at least determined by a net social benefit calculation and a ranking of those choices pervades most versions of the precautionary principle. Let us consider the European Treaty of Union, Maastricht 1992, Article 130r (Title XIX, now renumbered as Art. 174, which is legally enforceable) states the legislative mandate for a policy of environmental protection as:

"2. Community policy on the environment shall aim at a high level of protection ... [and] shall be based on the precautionary principle ...

3. In preparing its policy on the environment, the Community shall take account of:

(i)   available scientific and technical data;
(ii)   environmental conditions in various regions of the Community;
(iii)   the potential benefits and costs of action or lack of action;
(iv)   the economic and social development of the Community as a whole and the balanced development of its regions."

This Treaty also contains language about *sustainable growth, quality of life, environmental protection* and *economic and social cohesion.* It adds Article 130(u), which states that the policy on development *shall contribute to the general objective of developing and consolidating democracy and the rule of law, and that of respecting human rights and fundamental freedoms.*

Furthermore, Article 130(r)(1) states that the European Community environmental protection objectives include:

(i)   "preserving, protecting and improving the quality of the environment;
(ii)   protecting human health;

(iii)   prudent and rational utilization of natural resources and

(iv)   promoting measures at international level to deal with regional or worldwide environmental problems."

Acting notwithstanding uncertainty is implicit in (i), causation is explicit in (ii) and *rational* decision-making is explicit in (iii). An unspecified measure of risk aversion is explicit in (iii) through the term *prudent*. The European Union's, EU, precautionary principle points to decisions justified by risk, cost and benefit analysis because making rational decisions requires efficient choices based on the balancing of *the potential benefits and costs of action or lack of action*. In the EU, environmental damage *should be as a priority rectified at source*, as held by the European Court of Justice in the *Walloon Waste* case (Case 2/90). Other jurisdictions have adopted the precautionary principle. For example, in the case *A. P. Pollution Control Board* v. *Nayudu*, the Supreme Court of India stated the precautionary principle as follows:

"The principle of precaution involves the anticipation of environmental harm and taking measures to avoid it or to choose the least environmentally harmful activity. It is based on scientific uncertainty. Environmental protection should not only aim at protecting health, property and economic interest but also protect the environment for its own sake. Precautionary duties must not only be triggered by the suspicion of concrete danger but also by (justified) concern or risk potential"

Importantly to the ends of the principle, this Court also shifted the burden of proof from the plaintiff to the defendant. It stated that *the party ... maintaining a less-polluted state should not carry the burden of proof and the party who wants to alter it must bear this burden*. This shift is a major policy shift from traditional civil law where the plaintiff bears the burden of proving that the act or omission of the defendant caused the harm allegedly suffered by the plaintiff. Providing that level of proof requires scientific evidence of causation and the formal analysis of uncertainty (Ricci and Molton, 1981; Ricci and Gray, 1998).

1.3. VARIANTS OF THE PRECAUTIONARY PRINCIPLE: US law

US law contains several variants of the precautionary principle. Consider the statement that (US EPA, 2004) the *role of the US EPA is adequately to protect health, a goal that is assisted by risk assessments based on data and methods that, ideally, would not underestimate risks.* However, as the US EPA states (US EPA, 2004), even the very statement of goals is ambiguous:

"because there are many views on what 'adequate' protection is, some may consider the risk assessment that supports a particular protection level to be 'too conservative' (i.e., it overestimates risk), while others may feel it is 'not conservative enough' (i.e., it underestimates risk)."

As an example of the language and intent of Congressional mandates that govern the operations of the EPA (US EPA, 2004) variously state that:

"a) In the case of threshold effects an additional ten-fold margin of safety for the pesticide chemical residue shall be applied for infants and children ... (Federal Food, Drugs, and Cosmetics Act, FFDCA §408 (b)(2)(C))b),

b) The Administrator shall, specify, to the extent practicable: 1) Each population addressed by any estimate of public health effects; 2) The expected risk or central estimate of risk for the specific populations; 3) Each appropriate upper-bound or lower-bound estimate of risk ... (Safe Drinking Water Act, SDWA § 300g-1 (b)(3)),

c) The Administrator shall ... [add] pollutants which present, or may present, through inhalation or other routes of exposure, a threat of adverse human health effects ... or adverse environmental effects through ambient concentrations, bioaccumulation, deposition, or otherwise but not including releases subject to regulation under subsection (r) of this section as a result of emissions to air... (Clean Air Act, CAA §112(b)(2)),

d) Provide an ample margin of safety to protect public health or to prevent an adverse environmental effect (Clean Air Act, CAA §112(f))."

This issue is further complicated by the language of individual statutes (e.g., the Federal Fungicide, Insecticide, and Rodenticide Act, FIFRA; the Clean Water Act, CWA and so on) as to the appropriate level of protection (US EPA, 2004):

"a) To assure chemical substances and mixtures do not present an unreasonable risk of injury to health or the environment (the Toxic Substances Control Act, TSCA §2(b)(3)).

b) Function without unreasonable and adverse effects on human health and the environment (FIFRA §3).

c) Necessary to protect human health and the environment (Resources Conservation and Recovery Act, RCRA §3005 as amended).

d) Provide the basis for the development of protective exposure levels (NCP §300.430(d)).

e) Adequate to protect public health and the environment from any reasonably anticipated adverse effects (Clean Water Act, CWA §405(d)(2)(D))."

Even the language of a single statute, the Clean Air Act, used within the EPA's Office of Air and Radiation (OAR), shows differences, depending on the section of the Act:

a) Protect public health with an adequate margin of safety (Clean Air Act, CAA §109).

b) Provide an ample margin of safety to protect public health or to prevent an adverse environmental effect (OAR; CAA §112(f)).

c) Protect the public welfare from any known or anticipated adverse effects (OAR; CAA §109).

d) [Not] cause or contribute to an unreasonable risk to public health, welfare, or safety (OAR; CAA §202(a)(4)).

The 1990 CAA Amendments direct EPA to consider risk to the *individual most exposed* (IME) when determining whether a source category may be deleted from the list of sources of hazardous air pollutants, HAPs, (§112(c)(9)(B)(i)), and when determining whether residual risk standards are necessary (§112(f)(2)(A)). Additionally, §112(f)(2)(B) uses a two-step risk assessment for setting residual risk standards, as required under §112(f)(2)(A). (However, EPA has not yet proposed residual risk standards, but is working on a number of residual risk determinations and expects to issue the first proposal in 2004.) The two-step risk framework was articulated in EPA's 1989 Benzene NESHAP (Federal Register, 54: 38044) is as follows (US EPA, 2004):

"1) A first step, in which EPA ensures that risks are 'acceptable.' As explained in the Benzene NESHAP, in this step EPA generally limits the maximum individual risk, or 'MIR,' to no higher than approximately 1 in 10 thousand. The benzene NESHAP defines the MIR as the estimated risk that a person living near a plant would have if he or she were continuously exposed to the maximum pollutant concentrations for a lifetime (70 years).

2) A second step, in which EPA establishes an 'ample margin of safety.' In this step, EPA strives to protect the greatest number of persons possible to an estimated individual excess lifetime cancer risk level no higher than approximately 1 in 1 million. In judging whether risks are acceptable and whether an ample margin of safety is provided, the benzene NES In judging whether risks are acceptable and whether an ample margin of safety is provided, the benzene NESHAP states that EPA will consider not only the magnitude of individual risk, but 'the distribution of risks in the exposed population, incidence, the science policy assumptions and uncertainties associated with the risk measures, and the weight of evidence that a pollutant is harmful to health.' Therefore, decisions under §112(f)(2)(A) typically will include consideration of both population risk and individual risk, as well as other factors. Note that the ample margin of safety analysis in the second step of the residual risk framework also includes consideration of additional factors relating to the appropriate level of control, 'including costs and economic impacts of controls, technological feasibility, uncertainties, and any other relevant factors.'"

As the US EPA (2004) summarizes, another statute, CERCLA,[2] requires that actions selected to remedy hazardous waste sites be protective of human health and the environment. The National Oil and Hazardous Substances Pollution Contingency Plan, NCP, implements CERCLA (US EPA, 2004). The NCP establishes the overall approach for determining appropriate remedial action at Superfund sites with the national goal to select remedies that *are protective of human health and the environment,* and *that maintain protection over time.* One of the policy goals of the Superfund program is to protect a high-end, but not worst case, individual exposure: the reasonable maximum exposure (RME). The RME is the highest exposure that is reasonably expected to occur at a Superfund site, is conservative, but within a realistic range of exposure (US EPA, 2004).

The EPA defines the *reasonable maximum* such that only potential exposures that are likely to occur will be included in the assessment of exposures. In addition to RME individual risk, EPA evaluates risks for the *central tendency exposure* (CTE) estimate, or average exposed individual. EPA received comments implying that the RME represents a *worst-case* or *overly conservative exposure estimate.*Thus, the RME is not a worst-case estimate -- *the latter would be based on an assumption that*

---

[2] 42 U.S.C. 9605 (CERCLA, 1980), Superfund Amendments and Reauthorization Act of 1986 (SARA), P.L. 99-499 (SARA, 1986).

the person is exposed for his or her entire lifetime at the site. As the preamble to the NCP states (US EPA, 2004):

"The reasonable maximum exposure scenario is 'reasonable' because it is a product of factors, such as concentration and exposure frequency and duration, that are an appropriate mix of values that reflect averages and 95th percentile distribution ... The RME represents an exposure scenario within the realistic range of exposure, since the goal of the Superfund program is to protect against high-end, not average, exposures. The "high end" is defined as that part of the exposure distribution that is above the 90th percentile, but below the 99.9th percentile."

1.3.1. Implications of Different Approaches to Precaution

The implications of these different definitions and interpretations can result in lengthy and costly litigation. For instance, the controversy about the ambient ozone standard under the federal Clean Air Act (CAA) exemplifies a risk aversion that is similar to the precautionary principle. The US EPA had set the *primary* standard for ozone at *0.12* [ppm/hour], not to be exceeded more than once per year. Adverse effects were found at exposures to concentrations between *0.15* and *0.25* [ppm/hour]. However, a stakeholder, the American Petroleum Institute, sued the US EPA, resulting in the case *American Petroleum Institute* v. *Costle*. A key issue in this litigation was whether the magnitude of the factor of safety adopted by the agency to set the ozone ambient standard was *adequate* to protect members of the general population and particularly sensitive individuals. The court discussed safety factors as being appropriate when some individuals are more susceptible to injury than the rest of the population and when science demonstrates that there is no *clear threshold* in the exposure-response function linking ozone exposure to adverse response. The court then held that the US EPA had acted properly.

Example 1.1. Suppose that there are no adverse effects in an experiment at dose of *10* parts per million, ppm. Suppose that the environmental agency opts for a factor of safety that equals *100*. The adequately protective dose is calculated as *10/100 = 0.01*[ppm].

In another example of legal precaution involving occupational exposure to benzene, the US Supreme Court, in the case *Industrial Union Dept.* v. *American Petroleum Institute,*[3] placed the burden on OSHA (the US Federal Occupational Health and Safety Administration) to show (Ricci and Molton, 1981) that *on the basis of substantial evidence, that it is at least more likely than not that long-term exposure to 10 ppm of benzene presents a significant risk of material health impairment.*

The prevailing standard of proof in tort law is *the preponderance of the evidence*, understood to amount to the balancing of the evidence presented up to the point where *51%* of that evidence (proffered by one party) overcomes the remaining *49%* (proffered by the antagonist). In the US, however, the admissibility of epidemiological or other medical evidence can be governed by the less stringent *reasonable medical certainty* standard. In determining whether substantial scientific evidence supported the agency's decision, it was enough *that the administrative*

---

[3] 100 S. Ct. 2844 (1980).

record contained respectable scientific authority supporting the agency's factual determinations. In other words, procedures become critical: if the record is complete, it can meet the administrative law *substantial evidence* test, used in reviewing OSHA's regulatory action. The US Supreme Court stated that:

"It is the Agency's responsibility to determine, in the first instance, what it considers to be 'significant' risk. ...OSHA is not required to support its findings that a significant risk exists with anything approaching scientific certainty... Thus, so long as they are supported by a body of reputable scientific thought, the Agency is free to use conservative assumptions in interpreting the data with respect to carcinogens, risking error on the side of overprotection rather than underprotection."

The US Supreme Court then held that the OSHA may issue regulations *reasonably necessary* to reduce *significant risk* of *material impairment; but these regulations should be technologically and economically feasible.*

The case *Ethyl Corp.* v. *US Environmental Protection Agency* (1976) reminds us of the European Union precautionary principle. In this case, the legal theory is that *some* scientific evidence can be sufficient for environmental regulation when there is significant risk. This case involved the EPA's phasing-out of lead in petrol because such fuel additive resulted in a *significant risk of harm* to those exposed to it. The majority opinion of the court stated that:

"... more commonly, 'reasonable medical concerns' and theory long precede certainty. Yet the statutes – and common sense – demand regulatory action to prevent harm, even if the regulator is less than certain that harm is otherwise inevitable."

That opinion also stated that:

"Where a statute is precautionary in nature, the evidence is difficult to come by, uncertain, or conflicting because it is on the frontier of scientific knowledge, the regulation designed to protect...we will not demand rigorous step-by-step proof of cause and effect."

Reserve Mining Co. v. *EPA* (1975) exemplifies yet another aspect of a precautionary approach developed in US federal law. This case dealt with a mining activity that discharged asbestos material to the Great Lakes water and to the air. Those emissions could result in the ingestion of carcinogenic fibres by nearby residents, in the city of Duluth. The US EPA sought an injunction against that waste disposal activity by the company Reserve Mining to avoid what the EPA considered an immediate danger of cancer from asbestos exposure. The court held that the discharges and cancer were related through an *acceptable but unproved medical theory* of carcinogenesis. The court, which had hired an independent expert to determine the distribution of asbestos, stated that:

"In assessing probabilities in this case, it cannot be said that the probability of harm is more likely than not. Moreover, the level of probability does not readily convert into predication of consequences. ... The best that can be said is that the existence of this asbestos contaminant in air and water rises to a reasonable medical concern."

Regardless of the uncertainties about scientific causation, which were well understood by the court, it held that the contamination should be removed.

The CAA §112 deals with a relative large number of air toxicants, such as carcinogens, and is health-based. Specifically, the 1990 amendments to the CAA, which are stated in §112, list 189 hazardous pollutants. The US EPA must develop emission standards based on the maximum degree of control.[4] The Act now admits a health threshold (where scientifically recognized), and environmental standards are set at the *ample margin of safety*.[5] Technology-based standards are allowed under §111.[6] The US EPA was required to set air quality standards using an *ample margin of safety to protect public health*.[7] Such risk-based standards prohibit the US EPA from considering costs and technological feasibility.[8] In *Natural Resources Defense Fund* v. *Environmental Protection Agency,* the court held that *safe* is not *risk free* and that the EPA's *expert judgment* can determine the acceptable level of risk of cancer. After the initial determination of acceptability of risk, the ample margin of safety could be determined by reducing the statistical risk from exposure to vinyl chloride. Following this legal decision, the acceptable level of risk was one in ten thousand excess lifetime cancers and the ample margin of safety should protect *99%* of the individuals within *50* kilometers of the emitting sources.

The process of setting federal US environmental standards illustrates how the US EPA deals with partial information and cost-benefit analysis. In 1982, the US EPA reviewed the particulate matter (PM_{10}), under a different section of the CAA. However, in 1992, the EPA did not again review the PM_{10} standard, as it was commanded by the CAA to do. The American Lung Association sued the agency to force review and to make sure that such standard would now include $PM_{2.5}$.[9] The court ordered the EPA to review that standard. The peer review of the literature, the use of that literature and a US EPA-funded study that was mandated by Congress done by the Clean Air Scientific Committee, CASAC, were controversial. Nonetheless, although there was general agreement of the *need* to set a standard for $PM_{2.5}$ there was division among the members of CASAC about the adverse effect of different exposures on the lung. In the end, CASAC did not endorse the proposed EPA's $PM_{2.5}$ standard because it found that the *science was too weak to lend support for any of the specific levels chosen by the EPA* (Faigman, 2000). The reasons were weak empirical causation, uncertainty and the high cost of meeting the new standards.

This last issue was raised even though the law (the CAA) seems quite clear on the point: costs are not to be directly considered in setting primary standards for *criteria* air pollutants. On the theme of cost, in 2001, the U.S. Supreme Court affirmed that, under Section 109 of the CAA (Amended)[10], federal standard setting (for the

---

[4] Clean Air Act, §112(d)(2).
[5] Clean Air Act, §112(d)(4).
[6] Clean Air Act, §111.
[7] 42 U. S. C. §7401 *et seq*. as amended.
[8] 824 F.2d 1146 (D. C. Cir. 1987).
[9] *American Lung Assoc.* v. *Browner*, 884 F. Sup. 345 (D. Ariz. 1994).
[10] 42 U.S.C. Sect. 7409(a).

National Ambient Air Quality Standards, NAAQS, section of the CAA) for particulate matter and tropospheric ozone must be set *without* consideration of costs and benefits. First the Court held that Section 109(b)(1) of the CAA does not grant the US EPA legislative powers: an agency cannot use its own interpretation of a federal statute to gain what it cannot do.[11] Moreover, the US EPA cannot arbitrarily construe statutory language to nullify *textually applicable provisions* (contained in a statute) *meant to limit its* (the agency's) *discretion*. This case, decided by the US Supreme Court as *Whitman, Admin. of EPA, et al.*, v. *American Trucking Associations, Inc., et al.*,[12] addresses four issues, two of which are critical to understanding setting of ambient environmental standards, under the congressional mandate (Section 109(d)(1)) *to protect the public health* with an *adequate margin of safety*. Such congressional mandate is explicit: it is the standard *requisite to protect the public health*.

Specifically, the Supreme Court held that *section 109(b) does not permit the Administrator to consider implementation costs in setting NAAQS.*[13]

1.3.2. *De Minimis* Risk, Significant Risk, and Other Principles of Precaution

In the discussions of legal principles and tests, as well as the inevitable conjunction of scientific evidence and causation, the legal concept of *de minimis* risk becomes relevant because it provides a bound on risk below which no regulatory action, including precautionary ones, should be taken. The US experience with regulating small individual lifetime- of- cancer risk (such as one in a million increased lifetime probability of cancer from exposure to an ambient air pollutant) is informative. That experience has shown how, when the statutes contain terms such as *substantial release* and *significant amounts*, an Agency only needs to establish *a rational connection between the facts ... and the choices* to regulate. That agency is not required to perform a quantitative risk assessment to find the *de minimis* threshold.[14]

Earlier American decisions allowed a *de minimis* threshold to be *more than insignificant*.[15] In cases under the Clean Air Act and the Toxic Substances Control Act, TSCA,[16] the courts have accepted that regulatory agencies (here, the US EPA) can *overlook circumstances that in context may fairly be considered de minimis*.[17] However, ignoring an exposure level that can be dangerous requires determining that only trivial benefits derive from the regulation.[18] An agency, in reaching a conclusion that exposure causes the *minimum reasonable risk*, must set regulations that are not unduly burdensome. For instance, the US EPA's analyses indicated that it

---

[11] Under the CAA Section 307, a final standard by the EPA can be reviewed when that final action is an implementation policy.

[12] No. 99-1257, Feb. 27, 2000.

[13] *FindLaw Const. Law Center* (2001).

[14] *Chemical Manufacturers Assoc.* v. *EPA*, 899 F.2d 344 (5th Cir. 1990).

[15] *Monsanto* v. *Kennedy*, 613 F. 2d 947 (D. C. Cir. 1979), at 955.

[16] TSCA, § 6; *Chemical Manufacturers Association* v. *EPA*, 899 F.2d 344 (5th Cir. 1990).

[17] *Environmental Defense Fund*, v. *EPA*, 636 F. 2d 1267 (D. C. Cir. 1980), p. 1284.

[18] Ibid., p. 1284.

would cost from seven to eight million dollars to seventy-two to one hundred and six million, per statistical life saved.

Chemical Manufacturers Association v. *EPA* details the issues for judicial review. These include: the amount of the releases (being substantial), exposure being substantial or significant and a *rational connection between the facts ... and the choices* but the agency need not rely on quantitative risks and benefits. For example, the US Supreme Court, in *Industrial Union*, held that the agency had the burden of showing that it is at least more likely than not that long-term exposure to *10* ppm of benzene presents a significant (a qualifier that does not appear in the statute) risk of material health impairment. In this case, the Court required the OSHA to develop better evidence of adverse effects (leukemia) from occupational exposure to airborne benzene, and concluded that *safe* is not equivalent to risk-free. The *significance* of risk is not a *mathematical straitjacket* and OSHA's findings of risk need not approach *anything like scientific certainty*. As the Court stated:

"The reviewing court must take into account contradictory evidence in the record..., but the possibility of drawing two inconsistent conclusions from the evidence does not prevent an administrative agency's findings from being supported by substantial evidence."

Importantly, the Court refused to determine the precise value of *significant risk*. Although it noted that chlorinated water at one part per billion concentrations would not be significant, but one per thousand risk of death from inhaling gasoline vapor would be, it did not provide the risk acceptance (or tolerability) criterion. The absolute risk-based standard is *zero risk*.[19] It is most closely associated with a narrow set of regulations taken by the Food and Drug Administration (FDA) under the Federal Food Drug, and Cosmetics Act, FDCA, Sections 201(s) and 409[20]. This statute, in general, requires balancing the economic benefits with the costs of reducing exposure (21 USC §346), rather than zero risk. An example of the application of zero risk occurred in the controversy surrounding the ban of Orange Dye No. 17 and Red Dye No. 19, used in cosmetics. The FDA had found that these additives were carcinogenic in animals, and had estimated the individual lifetime estimates of human risk were, respectively for these dyes, $2.0*10^{-10}$ and $9.0*10^{-6}$.[21] The FDA then concluded that these risks, developed through what it considered to be the state-of-the-art but with conservative, or pessimistic, assumptions were *de minimis*, but not zero.

The FDA allowed these two colorants to be used. However, because the FDCA contained the express prohibition (the Color Additives Amendment of 1960) against all color additives intended for human consumption that *induce cancer in man or*

---

[19] *Alabama Power Co.* v. *Costle*, 636 F. 2d 323 (D. C. Cir. 1979). The *de minimis* principle allows judicial discretion to avoid regulations that result in little or no benefit. *Les* v. *Reilly*, 968 F.2d 985, *cert*. den'd, 113 S. Ct. 1361 (1992) affirms the Delaney Clause prohibiting carcinogenic food additives.

[20] The first section defines food additives and includes wording that takes out of §409 additives that are generally recognized as safe.

[21] 51 Fed. Reg. 28331 (1986) and 51 Fed. Reg. 28346 (1986).

animal,[22] in the litigation that follow the rules by the FDA, a court rejected, albeit with *some reluctance,* the FDA's conclusion of *de minimis* risk.[23] Although *zero risk* would appear to reduce the need to establish causation by imposing a threshold finding (e.g., cancer in two animal tests) by legislative *fiat,* such a principle is plausible only in a few practical instances. These include the situation where causation is not in general dispute *and* the net benefits to society from banning a substance are trivially small.

1.3.3. Dissemination of Information about Risk Assessments and Management

In 2003, the US Office of Management and Budget, OMB, concluded that precaution plays an important role in risk assessment and risk management, but precaution, coupled with objective scientific analysis, needs to be applied wisely on a case-by-case basis (US EPA, 2004). Following the 1991 Executive Office of the President document Regulatory Program of the United States Government (US EPA, 2004), risk assessment has specific requisites:

"a) Risk assessments should not continue an unwarranted reliance on 'conservative (worst-case) assumptions' that distort the outcomes of the risk assessment, 'yielding estimates that may overstate likely risks by several orders of magnitude.'

b) Risk assessments should 'acknowledge the presence of considerable uncertainty' and present the extent to which conservative assumptions may overstate likely risks.

c) EPA risk assessments must not 'intermingle important policy judgments within the scientific assessment of risk.' Rather, the 'choice of an appropriate margin of safety should remain the province of responsible risk-management officials, and should not be preempted through biased risk assessments.'"

The US Office of Management and Budget (OMB) has established a set of Final Guidelines, the *Information Quality Guidelines* that controls all US federal agencies regarding the collection, processing and dissemination of information that has to do with risk assessment.[24] The Guidelines apply to scientific information used by federal agencies. Specifically, the OMB refers to the Safe Drinking Water Act as the gold standard for justifying public decision-making based on risk.[25] The SDWA applies to what the OMB calls *influential* information, as follows:

"(A) Use of science in decision-making. In carrying out this section, and, to the degree that an Agency action is based on science, the Administrator shall use:

(i)   the best available, peer reviewed science and supporting studies conducted in accordance with sound and objective scientific practices; and
(ii)   data collected by accepted methods or best available methods (if the reliability of the method and nature of the decision justifies use of the data).

---

[22] PL 86-618; 21 U. S. C. §376(b)(5)(B), the Delaney Clause.
[23] *Public Citizens* v. *Young,* 831 F.2d 1108 (D. C. Cir. 1987).
[24] 67 FR 8452-8460 (Feb. 22, 2002).
[25] 42 USC Sect. 300g-1(b)(3)(A), (B).

(B) Public information. In carrying out this section, the Administrator shall ensure that the presentation of the information on public health effects is comprehensive, informative and understandable. The Administrator shall, in a document made available to the public in support of a regulation promulgated under this section, specify, to the extent practicable:

(i)   each population addressed by any estimate of public health effects;

(ii)   the expected risk or central estimate of risk for the specific populations;

(iii)   each appropriate upper-bound or lower-bound estimate of risk;

(iv)   each significant uncertainty identified in the process of the assessment of public health effects and studies that would assist in resolving the uncertainty; and

(v)   peer-reviewed studies known to the Administrator to support, are directly relevant to, or fail to support any estimate of public health effects and the methodology used to reconcile inconsistencies in the scientific data."

Influential information is defined to be that scientific, financial, or statistical information, which *will have or does have a clear and substantial impact on important public policies or important private sector decisions*. Some US federal agencies have adapted the SDWA guidelines for *influential* information. For example, the Centers for Disease Control and Disease Prevention, CDC, had adapted the OMB Guidelines (CDC can legally do so) to deal with judgments based on qualitative information as follow:

1.   The best available science and supporting studies conducted in accordance with sound and objective practices, including peer-reviewed studies,

2.   Data collected by accepted methods (if reliability of the method and the nature of the decision justify use of the data),

3.   Ensure that information disseminated to the public related to risk effects is comprehensive, informative and understandable.

The OMB Guidelines also allow an individual to bring civil law suits to challenge the value of the influential information, including risk assessments.

1.4. DIFFICULTIES WITH PRINCIPLES OF PRECAUTION

As discussed, most forms of principles of precaution involve cost-benefit balancing. Whether environmental management should be based on cost-benefit analysis, or not, cannot be resolved in this book because, before even engaging in that debate, causation and uncertainty must be explicitly assessed. On law and logic, it is difficult justifiably to act on behalf of society when cause and probability of the adverse outcome are unknown. Moreover, any action taken on behalf of society should require more than a scientific conjecture. Otherwise, the precautionary principle (with the due process and other fundamental guarantees) can be abused because of potentially forcing an action that can be even more costly to society than the one being prevented.

These are the horns of the dilemma: either to act prudently basing the act on a scientific conjecture or to act with minimum uncertainty and the full account of cost and benefits. Neither seems satisfactory because of the potential residual risks and potentially high cost. Fortunately, there are practical ways to deal with this dilemma. Let us see how. Kriebel et al., (2001) state that *the term precautionary principle has the advantage that it provides an overarching framework that links environmental science and public health.*[26] They also give three main points in opposition to the precautionary principle:

1. "current regulatory procedures are already precautionary; for example, the safety factors used in risk assessment…
2. the precautionary principle is not scientifically sound because it advocates making decisions without adequate scientific justification …
3. the precautionary principle would stifle innovation …"

These authors discuss uncertainty by concentrating on hypotheses, statistical independence, Type I (false positive), II (false negative) and III errors. This last error is the error of providing an *accurate answer to the wrong problem.*

The sciences and law intersect in the drafting and implementation of the precautionary principle. Their intersection becomes critical at the end of the process of applying principles of precaution, when a court or tribunal reviews the actions of an agency, authority or private party. The precautionary principle, as drafted in primary legislation and interpreted in case law, does not provide a formal framework necessary to achieve such linkage. This issue becomes increasingly apparent after examining other variants of the principle found in American law as well as the law of other jurisdictions. As shown, the precautionary principle needs a threshold of scientific knowledge that, when crossed, commands a public institution to provide a *high level of protection* in spite of uncertainty. This threshold is determined by linking probable events within a causal network. The output from this analysis is a distribution of values.

Unfortunately, for decision-makers, a probabilistic threshold cannot be a bright line, unless it is certain. Practically, the choice of any specific threshold requires scientific and political considerations. Thus, there can be a subset of values that does not trigger the precautionary principle. Should it be otherwise, the balancing of costs (including risks) and benefits might not be justified and consequently fail the legal challenges feared by the US Comptroller General, 3 decades ago.

The precautionary principle has two principal forms. In one, it requires the *explicit legislative* balancing of risks (as a form of cost measured by a change in the cumulative probability of an adverse outcome) with *monetized* cost and benefits. This is the *relative* form of the precautionary principle. In the other variant, the *absolute* form, the principle urges precaution when the magnitude of the potential adverse event is large or the adverse outcome is severe, even if its probability is small. In this form, it is akin to separating the magnitude of the adverse outcomes, e.g., *500* prompt deaths, from their probability of occurring, e.g., *1/100,000* per year. The absolute precautionary principle excludes calculating the expected value, $500*100,000^{-1}$. It instead asks for action based on the large number of potential deaths, disregarding uncertainty. It invokes action but does not require the balancing

---

[26] Kriebel, D, Tickner, J Epstein, P Lemons, J Levins, R Loechler, EL Quinn, M Rudel, R Schetter, M Stoto, 2001, The Precautionary Principle in Environmental Science, *Env. Health Perspectives*, 109: 871-876.

of risks, costs and benefits: the magnitude of those prompt deaths is intolerable. Under both forms of precaution, the representation of the uncertainty and the methods for supporting causal assertions made under it must be scientifically sound because uncertainty can be used either as a sword or as shield to justify a policy or to avoid taking action.

Often, however, uncertainty may neither be reduced nor eliminated. First, if the process is inherently random, there is no practical way to eliminate randomness through gathering more data or conducting more research, even when funds are available to do so. Second, the methods to deal with the various forms of uncertainty are complex and may not always be used in practical work. Third, although in some analysis uncertainty can be taken care of by using factors of safety, in many environmental decisions this option is not theoretically sound because it is ad hoc. It is therefore critical to characterize risks in a way that is replicable and sound through quantitative methods.

The policy maker or manager can make a conservative choice, but the legitimacy of such choice relates to the means for analysis and to the fairness of the outcome. The decision-maker often confronts a limited set of (seemingly) plausible actions and a set of situations that are not in her control (these states are often called states-of-nature, meaning that the decision maker cannot control what actually happens, regardless of what she chooses to do).

1.5. ELEMENTS OF RISK ASSESSMENT AND MANAGEMENT

A risk is generally measured by the probability of an adverse outcome of a specific magnitude. If the decision-maker had all of the relevant and deterministic representations of that information, then the optimal choice would be the action (out of a set of mutually exclusive and fully exhaustive set of actions) with the lowest deterministic net cost or the maximum net benefit. This criterion requires that the set of all possible actions is fully exhaustive and that each potential action, within that set, is mutually exclusive. Importantly, this approach can account for the cost of adding information. If that cost exceeded the value of the expected benefit, the decision-maker would not want to spend money, regardless of the decrease in the variability that would result from this action.

The issues to be addressed when making risky choices overview are shown in Figure 1.1.

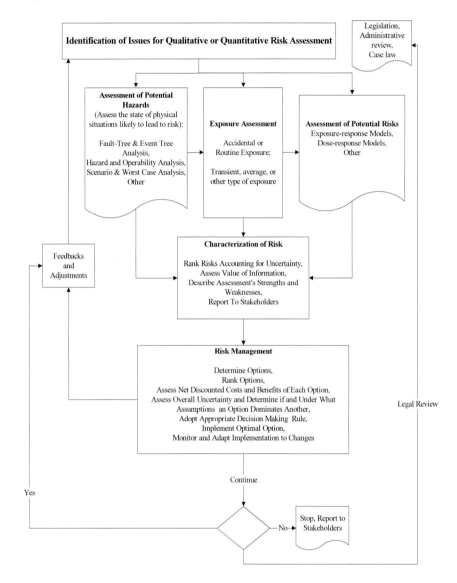

Figure 1.1. Linkages between the Components of Risk Assessment and Management

The components of risk analysis and management include:

1. Hazard identification (the source of danger),
2. Exposure characterization (the type and form of hazard),
3. Exposure assessment (the spatial and temporal extent of exposure),
4. Risk characterization (the causal relation between exposure and response),
5. Risk management actions (the choice and net benefits of actions to reduce or eliminate the hazard, including monitoring exposure and changes in choices as new information becomes available).

6.   Communication of hazard and risk to those affected (the sharing of knowledge and information).

The role of independent administrative and judicial reviews, as discussed in this chapter, is important to risk management because *the king can do wrong*. An example is provided by the US EPA guidance (US EPA 2003) about the risks associated with early (fetal and childhood) exposure. The emphasis of the US EPA Guidelines is to base risk assessment on *weights-of-evidence*, which use cellular and molecular *modes of action*, as well as interspecies comparisons, which can b used when epidemiological data is unavailable or insufficient. Guidelines are not *substantive, binding "rules" under the Administrative Procedure Act or any other law* ...; the EPA guidance is *a non-binding statement of policy*.

1.6. UNCERTAINTY AND VARIABILITY

Uncertainty and variability describe different states of knowledge. As we have shown, *uncertainty* and *variability* are often used interchangeably in legal and other arguments about risks, costs and benefits. These two terms are not equivalent. In situations typically encountered in risk assessment and management, uncertainty and variability describe a range of uses: from lack of knowledge to incomplete knowledge, from the variability of a sample or a population to that of a model's predictions. The concern with using uncertainty and variability interchangeably, and often without relation to a measure (such as probabilities), has reached the highest levels of decision-making. For example, the US National Research Council (US NRC, 1994)) discusses and defines some of these terms in the context of the Clean Air Act as *uncertainty analysis is the only way to combat the 'false sense of certainty,' which is caused by a refusal to acknowledge and (attempt to) quantify the uncertainty in risk predictions.*

For the NRC *uncertainty* is the *lack of precise knowledge as to what the truth is, whether qualitative or quantitative*. For example, the *validity* of alternative models is assessed through measures of uncertainty. Additionally, the US NRC (1994) stated that:

"the very heart of risk assessment is the responsibility to use whatever information is at hand or can be generated to produce a number, a range, a probability distribution – whatever expresses best the present state of knowledge about the effects of some hazard in some specific setting."

These and other considerations led to the NRC's admonition that the analyst should present separate assessments of the parameter uncertainty that remains for each independent choice of the underlying model(s) involved. In practical risk assessment and management, a key uncertainty is model uncertainty. It applies to the form and variables of any risk models (Cox and Ricci, 2000). The US NRC (1994) conclusion is somewhat hybrid. It states that:

"... given the state of the art and the realities of decision-making, model uncertainty should play only a subsidiary role in risk assessment and characterization, although it may be important when decision-makers integrate all the information necessary to make regulatory decisions."

The concept of *variability* is more commonly understood as a distribution of values: all are familiar with the variability of weather conditions, moods and so on. Although the concept is familiar, its treatment in risk assessment can often be complicated and technically demanding. The NRC concludes:

"As a result, the EPA's methods to manage variability in risk assessment rely on an ill-characterized mix of some questionable distributions, some verified and unverified point values intended to be 'averages,' and some 'missing defaults,' ... hidden assumptions that ignore important sources of variability."

The US NRC (1994) does not define *variability*, but it explains it as:

"individual-to-individual differences in quantities associated with predicted risks, such as in measures or parameters used to model ambient concentrations, uptake or exposure per unit ambient concentration, biologically effective dose per unit of exposure, and increased risk per unit of effective dose."

Such discussion points to the fact that a distribution of values represents variability. In particular, practical risk assessment and management can use *known* or *unknown* probability distributions and *subjective* probability distributions. A probability distribution (PDF, pdf), which can be continuous, discrete or a mix of the two, describes *variability*, encompassing objective or subjective probabilities, over all possible values taken by one or more random variables. The subjective assessment of probabilities, discussed later in this chapter, is *used for a risk assessment component for which the available inference options are logically or reasonably limited to a finite set of identifiable, plausible, and often mutually exclusive alternatives*. A distribution function (discrete or continuous) should be used to characterize variability. Variability pervades all of the aspects of risk assessment; namely all of the sub-processes leading to the calculations of emission rates, transport and fate of risky agents, exposure patterns and dose and response. Variability can practically be described by the distribution function of a random variable, or as the joint distribution function of two or more random variables. The pervasiveness of variability, and its importance for those concerned with making decisions based on health risk information, is that:

"The dose-response relationship (the 'potency') varies for a single pollutant, because each human is uniquely susceptible to carcinogenic or other stimuli (and this inherent susceptibility might well vary during the lifetime of each person, or vary with such things as other illness or exposure to other agents)."

In other words, once a relationship (or model) is selected as the most plausible one, given the current state of knowledge supporting its choice, then it is proper to assess the variability of its parameters. For example, in the instance of a simple, linear dose-response model, probability of response equals parameter times dose, $r = b*d$, the NRC's statement is correct if it is interpreted as a distribution of values about the mean value of parameter of the dose-response (namely, b, or potency). But it is often incorrect because a dose-response relationship, at ambient exposure levels that do not cause unambiguous adverse responses, does not yield a unique potency, unless the model is linear. In many situations, linearity can be a scientific conjecture that requires further analysis of the disease and the specific substance causing it before being biologically plausible and becoming a scientific fact. More specifically, the first derivative of the dose-response of the linear model is constant: this is what the US EPA labels a potency number, but it is variable in the sense that it is a random variable with a specific distribution. Estimation of parameters (using such statistical methods as the maximum likelihood, discussed in Chapters 4and 5, applied to a sample, such as an animal bioassay) yields central estimates (e.g., mean) and statistics that describes variability (e.g., variance), upper and lower confidence limits and other statistical information.

Example 1.2. The potency of arsenic from a data set of animals (hamsters) is estimated to be $3.5*10^{-5}$ [mg of arsenic/kg body weight of the person-day]$^{-1}$. The single-hit cancer model for lifetime dose d is:

$pr(d) = 1-exp(-3.5*10^{-5}d)$.

In this model, the probability is cumulative and lifetime, as explained in Chapter 5.

The NRC suggests that variability can be dealt with by either gathering *more data or rephrasing the question*. More specifically, according to the NRC, there are four ways that traditionally have been used to deal with variability: 1) *ignore the variability and hope for the best*, 2) *explicitly disaggregate* it, 3) *use the average*, and 4) *use a maximum or minimum* value. These "ways" raise the following issues:

- Ignoring uncertainty can be either a subterfuge or an excuse for choosing an option
- Disaggregating variability can require operations (such as deconvolutions) that may be unavailable in practical work
- The average, although mathematically correct, can be misleading when the distribution of values is wide, has "heavy tails," or is multimodal (e.g., the distribution of the mean aerodynamic diameter of particles is bimodal)
- Choices based on the minima and maxima of a variable require an analysis based on specific criteria (e.g., minimax) to avoid being ad hoc

The US EPA (2004) states that "uncertainty can be defined as a lack of precise knowledge as to what the truth is, whether qualitative or quantitative." The US EPA (2004) also states that:

"Uncertainty in parameter estimates stems from a variety of sources, including measurement errors (such as) random errors in analytical devices (e.g., imprecision of continuous monitors that measure stack emissions), systemic bias (e.g., estimating inhalation from indoor ambient air without ... the effect of volatilization of contaminants from hot water during showers), use of surrogate data ... (e.g., use of standard emission factors for industrialized processes), misclassification (e.g., incorrect assignment of exposures of subjects ... due to faulty or ambiguous information), random sampling error (e.g., estimation of risk to laboratory animals or exposed workers in a small sample), (and) non-representativeness (e.g., developing emission factors for dry cleaners based on a sample of 'dirty' plants)."

Furthermore, according to this agency (US EPA, 2004):

"Common types of model uncertainties in various risk assessment–related activities include: a) relationship errors (e.g., incorrectly inferring the basis of correlations between chemical structure and biological activity), b) oversimplified representations of reality (e.g., representing a three-dimensional aquifer with a two-dimensional mathematical model), c) incompleteness, i.e., exclusion of one or more relevant variables (e.g., relating asbestos to lung cancer without considering the effect of smoking on both those exposed to asbestos and those unexposed), d) use of surrogate variables for ones that cannot be measured (e.g., using wind speed at the nearest airport as a proxy for wind speed at the facility site), e) failure to account for correlations that cause seemingly unrelated events to occur more frequently than expected by chance (e.g., two separate components of a nuclear plant are both missing a particular washer because the same newly hired assembler put them together) f) Extent of (dis)aggregation used in the model (e.g., whether to break up fat compartment into subcutaneous and abdominal fat in a physiologically based pharmacokinetic, or PBPK, model). … This Agency concludes that the 'EPA's models are often incomplete and knowledge of specific processes limited. As a result, EPA relies on specific default assumptions as a response to uncertainty.'"

The recommendations made by the US NRC to the US EPA point to probabilistic and statistical reasoning as being essential to risk assessment and management. These recommendations are consistent with past choices of exposure scenarios such as the *maximally exposed individual*, the more recent *high-end exposure estimate* and the *theoretical upper-bounding exposure*. Those recommendations are also consistent (and often require) simulation techniques, such as Monte Carlo methods, to estimate the variability in the outcome variable of a risk model.

1.7. SCIENCE AND LAW: From scientific evidence to causation

First, the law coerces (through injunctions), corrects (by redistributing wealth) and punishes (through fines or confinement). Second, developing scientific causation demands formal reasoning about incomplete knowledge, variable data, scientific conjectures, mathematical models and empirical analyses. Third, probability (taken as an appropriate measure of uncertainty) and magnitude of outcomes (e.g., the number of prompt accidental deaths) should be considered for coherent decision-making. Fourth, deterministic thinking can be counter-productive by giving a sense of false security. Fifth, political, pressure-group or media-related statements about environmental risks may give far too great saliency to a problem that does not, after thoughtful consideration, warrant the attention and expenditures made to cope with it. Sixth, there are perceptual and cognitive issues. Finally, as Zeckhauser and Viscusi (1990) remarked, society may be *willing to spend substantial resources to reduce risk* but is *reluctant to look closely at the bargain it has struck.*[27]

The scientific evidence about cause (e.g., exposure to ionizing radiation from inhaling radon) and effect[28] (e.g., a response such as the increased risk of lung cancer) can consist of epidemiological studies, animal bioassays, molecular studies,

---

[27] They suggest that the decision making process *should be judged in terms of their effect on expected utility, the only well-developed prescriptive framework for choices under uncertainty.*
[28] There are legal cases in which causation is not a legal issue. For instance, the carcinogenic effects of DES (causing clear cell adenocarcinoma in the daughters of the mothers who took the synthetic hormone) was not disputed, as discussed in Note, Market Share Liability: An Answer to the DES Causation Problem, *Harv. L. Rev.* 94: 668 (1981).

tests on lower organisms (e.g., *in vitro*) and other work. Establishing causation from these heterogeneous forms of scientific evidence also involves reasoning with uncertain data using complex statistical and probabilistic models. The sources that release noxious agents, their transport and transformation in air, water, and soil also demand accurate modeling and monitoring to characterize exposure to agents likely to cause adverse effects in those exposed. Further *in vivo* transformations, involving pharmaco-kinetics and other biochemical and molecular mechanisms, (discussed in Chapter 11) can complicate causal analyses, particularly from exposures to low doses. Exposure can be controversial. The reasons include: 1) background exposure is often poorly characterized and 2) the reconstruction of past exposure is complicated by having to reconcile it with circumstances that were different (and often much worse) than those in existence when the disease is detected.[29] If exposure and response were determined by an epidemiological study that uses aggregate data, those results can be affected by Simpson's paradox. Of course, unlike animal studies, epidemiology does not require interspecies comparisons, responses can be age-race-sex adjusted, standardized to a specific population, and can account for confounders and effect-modifiers using the statistical methods discussed in Chapters 4-11. On the other hand, animal bioassays use strains of animals that are homogeneous, there is strict control over exposure, diet and other factors likely to affect response. However, the results from these experiments require interspecies comparison to be applicable to human risk assessment.[30] Moreover, some strains are developed for reasons unrelated to risk assessment, and have organs not found in humans (such as the zymbal gland in rodents) that can become cancerous.

Neither scientific nor legal causation can hope for complete certainty.[31] Statistical evidence alone, even when based on well-conducted studies, can be rebutted by skillful legal interrogations using contradictory scientific theories and data. The strength of the legal causal link is mediated by evidence and standards of proof. Thus, when the scientific uncertainty about causation (e.g., that a particular dose-response model was biologically accurate) was stated as a *possibility*, it was

---

[29] In *Miller* v. *National Cabinet Co.*, 204 NYS2d 129 (1960), the court held that, even though the deceased plaintiff had been exposed to the solvent benzene for approximately 25 years, the fact that the leukemia that killed him had occurred several years after employment had terminated did not allow a finding of *legal* causation. In 1960, according to an expert witness from the National Cancer Institute, ... *no country has officially recognized benzol (benzene) leukemia as a compensable occupational disease.* The court differentiated between possibility and probability. It held against the plaintiff because her testifying physician stated that it was *possible* that the leukemia had been caused by exposure to benzene, rather than *probable*.

[30] Courts do not often accept results from animal bioassays and their statistical analysis to show causality for humans relative to the incidence of diseases.

[31] *Allen* et al., v. *U. S.*, 588 F. Supp 247 (1984), rev'd on other grounds, 816 F.2d 1417 (10th Cir. 1988), cert. den'd, 484 U. S. 1004 (1988) illustrates how a modern court deals with uncertainty. The court held that each plaintiffs should meet at the *preponderance of the evidence* test for i) probability of exposure to fallout resulting in doses in significant excess of *background*, ii) injury is consistent with those known to be caused by ionizing radiation; iii) and that the individual claimants resided near the Nevada Test Site for at least some, if not all, of the years when the tests were conducted. The *Allen* court also recognized that lack of statistical significance does not mean that the adverse effect is not present.

insufficient[32] to demonstrate legal causation by the *preponderance of the evidence*.[33] These difficulties, and the often-conjectural model of dose-response, increasingly force the legal system, particularly in tort law, to ask science for better proof of causation and methods for dealing with inevitable uncertainty and variability.[34]

1.7.1. Scientific Evidence: What can be legally admissible?

As a process for environmental decision-making, risk assessment combines reasoning with uncertain data complex mathematical and statistical methods to describe and extrapolate far from the relevant range of the observations, often basing those extrapolations on incomplete or limited theoretical bases. The difficulties arise from having to base legal or policy decisions on scientific information and conjectures that can mystify, rather than clarify, the causal link sought. The discussions that follow use US law as examples. The question that we are addressing is: What is the legal test that governs the admissibility of scientific evidence? In 1993, the U. S. Supreme Court, in *Daubert*,[35] addressed the extent to which scientific results could be allowed in court. Before Daubert was decided, it had been held, for instance given epidemiological evidence of the effect of Bendectin,[36] that scientific evidence based on chemical structure activity, *in vitro*, animal tests, and recalculations that had not been peer-reviewed, did *not* meet the test that governs the admissibility of the evidence, generally known as the *Frye* test.[37] The unanimous Supreme Court disagreed holding that *general acceptance* of a scientific finding is not *required* for that finding to be admissible into evidence, thus rejecting *Frye*. The *Frye* test states that:

"(j)ust when a scientific principle or discovery crosses the line between the experimental and the demonstrable stages is difficult to define. Somewhere in this twilight zone the evidential force of the principle must be recognized, and while the courts will go a long way in admitting expert testimony deduced from well-recognized scientific principle or discovery, the thing from which the deduction is made must be sufficiently established to have gained general acceptance in the particular field in which it belongs."

---

[32] In US practice *sufficient evidence* is evidence that can proceed to be heard by the jury or by a judge acting as the fact-finder.

[33] *Green* v. *American Tobacco Co.*, 391 F.2d 97 (5th Cir. 1968), 409 F. 2d 1166 (5th Cir. 1969). *Bowman* v. *Twin Falls*, 581 P.2d 770 (Id. 1978) held that *to require certainty when causation itself is defined in terms of statistical probability is to ask for too much.*

[34] Courts often do not examine the details of a scientifically-based causal assertion, as the cases *Karjala* v. *Johns-Manville Products Corp.*, 523 F. 2d 155 (8th Cir. 1975) and *Bertrand* v. *Johns-Manville Sales Corp.*, 529 (F. Supp. 539, D. Minn. 1982) show.

[35] *Daubert et al.*, v. *Merrell Dow Pharmaceuticals, Inc.*, *L. Week* 61:4805. In *Christophersen* v. *Allied-Signal*, No. 89-1995 (5th Cir. 1991), cert. den'd, 112 S. Ct. 1280 (1992), the scientific testimony about the causal association between exposure to nickel and cadmium and colorectal cancer was excluded because it was not mainstream science. And yet, some American state courts have allowed into evidence scientific explanations that would fail *Frye*-type tests if those explanations are *sound, adequately founded ... and ... reasonably relied upon by experts*, *Rubanick* v. *Witco Chemical*, 593 A. 2d 733 (NJ 1991).

[36] In *Brock* v. *Merrell Dow Pharmaceuticals*, 874 F.2d 307 (5th Cir.), 884 F.2d 166 (5th Cir.), reh'g en banc den'd, 884 F. 2d 167 (5th Cir. 1989).

[37] *Frye* v. *U. S.*, 293 F. 1014 (D. C. 1923).

The US Supreme Court held that:

"(t)o summarize, the 'general acceptance' test is not a necessary precondition to the admissibility of scientific evidence under the Federal Rules of Evidence, but the Rules of Evidence -- especially Rule 702 -- do assign to the trial judge the task of ensuring that an expert's testimony both rests on a reliable foundation and is relevant to the task at hand. Pertinent evidence based on scientifically valid principles will satisfy those demands."

The *Daubert* Court also discussed scientific knowledge. It stated that:

"(t)he adjective 'scientific' implies a grounding in the methods and procedures of science. Similarly the word 'knowledge' connotes more than subjective belief or unsupported speculation."

It then added that *(o)f course, it would be unreasonable to conclude that the subject of scientific testimony must be 'known' to a certainty; arguably there is no certainty in science.* The Court explains the role of peer-review as *relevant, though not dispositive, consideration in assessing the scientific validity of a particular technique or methodology on which a particular technique is premised.*

The Court went on to discuss the difference between legal and scientific analysis, and gave the reason for the difference, stating that *scientific conclusions are subject to perpetual revision. Law, on the other hand, must resolve disputes finally and quickly.* The former is unquestionable; the latter is troublesome because the judicial process is notoriously slow. The proper inquiry into the scientific basis for any subsequent assertion of *legal* causation must explore often-uncharted paths through the maze of heterogeneous evidence. The weight of each item into evidence as part of the proof must be determined through methods that are reproducible and coherent. It is appropriate to assign a higher degree of belief to peer-reviewed evidence than to other evidence. Thus, *Daubert*'s approach is a sensible step to determine legal causation from scientific data and theories, but it must be based on formal ways to establish the factual causation, in the context of legal causation.

1.7.2. Scientific Evidence in Legal Causation

In *Usery* v. *Turner Elkhorn Mining Co.*, the US Supreme Court made it clear that uncertain causation (between exposure and cancer) does not prevent compensating the workers who are exposed to dangerous ionizing radiations.[38] Moreover, a negative finding of cancer in a radiological exam is insufficient to deny a worker's compensation claim.[39] *Usery* stands for the principle that Congress can develop probabilistic standards of compensation. Those are necessary to provide just compensation when uncertain etiological factors are associated with dreaded, fatal and causally complex diseases. The Court permitted recovery as *deferred* compensation for the plaintiff's widow.

---

[38] 96 S. Ct. 2882 (1976).
[39] 428 U. S. 1 (1976). The statute is found at 33 U.S.C. § 920 (Supp. IV, 1974).

Direct legal evidence does not require inference: the facts speak for themselves. This simplicity is appropriate when causation is well established and the statistical problem is the measurement error only. Simplicity can fail in most environmental situations in which exposure is low and there are several causal factors: uncertain and multifactorial causation is the rule. A way to deal with causation is to take the *strong* version of the preponderance of the evidence. This test does not admit statistical correlations, such as regression models, without additional *direct and actual knowledge of the causal relationship between the defendant's conduct and the plaintiff's injury.*[40] The alternative, consistent with the precautionary principle, is the *weak* version of the evidence where statistical correlations can determine causation. However, the weak version only suggests hypothesis generation, rather than causal association, such as that found in an *ecological* epidemiological study. Uncertainty and multiple and possibly heterogeneous factors[41] can become important issues, as the law has long recognized.[42]

The traditional rule in US administrative law makes the *proponent of the rule* bear the burden of proof; but the evidentiary standards are less demanding than in tort law.[43] In general, judicial review of agency actions is based on the liberal (i.e., giving much latitude to an agency) *arbitrary and capricious* test. However, federal courts can review the record, including the scientific evidence, under the *hard look* theory[44] Determining whether the agency's decision is supported by substantial evidence requires finding that *the administrative record contain(s) respectable scientific authority* supporting the agency's factual determinations. Under the *hard look*, agency rulemaking will be sustained if: i) there is a reasoned explanation of the basis of fact, ii) its decision is supported by substantial evidence, iii) other alternatives were explored, and given reason for their rejection, and iv) the agency responded to public comments. For instance, medical evidence of exposure to lead and resulting anemia, other adverse effects on the red blood cells and neurological effects led the court to find that the US EPA had not acted unreasonably when the uncertainty of the adverse effects was large.[45] The *hard look* has been used to ascertain the validity of

---

[40] See *Symposium on Science - Rules of Legal Procedure*, 101 F. R. D. 599 (1984).

[41] *Environmental Defense Fund* v. *U. S. Environmental Protection Agency*, 489 F2d 1247 (D. C. Cir. 1973) held that in the *great mass of often inconsistent evidence* there *was sufficient evidence to support the Agency's ultimate decision.*

[42] *Environmental Defense Fund* v. *U. S. Environmental Protection Agency*, 548 F2d 998 (D. C. Cir. 1976) held that *causation is key in environmental law.* Moreover, the court held that *substantial evidence, including experimental evidence ... (from) laboratory animals* is appropriate in suspending the registration of two pesticides. The registrant, not the Agency, had the burden of proof because Congress had *made it clear that the public was not to bear the risk of uncertainty concerning the safety of a covered poison.*

[43] Federal Administrative Procedure Act, U.S.C. Title 5, Chapter 5, Subchapter II, § 556 (d).

[44] *Marsh* v. *Oregon Natural Resources Council*, 490 U. S. 360 (1989). The hard look review indicated that the allegedly important information was in fact less important than suggested by the claimants and that, although some other decision-maker might have reached a different conclusion, the agency did not make a *clear error in judgment* by not conducting more assessments.

[45] *Lead Industry Assoc.*, Inc. v. *Environmental Protection Agency*, 647 F.2d 1130 (D.C. Cir. 1980)

an agency's choice of mathematical models,[46] challenging the technical assumptions made by another agency[47] and requiring documentation of the health effects of *428* toxic substances.[48]

The opposite of the hard look is the *soft glance* in which:

"because substantive review of mathematical and scientific evidence by technically illiterate judges is dangerously unreliable, I continue to believe that we will do more to improve administrative decision-making by concentrating our efforts on strengthening administrative procedure." (Judge Bazelon, in *Ethyl Co.* v. *US Environmental Protection Agency*).[49]

We should remember that there is a significant difference between tort law and environmental law. Specifically, the courts are traditionally deferential towards rules and regulations issued by public agencies. For instance, in *Chevron* v. *Natural Resources Defense Fund*,[50] involving the interpretation of a Section of the Clean Air Act, the Court stated that:

"the Administrator's interpretation represents a reasonable accommodation of manifestly competing interests and is entitled to deference: the regulatory scheme is technical and complex, the agency considered that matter in detailed and reasoned fashion, and the decision involves reconciling conflicting policies... Judges are not expert in the field, and are not part of either political branch of the Government...When a challenge to an agency construction of a statutory provision, fairly conceptualized, really centers on the wisdom of the agency's policy, rather than whether it is a reasonable choice within a gap left open by Congress, the challenge must fail."

Deference can result in a judicial unwillingness to deal with uncertain causation. For instance, the US Nuclear Regulatory Commission was held to be free to adopt conservative assumptions by *risking error on the side of over-protection rather than under-protection... when those assumptions have scientific credibility.*[51] The estimates of carcinogenic risk of death from radon, from uncontrolled uranium mine tailings, was calculated by the Commission to be one in fifty million, compared to the background cancer risk of death from ionizing radiation, which is approximately fifty per million. However, for residents nearby a uranium mine with uncontrolled tailings, the radon-related risk had been calculated to be one in two-thousand six-hundred person-lifetimes exposure. This result was a *satisfactory basis* for finding an unreasonable public health risk: *unreasonable risk* (15 USC §2058(f)(3)(A)) is a quantitative measure of excess risk over background. The court accepted the Commission's bounding of that risk *somewhere between one in two thousands and one in fifty million, is appropriately left to the Commission's discretion, so long as it was reasonable.* Deference to agency *rulemaking* remains strong. For instance, in *Baltimore Gas and Electric, Co.* v. *Natural Resources Defense Council,* the Supreme Court unanimously reversed the DC Court of Appeals, finding that the US Nuclear

---

[46]*Ohio* v. *EPA*, 784 F. 2d 224 (6th Cir. 1986).

[47]*Gulf South Insulation* v. *CPSC*, 701 F. 2d 1137 (5th Cir. 1983).

[48]*AFL - CIO* v. *OSHA*, 965 F.2d 962 (11th Cir. 1992).

[49] 541 F.2d 1 (D. C. Cir. 1976).

[50] 467 U. S. 837 (1984).

[51]*Kerr-McGee Nuclear Corp.* v. *U. S. Nuclear Regulatory Commission*, 17 ERC 1537 (1982).

Regulatory Commission had acted arbitrarily and capriciously.[52] The Court held that a *most deferential* approach should be given an agency engaged in making legitimate predictions of risks that fell *within its area of expertise at the frontiers of knowledge*, and when the *resolution of ... fundamental policy questions lies ... with ... the agency to which Congress has delegated authority.*

However, there must be a *rational connection* with Congressional intent. That is, an agency rulemaking is limited by the objectives dictated by the legislation by the substantial evidence test applied to the facts, and by an arbitrary and capricious test that applies to policy judgments and informal rulemaking. If an agency fails to comply with its own procedures, as in not submitting its findings to peer review, the court may find this error to be harmless, as the DC Circuit concluded in the review of the setting of the ozone standard, under the Clean Air Act.[53] The court stated that safety factors are appropriate when certain groups of individuals are less resistant, and thus more susceptible to injury from ozone exposure, than the rest of the population and when there is *no clear threshold.*

Setting occupational health and safety standards generally requires *substantial evidence*, which is a more rigorous requirement than the general *arbitrary and capricious* standard of review adopted by courts to review an agency's rulemaking. This view was developed in *Gulf South Insulation, Inc.* v. *Consumer Product Safety Commission.*[54] The Consumer Product Safety Commission, in the rulemaking to reduce indoor exposure to formaldehyde, proposed to ban it because the evidence from an animal study involving approximately *250* rats, suggested that the risk to humans of nasal cancers could be up to *51* carcinomas per million individuals exposed. The Commission had used the multistage model of cancer and found that the approximate 95% upper confidence limit risk was sufficiently high to be unreasonable.

The court reviewing this Agency's rulemaking held that *the Commission cancer prediction of up to 51 in a million provides ... no basis for review under the substantial evidence standard.* The court struck the ban down because it held that the *record as a whole* did not support the agency's finding under the substantial evidence standard. This standard, the court held, is more restrictive than the general standard of judicial review. A more complete statistical study (such as a randomized study) and a more thorough investigation of the biological basis associating formaldehyde exposures with cancer might have met the substantial evidence test.[55]

Thus, the US federal agency that has control over occupational risks, OSHA, may issue regulations *reasonably necessary* to reduce a *significant risk* of material *impairment.* These regulations should be technologically and economically feasible.[56] Several authors have discussed the regulation of formaldehyde by US

---

[52] 426 US 87 (1983).

[53] *American Petroleum Institute* v. *Costle*, 16 ERC 1435 (D.C. Cir. 1981).

[54] 501 F.2d 1137 (5th Cir. 1983).

[55] The court found that eleven epidemiological studies were negative: this strongly influenced the striking of the ban under the substantial evidence test.

[56] Substantial evidence *requires substantive review of the evidence in the agency record and*

federal agencies (Ashford et al., 1983). They found that there was a *failure* on the part of those agencies involved in regulating this chemical, to follow established federal cancer policy. Moreover, the EPA apparently held meetings with industry representatives and allowed certain appointments of individuals who may be considered to have a conflict of interest in their duty to the agency. OSHA apparently denied an emergency temporary standard (ETS) for formaldehyde. It had calculated that exposure to *three* parts per million would result in an excess risk of *four* cancers per *1,000* workers. Because the average cancer mortality in workers is *4/1,000*, OSHA concluded that formaldehyde did not pose a grave risk of cancer such as to justify the ETS. *Grave danger* had been interpreted to mean that:

"(w)hile the Act does not require an absolute certainty as to the deleterious effect of the substance in man, an emergency temporary standard must be supported by evidence that shows more than some possibility that a substance may cause cancer in man."[57]

The past discussions indicate that precautionary legal principles have a fundamental link to scientific evidence and scientific causation. Thus, the combination of uncertain evidence and causation with legal tests balancing risks, costs and benefits points to the need for a framework for dealing with uncertainty and variability that is also consistent with any reasonable principle of precaution. One such framework is outlined next.

1.8. PROBABILITIES TO EXPRESS UNCERTAINTY AND VARIABILITY

As the US EPA (2004) states:

"Probabilistic methods can be used in ... exposure assessment ... because the pertinent variables (for example, concentration, intake rate, exposure duration, and body weight) have been identified, their distributions can be observed, and the formula for combining the variables to estimate the lifetime average daily dose is well defined (US EPA, 1992a). Similarly, probabilistic methods can be applied in dose-response assessment when there is an understanding of the important parameters and their relationships, such as identification of the key determinants of human variation (for example, metabolic polymorphisms, hormone levels, and cell replication rates), observation of the distributions of these variables, and valid models for combining these variables. With appropriate data and expert judgment, formal approaches to probabilistic risk assessment can be applied to provide insight into the overall extent and dominant sources of human variation and uncertainty. In doing this, it is important to note that analyses that omit or underestimate some principal sources of variation or uncertainty could provide a misleadingly narrow description of the true extent of variation and uncertainty and give decision-makers a false sense of confidence in estimates of risk. Specification of joint probability distributions is appropriate when variables are not independent of each other. In each case, the assessment should carefully consider the questions of uncertainty and human variation and discuss the extent to which there are data to address them. Probabilistic risk assessment has been used in dose-response assessment to determine and distinguish the degree of uncertainty and variability in toxicokinetic and toxicodynamic modeling. Although this

---

not merely procedural review of the agency's genuine considerations of diverse informed opinions, as was held in *Texas Ind. Ginners Assoc., Inc.* v. *Marshall*, 630 F. 2d 398 (5th Cir. 1980).
[57] *Dry Color Manufacturers Assoc.* v. *Dept. of Labor*, 486 F.2d 98 (3rd Cir. 1973).

field is less advanced than probabilistic exposure assessment, progress is being made and these guidelines are flexible enough to accommodate continuing advances in these approaches."

As stated in EPA's Policy for Use of Probabilistic Analysis in Risk Assessment (1997, in US EPA 2004):

"For human health risk assessments, the application of Monte Carlo and other probabilistic techniques has been limited to exposure assessments in the majority of cases. The current policy and associated guiding principles are not intended to apply to dose-response evaluations for human health risk assessment until this application of probabilistic analysis has been studied further."

The EPA's Science Advisory Board (SAB) suggested that the Agency reconsider this science-policy opinion (US EPA (2004), citing US EPA, 2000e) for dose-response evaluations for human health risk assessments. The EPA (US EPA, 2004) states that:

"probabilistic analysis" is a means for describing the uncertainty in risk estimates by characterizing the uncertainty and population variability in the individual steps by probability distributions. Thus, the likelihood of each risk is quantitatively characterized in the resulting estimates. This is generally implemented by a 'Monte Carlo' approach, which performs a computer simulation to produce the estimates. In addition, probabilistic analysis may avoid some potential problems of apparent overestimation of risk estimates from multiplying UFs (uncertainty factors)in a deterministic risk assessment. ... We recognize that probabilistic analysis is no panacea. It will not necessarily result in different outcomes or decisions, nor are its implementation and use simple processes. The accuracy of probabilistic analysis will still depend upon the quality of the data used for the analysis. Moreover, in some situations, a probabilistic analysis may not be appropriate. For example, at cleanup sites (such as Superfund, Resource Conservation and Recovery Act, or Brownfields sites), the resident populations may not have the wide-ranging susceptibilities that would warrant conducting a probabilistic risk assessment."

Dealing with risk is a task that requires representations of uncertainty and variability that are often represented by probabilities and distributions. Although there are other measures that can be used to describe and account for uncertainty, we take probability measures and their calculus as a suitable means to accomplish this science-policy task. To begin with, there are different views of probabilities. They have been understood as either subjective measures of belief or objective as the result of the long-term replication of the same physical experiment. A *belief*, as Frank Ramsey's observed in 1929, is *obtained by a reliable process* having a causal basis. It is knowledge.[58] A common way to think of probabilities is as relative frequencies: the ratio in which the numerator is a number representing one type of events (positive counts, for instance) and the denominator is the sum of all events (positive and negative counts).

Probabilities, *pr*, are dimensionless numbers between *0* and *1*, including these two limits. Frequencies, *fr*, are also dimensionless numbers between *0* and *1*, and include these two limits. Both of these numbers are ordinal and use the same calculus. Ordinal means that it cannot be said that the probability of *1* is twice the probability

---

[58] Ramsey FP, *Philosophical Papers*, (DH Mellor, Ed), Cambridge Univ. Press, Cambridge (1990).

of *0.5*. When such comparisons are needed, probability numbers can be transformed to yield a cardinal number that permits comparisons such as *twice as large*. The transformation, for exponential failure rates, is a hazard rate:

r(pr) = {-[natural logarithm(1-probability number)]} = -ln(1-pr).

At low probability numbers, the transformation is unnecessary because *r(pr)* is approximately equal to the probability number *(pr)*. As Cox (2002) discusses, *r(pr)* is the value of the continuous-time hazard rate which, if constant over the period of time of interest to the assessment, yields the instantaneous probability of adverse response.[59]

Example 1.3. Let *pr = 0.001*, then: *r(0.001) = [-ln(1-0.001)] = 0.001005*. If we let *pr = 0.15*, then *r(0.15) = [-ln(1-0.15)] = 0.16*.

Example 1.4. Suppose that activity *A* has a *75%* probability of failure per year, while activity *B* has a *95%* probability of failure per year. For activity *A* the constant hazard rate is: *[-ln(1-0.75)] = 1.34*, while *B* has the constant hazard rate of approximately *3.00*. In other words, *B's hazard rate* is more than twice that of *A*.

Example 1.5. Suppose that activity *1* has a *0.75* probability of failure and activity *2* has twice the probability of failure of activity *1*, per year. The hazard rate for activity *1* is: *[-ln(1-0.75)] = 1.34*. Twice *1.34 = 2.68*: the probability of failure for activity *1* is calculated as: *1-exp(-2.68) = 0.93*.

In the next sub-section we will discuss how probability measures can have different interpretations. These are briefly discussed in the next sub-section, which is provided as an introduction. The important matter to keep in mind is that degrees of personal belief can measure uncertainty about specific events or propositions, without having to use the probabilistic interpretations via frequencies, but including them.

1.8.1. Three Views about Probability

It is important to understand some of those views because, in many circumstances, risk assessors and managers must deal with different justifications of uncertainty or variability that are represented by probabilities. Among the views, the following three stand out.

View 1. According to von Mises, a probability measure is defined as an analytical proposition, which is associated with some *physical* meaning. That is, one speaks of *probability of...*, not of *probability*. Moreover, in this view, there no probability that

---

[59] The terms are explained and illustrated in the chapters that follow. For completeness, consider a random variable that has a Poisson distribution. It can be shown (see Chapters 5 and 11) that this distribution has a constant hazard rate, symbolized by λ. The probability of *surviving* for *T* periods of time can be demonstrated to be *exp(-λT)*; therefore, the probability of *failure* is the complement of the probability of λ success, *1-exp(-λT)*. For a unit time period, *T = 1*, the probability of at least one failure is *[1-exp(-λ)]*. It follows that $\lambda = [-ln(1-pr)]$.

can be attached to a metaphysical question, nor is there any true probabilistic meaning in the statement *the probability of Switzerland attacking Austria in the year 2006 is....* Thus, a probability has a long-run characteristic: it is the limit of a relative frequency. This limit tends to infinity and the relative frequency tends to a constant value. The critical element in this view is the uniformity of the ensemble (the set of elements) over which the probability is calculated and the independence of the limit from the placement of the elements within the ensemble. This is what classical statistical analysis deals with, as studied by Fisher, Kolmogorov, Neyman, Pearson and others.

View 2. Ramsey and de Finetti (de Finetti, 1970) developed the subjective view of probabilities in the 1920s. Subjective probabilities are qualitative because they are formed through human experiences but are measured by numbers. These *qualitative probabilities* require coherence, which is the condition that no one would opt for a bet that results in a sure loss, given the choice of a lottery with at least one positive outcome. If this (rational) requirement is met, de Finetti demonstrated that the numbers associated with the lottery are consistent with the axioms of finitely additive probability measures. The interpretation of subjective probabilities is that *1* is full belief, *0* is its contradiction; numbers in between these are statements of a degree of uncertainty (these numbers are *partial* beliefs). As Ramsey stated (1990), *it is not so easy to say what it is meant by a belief 2/3 of certainty, or a belief in a proposition being twice as strong as in that of its contradictory (proposition).* For Ramsey, a probability (as a degree of belief) is a causal property, measured by a willingness and ability to act on a choice plus specific, but possibly hypothetical, circumstances. The odds of an outcome measure the intensity of belief associated with that outcome, which can be prospective (or yet unrealized). The fact that a hypothetical situation can be considered is consistent with the nature of choices. That is, a choice does have to be made. An issue with such personal but informed basis for belief is that bets must be comprehensible and must be backed by sufficient wealth to pay-up in the event of a loss. This view of probability has lead to Bayesian analysis, discussed later in this chapter.

View 3. In physics, the early deterministic view of events, changing from the early 1900's transition probabilities of atomic physics (Bohr and Einstein), changed even further with quantum mechanics (Heisenberg, Dirac, and Born, in the late 1920s). For example, there is no indeterminacy in classical physics. In other words, the laws of physics determine exactly future outcome, given the initial conditions and laws of motion. Although an outcome as the sudden collapse of a walk-way that kills tens of people can be predicted using calculations involving strength of materials, dynamics and so on, reversing that failure is extremely improbable. This is the domain of statistical mechanics, developed by Maxwell, Boltzman and Gibbs, in which the probability of events is based on generalizations and irreversibility, but can also account for reversibility if a lower level of generalization is adopted (Guttman, 1999). On the other hand, indeterminacy can governs, as we know from the indeterminacy (or uncertainty) principle of Heisenberg. Von Neumann's demonstration that quantum mechanics is irreducibly *acausal* is important as it contrasts with de Finetti's idea that probabilities are epistemic. That is, an epistemic probability describes the level of ignorance about a phenomenon, but a quantistic probability does not serve this purpose because it is a description of a probable state (von Plato, 1994).

1.9. PRACTICAL ANALYSIS OF UNCERTAINTY AND VARIABILITY

The uncertainties that confront environmental risk managers include scientific conjectures, lack of knowledge of a model's structure (leading misspecification of the model), and statistical uncertainty and variability. For the moment, we can discuss uncertainty as the combination of:

Scientific conjectures. These arise when causal models are unknown (a conjecture is a theory that has an equal chance of being right or wrong).

System (model) misspecification. It consists of the exclusion of fundamental components in a multi-component model. Specification refers to the choice of mathematical form (linear, polynomial, and so on) and the variables excluded: relevant explanatory variables that cannot be accounted for by random error. For example, a choice of dose should account for biochemical changes that occur after exposure as the chemical moves through physiological and biochemical pathways to reach the target organ.

Statistical uncertainty. This is a familiar form of uncertainty that generally refers to the natural variability of data (e.g., sampling variability, heterogeneity, and so on).

Probabilistic reasoning helps to resolve these difficulties. A form of such reasoning is known as Bayesian reasoning. Its central aspect is reliance on the combination of expert judgment (expressed as prior probabilities or distributions) and experimental data (the likelihood) to for a posterior probability or distribution. Specifically, four concepts that characterize Bayesian analysis are:

a) *Prior knowledge is represented by probabilities.* Specific knowledge, information, and scientific judgments are represented by prior (before the experiment) probability distributions for explanatory variables and by conditional probability distributions for independent variables.

b) *Likelihoods formally represent empirical evidence that updates prior knowledge.* This (conditional) probability summarizes the role of the data for a given hypothesis. The likelihood determines the empirical support, if any, for the prior belief that is stated as a probability or a distribution.

c) *Prior beliefs are updated by conditioning on observed evidence, such as the result from experiments.* Given the prior beliefs about uncertain quantities and the evidence, measurements, model, and data, the posterior distribution of one or more parameters is computed by Bayes' theorem. Many new methods and special-purpose graph algorithms for factoring high-dimensional joint distributions now make practical the computations required for Bayesian inference (and efficient knowledge representation). Some of these methods are Gibbs sampling and other Markov chain Monte Carlo (MCMC) methods.

d) *Uncertainty about the correct model, out of several alternatives, can be treated in the same coherent framework.* To include such uncertainty, model-based predictions or inferences are calculated for each of a set of alternative models that are considered mutually exclusive and collectively exhaustive. Bayesian model-averaging (BMA) uses the predictions from these multiple plausible models, weighted by *Bayes' factors* (ratios of posterior odds to prior odds for different models) that reflect their relative plausibility conditional on the data. These factors can give more realistic estimates than selecting any single model on the basis of goodness-of-fit criteria, which we discuss in later chapters.

1.9.1. Bayesian Analysis

The central aspect of Bayesian analysis, which uses joint, conditional and unconditional probabilities or distribution functions, is derived from the formula (in which "|" means conditioning):

pr(population parameter and data) = [pr(population parameter)][pr(data|population parameter)].

From elementary probability theory we know that:

[pr(data and population parameter)] = [pr(data)][pr(population parameter|data)].

Bayes' theorem, which yields the posterior probability, follows as:

[pr(population parameter|data)] = {[pr(data|population parameter)][pr(population parameter)]}/[pr(data)].

In this formula, a *parameter* can be understood as the population mean; its estimate is the sample mean. Bayes' theorem states that the *prior*, or subjective, belief of the *researcher* or *the decision-maker* is updated by empirical (experimental or observational) information (the likelihood function) that is new and additional to that prior belief. The prior belief can be a probability of an event, such probability being subjective in the sense that it is a coherent statement of the prior information that the researcher has available. Bayes' theorem extends to more than two random variables, as will be discussed in Chapter 2, and to further updates, given new information

Deterministic models (such as a quadratic equation studied in high school) do not include probabilistic measures. The selection and use of several plausible values for one or more coefficients in a system or model does not necessarily mean that such choice is truly probabilistic. Rather, it can be a practical attempt to approximate one or more forms of uncertainty through sensitivity analysis. It consists of determining changes in the output from its *base case*, by varying the value of the coefficients of a model either singly or in combinations. The choice of the values input in this sort of sensitivity analysis requires technical judgment, is consistent with deterministic reasoning, and can use simple statistical measures such as the mean of the parameters being changed in the analysis. The preferable approach, consistent with Bayesian reasoning, is to treat all uncertain quantities as random variables. Uncertainties about functions, values of parameters or variables, and stochastic or sampling uncertainties are handled in a uniform computational and methodological framework based on conditional probabilities, allowing for coherent beliefs. Bayesian analysis is a theoretically sound and practical way to deal with the applications of the precautionary principle and its variants. First, the fewer the assumptions, the clearer the analysis. Second, risk assessors and managers are familiar with probabilities, their calculus and with making inferences. Third, Bayesian conditioning makes explicit the dependence of empirical data on the hypothesis. Howson and Urbach (1993) discuss additional issues relevant to Bayesian analysis.

The probabilistic framework has direct application to making environmental decisions about risk through legal principles. An important legal test that used probabilities is *Learned Hand's* test. It is found in US tort law to determine if a defendant exercised reasonable care. The case where this probabilistic approach was used is *United States* v. *Carrol Towing Co.*[60] Under this test, a person can be liable if the burden of care (measure in dollars) is less than the probability of harm multiplied by the magnitude of the injury. In *Moisan* v. *Loftuss*, Judge Learned Hand discusses the complexities of his test fully recognizing the difficulties in estimating probability and magnitude of the risk, as well as the costs of care.[61] This test is neutral. It neither favors nor penalizes the magnitude of the harm. In other words, it uses the expected value of the cost of protection, is calculated as [*(probability)* times *(magnitude of the cost)*], as a threshold for the duty of care. If the damage is less than the expected cost, there is no liability; otherwise, the person or concern that caused damage can be liable. The expected value, $\sum [pr_i(M_i)]$, M is a magnitude, can be used in making decisions about catastrophic events. However, correction for events characterized by large magnitude can be introduced by restating the expectations as follows: [*(probability)*(magnitude)^k*], omitting the summation. The expected value has $k = 1$: if the magnitude is *10*, then $10^1 = 10.00$. Otherwise k can be either greater or smaller than unity; the value of $k > 1.0$ implies that the magnitude is amplified; $10^{1.5} = 31.62$ or reduced, viz., $10^{0.5} = 3.16$. This change biases the calculation, without explicitly providing a sound theoretical explanation for the choice of bias. Theoretically defensible alternatives to arbitrary exponents are discussed in Chapter 3.

1.10. RISK ASSESSMENT AND RISK MANAGEMENT

Our discussions have dealt with risk assessment and risk management in a number of contexts characterized by precautionary choices. Risk assessment and management consist of a process with several sequential steps taken over time:

{calculate: sources of risk and exposure}$_t$ $\rightarrow$ {calculate: exposure & response}$_{t+k}$ $\rightarrow$ {calculate: net costs of control and risk minimization} $\rightarrow$ {develop: set of choices}$_{t+k+n}$ $\rightarrow$ {calculate: optimal choice} $\rightarrow$ [select and implement optimal choice] $\rightarrow$ {monitor outcomes} $\rightarrow$ [feedback to beginning of process] $\rightarrow$ {modify choice, if needed} $\rightarrow$ {iterate again, if needed} $\rightarrow$ [end].

In this process, *t* is a unit period; *n* and *k* are additional periods. This process requires some knowledge of one or more of the following elements:

Time: events leading to adverse effects are disclosed over time.
Threshold: the value that triggers legal or managerial action to study or to act.

---

[60] 159 F.2d 169 (2d Cir. 1947).
[61] 178 F.2d 148 (2d Cir. 1949).

Source of risk: the hazard.
Exposure: the event that originates from the source of risk (the hazard) and can place individuals at risk.
Magnitude: the number characterizing the numbers of adverse outcomes for those at risk.
Severity: increases in the degree of harm for an outcome on those at risk.
Net discounted social benefits: the algebraic sum of the net direct, indirect, tangible and intangible benefits and costs, discounted to account for the effect of time on the value of money or other discountable quantity.
Uncertainty: It affects each of the elements of the process because they contain random variables. Thus, time (as the arrival of a failure, a random variable), threshold, severity, magnitude, costs and benefits are all characterized by distribution functions.

Uncertainty and variability affect the process of risk assessment and management by causing false positives and false negative errors. Two mutually exclusive, complete, and fully exhaustive choices, which can be affected by chance (the state-of-nature), can result in two errors:

- *Type I*: rejecting the statistical null hypothesis when it is in fact true.
- *Type II*: accepting the null hypothesis when it is in fact false.

A basis for rejecting the null hypothesis is the small probability that it has occurred because of chance alone. This probability is the level of *significance*; it is a probability most often stated as α, ($\alpha = 0.05$, for instance). The implication of the type *I* error is that the researcher mistakenly finds an association, when there is none. Hence, she wants to keep α low. β is the probability of making the type *II* error: accepting an hypothesis (generally, the null hypothesis) when it is false. The *power* of the test is determined by subtracting, from *1*, the probability of committing a type *II* error, ($1-\beta$): power is the probability of rejecting the null hypothesis when it is false. In the context of precaution, the trade-off between false positive and false negative errors suggests focusing on policy (and managerial) choices that minimize false negatives. What matters is that those at risk are correctly identified (at least in terms of their magnitude): if so, the false negative rate is what matters. The language of precautionary principles can be interpreted as attempting to minimize the false positive error rate. It could be modified by letting the minimization of the false positive rate continue to be the critical issue for the risk manager. Therefore, at time *t*, he opts to minimize the type *I* error. However, by adopting a resilient policy that allows updating the initial decision with new information and re-assessing the risks, at time *t+k*, uncertainty is generally reduced: *both* probabilities of error change and have to be assessed as new information becomes available.

1.11. CONCLUSION

Risk assessment develops choices that risk managers can rank according to risk-cost-benefit analysis or other criteria, and implement, monitor and change as new knowledge becomes accepted. In the risk assessment of air pollution, for example, an empirical causal link relates emission of noxious substances, such as particulate matter, gases and so on, to such adverse outcomes as lung cancer, damage to plants and monuments and reduction in visibility. Establishing these links relies on information and knowledge from an equally varied range of disciplines, from molecular biology to epidemiology, and from plant genetics to pathology.

Environmental and health protection rely on scientific evidence to justify legislative, administrative or judicial processes that result in air, water, soil and other environmental standards or guidelines. Applied risk assessment and management are either directly created by law or influenced by concern with liability. The law attempts to prevent or cope with routine and non-routine hazardous conditions likely to affect the environment in which we live. Several different legal principles, from the precautionary principle to *zero risk*, currently exist. Each of those can have equal plausibility, depending on the context within which they are applied. However, the law looks and uses scientific evidence more qualitatively than science. Because of this difference, an environmental risk assessor or manager should understand how the judicial, administrative and legislative branches interpret and use scientific information when commanding precautionary actions. The discussions point to the conclusion that risk, cost and benefits must be balanced in ways consistent with societal preferences. However, there are areas of risk management where society has decided that economic costs are not directly relevant to pollution prevention or when safeguarding health. The heterogeneity of those at risk, from infants to people with AIDS and the ubiquity of the toxic agents, can point to a societal choice of options based on risk minimization or avoidance alone. Nonetheless, the general rule ranks and selects choices by balancing risks, cost and benefits. Social and private decision-makers or stakeholders rely on the representation of uncertainty and variability to make plausible and defensible decisions. A rational, causal link between exposure to a hazard and response is the general legal requirement for both. Such causal links are developed through qualitative and quantitative methods that require a mechanistic basis for building them. The rationale for using probabilities as representations of risk is common to both scientific and legal reasoning. Because risks are practically treated as probabilities, both in science and in management, we have also outlined a formal method for assessing them using Bayesian reasoning. The reasons include the fact that data is often unavailable or partially available, thus preventing frequencies from being calculated or being reliably used. Moreover, causation can be dealt with using such methods as Bayesian networks, which is discussed in Chapter 4.

CHAPTER 2.

SUSTAINABILITY AND MAKING DECISIONS UNDER UNCERTAINTY

Central objective: discuss and exemplify sustainability and risk assessment and management, the legal bases for risk, cost and benefit analysis and criteria for making optimal choices under risk or uncertainty

This chapter is a bridge between Chapter 1 and the more technical chapters that follow. It introduces sustainability as a criterion for guiding human development, specifically in terms of the social and economic aspects of sustainability. Sustainability, which is defined in different ways by different disciplines, generally stands for potential long-term viability and equilibrium, and links with environmental and health risks because it requires assessing social, economic, ecological, health, and other concerns. In this sense, risk assessment and management are used to determine if options are sustainable.

In this chapter, consistently with our overall risk management approach to dealing with uncertain or variable information and causation, we generally think of sustainability as a means to achieve a fair and just distribution of scarce resources, while protecting the environment to the fullest extent possible. The reason for their discussion is that these terms are, at least by implication, value laden because they add respectability to results that may in fact be unsound. Therefore, their usage can bias the understanding and the communication of results that are important to managing risks.

Thus, with sustainability, we also discuss a number of risk-related aspects of decision-making under uncertain conditions, focusing specifically on decision under risk (namely, situations where the distributions are either known or can be assumed). We thus deal with:

- Criteria for making social choices, such as the maximization of expected utility, minimax, maximin, and Pareto's principle
- Scales of measurement for risk factors and outcomes
- Bayesian reasoning and the *policy* basis of cost-effectiveness, and risk-cost-benefit analysis
- Verification, validation and confirmation of a model's structure and results, relative of the *real world* it seeks to represent, as an informal assessment of that model's characteristics

The flow chart below summarizes the principal linkages between these elements:

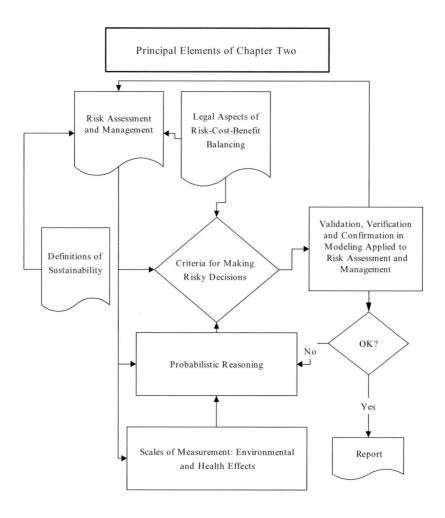

We begin with a review of sustainability. Those definitions set up the factors that are relevant to understanding the outcomes of sustainable actions. Qualitative and quantitative relations between and within them must be set up so that the set of potential outcomes can be described, predicted and used in making the choice deemed to be sustainable. The factors that represent sustainability and risky outcomes are measured on different scales. These scales include the nominal, ordinal, interval or ratio scales. Moreover, uncertainty and variability can affect each factor. Because measurements and operations on these measurements must account for uncertainty and variability, Bayesian analysis is discussed further, continuing and expanding from Chapter 1. However, the eventual acceptability of a model's result by the stakeholders uses terms such as *validation, verification* and *confirmation*.

2.1. SUSTAINABILITY AND RISK MANAGEMENT

There are several definitions of sustainability. The discussions that follow, although not exhaustive, show the breadth and variety of those definitions.

1. *Economic-Demographic. Stabilization of population through increases in the standard of living* (Cohen, 1995).

2. *Deterministic-Mathematical.* The solution to the deterministic logistic differential equation $dN/dt = rN(K-N)$. Sustainability is defined as *the upper level, beyond which no major increase can occur ... is the upper asymptote carrying capacity*, N is population, K is the carrying capacity and r is the (constant) growth rate. The actual model can include crowding, competition and environmental effects.

3. *Regulatory.* The *level of human activity (including population dynamics and economic activity) that a region can sustain at an acceptable quality of life in the long-term* (US EPA, 1973).

4. *Biological(II).* The preservation of global diversity, through the regulation of biological resources, and the identification of activities likely to affect growth rate and significantly impact the *conservation of biodiversity.* The adoption of *economically and socially sound measures that act as incentives for the conservation of biological diversity.* (Wiggering and Rennings, 1997)

5. *Legislative.* "Public lands (should) be managed in a manner that will protect the quality of scientific, scenic, historical, ecological, environmental, air and atmospheric, water resources, and archeological values; ... where appropriate, will preserve and protect certain public lands in their natural condition, ... provide food and habitat for fish and wildlife and domestic animals, and that will provide for outdoor recreation and human occupancy and use." 43 USC Sect. 1701(a)(8), (1988).

6. *Socio-economic.* Sustainable income developed through consensus in which natural resources should remain intact (Hueting and Bosch, 1996).

7. *Physical.* The *sustainable yield* that, for a single resource (e.g., fisheries threatened by depletion), is its *minimum sustainable yield.*

8. *Genetic.* A part of the *essential ecological processes and life support systems* of a community (International Union for the Conservation of Nature, 1980).

9. *Economic-ecological*. "Sustainable scale of economic activity relative to its ecological life support system, not only between the current generation, but also between present and future generations; and an efficient allocation of resources over time." (Daly, 1992).

A comprehensive example of what is meant by *sustainability* can be summarized as (Resources Management Act (NZ) (1991), section 5(2)):

"managing the use, development, and protection of natural and physical resources in a way, or at a rate, which enables people and communities to provide for their social, economic, and cultural well-being and for their health and safety while

- a) sustaining the potential of natural and physical resources (excluding minerals) to meet reasonably foreseeable needs of future generations
- b) safeguarding the life-supporting capacity of air; water, soil and ecosystem
- c) avoiding, remedying, or mitigating any adverse effects and activities on the environment."

Operationally to define sustainability, in the context of the management of risks, requires representations through a dynamic and probabilistic process. Those representations are formal because risk assessment is a formal system of analysis. Consistent with risk assessment, the *physics* (e.g., the ecology of the system under analysis) should be understood so that the effects of policy undertaken to enhance sustainable development can assessed in terms of being able to cope with unforeseeable events. The example that follows explains why.

Example 2.1. Consider the deterministic logistic equation. When the equation is stated as a finite difference (Beer, personal communications, 2002) it is written as:

$$N_{t+1} = rN_t(K-N_t).$$

It represents a population of size N that grows exponentially when it is small but whose growth rate declines as the population increases until it reaches a steady state, when $N_{t+1} = N_t$, or when $dN/dt = 0$ in the continuous case, with solution $N_{t+1} = N_t = K$. However, just because a mathematical model has a steady state solution does not mean that, in practice, that solution occurs. May (1976) used this equation to illustrate that even simple mathematical models can have complicated dynamics, that are often unexpected. This is due to particular values of r and K for which the population does not steadily and monotonically grow to an asymptotic carrying value (the carrying capacity of the system, K). Instead, it becomes chaotic: it results in apparently unstable oscillations. Therefore, even with a simple, deterministic equation, it can be very difficult to determine whether there is a stable solution (representing sustainability) or whether the system is unstable (and thus be unsustainable*).*

To some, sustainability means ensuring that a supply of resources is available for use. To others, sustainability means conserving the existing supply of natural or exhaustible resources, either by forbidding their use or by creating banks or using a combination of methods. The many definitions of sustainability point to the complexity of its physics, factors and linkages. Sustainability, in terms of risk assessment and management, uses indicators to simplify the understanding of that physics. Indicators and indices, however, provide useful information. For example, financial indicators omit social cost of pollution and thus underestimate negative or positive contributions to the balance sheet. Other indicators are linked to the depletion of resources and other significant impacts that are difficult to quantify can mask some effects (Milne, 1996). Enhancing them by accounting for omitted environmental services is critical to proper decision-making and to sustainability.

Sustainability, in the general context of risk management, is an overarching concept that includes some of the essential elements inherent to such accounting. These elements include:

- *Economic Growth*. It is measured by indices such as the gross domestic product, GDP, gross national product, GNP, productivity, and other economic factors.
- *Sustainable Economic Growth*. It is also measured by economic indices, such as the GDP, but accounts for resource depletion, social and environmental constraints, and the potential for irreversible effects on the environment and the interventions that can be taken to change degradation or exhaustion.
- *Assimilative Capacity*. The amount that can be retained without producing short- or long-term adverse impact on the receiving body.
- *Rules or Limits to Resources Exploitation*. For instance, "[h]arvest rates should not exceed regeneration rates; waste emission should not exceed the relevant assimilative capacities of ecosystems; non-renewable resources should be exploited in a quasi-sustainable manner by limiting the rate of deletion to the rate of creation of renewable substitutes." (Wiggering and Rennings, 1997)
- *Causation*. Causation represents the *physics* of the system. It consists of: initiating events(s) resulting in adverse, neutral or positive outcomes, their magnitude and severity, accounting for feedbacks, variable inputs, outputs, and different degrees of knowledge

Because of the number of causal factors, it can be useful to develop indicators and indices to capture the essential elements of sustainability. For example, the GDP is an economic indicator (a composite of several economic factors) that is an index of macroeconomic performance of the economy. Similarly, air quality or water quality indices aggregate several individual air or water quality parameters (Ricci et al., 1977; Ricci, 1979) in a single number.

2.1.1. Indicators and Indices

Indicators and indices are quantitative, single representations of two or more factors that are calculated by a specific formula. Indicators summarily and concisely attempt to represent the underlying characteristics of a complex economic and ecological system. Monetary indicators, for example, measure a set of unique and salient economic factors. For example, the price and quantity of two or more goods or services are estimated at different points in time, and then are related by letting one of them be the base-year for the comparison. Constructing indicators includes considering their variables in terms of:

- The completeness of each indicators or indices selected
- The measures of uncertainty and variability to account for events and relationships summarized by the indicator

- The representations of short-term, intermediate and long-term effects from alternative hypotheses about the structure of the relationships between sustainable development and its outcomes

Indicators are critical in assessing feedbacks by comparing the output of a process to its desired (and evolving) standard of performance. In this context, indicators of performance can compare an actual outcome to the desired outcome, accounting for economic, biological and other dimensions in a way that can give early warning to the stakeholders.

Example 2.2. The Dutch Advisory Council for Research on Nature and the Environment (1992) accounts for ecological capacity, which includes sustainability. Specific sustainability criteria have been suggested for resources depletion and pollution. An indicator system measuring species diversity and performance on a relatively large scale (Weterings and Opschoor, 1997) includes:

1. A sustainable ecological level (e.g., extinction of less than five species/year),
2. An expected ecological level at a specific year (e.g., 2040),
3. A minimal reduction in species loss as a percent of a baseline value, and
4. A minimal amount of contamination from human sources, as a percent of mass to be achieved by a specific year (e.g., 2040).

It follows that definitions of sustainability that have specific relevance to risk assessment and management should describe and summarize:

- The inputs and outputs of the processes involved from the cradle to the grave
- Technological forecasts for processes in which new technology is planned for use
- The reproducibility and consistency of experimental results in different ecosystems and for different species
- Knowledge about and representation of cause and effect relations, including uncertainty and variability
- The experimental and statistical protocols for data development and methods for analysis

These points are pre-requisite to the broad German Council of Environmental Advisors' *essential criteria* for assessing sustainability (Wiggering and Rennings, 1997):

- A reference of actual environmental quality to some sustainable standard
- Describe the temporal and spatial characteristics of the process under study from an ecological perspective
- Account for all ecological functions
- Synthesize the effect of changes
- Be capable of replication and generalization

Because uncertainty and variability affect the assessment of sustainable choices, the results of an assessment should be characterized by distributions, not by single numbers. Moreover, suitable methods, such as Monte Carlo simulations, propagate and combine the uncertainty of the output, when the inputs are variable. The methods to accomplish this task are discussed in Chapter 6.

> **Example 2.3**. Indicators can be used at the local level as well as at the international level to measure the same functions. International macroeconomic indicators can include sustainability (Pearce and Atkinson, in Weterings and Opschoor, 1992). Their indicator includes macroeconomic factors, such as a country's national *saving*, S, national income, Y, depreciation and a measure of *damage to natural resources and the environment*. The resulting Z-indicator is $Z = \{(S/Y)/[(depreciation\ of\ man\text{-}made\ capital)\text{-}(depreciation\ and\ damage\ to\ natural\ resources\ and\ the\ environment)]\}$. Positive values of Z mean sustainability, $Z = 0$ means that the countries so characterized are marginally sustainable, negative values of Z mean that the national economies are not sustainable. For instance, the US's $Z = 2$, Japan's $Z = 17$, Mexico's $Z = 0$ and Ethiopia's $Z = -7$.

Some indicators capture the efficiency of the system. Efficiency is critical because if option A is more efficient (read: more productive for the same level of effort and for the same conditions) than B, all other things being equal (the *ceteris paribus* assumption), then A is preferred to B. These considerations are common to risk management.

> **Example 2.4**. The operation of a power plant is optimized for thermodynamic and other aspect of technological efficiency. Therefore, it is also optimized over the choice of engineering materials, fuels (oil, coal or other), operation and environmental controls. The fact that pollution removal equipment requires electric energy to run, results in some (say, *10%*) loss on the overall efficiency of production and therefore can cost more money to the consumers of energy, relative to an unregulated plant with the same efficiency. *Derating* should be explicitly captured by an indicator.

There are at least two challenges to building indicators. The first is to make them comprehensible by including a few demonstrably essential factors and avoiding complex units. The second challenge is to make them predictive. There are some obstacles to successfully meeting these two challenges. First, there are inconsistencies between the means to analyze sustainability and the definitions of what it is. Second, avoiding complexity and yet being comprehensive can result in a weak indicator of sustainability. The *curse of heterogeneity* can plague the builders of indicators because:

- Heterogeneity among the factors captured by an indicator can be high
- Individual values are known to be heterogeneous
- Social and economic heterogeneity is high

Although the *maximand* (e.g., the objective to maximize *ecological sustainability*) is generally easy to state, the system that is associated with the maximization is often difficult to model because the factors to be included are either not available or, when available, may not be measured with the accuracy required by the decision-maker. The difficulties of measurement and representation that affect implementation of the precautionary principle also affect the assessment of sustainability. It follows that it may be preferable, and consistent with the discussions in Chapter 1, to use analytical criteria to justify either probabilistic or deterministic choices. Moreover, risk assessors and managers use data and models to make decisions and defend their choices using terms such as *verification*, *validation*, and *confirmation*. Some of the controversies that arise in risk assessment and affect risk management can be related to the use of these and other terms that justify why one choice is more credible than another. For example, in risk management, in a public meeting, a group can seek to influence the outcomes of a risky decision by asserting that the models used in risk assessment are *valid,* that the results have been *confirmed,* and that all assumptions have been *verified.* The sections that follow are an introduction to some of them, focusing on verification, validation, and confirmation.

2.2. VERIFICATION, VALIDATION, AND CONFIRMATION

These terms apply to scientific theories, methods for analysis and results (Suppes, 1999). We begin by suggesting that verification, validation and confirmation can be value-laden. Yet, the consequence for a theory or result that lacks such desirable properties can be disparaging or fatal to the theory or results propounded. It is therefore important for risk assessors and managers to understand their limitations. The question of whether a theory is *true* is resolved through *verification.*[1] A system can be verified, meaning that its truth can be ascertained, if it is *closed.* For such systems, it can be formally demonstrated that when the premises are true, the conclusions cannot be false (Oreskes et al., 1994). For example, verification is the correct term to use when demonstrating a theorem and for testing some numerical algorithms, often part of a computer code. However, verification does not apply to the forms of complex causation discussed in legal proceedings and in most risk assessments. The reason is that *inductive* inference, consisting of statistical analysis cannot be deductively verified as theorems, but can be validated, corroborated or confirmed (Boyd, 1991). Specifically, in risk assessment, decision-makers must deal with open systems that are characterized by assumptions that cannot be verified *a priori* (Oreskes et al., 1994). Verification in risk assessment and management is therefore a strong requirement that should be avoided in favor of less demanding terms.

Validation stands for the inductive, independent replication of results. Accordingly, validation focuses on the consistency of the system relative to what it attempts to portray, and includes empirical corroboration from other sources. Validation can include auxiliary assumptions (premises) that enhance it. *Sense-I* validation that meets logical, first principles and internal consistency is weaker than verification, which is truth determining and deductive, but is consistent with assessing risk assessment methods. Validation, in this sense, *corroborates* empirical results through the independent application of first principles and appropriate mathematical procedures. Statistical analysis of environmental systems provides, in conjunction with sound modeling practice, *sense-II* validation, unless the statistical analyses are exploratory or just hypothesis generating. In the latter case, there is no validation, just an exploratory assessment based on empirical findings. A reason is that mere agreement between alternative sources of data *can* provide increasing levels of approximations of the actual system being measured, but cannot guarantee an accurate representation of the physical system. An example is the choice of empirical, inductive models through statistical criteria such as the loss function (Judge et al., 1980). Such a function is related to the error (or bias) made in the

---

[1] The *verification theory of meaning* is due to Wittgenstein, (1922).

estimation of statistical values, relative to the true population parameter of interest, thus leading to choosing the model with the smallest expected loss. Consistency of results, in other words, does not mean that results are accurate, but they can certainly be precise. Models describe and predict dispersion, diffusion and transport of contaminants and bacteriological agents. Statistical models of dose and response often require numerical methods to obtain solutions because their analytical solutions are either too difficult, or are too cumbersome, to achieve results in a reasonable amount of time. *Verification*, in this context, is the congruence of a set of empirical results and a corresponding set of analytical results. When a numerical solution approximates an analytical solution, the approximation does not *verify* the numerical solution. The preferable term for this form of assessment is *bench-marking*.

Confirmation is the process of deductively assessing empirical results that are generated by a scientific law. The concept that *science requires that empirical observations be framed as deductive consequences of a general theory* is the basis for what is generally understood as the confirmation of a theory (Oreskes et al., 1994). Confirming a scientific law means that the observations are empirically true. The strength of the confirmation increases as the number of very similar independent empirical results (stated as magnitude and direction of the relationships) increases. A classical issue with confirmation is that if *a model fails to reproduce observed data, then we know that the model is faulty in some ways, but the reverse is never the case* (Popper, 1959). There are two meanings for the term *reverse*. One refers to Popper's falsification: Popper's tenet that it is not possible, even in principle, to prove that a theory is true. However, it is possible to prove that a theory is false. Thus, if a model reproduces observed data, we still do not know whether the model is correct in all respects. The second meaning of the term *reverse* is that models and data are uncertain. When a model fails to reproduce observed data, it is possible that the model is valid but that the observed data is faulty.

Calibration is *the manipulation of the independent variables to obtain a match between the observed and simulated distribution of a dependent variable* (Oreskes et al., 1994). Understood this way, calibration is a means, but not the ultimate means, to provide *empirical adequacy* to a particular theory (Van Fassen, 1984). Refinements are generally required to achieve an acceptable level of empirical adequacy. In practice, those refinements occur through sensitivity analysis of a model's results. In Chapter 4, which deals with statistical models, it will become clear that variables can also be studied through sensitivity analysis to determine whether they should be linear, be multiplied or take some other form. For example, an independent variable can be changed from linear, X, to quadratic, X^2, from diagnostic analyses.

An important aspect sought when trying to establish empirical causation in risk assessment is the *empirical sufficiency*[2] of the scientific evidence. It depends on the state of the information and on methods to determine it, at the time that the risk-modifying actions are being studied. Empirically sufficient evidence, such as the combination of relevant data and models, must be at least sense-II valid. The results

---

[2] This follows the concepts developed by van Fassen, summarized by Oreskes et al., (1994).

from exercising models must be capable of confirmation, if the mathematical structure of the model has been assessed for errors and omissions.

Empirical sufficiency requires probabilities, or other measures of uncertainty, and expert evidence to construct plausible models. Oreskes et al., (1994) have stated that *confirmation is a matter of degree* and that the *central problem with the language of validation and verification is that it implies an either-or situation.* However, empirical knowledge is probabilistic and conditional on what is known at the time. Perhaps, the *generalizability* of results and *necessary-for-purpose* may be all that is needed (and thus be *sufficient*) for some risky decisions that must be taken under the precautionary principle. For risk assessment and management specifically, empirical sufficiency is a plausible basis for accepting the conceptualization, design, testing and generalizations of a model and its results.

2.3. LEGAL BALANCING OF RISKS, ECONOMIC COSTS, AND BENEFITS

There are three conceptually different ways to balance cost and benefits. These are:

- Cost-effectiveness analysis
- Cost-benefit analysis
- Feasibility analysis

In this section, we briefly summarize them in the context of alternative forms of risk, cost and benefit balancing found in managing the risks of pollution through their legal basis. Chapter 3 contains a discussion of the techniques and theoretical basis of cost-benefit analysis, including the methods to deal with intangible costs and benefits. The examples are American, but the reasoning can be similar to that of other jurisdictions.

Risks can be stated as costs, or health burdens, but they are measured by the change in the probability of an adverse outcome, relative to background risks, from one or more activities. Traditional economic costs include economic, social, and any other impacts that can be stated in monetary units.[3] Those costs and benefits are *monetized*. However, some costs and benefits cannot be directly monetized because they are intangible. Changes in visibility are an example of an environmental good generally not priced by the operations of the market.

Cost-Effectiveness. Economic efficiency drives this form of balancing. It consists of calculating, for alternative pollution control technologies, the efficiency of pollution

---

[3] Coase (1960) has demonstrated that a *frictionless* market allocates production at the point where marginal costs equal marginal benefits, regardless of legal liability rule (if liability is known). This is the optimal allocation between those that supply and those who consume goods and services. In practice, the market imposes considerable costs (including the costs associated with civil liability, such as damages in tort or environmental law) for its smooth functioning. Nevertheless, market operations are not frictionless and therefore the efficiency of production and exchange can be less than optimal (Cooter, 1982).

removal, the total costs of operation and maintenance of each alternative and making the best (least cost) choice allowed by law.

In risk assessment, technological choices and risk reductions can be represented as depicted in Figure 2.1, noting that the efficiency of removal is bounded between *0%* and *100%*.

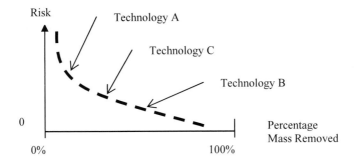

Figure 2.1. Technological Trade-off Analysis, Based on Risk Reduction

The increase in cost in going from the Best Practicable Control Technology (technology A) to the Best Available Control Technology (technology B) may involve an intermediate technology, the Best Conventional Control Technology. Legally, costs were found by US courts to be important, but not likely to limit, the use of expensive technology, unless those costs were disproportionate to the benefits achieved by the selected control technology. The US Federal Water Pollution Control Act imposes a limited form of cost-benefit analysis for water pollution control depending on the level of effluent control.[4] Marginal cost analysis of control technology plays an important role in determining the level of effluent discharge permitted. *Chemical Manufacturers Association* v. *EPA*, which dealt with changes in the marginal cost of controlling effluent discharges, exemplifies how cost-effectiveness can be used in risk-based regulations.[5]

The US EPA can impose a standard unless the reduction in effluent is *wholly out of proportion* to the cost of achieving it. In the US case *Portland Cement Assoc.* v. *Ruckelshaus*, the court held that Section 111 of the Clean Air Act explicitly requires taking into consideration the *cost of achieving emission reductions* with the *non air quality health and environmental impacts and energy requirements*.[6] The court then

---

[4] 33 U. S. C. Section 1251 *et seq.*
[5] 870 F.2d 177 (5th Cir. 1989). The marginal cost is the cost per unit of mass of pollution removed.
[6] 486 F.2d 375, cert. den'd, 417 U. S. 921 (1973).

found that the US EPA met the statutorily required consideration of the cost of its regulatory actions.[7]

Cost-Benefit Balancing. Cost-benefit analysis consists of calculating the present discounted value of the difference between the total *social* costs and the total *social* benefits of one or more actions, discounted at the appropriate social discount rate. Each action has the same objective. Discounting occurs over the economic lifetime of the alternatives considered; it is the means to account for the effect of time on the value of money. The alternative yielding the highest positive difference (B-C) is optimal. The calculations include direct costs and benefits, such as employment and employment effects, as well as indirect costs and benefits, whether tangible or not. This type of analysis also includes anti-competitive effects, if the alternatives being studied cause such effects. Hence, distortion to the labor market, price-induced effects on wages and so on are part of cost-benefit analysis, if relevant to the specific application. For example, the US National Environmental Policy Act (NEPA), one of the earliest modern environmental protection acts, requires the balancing of the environmental costs of a project against its economic and environmental benefits. Although a numerical balancing is often involved, NEPA does not require the calculation of a numerical cost-benefit ratio or the computation of the net cost-benefit number for each alternative taken to mitigate an environmental impact. It should come as no surprise that cost-benefit analysis has generated many disputes between stakeholders such as environmental organizations, industry and the US EPA. A recent legal case exemplifies the issues.

In 2001, the US Supreme Court affirmed that, under Section 109 of the Clean Air Act (CAA, amended)[8], federal standard setting (for the National Ambient Air Quality Standards) for particulate matter and tropospheric ozone must be set *without* consideration of costs and benefits. The new ozone NAAQS was being lowered to *0.08* ppm, and for the first time the EPA was regulating *2.5* microns particulate matter (lowered from *10* microns). The sources of particulate matter being regulated were principally trucks and other diesel engines, as well as mining. The scientific basis for lowering the particulate matter standard was that such reduction would reduce the number of premature deaths by *15,000* per year and serious respiratory problems in children by *250,000* per year. According to the US Chamber of Commerce, to implement these regulations would cost industry approximately *46* billion US dollars per year.

This case, decided by the US Supreme Court as *Whitman, Admin. of EPA, et al.,* v. *American Trucking Associations, Inc., et al.,*[9] addresses four issues, two of which are critical to understanding setting of ambient environmental standards under Section 109(d)(1). This legislative mandate is the *attainment and maintenance ... requisite to protect the public health* with an *adequate margin of safety*. It is explicit. The Supreme Court held that *section 109(b) does not permit the Administrator to consider implementation costs in setting NAAQS. Whitman* stands for the proposition

---

[7] *Sierra Club* v. *Costle*, 657 F.2d 298 (D. C. Cir. 1981) also explains these issues.
[8] 42 U.S.C. Sect. 7409(a)
[9] No. 99-1257, Feb. 27, 2000.

that the requirement by Congress must be clearly stated: if costs and benefits are considered, the Clean Air Act must specifically allow it.

The legislative *adequate margin of safety* does not trigger cost-benefit analysis by itself, as discussed in Chapter 1. It only commands the Administrator of the EPA to adopt an *unspecified* degree of conservatism, as can be represented and achieved through factors of safety. Moreover, not all sections of the Clean Air Act are so stringent, as was held in *Union Electric Co.*, v. *EPA*, a US Supreme Court case dealing with air pollution.[10] This case illustrates the issue of having to consider economic factors (e.g., costs of compliance) when attempting to safeguard human health. Specifically, costs *are* excluded under Section 109, the section governing National Ambient Air Quality Standards, NAAQS, but risks are not.[11] This result was established in *Lead Industries Assoc.*, v. *EPA*, a case that was decided in a lower court.[12] The *Whitman* decision appears to settle this issue: the clear absence of a command to account for costs was not an omission by Congress. Section 109 is protective of human health and welfare. The American Trucking Association argued that imposing stringent air pollution standards would increase the risk to health by increasing unemployment because industry would have to lay-off workers. The residual risks can be the increased number of suicides, violence at home, and other social problems such as those due to unemployment. After reviewing the record from the early days of the Clean Air Act, from 1967 to Feb. 27, 2001, the Court determined that Congress was quite explicit when it wanted to include any consideration of cost. It had done so in other cases involving water and air pollution. So far as the CAA, Section 109, the Court found that Congress clearly meant to be protective: [13]

"the health of the people is more important than the question of whether the early achievement of ambient air quality standards protective of health is technically feasible. ... (Thus) existing sources of pollutants either should meet the standard of the law or be closed down ...."

The Supreme Court rejected the American Trucking Association's contention that the phrase *adequate margin of safety* does not provide *an intelligible principle* of protection, therefore being vague and thus unconstitutional. The CAA requirement is that the EPA must establish:

"(for the relatively few pollutants regulated under Sect. 108) uniform national standards at a level that is requisite to protect public health from the adverse effects of pollutants in the ambient air."

The term *requisite* legally means *sufficient, but not lower or higher than necessary*. This is the fundamental constraint that bounds the US EPA's ability to set standards: if it is exceeded, then the EPA has violated the Constitution because it now has *legislated* change. The Court also held that Section 109(b)(1) of the CAA does not grant the US EPA legislative powers: an agency cannot use its own interpretative

---

[10] 427 U.S. 246 (1976).
[11] It was established by Congress under Section 109(b)(1).
[12] 647 F.2d 1130, 1148 (CA DC, 1980).
[13] *Senate Report* No. 91-1196, pp. 2-3 (1970).

discretion to somehow gain what it cannot do. Moreover, the US EPA cannot arbitrarily construe statutory language to nullify *textually applicable provisions meant to limit its discretion.* The US Constitution prohibits the Executive branch of the government from legislating. Legislation is the prerogative of the Congress of the United States. Justice Breyer agreed that the CAA is unambiguous on this point: costs are not included in the determination of the NAAQS. He concluded that *a rule likely to cause more harm to health than it prevents is not a rule that is requisite to protect human health.* Section 109(a) does not preclude comparative risk assessments.

Feasibility Analysis. Feasibility analysis determines the extent to which some firms in an industry can fail, if that industry as a whole is not threatened. The phrase *to the extent feasible* was held to mean *capable of being done* by the US Supreme Court, when it found that Congress had done the appropriate cost-benefit balancing. This fact excluded an agency of the government from doing such balancing, particularly when Congress intended to place *the benefit of the workers' health above all other considerations, save those making the attainment of the benefit unachievable.*[14] The federal agency involved in this dispute was the federal Occupational Health and Safety Administration, OSHA, which had estimated the cost of compliance with the new standard, determined that such costs would not seriously burden the textile industry and would maintain *long-term profitability and competitiveness* of the industry (Ricci and Molton, 1981). The relevant section of the Occupational Health and Safety Act, 29 USC §§655(b)(5) and 652(8), read as follows:

"The Secretary, ... , shall set the standard which most adequately assures, to the extent feasible, on the basis of the best available evidence, that no employee will suffer material impairment of health or functional capacity even if such employee has regular exposure to the hazard dealt with by such standard for the period of his working life."

We have discussed the meaning and implications of *most adequately* and *to the extent feasible*, as legal phrases that determine the acceptability of risk management choices. This quote contains the phrase *best available evidence* and further exemplifies the inevitability of the intersection of science and law.

2.4. CRITERIA FOR MAKING CHOICES IN RISK MANAGEMENT

Making decisions about risky activities requires considering one or more criteria to justify the selection of one choice over another. For all of those criteria, the decision-maker knows and uses the set of mutually exclusive, fully exhaustive and complete set of possible choices. Some of the criteria apply to the single decision-maker, other to more than two decision-makers. Some of the best-known analytical criteria for making choice in risk management include the following.

Maximum expected utility criterion. If certain axioms are accepted, then the best choice that the single decision-maker can make is the one that has the highest expected utility. The choices are described as lotteries with pay-offs associated with

---

[14] *American Textile Manufactures Assoc.* v. *Donovan, U. S. L. Week* 49: 4720 (1981).

them. The optimal choice is guaranteed by a theorem that demonstrates the decision-maker consistency with the axioms (the axioms of rationality).[15]

Maximin, maximax and minimax criteria. The *maximin* criterion consists of selecting the action that maximizes the minimum payoff between the actions open to the decision-maker. The *maximax* criterion selects the maximum of the maxima. An alternative criterion is to minimize the maximum loss: this is the *minimax* criterion.

	Potential, Mutually Exclusive and Fully Exhaustive Options (ranks: 0 = least preferred, 100 = most preferred)		
Severity of Outcomes	*Ban Exposure*	*Limit Exposure*	*Do Nothing*
High	90	30	0
Medium	60	70	40
Low	0	30	100
Maxima	*90*	*70*	*100*
Minima	*0*	*30*	*0*

Example 2.5. Consider the following situation (the units are arbitrary and the values are deterministic). A single decision-maker who faces three options with three outcomes:

The minimum of the maxima is *70*, which is associated with the option *limit exposure*. The maximum of the minima is *30*, which is associated with *limit exposure*. The maximum of the maxima is *100*, which is associated with *do nothing*. The maximin criterion leads to choosing the option *limit exposure*. The minimax criterion leads to choosing the option *do nothing*. Additional discussions and analyses are found in Anand (2002) who inspired this example.

The maximax criterion is the most optimistic; the maximin is conservative while the minimax is optimistic. These three criteria are not based on uncertainty or risk. Although they can be considered to be deterministic, they may actually be used in situations where there is considerable uncertainty and the decision-maker feels uncomfortable in assigning probability numbers.

The pessimism-optimism criterion (Hurwicz). This criterion uses an index of *pessimism-optimism* (Luce and Raiffa, 1957) to represent a single-decision-maker's attitude. The index is:

$$H = \{(kM) - [(1-k)m]\},$$

Where m is the minimum, M is the maximum monetary value of the ith action and the term k is contained in the interval $(-1 \leq k \leq 1)$. The preferred option is one for which the index is the largest. This criterion includes the *maximin* and *maximax* criteria and can use utilities rather than monetary or other measures of value. The meaning of the constant k is as follows. If k is close to zero, then the importance of the maximum pay-off is reduced, relative to the importance of the minimum pay-off. Otherwise we obtain the opposite results, as the reader can easily calculate.

---

[15] The axioms, based on binary preferences, are complete ordering, continuity, substitution and probability ordering between lotteries, and indifference between equally preferred lotteries.

Thus, if alternative A is characterized by $k = 0.8$, with the minimum $10,000 and maximum at $100,000, we obtain: $H_A = (0.8)(100,000)+(0.2)(10,000) = 82,000.00$. If H_A has the highest value it should be selected from its alternatives.

Minimax opportunity loss (regret) criterion. We begin by introducing the following decision table:

	Consequence or Outcomes	
Actions	*Positive*	*Negative*
A	200	-180
B	100	-20
C	0	0

We then define the opportunity loss, the regret, as the maximum value associated with not selecting the best action, *given the outcomes*. Thus, for a positive outcome, the regret is *200*. For a negative outcome, the regret is *0*. It should be clear that the decision-maker does not know which outcome is going to occur: he has no control over them but he does have control over which actions he can take. The table of regrets for all consequences is:

	Consequence or Outcomes	
Actions	*Positive*	*Negative*
A	200-200 = 0	0-(-180) =180
B	200-100 = 100	0-(-20) = 20
C	200-0= 200	0-0 = 0

The minimax *regret* is 100, that is: *min(0, 1, 20, 100, 180, 200)* = *100*: choose action *B*.

Pareto's criterion. The Pareto efficient solution is a point where the welfare of one of the parties cannot be improved without damaging that of the other. The analysis is based on a *contract* curve, which shows the preferences of two (or more) individuals for two goods or services. A Pareto improvement is a reallocation of resources that will make one party better off without decreasing the well-being of the other.[16]

Example 2.6. Consider the following situation: A and B have a set of preferences for butter (in kilograms per year) and for bread (in kilograms per year). Those preferences are depicted by the curves labeled $a_1, \ldots$, and $b_1, \ldots$.

Note that this diagram has two sets of axes and that the third, the utility axis, is not shown because it is perpendicular to the diagram:

---

[16] The necessary condition is that the marginal rates of substitution for the goods or services are the same for the two parties contemplating the exchange. The fundamental theorem of welfare economics guarantees this result for a perfectly competitive economy (Baumol, 1977).

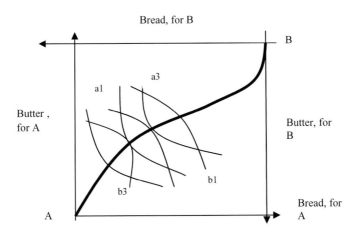

Bread, for B

Preference curves are slices taken horizontally to the axes shown and projected on the two-dimensional plane: these curves are a contour map. It shows increasing preferences, measured by *utilities*, are constant along any single preference curve, but increase in the NE direction for *A* and in the SW direction for *B*. For example, curve b_1 is less preferred than curve b_3, curve a_1 is less preferred than curve a_3. Each preference curve identifies the combinations of kilograms of butter and kilograms of bread that have the same utility and therefore have the same preference level, for instance a_1. An efficient solution to making the exchanges that can be represented in this diagram (known as the Edgeworth diagram) occurs where two preference curves are tangent to each other. The contract curve is the thick black line, beginning at *A* and ending at *B*. It identifies all possible sets of efficient solutions to the potential exchanges of bundles of bread and butter between *A* and *B* (Katz and Rosen, 1998).

Example 2.7. Following Straffin (1993), consider the following situation:

		Individual A	
		Strategy A1	*Strategy A2*
Individual B	*Strategy B1*	(3, 3)	(-1, 5)
	Strategy B2	(5, -1)	(0, 0)

This situation illustrates that there is only one strategy that is not Pareto optimal: the values in cells *A2, B2*.

Nash Equilibrium. The optimal solution is the point where product of the gain in utility is the highest, if the point of no agreement is the least preferred. In other words, at a Nash equilibrium, the stakeholders would not rationally change their equilibrium positions unilaterally. Utility theory applies and the choice of contract is independent of the scale of measurement. There are several axioms that have to be satisfied to guarantee Nash's result. Recent work has dealt with the lack of an appropriate process of negotiation, situations with incomplete information, and a process of negotiations that include the dynamics of exchange (Samuelson, 1984; Roth, 1985; Katz and Rosen, 1998).

Nash's work helps to resolve the issue characterizing the equilibria, represented by the outcomes of different strategies, including probabilistically weighted ones. The optimal strategy of each individual must be identified, given the corresponding strategy of other individuals. Nash equilibrium or equilibria, depending on the situation, are the result of having identified strategies that are at a particular type of equilibrium. One or more Nash equilibria may not be optimal for a stakeholder.

Example 2.8. Consider the following table of pay-offs between two individuals (after Straffin, 1993):

		Individual A	
		Strategy A1	*Strategy A2*
Individual B	*Strategy B1*	(2, 4)	(1, 0)
	Strategy B2	(3, 1)	(0, 4)

Suppose that both decide to use probabilities. Then, *B*'s strategy is to combine her strategies (*0.43* act, *0.57* do not act). This mixed strategy, if adopted by *B*, yields the expected value *2.28*, which does not depend on what *A* does. Alternatively, *A* can adopt the mixed strategy (*0.5* act, *0.5* do not act) which ensures that *B*'s expected value is *1.5*, regardless of what *B* opts to do. These strategies are at equilibrium, although this equilibrium may not be optimal. Notably, these strategies are such that neither *A* nor *B* can gain by adopting another: this is the Nash equilibrium.

Once we adopt a criterion, it can be useful to guide risk-based decisions as follows:

- Rank and justify how an act was chosen, relative to all other options available
- Monitor the outcomes of a decision after the choice is made and the results are available
- Provide a re-ranking of options as new information becomes available, through accounting for changes in the information and preferences that may occur over time
- Decide whether to continue research, or stop or take action by accounting for the value of new information and its cost.

Typically, several attributes (or factors) characterize each decision; each factor can be measured on a different scale. For example, attributes of a decision include *dollar, mass, severity, sex,* and so on. Some attributes are qualitative; others are quantitative. The differences in the units of measurement suggest that there are different scales over which those outcomes are measured. If so, a plausible question is: What mathematical and statistical operations can be made on those scales of measurements?

Another reason for our interest in scales of measurements is that it is important to determine how to represent different observations through a formal system and whether the scales over which the measurements are obtained are unique (Luce and Narens, 1990). We will describe four such scales: nominal, ordinal, interval and ratio because these are often encountered in risk assessment and management. As briefly described in this chapter, risk management can consist of either a single or multiple decision-makers. The single decision-maker can be enlightened and account, in her deliberations and choices, for the preferences of several stakeholders. She can also act as a despot. Often, in many practical situations there will be more than a single decision-maker. In the rest of this chapter, we focus on the single decision-maker.

2.5. A PARADIGM FOR DECISION-MAKING: The single decision-maker

A generic decisional problem concerns a decision maker who faces a number of possible actions, the ultimate acceptability of the choice depending on chance. The issue is that the decision-maker cannot control the outcome of the action chosen because chance (which we call *nature* or *states-of-nature*) can prevent the desired outcome from occurring. This is the *decision problem under uncertainty* (Luce and Raiffa, 1957).

Example 2.9. In this example, *Nature* can yield outcomes that are different from those the *Decision-maker* calculates to occur from his three possible, but mutually exclusive, options (Act *1*, Act *2* and *Act 3*). The outcomes are represented by the notation (5, -1), ... , (2, -8):

		Nature	
		Yes	*No*
Decision-	*Act 1*	(5, -1)	(10, 40)
maker	*Act 2*	(1, -1)	(3,-10)
	Act 3	(0, 0)	(2,-8)

Specifically, the entry (5, -1) means that if the decision-maker chooses act *1* he can gain 5 units of value. If chance (nature) prevails, he loses the *1* units of value, namely *-1*. In other words, a benefit from a decision can have a positive outcome but nature can produce a completely different outcome.

This example does not make uncertainty explicit. To do so, we can use probabilities. In this case, the problem is one of decision-making under risk. Those who engage in practical risk assessment and management have several ways to think about how uncertainty affects the process of making decisions. One is based on probabilities as a measure of uncertainty. This leads to using probability theory for problem solving because it can be used to deal with both uncertainty and variability in transparent, consistent and coherent ways. Another practical way for reasoning about uncertainty is more intuitive, can include probabilities to represent specific aspects of uncertainty, but need not always do so. That is, uncertainty is represented by qualitative statements such as *it is possible that* ..., or similar phrases.

In risk assessment, as we discussed, the prevalent measure of uncertainty and variability is the probability. It can be used within the protocol that follows:

1. Define and identify the boundaries of the risky choice or problem,
2. Define working hypotheses, scientific conjectures and *transcientific* issues as well as by giving a qualitative description of the processes leading to all relevant outcomes,
3. State *statistical* hypotheses,
4. Determine the state-of-knowledge and represent it through literature reviews and prior distributions,
5. Assess the need for additional experimental work, develop stopping rules for limiting the gathering of costly information, based on value of information calculations and explicitly include the cost of adding information,

6.   Conduct an experiment or experiments designed to provide likelihoods (conditional probabilities),
7.   Model empirical, causal relations between source of hazard and adverse effect; account for uncertainty through measures of uncertainty (e.g., upper and lower probabilities, intervals, possibilities or in some other ways),
8.   Make inference from a sample to the population from which the sample came (e.g., using confidence bounds),
9.   Establish formal expression of the relations between initiating events, outcomes, policy interventions, costs, benefits and risks, including the discounting of economic costs and benefits as well as the uncertainty about these,
10. Choose the criterion that is consistent with risk-reduction (e.g., minimax cost, minimax regrets, maximize expected monetary value or utility) and explain the rationale for the choice of criterion for selecting an alternative,
11. Identify residual risks (those risks that are corollary to the main risk-reduction activity),
12. Choose the optimal strategy: interventions (e.g., mitigation works, monitoring exposures indoors), policy (e.g., lawsuit) or political (e.g., repeal or lack of reauthorization of a statute by legislative means) or other,
13. Communicate risks and the net benefits, include monitoring and feedback from and to the stakeholders to the decision.
14. Resolve outstanding issues,
15. Closure.

In these steps there is an implicit symmetry among those who are at risk and those who benefit because the process for making risk-based decisions should be, at least, fair on the average.

2.6. PRACTICAL PROBABILISTIC REASONING: An elaboration

To expand from Chapter 1, the key concepts are as follows. The prior probability is a subjective degree of belief, based on knowledge, encoded in either a probability or a distribution. The conditional probability describes the effect that the probability of an event that can occur will have on another event yet to occur: the two events influence one another. For example, for events *E that has occurred* and *A that* has not, then:

$$pr(A|E) = [pr(A \text{ AND } E)]/pr(E).^{17}$$

It follows that, if *A* and *E* are (statistically) independent, *pr(A AND E)* is simply the product *pr(A)pr(E)*. Bayes' theorem also applies to probability density functions and mass functions discussed in later chapters. Recollect that Bayesian probability theory updates subjective belief (which are expressed as a prior probability or distribution based on past information) through experimental results, the likelihood. Given a prior probability and the likelihood (which is a conditional probability), Bayes' theorem[18] yields the posterior probability, as exemplified next.

---

[17] This is a key axiom of probability theory.
[18] A demonstration of Bayes' theorem follows. From *pr(A AND B) = pr(A)pr(B|A)* and *pr(B AND A) = pr(A)pr(A|B)* equate the right hand sides to obtain *pr(A|B) = [pr(B|A)pr(A)]/pr(B)*.

Example 2.10. Suppose that historically *10* individuals in a set of *1,000* are known statistically to have bladder cancer. Of these *10* individuals, *eight* actually test positive for a gene known to predispose to bladder cancer. Therefore, *two* who have the bladder cancer test negative for that gene. Of those *990* individuals who are statistically without the bladder cancer, *100* will test positive for the gene, diagrammed as follows (following Petitti, 1999):

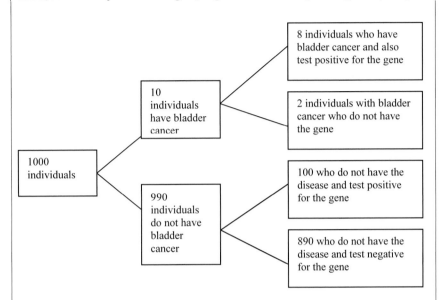

Thus, the probability of bladder cancer, *pr(cancer)*, is *10/1,000 = 0.01*. The probability that an individual tests positive for the gene given that he has cancer, *pr(positive|cancer)*, is *8/10 = 0.80*. The probability that an individual tests positive for the gene but does not have the cancer, *pr(positive|no cancer)*, is *100/990 = 0.101*. The probability of no disease is *990/1,000 = 0.99*. The application of Bayes' theorem yields the posterior probability, *pr(bladder cancer|test+)*:

pr(bladder cancer|test+) = pr(bladder cancer)pr(test+|bladder cancer)/[pr(bladder cancer)pr(test+|bladder cancer)+pr(no bladder cancer)pr(test+|no bladder cancer)].

Using the data in the example the posterior probability is:

[(10/1,000)(8/10)]/[(10/1,000)(8/1,000)+(990/1,000)(100/990)] = (0.008)/(0.018+0.10) = 0.067.

Bayesian methods are independent of the area of application. Thus, they have been applied to medical decisions, risk assessment of radioactive soil contamination, water resources analysis, risk management and economic analyses. It is important to be able to capture subjective beliefs from such sources as the literature, experts or other appropriate sources about the distribution of events before the experiment. Four key concepts of Bayesian analysis relevant to risk assessment and management are expanded next.[19] Recollect that: 1) a prior distribution is the distribution that a risk assessor *believes* to be true, for the particular problem she is assessing; 2) the experimental data is used to form the likelihood. Let *F(b)* denote the prior (subjective) joint distribution of uncertain quantities in the argument, *b*. Then, *(F|L)* is the posterior probability distribution (developed by applying Bayes' theorem to the prior distribution and the likelihood function) for the uncertain quantities in *b*. When the calculation of the updated probabilities *(F|L)* becomes computationally burdensome, methods based on belief networks, discussed in Chapter 6, make the calculations manageable (Pearl, 1988).

(a) *Probabilities or probability distributions represent knowledge and beliefs.* Prior probability distributions (such as density functions, for continuous data and probability mass functions, for discrete distributions) represent prior knowledge, information, judgments and beliefs for the independent variables. This is an aspect of analysis that cannot normally be resolved by classical statistical methods based on frequencies.

(b) *Likelihood functions represent empirical evidence.* All empirical data and modeling information are summarized by likelihood functions, also discussed in chapter 4. Given a probability model *pr(y; x, b)*, the corresponding likelihood function for the parameter *b* (possibly a vector, hence the bold letter) is simply *pr(y; x, b)*. It is a function of *b*, instead of a function of X and Y. The pairs of observed *x* and corresponding *y* values are data from the X and Y random variables. Thus, representing data as vectors *(x, y) = (x$_i$, y$_i$)*, *i = 1, 2, ..., N*, the likelihood for a value *b* of the parameter vector is the conditional probability of observing the data *y*, conditioned on the data *x*. The quantity *L(b; x, y, pr)* denotes the likelihood function for *b* based on observed data *x* and *y* of *n* individual *(x, y)* pairs, assuming that the probability model is correct. The maximum likelihood estimate, MLE, of *b* is the value of *b* that maximizes *L*, given the observed data and the model that has been assumed to be correct and used in the analysis.

Example 2.11. Suppose that a study consists of *six* individuals, *two* of whom fail, $F = 2$, leaving *four* successes, $S = 4$. The likelihood function for binary outcomes, assuming independence, is calculated from the formula $pr^F(1\text{-}pr)^S$. This is what is meant by *assuming a model*. What value of *pr* maximizes its value? The value that does this is the maximum likelihood. The expression $pr^F(1\text{-}pr)^S$, in logarithmic form, is: *(F)ln(pr)+(S)ln(1-pr)*. Using several arbitrary values of *pr*, we obtain:

Probability of failure.	Log-likelihood Function and Result
0.01	$2\ln(0.01)+4\ln(0.99) = 2(-4.605)+4(-0.010) = -9.25$

---

[19] The discussion uses the term *vector* which is a mathematical quantity characterized by direction and magnitude. A set of *n* observations on the random variable X, that is *[x$_1$, x$_2$, ..., x$_n$]*, is a vector. For example, a sample of concentrations of SO_2 is a vector, a sample of counts of daily hospital admissions for September 2001 is also a vector (with *30* values, *n = 30*), and so on. The data for a problem may require more than one vector. In this section we bold the letters representing vectors: *x = [x$_1$, x$_2$, ..., x$_n$]*. Two or more vectors can be represented by a matrix, introduced later.

0.10	$2\ln(0.10)+4\ln(0.90) = 2(-2.30)+4(-0.105) = -5.02$
0.20	$2\ln(0.20)+4\ln(0.8) = 2(-1.609)+4(-.22) = -4.10$
0.40	$2\ln(0.4)+4\ln(0.60) = 2(-0.92)+4(-0.51) = -3.88$
0.50	$2\ln(0.50)+4\ln(0.50) = 2(-0.69)+4(-0.69) = -4.14$

The maximum occurs approximately at $pr = 0.40$; the search for the maximum can be refined using smaller probability numbers near 0.40, but this is unnecessary for this example. Also note that we show only a part of the full likelihood; the full formula is discussed in chapter 4.

(c) *Updating prior beliefs with observed evidence using likelihoods.* Given the prior beliefs $F(b)$ about the uncertain quantities b and the evidence, measurements, model, and data summarized in L, posterior beliefs (meaning the combination of the prior and the likelihood) about b are computed by Bayes' theorem.

Example 2.12. Suppose that there are F failures for a total number of time intervals N, such that each interval has duration t; t is sufficiently small relative to N. The total length of time is the product $(N)(t)$, shortened to Nt. Let π be the conditional probability of failure in a period of time t. This means that the individual at risk has survived until the previous period. The likelihood function (omitting the combinatorial term, see Chapter 10) is: $[\pi^F(1-\pi)^{N-F}]$, because the distribution of binary outcomes is binomial. The rate parameter is, under the assumption that t is small, $\lambda = \pi/t$. The (simplified) likelihood for the rate parameter λ is $[(\lambda t)^F(1-\lambda t)^{N-F}]$. The following transformation is generally used: $[(F)\ln\lambda+(F)\ln(t)]+[(N-F)\ln(1-\lambda t)]$. In continuous time, the likelihood becomes $\lambda^F \exp(-\lambda Nt)$, which is called the Poisson likelihood.

Example 2.13. Suppose that exposure to air pollution is known to cause $F = 10$ cases of asthma, and that the total observation time is 1000 person-years. The log-likelihood function, using the Poisson distribution, is $[(10)\ln(\lambda 1000)-1000\lambda]$. Then, the rate is $\lambda = 10/1000 = 0.01$ cases of asthma per person-year of exposure to the pollutant. Suppose that $F = 1.0$ and that the total observation time is 500 person-years, then $\lambda = 1/500 = 0.002$. Suppose that $\lambda = 15/100 = 0.15$, and so on. The question is: Does the log-likelihood reach a maximum for a specific value of the rate parameter λ? The plot of the log-likelihood as a function of the rate parameter λ can be used to investigate where the function reaches a maximum, noting that the log-likelihood will be negative. The rate of failure can be used to calculate cumulative probability of survival over a specific period of time $T = Nt$. Assuming that λ is constant over time, then the conditional probability of failure in any interval of time is λt; the probability of surviving is $(1-\lambda t)$. If probabilistic independence holds, then the cumulative conditional probability of surviving N time periods is $(1-\lambda t)^N$. When the log-form of the conditional cumulative distribution, namely $[N\ln(1-\lambda t)]$, is small because t is sufficiently small, that quantity can be replaced by $-\lambda(Nt) = -\lambda T$, which is the cumulative rate of failure. For rare events, the expression $\ln[\exp(-\lambda T)]$ equals $\ln(1-\lambda T)$; the latter is often called the cumulative risk.

(d) *Managing alternative models and assumptions.* An important step in Bayesian analysis is the calculation of:

1. The posterior beliefs $(F|L)$ from prior beliefs that are encoded in F, and
2. The assumptions and evidence encoded in L, which depend on the probability model, pr, short for $pr(y; x, b)$, and on the observed data (namely, x and y).

In practice, uncertainty about the correct model, pr_i, out of several alternative models, is often the largest affecting the analysis. To account for it, let $\{pr_1,...,pr_n\}$

denote the set of *alternative models* that are known to be (or are considered) mutually exclusive and collectively to exhaust all possible probability models. Let $L_1,...,L_n$ denote the likelihood functions for alternative models, and let $w_1,...,w_n$ be the corresponding judgmental probabilities, also called *weights of evidence,* that each model is correct. If the models are mutually exclusive and collectively exhaustive, these weights must sum to *1*. The posterior probability distribution that is obtained from the prior *F*, data *(x, y)*, model weights of is the weighted sum: $w_1(F|L_1)+...+w_n(F|L_n)$. The advantages of the probabilistic reasoning as just set out include the following:

1. Showing the *distribution of values* for the true parameter *b*, (or vector *b*) based on available information and scientific judgments. The probability distributions show how much statistical *variability* can be reduced by further research (which would narrow the distributions, but not generally result in a single deterministic value).
2. Distinguishing the *contributions of different sources of evidence,* identifying specific areas where additional research is most likely to make a significant difference in reducing final uncertainty. This deals with the *uncertainty* of model building.

Despite the advantages, probabilistic methods have limits to their ability to represent uncertainty. The approach is appropriate for a single decision-maker concerned with a single decision because all that matters to this manager is the joint posterior distribution function for all uncertain quantities. However, when more than one decision-maker is involved, the unified probabilistic presentation of different *types* of uncertainties (e.g., observer-independent stochastic characteristics with theoretical assumptions, subjective judgments, and speculations about unknown parameter and variable values) is as much a liability as an asset. Other limitations include the following.

Sources of prior probabilities. It is often unclear *whose F* should be used, if different experts have different opinions. Two individuals with identical but incomplete objective information might express their beliefs with different prior probability distributions. Moreover, probability models cannot adequately express ambiguities about probabilities. For example, an estimated probability of *0.50* that a coin will come up *heads* on the next toss based on lack of information may not be distinguished from an estimated *0.50* that is based on *10,000* observations.

Unknown and incomplete probability models. A problem related to the need to specify a prior probability is that the assumed probability model *pr(y; x, b)* often requires unrealistically detailed information about the probable causal relations among variables. The Bayesian view is that an analyst *normatively* should use either his or her own knowledge and beliefs to generate a probability model when objective knowledge is either incomplete or even inadequate. The opposing view is that the analyst has no justification, and should not be expected or required, to provide numbers in the absence of substantial and relevant knowledge. This point of view has led to *Dempster--Shafer belief functions* to account for *partial knowledge.* Belief functions do not require assigning probabilities to events for which there is a lack of relevant information. Specifically, part of the probability mass representing beliefs can be uncommitted. The result is that instead of the posterior *(F|L)*, the analyst obtains a posterior belief function from which only upper and lower bounds on posterior probabilities can be derived.

Posterior probabilities based on weights of evidence are ambiguous. When the correct model is unknown and multiple models and weights are used, or when multiple sources of evidence giving partially conflicting posterior probabilities are combined, the resulting aggregate posterior probability distribution is *ambiguous.* An infinite variety of alternative models and weights are mapped by the formula $[w_1(F|L_1)+,...,+w_n(F|L_n)]$ onto the same aggregate posterior probability distribution for the risk estimate. A partial solution to these problems is to present posterior distributions and corresponding weights for each model separately.

Example 2.14. Take two competing models, Model *1* and Model *2*, and two sets of data, Experiment *1* and Experiment *2*. The sets of data are used to estimate the parameters of those two models. Because there are two alternative models and two sets of data, the risk assessor develops a decision tree to clarify the situation he confronts. The circles represent chance nodes:

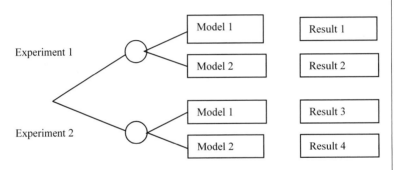

An assessor's degrees of belief (probabilities) are assigned to the branches of the tree. Suppose that the distribution of the final results for that assessor is:

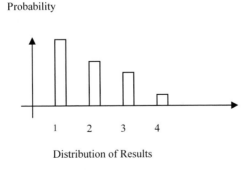

Distribution of Results

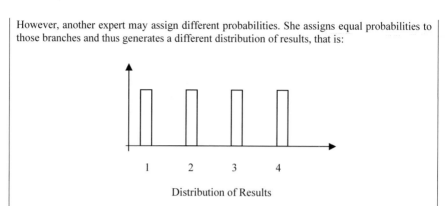

However, another expert may assign different probabilities. She assigns equal probabilities to those branches and thus generates a different distribution of results, that is:

Distribution of Results

For this assessor, her distribution is rectangular: all four results are equally probable.

Some relevant scientific knowledge cannot be expressed probabilistically. Some of the knowledge used to draw practical conclusions about risks can be abstract or non-quantitative. Other aspects of qualitative knowledge can be used to constrain probabilistic calculations, but cannot be represented by probabilities. For example, the statement *the dose-response function is smooth and s-shaped* can be used to constrain non-parametric estimates of a statistical model. However, this statement is not about the probable value of a variable conditioned on the values of other variables. Hence, it cannot be explicitly represented in these terms.

A set of mutually exclusive, collectively exhaustive hypotheses about the correct risk model is seldom known, making the use of probabilistic *weights of evidence* for different possible models inexact. Probability models inherently make the *closed world assumption* that all the possible outcomes of a random experiment are known and can be described (and, in Bayesian analysis, assigned prior probabilities). This assumption can often be unrealistic because the true mechanisms may later turn out to be something entirely unforeseen. Conditioning on alternative assumptions about mechanisms only gives an illusion of completeness when the true mechanism is not among those considered. Notwithstanding these issues, the usefulness of Bayesian analysis stems from several considerations. First, it provides a formal method for updating prior scientific knowledge by requiring the researcher to think in probabilistic terms. This requires the risk assessor having to specify *how* past information can be folded into a distribution function. Second, the analyst deals with distributions, rather than with single numbers. Third, the likelihood links the sampling design to the structure of the model most likely to be determined by the sample. Fourth, the assessor must disclose the reasons for her choice of prior distribution, which is shown through the form of the prior distribution function adopted and clarified by the reasons for adopting that distribution.[20] Finally, the analysis is transparent and can be discussed for lack of completeness, arbitrariness of assumptions, and adequacy of the experimental results. Discussions, aided by a

---

[20] As discussed in Chapter 11, distribution functions often have a theoretical basis, not just data-driven justification.

formal framework, provide a common and verifiable basis for assessing alternative assumptions, defaults and the adequacy of the data.

2.6.1. Comment

Traditionally, risk managers can account for uncertainty by selecting statistical decision rules based on hypothesis testing and confidence intervals. An important aspect of using the likelihood in the context of risk assessment is that it can be used to study the performance of estimation. For example, confidence limits can show the degree of empirical support for a parameter value by using the likelihood ratio. It is the difference between the maximum of the log-likelihood and another lower log-likelihood value. The interpretation of confidence intervals in Bayesian analysis differs from the classic confidence intervals. The frequentistic interpretation (meaning that probabilities *are* frequencies and *not* degrees of belief) is predicated on large samples or exact computations based on combinatorial analysis. Using frequencies, the analyst has sound theoretical reasons to expect that the confidence interval can contain the unknown population parameter, given the structure generating the data and the sample subsequently taken, with a specific probability (say, *95%*). The Bayesian interpretation of confidence intervals is a probability, relative to the evidence, that an unknown parameter is found within a specified interval (Houson and Urbach, 1993). De Finetti states that:

> "we would give an interval having some stated probability of containing X ... (in general *100β%*)... following Lindley, we could refer to this interval *[x', x"]* as a *100β%* (Bayesian) confidence interval for X."

The Bayesian confidence interval is the *100β%* probability that the true value of the population parameter lies within that interval. Contrast this view with the frequentistic interpretation of a confidence interval that is valid only if (infinitely) many samples were taken. When so, then the probability that those intervals would bracket the true population parameter is *100β%*. It follows that, from a single sample, the analyst will not know whether the confidence interval that is constructed about the unknown parameter actually covers it, unless the large sample assumption holds. In most jurisdictions, setting environmental standards and guidelines requires some form of risk, cost and benefit balancing, as well as their characterization through probabilistic methods. Chapter 3 provides the details of the methods used in making and using risk-cost-benefit calculations.

2.7. SCALES OF MEASUREMENT AND IMPLICATIONS FOR RISK ANALYSIS

Risk assessors measure different objects (age, sex, mass and so on) and thus use different scales of measurement. The principal scales are as follows.

Nominal scale. There is no ranking or order in the measurements made on this scale. An example is the letters and numbers on automobile number plates. These are simple labels that uniquely identify a car but do not give an order (from small to high) to those plates. A statistical measure of central tendency for nominal data is their *mode*, the most frequent value. The dispersion about the mode is estimated by the *variation ratio*, which is calculated as the percentage of categories that do not follow the one selected. Nominal data can only be equivalent, they cannot be equal; a nominal scale is unique up to a one-to-one transformation, which means that the order of the items measured can be changed relative to another nominal scale.

Ordinal scale. This scale is based on ordering the measurements. Ordering can be by increasing magnitude, increasing frequency and so on. An example of a measurement on this scale is the severity of an injury: an injury can be ranked as having low, medium or high severity. Although there is order between these three values, the distance between each ordered item is not specified and can thus be different. Equivalence and operations such as *greater than* are admissible. This scale is unique up to a monotonic transformation because that transformation preserves order. The appropriate statistical measure of central tendency is the median. The measure of dispersion is the percentile. Several non-parametric statistical tests can be performed on ordinal data (Siegel, 1956).

Interval scale. Both order and difference are equal for equal intervals, but the origin of the scale is arbitrary. An example of this scale is chronological age. Admissible operations must preserve both order and distance. The interval scale is unique up to a positive linear transformation. An appropriate measure of central tendency is the arithmetic average; the standard deviation measures dispersion. In general, parametric and non-parametric statistical tests can be performed on data measured on interval scales.

Ratio scale. This scale is characterized by an absolute origin, which has theoretical meaning. The reason why this scale is called ratio is that if the mass of one person is *80* [Kg], and that of another is *40* [Kg], then the ratio has physical meaning. Zero temperature (absolute zero, when measured in degrees Kelvin) and zero mass have theoretical meaning.[21] All arithmetic operations are permissible on this scale, which is unique up to a positive linear transformation. Statistical parametric and non-parametric tests are permissible on measurements taken on this scale.

Example 2.15. Take the attribute to be loss of wetland. The value and the description of the impact associated with that value is:

Value (on the Ordinal Scale of Measurement)	Physical Dimension of the Attribute	Attribute (Description of Impact)
0	Square meters [L^2]	No loss of habitat
1	Same	Loss of 100 square meters of habitat
...	...	...
10	Same	Complete loss of habitat

---

[21] We will often use physical units and generally place them within square brackets; L stands for length, M for mass, and T for time.

> Because the scale of measurement is ordinal, it can arbitrarily begin at *0* and end with *10*, enumerating *11* levels of ranked values of the attribute *X*, loss of a wetland in square meters.

Professional judgment, *a priori* beliefs, consensus opinions, the literature and scientific experiments can be used to establish which attributes (e.g., age, visibility) are measured and on what scale. For instance, a scale may show the several levels of severity assigned to a particular impact. Thus, a level *1* impact may be the loss of *1* hour of work, and the level *10* impact (the values on scale being from *1* to *10*) might measure the closure of a plant for a month. Other examples of scales of measurement include measures of public attitude, for example: *1* denotes complete support, *0* denotes complete neutrality and *-1* complete rejection. Some measurements can mask important aspects inherent to what is measured. Consider a scale measuring death counts (e.g., *1* death per day, *5* deaths per day). On that scale, the death of a child and that of an octogenarian are counted as a single event. However, the loss of life expectancy of the child's death is much greater than that of the *80* year old person (about a factor of *10*). Death counts do not reflect the natural consequence of loss of life expectancy due to premature death. Although an analyst should differentiate between these counts, this may not always be possible. For instance, scales are often developed to measure *proxy* factors or variables, rather than the actual factors or variables of interest themselves (Keeny, 1980; Keeney and Winkler, 1985). A way to measure the value of alternative actions is based on lotteries and *sure* bets. Such a method uses elicitations to discover the expressed preferences of individual stakeholders in the decision process. A brief explanation of the method is as follows. The respondent to the elicitation procedure sets a certain (*pr = 1.0*) value of action *X*, the simple lottery in which *Y* is the most preferred alternative, with probability *pr*, and *W* is the least preferred alternative, with probability (*1-pr*). The individual would choose the most preferred alternative, *Y*, if *pr* is large enough and, logically, prefer *X* when the probability of *W*, (*1-pr*), is small enough. A search over the probability space yields a point where the respondent will be indifferent between a lottery with probability values *pr = pr**, *1-pr** and the sure outcome *X*. The probability value *pr** is the *utility* of the alternative *X* for the individual respondent. The desirability of any alternative is mapped on the utility scale. In other words, if *Y* is the most desirable alternative, then *U(Y)* would be approximately *1.00*, *U(Y)* is the probability rate at which obtaining *Y* for certain is traded against the most preferable alternative. The prescription that follows is that those individuals who adhere to this way of reasoning *should* act to maximize their individual expected utility (Luce and Raiffa, 1957; Pratt, Raiffa and Schleifer, 1995). The requirement for rationality makes subjective utility theory normative. A critical issue that blunts the theoretical appeal of this approach is, as Cox states, that ... *real preferences and subjective probabilities systematically violate the assumptions of subjective utility theory* (Cox, 2002). Theory is often unmet by practice. Nonetheless, the importance of stating that expected utility should be maximized is that the form a basis for discussion. There are two aspects of evaluation. The first establishes a utility function. The other determines a value function. The evaluation of subjective utility is generally done either by direct elicitations of individuals, such as experts, or through surveys (Currim and Sarin, 1984). A value-function serves the purpose of determining the value (e.g., expressed as money) of changes between different levels of the attributes of a particular action. This discussion continues in Chapters 3 and 12.

Example 2.16. A value-function is $V(X)$, measured on the closed interval (*the least desired* = 0 < $V(X)$ < *the most desired* = 1), such that a higher value corresponds to a correspondingly larger monetary magnitude. The strength of preferences between outcomes can be measured using the differences in the value of two or more functions. That is $[V(Z)-V(Y)] > [V(X)-V(T)]$, if and only if Z is preferred to Y more than X is preferred to T (Dyer and Sarin, 1979; Krantz, Luce, Suppes, and Tversky, 1971; Kreps, 1988). Chapter 12 provides more details and examples.

2.8. CONCLUSION

In this chapter we reviewed and discussed several definitions of sustainability and linked these to making choices in risk management. We also discussed some of the principal scales of measurement and completed the exposition of Bayesian reasoning, which provides a useful way of dealing with both uncertainty and variability. This connects to cost-effectiveness, risk, cost, and benefit analysis. In risk management, the law governs the extent to which cost-benefit analysis can be used in setting environmental or health and safety standards. Setting those standards requires models and data. The last section of this chapter also dealt with model verification, validity and confirmation, in terms of the role of scientific theory and data on building such models, testing, and using them for practical applications.

Some of the issues discussed throughout this chapter stem from the fact that legal principles and legal tests having to do with scientific evidence and scientific causation used in risk analysis and evaluations are often incomplete. Therefore, they cannot lead to replicable or coherent results. Sustainability, specifically, can benefit from risk methods because they provide a consistent way to address the difficulties that arise from multifactorial and uncertain causation and heterogeneous data. The task of making sound decisions is aided by such principles as Pareto's. However, some of these principles require applying utility theory. This theory is discussed in Chapter 3 and made operational in Chapter 12, in terms of single and multiple attributes, using models that are directly applicable to understanding the implications of risky choices.

QUESTIONS

1. Develop a definition of sustainable development that includes a *list* of what you consider to be the relevant factors in sustainable development. Assume that you are developing that definition for a regional or national environmental agency. Do not exceed 300 words in total.
2. What do you understand to be the principal differences between cost-effectiveness and risk-cost-benefit analysis? Use approximately 600 words and give an example of each from your own work, experience or using any facility you can access, such as the World Wide Web.
3. Discuss your understanding of probabilistic conditioning and critically assess its importance in practical decision making, using no more than 400 words.
4. Provide simple examples of different scales of measurement that can be used in environmental management.

5. In no more than 400 words, exemplify an environmental law case that deals with an environmental pollution problem using cost-benefit analysis. Please use an example from your own jurisdiction and give the full and proper citation to the case.

6. How does Learned Hand's test address uncertainty? Does this legal test address the variability about the magnitude of the loss? If you think it does not, suggest a simple way by which it can do so. Use approximately 400 words.

CHAPTER 3.

RISK, COST, AND BENEFIT ANALYSIS (RCBA) IN RISK ASSESSMENT AND MANAGEMENT

Central objective: to develop and discuss the theoretical basis, techniques and limitations of applied risk-cost-benefit analysis, including cost-effectiveness analysis

This chapter deals with risk-cost-benefit analysis, RCBA, which consists of a set of methods for assessing and managing risks associated with routine and non-routine hazards. In this chapter, we review and discuss:

- Methods that can be used to analyze the costs and benefits of risky decisions, including the evaluation of indirect effects and intangible environmental goods and services, such as visibility
- Uncertainty and variability
- Practical implications of evaluations of risk-based actions that use such measures of outcome as deaths or illnesses averted
- Economic instruments, such as emission trading and banking, which are used in addition to environmental standards to optimize the use of resources with the least amount of pollution emissions
- Equitable distribution of risks, costs, benefits and some of the intergenerational implications of environmental decisions

Although principles of precaution explicitly include RCBA, its uses have raised ethical objections. For example, RCBA is predicated on economic efficiency, which some find unacceptable for certain societal decisions. Ethical issues cannot be resolved in this textbook because resolving them requires arguments that go beyond our scope. Practically, nonetheless, the limitations of RCBA do not invalidate its use in making risky decisions. Rather, each limitation must be explicitly addressed and demonstrably accounted for in assessing and managing risks. Not discussing them makes the results of an RCBA suspect and damages the acceptability of those analyses.

Risks *are* costs measured by probabilities, hazard rates or, in basic financial management, by the variance of an investment. Scarce resources can be either public, private or of mixed ownership. Economic costs and benefits are either monetary or can be *monetized* (given a monetary value) and discounted to account for the effect of time on the value of money. In many instances, some economic costs and benefits cannot be determined through the price mechanism. Therefore, their assessment requires techniques that value them indirectly. One such method is travel-cost; another is contingent valuation. These and other methods are discussed in later sections of this chapter. RCBA relies on economics, statistics and other sciences to provide a formal balancing of the risks, economic costs and benefits of private and public actions geared to minimize risks. The RCBA balancing, based on damage and benefit functions, consists of finding the point where the marginal costs of an action equal its marginal benefits. Because different combinations of marginal costs and benefits characterize each action, the optimal choice is defined at the point where the maximum *social* benefit is achieved at the least *social* cost. This is the condition where these marginal costs equal marginal benefits. The linkages between the principal aspects of the discussions included in this chapter are depicted as follows.

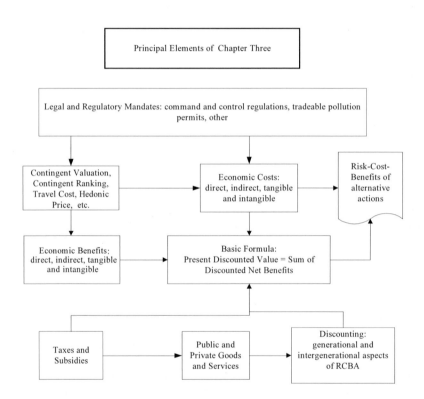

3.1. INTRODUCTION

Today's environmental regulations (e.g., those devised by the Organization for Economic Cooperation and Development, OECD, countries) place the cost of environmental protection on the entity causing environmental harm. This is the *Polluter Pays* principle. Additionally, environmental policy can use economic instruments, such as emission offsets and tradable permits, to achieve economically efficient levels of environmental and health protection. More generally, the trade-offs between setting risk-based technological standards, regulating by environmental standards or allowing the operation of a regulated market of emission permits and banking can be assessed using RCBA (Tietenberg, 2000; Adler and Posner, 2001; Sunstein, 2002). For example, regulations based on emission trading can reduce exposure as well as decrease the need for unwarranted or hidden subsidies. To justify environmental policy, society should have a baseline understanding of the risks, costs and benefits of an option (for instance, *do nothing*) and its importance relative to other options. This can be achieved through the application of theoretical and empirical methods for ranking alternatives: RCBA provides the guidance and methods that are consistent with the precautionary principle and its variants. The adverse effects of pollution, the burden of a disease and the economic costs of coping with that disease, often affect more than one generation. Cancer is an example because the latency of some cancers can be greater than twenty years, thus making intergenerational considerations important to risk management. Even more generations may be at risk from past or current emissions of pollutants, as is the case for the adverse effects of climate change from gases such as CO_2. In addition to the intergenerational effects, the RCBA of actions that are designed to minimize risks must account for the equitable distribution of the risks, costs and benefits of each action because those who benefit and those who pay and are at risk may be different from one another. Finally, the determination of tolerable risks depends on how society looks at the allocation of scarce resources to reach those risk levels and at other conflicting or competing objectives demanding those scarce resources.

As discussed, there often is a legal command to make decisions based the maximization of the net discounted benefits of social or private actions that are designed to reduce or eliminate health burdens or environmentally adverse outcomes. The actions designed to do so can produce difficult trade-offs. For instance, allowing (permitting) a private development in a wetland clashes with the protection and preservation of wetlands. Such protection implies a *social cost* because the value of preserving the wetland is the opportunity cost of not developing it. Public policy and laws that create and protect wetlands offer the protection that society has elected to have, possibly from considerations *other* than cost-benefit analysis. For instance, a wetland is protected because a wetland *is* a wetland. Society makes an ecological gain that escapes traditional financial accounting. Nonetheless, an economic trade-off has been made implicitly because the natural resource may be privately owned. Someone's property rights may have been reduced. In that situation, some jurisdictions have a constitutional requirement of compensation, if a privately owned wetland cannot be exploited for economic use (Ricci, 1995, Ricci et al., 1998). The economic effects of not developing it range from foregone earning to decreases in tax revenues. The societal benefit is that the wetland, and the natural services that it provides, are preserved for present and future generations.

RCBA provides a theoretically sound way to set priorities by considering the integrated aspects of the risks, costs and benefits relative to the boundaries of an objective. It includes temporal and spatial consequences. RCBA formally, qualitatively and quantitatively accounts for the total risks from a choice, any residual risks after the preferred choice has been selected, and the distribution of the risks and the net discounted benefits. Furthermore, because of the means-ends nature of risk-cost-benefit analysis, it yields a net benefit balancing that is essential to public decision-making. RCBA, explicitly or implicitly, subsumes risk assessment, RA, and risk management, RM. Risk management requires inputs from RCBA because each component of the RCBA, namely the risks, costs and benefits must be disclosed to the decision maker and to the stakeholders. Figure 3.1 depicts the flow of information within RCBA for air, water or soil pollution, assuming that the goal of RCBA is to reduce individual and aggregate risks from environmental pollution. RCBA can be constrained by the budget because resources are limited and costly. Moreover, the time allowed for a remediation, the potential delays and the fact that new information may become available could change the choice made before the remediation. Decision-making also requires considering the stakeholders' goals and objectives. Goals are broad statements that define objectives. The latter are specific statements of achievements measured and monitored by indicators or indices. Objectives are determined by the stakeholders' preferences. The context of the RCBA is the entirety of the *states-of-nature* and the set actions designed to meet the objectives of the decision-maker.

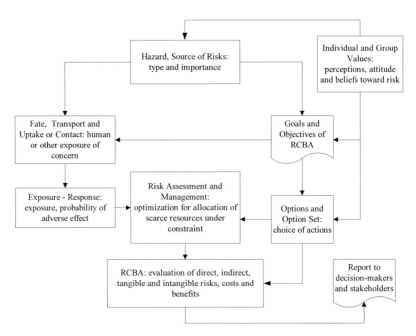

Figure 3.1. Simplified Risk Management Framework for Risk-Cost-Benefit Analysis

Example 3.1. In this example, the goal is to *eliminate* indoor exposure to unvented gases from indoor sources. An achievable objective that approximates the goal may be to *decrease* exposure to volatile organics from cans of paint in a storage room to a safe level. The set of options includes changing the formulation of the paint, changing the seals on the containers, and so on. The minimization of exposure to volatile organic compounds, VOCs, relates causally their concentrations in indoor air and to the adverse health effects that result from inhaling those substances. The achievability of the objectives relates to the constraints placed on the means to achieve those objectives.

As a passing comment, an objective can be stated as a function that captures the analytical aspects of what is being sought: for example, maximize net profits (Rosen, 1992; Baumol, 1977). The constraints are generally inequalities that must be considered in maximizing the net benefit function.[1] Constraints include funds, personnel, education and training, equipment and so on. The constraints, for example, include the fact that mass cannot be negative and account for limits on resources and operations. It should be remember that methods from operations research that use linear, quadratic, integer and other forms of mathematical programming can be useful, particularly in those areas of decision-making that are amenable to physical production (Rosen 1992). It should also be noted that *cost-effectiveness* analysis deals with the direct costs and benefits of production and use. If it costs *$10,000* to buy car model XYZ, and it costs *$11,000* to buy car MNO, then, for the same user and everything else being equal, buying car XYZ is more cost-effective than buying car MNO. If these costs do not fully include (or *internalize*) such additional costs as the cost of increased illnesses and mortality due to exposure to the air pollutants emitted by these two cars, the direct costs are biased in favor of the producer. Cost-effectiveness analysis is narrower in scope than cost-benefit analysis, CBA, because CBA accounts for the full social costs associated with producing these cars. In this textbook, cost-benefit analysis includes cost-effectiveness analysis, not the other way around. Cost-benefit analysis is ethically utilitarian. Although there is no guarantee that a cost-benefit analysis can satisfy all the stakeholders involved, the inclusion and the explicit exclusion of those factors that affect the outcome can help the stakeholders assess the fairness of the decision-making process and enhance its final acceptability. The decision maker's search for optimality in her decision-making, maximize the magnitude of the net discounted benefit over the set of possible actions, becomes transparent and can be used to identify weak assumptions and issues. Moreover, the value systems affected and affecting the decision-making process can be characterized through formal analysis.

3.2. COST-BENEFIT ANALYSIS, CBA

Cost-benefit analysis is used to balance direct, indirect costs and benefits that can be measured (or approximated) by a monetary unit of value through a specific and well-established set of formulae. The theory of cost-benefit analysis arises from a number of economic sub-disciplines that include microeconomics, public finance, financial management, welfare, labor and environmental economics. The methods for analysis are essentially statistical (more specifically, econometric) and use data from questionnaires, experiments or observational studies. Probability theory provides the

---

[1] The objective function describes what is being maximized (or minimized), the constraints limit the amount of the inputs that can be used to maximize the objective function; for example, *Maximize* $Z = 3.5X + 0.11 + Y + 3.1W$, with $X + Y \leq 10$ and $W > 0$. This is an example of a linear programming optimization (Rosen, 1992).

basis of uncertainty and variability analysis. The process of CBA can be summarized as a sequence of activities:

Define an Objective → Identify Alternative Options to Reach that Objective → For Each Alternative, Measure the Costs and Benefits of Achieving that Objective → Calculate the Net Present Discounted Value (using the same discount rate) of Each Alternative → Rank Each Alternative According to Its Net Present Discounted Value → Recommend the Choice of Alternative, Basing such Choice on the Maximum Net Present Discounted Value.

Viscusi (1996) exemplifies ways in which a risk-cost-benefit assessment can put into perspective government interventions using a common metric: the costs of saving lives. An abstract of these numbers (in 1984 $US) is given in Table 3.1.

Table 3.1. Selected Cost per Life Saved (in millions of 1984 $US; Viscusi (1996))

Action	Agency, Year of Regulation	Annual Risk	Expected Number of Annual Lives Saved by Regulation	Cost per Life Saved (Millions 1984 $)
Passive Restraints/Belts	NHTSA, 1984	$9.1*10^{-5}$	1,850	0.30
Occupational Benzene Exposure	OSHA, 1987	$8.8*10^{-4}$	3.8	17.10
Occupational Formaldehyde Exposure	OSHA, 1987	$6.8*10^{-4}$	0.01	72,000

The National Highways Traffic Safety Administration, NHTSA, and the Occupational Health and Safety Administration, OSHA, save statistical lives at the costs depicted in Table 3.1. The range of *costs per life saved* between mandating passive restraints and formaldehyde exposure in the work place is approximately *$72,000* million. Abelson (1993) summarized the cost of regulating pollution as approximately averaging *$115* billion (in $1990).[2] A deterministic expression for CBA derives from the formula for the future simple discounted value of an amount of money, *A*, which is held today, calculated for *t* uniform periods of time at a (constant) interest rate, *r* (dimensionless). The formula is:

Future Compounded Value, $FDV = A(1+r)^t$.

This formula yields the present discounted value, PDV, as a back-calculation in which *A* is changed to (B_t-C_t):

Present Discounted Value, $PDV = \sum_t (B_t-C_t)/(1+r)^t$.

The present discounted value formula provides today's ($t = 0$) monetary value of a discrete stream of net benefits, calculated as the difference between costs, *C*, and benefits, *B*, discounted at a rate *r*, over a number of periods of time (each of which is

---

[2] Using 1984 dollars means that all values expressed in dollars are transformed into constant 1984 dollars. 1984 is the *base-year*. Values in dollars in other periods, say 1986, are *nominal*. This transformation uses an index such as the *price index*. Rosen (1992) gives the details.

symbolized by t, namely $t = 1$, $t = 2$,...), beginning at time $t = 0$. If a single, lump sum value (a single sum without prior payments) of the net benefits were sought, the formula simplifies to:

$PDV = [(B_t-C_t)/(1+r)^t]$.

Example 3.2. Let the value of the benefit equal $10,000.00 and the cost equal $5,000.00, the number of yearly time periods is *10*. Assume a constant interest rate of *5%*. Today's value of the net benefit that will be generated only at the end of year *10* is:

$PDV = 5,000/(1+0.05)^{10} = \$3,069.57$.

In this calculation the interest rate is *nominal*, which means that inflation and the risk of default is included in the *5%*. The *real* interest rate might be about *2%*.

Example 3.3. Let the net benefit stream be $1,000.00, $5,000.00, $9,000.00, and $15,000.00; the interest rate equals *10%*. The PDV is:

$PDV = 1,000+5,000/(1+0.10)+9,000/(1+0.10)^2+15,000/(1+0.10)^3 = \$24,253.19$.

The formula is not sensitive to monetary unit used. In other words, using the Euro, dollar or pound does not affect the rank order. However, the choice of the base year can affect the rank-order of the alternatives. Therefore, a constant-dollar should be used in the calculations in part because inflation and financial risk may affect the nominal value of the currency used, as can several other factors, including trade deficit, exchange rates and so on. The formula for rank-ordering potential actions through the CBA is as helpful as it is deceptive. The decision rule is to choose the alternative that yields the largest positive PDV. The formula, through to n time periods measured in years, is:

Net Present Discounted Value ($), PDV = $(B-C)_0+(B-C)_1/(1+r)^1+...+ (B-C)_n/(1+r)^n$.

The PDV states that what is being discounted are the net benefits, measured by a suitable monetary unit, e.g., dollar. A negative net benefit, a net *disbenefit*, occurs when the difference between the costs and benefits is negative. The interest accrues at the end of each period and does not change over the periods used in the CBA. If compounding occurred more frequently within a year, for example quarterly, the formula would have to be changed slightly to reflect that situation.[3]

3.2.1. Discussion

Both cost-effectiveness and CBA are predicated on the principle that the maximum net benefits occurs when the *marginal* benefits of a choice equal the *marginal* costs

---

[3] The cumulated discounted value of an initial amount, $(B-C)$, in which the interest rate per year is r, and k is the frequency of compounding per year for n years. A sum can be compounded k/r times per year as follows: *(Initial Amount)*$*[(1+r/k)^k]^n$. Consider the simple interest payments for $1,000 for *2* years at *10%* per year: $1,000*(1+0.10)^2 = \$1,210$. The effect of compounding interest *4* times over *2* years yields $1,216: $1,000*(1+0.10)^4 = \$1,216.00$.

of achieving that choice. The difference is the type of costs and benefits included in the calculations. Marginal, in economic analysis, means the rate of change (first derivative) of a total cost or total benefit function.

The marginal conditions for an economic optimum (an efficient solution) are necessary, but not sufficient, for the full optimization. The sufficient condition requires calculating the second derivative of the cost and benefit function. Marginal costs and benefits can be calculated directly from data sets, as exemplified next.

Example 3.4. A typical discrete calculation of marginal cost is as follows:

Output (Quantity)	Variable Cost ($)	Marginal Cost ($/unit of output)
1	100	---
2	160	60
3	210	40
4	230	20

In this example, variable cost is the schedule of costs that are not fixed.

The marginal cost is the cost of adding an additional unit of output or the cost of reducing pollution by one additional unit. Figure 3.2 depicts the marginal cost curves for technologies 1 and 2.

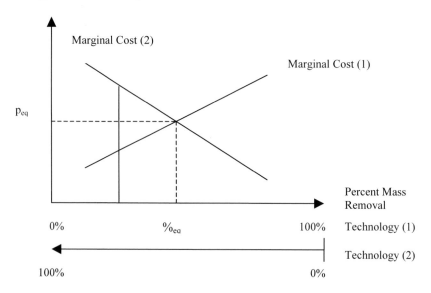

Figure 3.2. Marginal Cost Curves for Two Different Technologies of Pollution Control

The horizontal axis depicts the percentage of the mass of a pollutant removed; the vertical axis depicts the marginal price (price per unit of mass of pollutant removed). For instance, for air pollution, these can be a bag-house and an electrostatic precipitator. Note that, in Figure 3.2, there are two *x*-axes, one for polluter *1* and one for polluter *2*: the horizontal axis shows removal efficiencies from *0%* to *100%* and from *100%* to *0%*, respectively, for these two. These marginal cost functions are consistent with theory and practice: as more mass of an air pollutant is removed, the higher the price of such removal.

Equilibrium between the two technologies occurs at the intersection of the two marginal cost curves: any choice away from that equilibrium is economically inefficient. At a specific % removal away from equilibrium, emitter *1* has much lower marginal cost of removal than emitter *2*. Both would benefit, will be more efficient, by moving toward equilibrium. If this is not possible, emitter *1* could (in theory and if the law allowed such transfers) transfer some pollution credits to emitter *2*. The static marginal cost, marginal precaution underlying the discussion is depicted in Figure 3.3 (Tietenberg, 2000). As Figure 3.3 depicts, equilibrium occurs where the two marginal curves meet: the marginal cost of control equals the marginal level of precaution. This is the basis for using economic instruments, such as marketable credits for emissions, discussed later in this chapter.

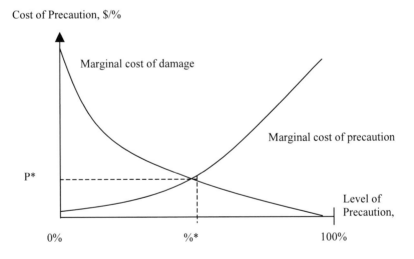

Figure 3.3. Cost-effective Level of Precaution and Marginal Cost of Precaution

The scope of CBA can be project specific, but can also be applied to regional, national or even international analyses, if the data is available and reliable. An example of nation–wide cost and benefit analysis, used in part to justify the regulations of air pollution under the US Clean Air Act, has included such social benefits as death and morbidity reduction, visibility improvements, and other factors due to air pollution. The results of this analysis, using the base year 1990 (Tietenberg, 2000) are given in Table 3.2.

Table 3.2. Benefits and Cost of Air Pollution Controls (Billions, $US 1990)

Benefits and Costs of Controlling Air Pollution, under the Clean Air Act, in billions $US (1990)				
Year	**1975**	**1980**	**1985**	**1990**
Benefits	355	930	1,155	1,248
Costs	14	21	25	26
Net benefits	341	909	1,130	1,220

The data in Table 3.2 is adapted from Tietenberg (2000, p. 29, Table 2.1), who developed these data from *The Benefits and Costs of the Clean Air Act*, 1970 to 1990 (US EPA, 1997). The costs are annualized over the life of the capital investment; the benefits are the *average* benefits. The direct costs associated with achieving those benefits include capital outlays, the costs of operating and maintaining pollution control equipment required by law to reduce exposure, as well as other costs of compliance, such as monitoring, and so on. If the marginal cost of precaution internalizes all of the factors that are needed properly to characterize it, then Figure 3.3 can represent a form of cost-benefit analysis. The balancing of risks of morbidity and death, economic costs just described and benefits can be used to:

1. Set priorities through a formal method,
2. Develop or justify actions (including regulations) by clarifying who benefits, who bears the risks, when those affected benefit and at what cost,
3. Provide the risk, benefit, and cost profiles for potential hazards; identify ways for their optimization through technological or other engineering choices, including changes in policies and procedures.

Several different costs are involved in cost-benefit calculations. For example, the *time cost of capital* is the value of the good or service in use, less the costs associated with reaching that productive use. Another is the *opportunity cost* of resources: the value of the forgone opportunity. The cost of unforeseen delays is another. Yet another cost is the contractual penalty likely to be included in the failure to meet a specified performance goal. Additionally, costs can be direct, such as the cost of labor (measured by the *wage* rate), and indirect, such as the cost of inflationary pressure put on the wage by the scarcity of the labor. Costs can also be tangible: for example, wage is tangible. The cost of bad corporate citizenship is an example of an intangible cost. Some costs can be fixed (capital equipment such as machinery, for example). Variable costs (the cost of labor measured by a wage, for example) can be changed (in a free market) in the short-run. In the long-run, both fixed and variable costs can be changed. Recollect that a risk is the probability of an adverse effect measured over the appropriate period. For example, the risk of death from a specific cause may be *1/100* per year (or one death, due to a specific event, per one hundred deaths from all other causes). This means that, on the average, an individual has a chance of death equaling *0.01* per year, from that cause alone.[4] Economic costs and

---

[4] This example assumes that the 100 individuals respond similarly (are homogeneous) with respect to that hazard.

benefits analysis can, and generally do, include probabilistic weights to establish how likely they are before (*ex ante*) a management's decision is implemented. Specifically, costs and benefits are random variables characterized by the appropriate distribution functions. The three examples show how uncertainty affects risk, cost and benefit balancing (Wilson and Crouch, 1987).

Example 3.5. A rare disease invariably follows from the exposure to a specific substance. Let the risk ratio be *500*. That is, exposure yields *500* times more cases than no exposure. There is, because of the rarity of the disease, no great difficulty in (at leas empirically) determining the causal relationship between exposure to the substance and the much larger response in those exposed, relative to the unexposed. Hence, not knowing the biological basis for the exposure-response relationship is not troublesome. The decision-maker can act to eliminate exposure (assuming that the substance can be eliminated and that the elimination is not too costly to society in dollars and risks).

Example 3.6. Consider a ubiquitous substance, deemed to confer some benefit on the many users and costing very little relative to the cost of the products within which it is found. Assume also that, if the substance were withdrawn from production, no cheaper alternative would be available. Therefore, the cost of the product to the consumer would increase 5-fold. However, that ubiquitous substance has been shown to be carcinogenic, but under unrealistic exposure patterns and only when it is administered to a very sensitive animal species. Interspecies conversions (from animals to humans) yield an expected number of cancers in humans equaling *100* cancer cases per year.

Example 3.7. Assume that a practical definition of worst-case (suitably defined) is a risk that equals $1*10^{-3}$. The tolerable level of risk established by regulation is $1*10^{-6}$, or less. This regulatory risk level can be achieved by installing a system of emission control equipment costing $\$5*10^{9}$ to reduce exposure to the concentration that yields that tolerable risk level. Assume that the exposed population is $2*10^{6}$ individuals. The change in risk (risk reduction) required is *0.00999*, which is rounded to $1*10^{-3}$. For a population of $2*10^{6}$ persons, the expected number of events, $\overline{N}$, is $\overline{N} = R*P$. That is, $\overline{N} = (1*10^{-3})*(10^{6} persons) =$ expected 10^{3} adverse outcomes. Using a total cost of $\$5*10^{9}$ for the engineering controls, the average total cost per adverse outcome is:

(Total Cost)/(# adverse outcomes) $= (5*10^{9})/(10^{3}$ *adverse outcomes*) $= (\$5*10^{6})/(adverse$ *outcome*).

As discussed, costs must be discounted if more than one period is considered. Discounting reflects the preferences for an amount of money foregone in a period for a (larger) amount in the next or later periods. It includes the effect of inflation and the probability that the borrower does not live up to expectations of the lender by, for example, by defaulting on the loan. Some economic instruments, such as US Treasury Bills, have negligible risk when backed by the full faith and credit of the issuing government.

3.2.2. Discounting: Social rate of discount and the interest rate

The analysis of costs and benefits should include, quantitatively and qualitatively, the implications and effects of changing attitudes and values: namely, individual and societal preferences for money over time. The discount rate accounts for individual attitudes (preferences) for money. It is the ratio of preferences for consumption of a good or service in a time-period relative to the preferences for the same good or service in a later time-period. The *private* discount rate is the interest rate familiar to most people.

Example 3.8. I lend you *$10.00* on 1/1/2001 and expect that you to return the *$10.00*, lent on 1/1/2002, and compensate me with *$1.00* for this loan. The rate of interest is *1.00*[$]/*10.00*[$] = *0.10*. This calculation shows that behavioral aspect of the interest rate. That is, I do not use the *$10.00* in the year, but you do.

The cost of private investment foregone equals the foregone rate of return on that investment. The *social* rate of discount reflects the collective value of future net benefits and is often different from the private rate of investment. The social discount rate can be measured by the opportunity cost of capital at a risk-free rate. That is, because the government makes public expenditures on behalf of its citizens, there is no risk premium to pay for the risk of a government's default, at least in theory. However, the government competes in the open market for funds, thus affecting the nominal value of the interest rate. The social discount rate is generally lower than the private discount rate. On this argument, the discount rate can be positive, zero, or even negative. The rationale for a positive discount rate is that most individuals would prefer to consume today, rather than tomorrow. If the discount rate were zero, then the value of a sum of money today is the same as it is in the future. A large discount rate would make future public investments prohibitive; a low one makes most such investments justifiable.

Example 3.9. *The Economist* (June 26, 1999) gives an example of using the appropriate *discount* rate for public projects with an intergenerational time horizon. Let the global Gross Domestic Product (GDP) grow at *3%* per year, after *200* years that value is *$8,000,000,000,000,000*. Suppose that the value of a social project is discounted at *7%*, assume that the benefit will arrive after *200* years. The PDV of *$8,000,000,000,000,000* discounted at *7%* per year is about *$10,000,000,000*. On these calculations, should society spend more than ten billion dollars today by exercising an option that would prevent the loss of the entire earth's production *200* years from now?

For a project that has social consequences, a weighted average of the rates of return of the opportunity cost rates of return for the area of the project can be practical (Peskin and Seskin, 1975; Marglin, 1963). Public concerns require a socially justifiable balance between the well-being of the present and future societal demands, an aspect of economic efficiency that is studied by welfare economic analysis. Concerns with the intergenerational impacts of an action are also accounted for by the appropriate choice of the social discount rate.

3.2.3. Variability in CBA

Consider a private concern that, for simplicity, seeks to maximize its net profits. Assuming constant dollars throughout the calculations, the variables are:

R = *gross yearly revenues,*
c = *unit cost,*

N = *number of units produced,*
F = *fixed capital costs,*
V = *variable capital costs,*
W = *total yearly wage,*
H = *total overhead,*
t = *tax rate* (assumed constant, for net profits and from a single taxing authority).
The profit-maximizing firm will have net revenues, NR, before taxes:

$$NR = R-(cN+F+V+W+H).$$

Applying a constant tax rate, t, to the net revenue, $t(NR)$ and subtracting from NR yields the net yearly profits, Π. Let the variables F, V, W be random. (The other variables are also random, but the total probability mass is assigned to a single number, that is, the number has probability 1.00). Therefore, Π will also be random (because a linear combination of random variables is also a random variable). The variability of Π could be accounted for by the normal distribution. The reason for using the normal distribution is that profits can be negative, namely, the firm can be operating at a loss.

Example 3.10. Consider the following hypothetical sample data:

Alternative 1 (current year $)		Alternative 2 (current year $)	
Probability, *pr*	Net Revenues ($*1,000)	Probability, *pr*	Net Revenues ($*1,000)
0.10	1,000	0.10	2,000
0.20	3,500	0.25	3,000
0.40	4,000	0.30	4,000
0.20	4,500	0.25	5,000
0.10	10,000	0.10	8,000

The sample mean net revenue is $\overline{NR} = \sum NR_i(pr_i)$. The sample standard deviation of the net revenue is $(VarNR)^{1/2} = \sum [pr_i(NR_i - \overline{NR})^2]^{1/2}$. These are calculated as:

$$\overline{NR}_1 = (0.1)(1,000)+(0.2)(3,500)+(0.4)(4,000)+(0.20)(4,500)+(0.10)(10,000) = \$4,300.00,$$

and:

$$[Var(NR_1)]^{1/2} = [(0.1)(1,000-4,300)^2+(0.2)(3,500-4,300)^2+(0.4)(4,000-4,300)^2+(0.2)(4,500-4,300)^2+(0.1)(10,000-4,300)^2]^{1/2} = \$2,124.$$

Similarly, we can calculate:

$$\overline{NR}_2 = (0.1)(2,000)+(0.25)(3,000)+(0.3)(4,000)+(0.25)(5000)+(0.10)(8,000) = \$4,200.00.$$

$$[Var(NR_2)]^{1/2} = [(0.1)(2,000-4,200)^2+(0.25)(3,000-4200)^2+(0.3)(4000-4200)^2+(0.25)(5000-4200)^2+(0.1)(8,000-4200)^2]^{1/2} = \$1,568.00.$$

The coefficients of variation, CV (a dimensionless quantity), are: $CV_1 = [var(NR_1)]^{1/2}/ \overline{NR}_1$ $= 2,124/4,300 = 0.49$, and: $CV_2 = [var(NR_2)]^{1/2}/ \overline{NR}_2 = 1,568/4,200 = 0.37$.

3.2.4. Supply and Demand Analysis

The difference between private and public goods and services is the right that the private owner has to the lawful and exclusive enjoyment of her property. Roughly, if you (a private citizen) can legally exclude me from your land, that land is privately owned by you. If you cannot, then that good or service is publicly owned. Private goods and services are valued within the market operations. Public goods are not (entirely) valued by the market, and must be valued indirectly because there is no market mechanism to send signals of relative value based on availability, price and the right of exclusion. For most goods and services, their individual *demand* function, *price = f(quantity)*, is downward sloped: more is demanded as the unit price of the good is lowered. For instance, assuming a linear model:

price = a-b*(quantity of the good or service demanded),

that is:

$P = a-bQ_d$.

Private individual demand functions are aggregated horizontally to obtain total demand. For instance, the aggregate demand of *n* firms in an industry must be aggregated horizontally to yield the industry's demand function. On the other hand, the aggregation of individual demand functions of public goods or services is vertical.

The horizontal aggregation of individual demands (for a private good) is calculated (for two individuals) as shown in Table 3.3.

Table 3.3. Individual Demand for Private Goods

Individual	Price/unit	Total Quantity
1	6	2
2	6	5
Aggregate	6	7

The vertical aggregation of individual demands (for a public good) is shown in Table 3.4.

Table 3.4. Individual Demand for Public Goods

Individual	Price/unit	Total Quantity
1	25	10
2	15	10
Aggregate	40	10

The *supply* of private goods is positively sloped:

price = d+c*(quantity of the good or service),

that is:

$P = d+cQ_s.$

The reason for making explicit the form of the supply equation is that producers would be willing to provide more quantity at higher price. From these two equations, it is clear that demand and supply functions are linked (they are simultaneous), as can also be seen from the two linear functions depicted in Figure 3.4. Static equilibrium between quantity demanded and supplied occurs at the point where these two equations intersect. Static means that time affects neither the relationships nor their equilibrium. When time is considered, we need a system of differential (or difference) equations, which can be studied using dynamic models that are outside the scope of this chapter. Using substitutions, we can find an expression for the solution of the system of static equations: that is, equilibrium price (P_{eq}) and equilibrium quantity (Q_{eq}).[5] The static relationship and equilibrium between supply and demand, expressed as two linear equations, is depicted in Figure 3.4.

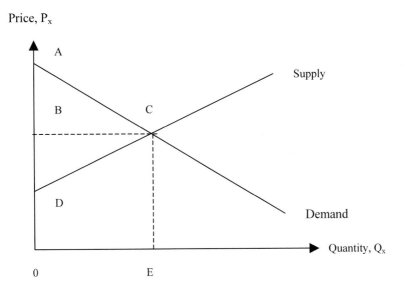

Figure 3.4. Static Supply and Demand Model

---

[5] Mathematically, $Q = a-bP$ and $Q = -c+dP$. Therefore, $P_{eq} = (a+c)/(b+d)$, provided that $P_{eq} > 0$. $Q_{eq} = (ad-bc)/(b+d)$, $Q_{eq} > 0$, provided that $ad > bc$. The subscript eq means equilibrium.

Price is in *$/unit* of the quantity supplied or demanded. At static equilibrium, point C, supply equals demand and therefore marginal cost (the first derivative of the cost function) equals marginal benefit (the first derivative of the demand function). The willingness-to-pay, WTP, function is the demand curve and it can be developed from questionnaires that ask respondents to state how much they are willing to pay for a particular good or service. WTP has been used to evaluate market and non-market goods and services, beginning in the 1940s with Hotelling's work for the US Forest Service. Once the WTP function is known, it becomes possible to calculate the consumer surplus, which is measured by area ABC. This area represents the gain made by the consumers when price and quantity are at equilibrium. In other words, because consumers do not pay high prices for less quantity, they gain. The producer also gains: the area BDC measures the amount of gain. Finally, the total revenues equal the area 0BCE and the total cost is measured by the area 0DCE. In practice, the data for developing demand and supply functions are *schedules* (or vectors) of data. Table 3.5 contains a hypothetical example of data that can be used to estimate the parameters of the supply and demand equations, namely *a, b, d,* and *c.*

Table 3.5. Aggregate Supply and Demand Schedules

Aggregate Demand		Aggregate Supply	
Price/unit	*Total Quantity*	*Price/unit*	*Total Quantity*
40	200	40	950
35	350	35	825
30	500	30	550
25	650	25	650
20	800	20	500
10	1000	10	300

To obtain estimates of the parameters *a* and *b*, we *draw* a simple linear function, as depicted in Figure 3.5.

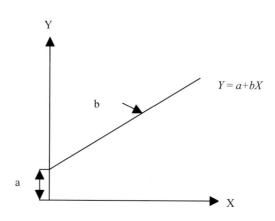

Figure 3.5. Simple Linear Function

Example 3.11. Suppose the risk assessor wants to model a simple relationship between two variables, Y and X. He believes, from having read the literature, that the relationship between these two variables is linear. The mathematical expression is $Y = a+bX$, for $a > 0$. A typical application is between price and quantity, $P = a+bQ$. Suppose that the literature has developed the following values: $a = 10$ and $b = 0.5$. The actual model, substituting P for Y and Q for X, is the equation $P = 10+0.5Q$. It is linear because the independent variable is not squared or otherwise modified. The equation could also be more complicated, but that complication is relegated to chapters 4 and 5. Given values of Q, we can determine the corresponding values of P. So, for $Q = 10$, $Y = 10+0.5*10 = 15$.

A quick way to obtain a fit is to eyeball a straight line to the data. For me, it is roughly as depicted in Figure 3.6 (the points are shown as small circles for ease of representation).

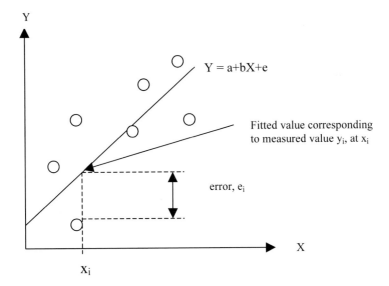

Figure 3.6. Eye-ball Best Fit of the Line $Y = a+bX$ to a Sample of Data

We use a set of data (sample or population) for X and Y. We can plot the paired observations, $\{(x_1, y_1), (x_2, y_2), \dots, (x_n, y_n)\}$. In Figure 3.6, y_i is the actual observation; the hand-drawn line provides the *fitted* value of y_i for a fixed value of x_i. The error that is made in fitting the line to the data is e_i. We then read on the Y-axis the value of a; the slope of the line is the value of b (Discussed in Chapter 4).

We need more than just a pair of eyes to obtain accurate and reproducible estimates of a and b. How are these two estimates obtained? The formula for estimating b is, noting that y_i and x_i must be paired observations on X and Y, is:

$$b = [\sum_{i=1}^{n} (x_i - \bar{x})(y_i - \bar{y})] / \sum_{i=1}^{n} (x_i - \bar{x})^2 .$$

In this formula, x_i is one of *506* values of X and y_i is one of *506* values of Y; $\bar{x}$ and $\bar{y}$ are the sample averages of these two sets of observations on X and on Y, respectively. The formula for estimating a, which requires having estimated b, is:

$$a = \bar{y} - b\bar{x} .$$

This is what standard estimation is fundamentally all about: obtaining the *best fit* of a model to the data, which is obtained by minimizing the squared errors, relative to each coefficient of the model. Estimation means that we use one or more formulae, applied to sample data, to approximate the unknown value of the corresponding population parameter. For example, we use the formula for the sample mean to estimate the unknown mean of the population from which the data is taken. A formula of dispersion of the data for the sample mean, $\bar{x}$, is the sample variance, *var(x)*:

$$var(x) = 1/(n-1) \sum_{i=1}^{n} (x_i - \bar{x})^2 .$$

Example 3.12. We use the data set *Boston housing*, JUMP5, to study the linear relation between housing values and concentrations of NO_x. The actual sample consists of *506* data (*x* is a value of X and *y* is a value of Y) points that plot as follows; that is, there are *506* paired values for X (labeled *nox*) and *506* values of Y, (labeled *mvalue*):

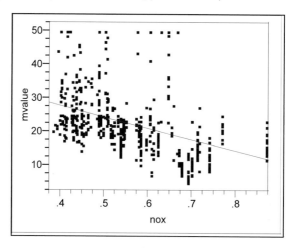

In theory, we would expect that higher concentrations of NO_x lower the median housing values: this relation is a simple damage function. The damage function is:

mvalue = *a*+*b**nox,

in which we expect that b has a negative sign $(-b)$ for *nox*. The estimated coefficients of the model that are obtained using the estimation procedure (called the ordinary least squares, OLS) replace a and b in the damage function model, obtaining:

mvalue = 41.35-33.92nox,

that is:

Median Housing Values = 41.35-33.92*NO$_x$.

Estimation confirms that, as the concentrations of NO_X increase, median housing values decrease, with a loss of approximately *$34* per unit concentration of *NO$_x$*.

The sample information can be studied in additional detail, as shown next.

Example 3.13. From the Boston housing sample of *506* observations I take and show eleven data points, median housing value is *MHV*:

MHV = 23, 23.7, 25, 21.8, 20.6, 21.2, 19.1, 20.6, 15.2, 7, 8.1.

NO$_x$ = 0.532, 0.532, 0.532, 0.532, 0.583, 0.583, 0.583, 0.583, 0.609, 0.609, 0.609.

For these *11* data points, the mean for *MHV = 18.66* and the mean *NO$_x$* concentrations equal *0.572*. This example conveys the essential ideas necessary to continue with the discussions that follow.

Because we are dealing with a sample, we must have a way to make inference from the sample to the population from which the sample comes from. The essence of the idea, discussed in more details in subsequent chapters, is described in the example that follows.

Example 3.14. The confidence interval is a specified confidence level, a probability, often stated as *0.95*. It implies a level of statistical significance, also a probability, which equals *0.05*. The confidence interval is built as follows.

Take the normally distributed population, with population mean μ and standard deviation σ. The value $z = 1.96$ is the value of the standardized normal distribution (a normal distribution in which $\mu = 0$ and standard deviation $\sigma = 1$, called the z-distribution).   The   *(1-α)*   confidence   interval   for μ is: $pr[-1.96 \leq (\bar{x} - \mu)/(\sigma/\sqrt{n}) \leq 1.96] = 0.95$.   The   values   of   the z-distribution   are discussed in a later chapter.

This framework is depicted for *($\alpha = 0.05$)*, and a two-tailed test, in Figure 3.7.

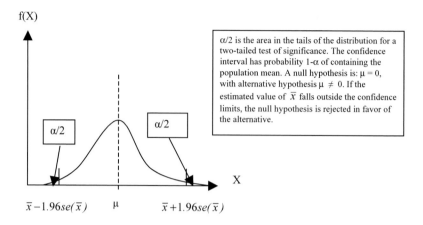

f(X)

α/2

α/2

α/2 is the area in the tails of the distribution for a two-tailed test of significance. The confidence interval has probability 1-α of containing the population mean. A null hypothesis is: μ = 0, with alternative hypothesis μ ≠ 0. If the estimated value of $\bar{x}$ falls outside the confidence limits, the null hypothesis is rejected in favor of the alternative.

X

$\bar{x}-1.96se(\bar{x})$ μ $\bar{x}+1.96se(\bar{x})$

Figure 3.7. Two-tailed Test of Significance for the Population Mean, from a Sample Mean

We can now extend these ideas to statistical modeling. If we take the regression model $Y = a+bX$ to be appropriate for the entire population of data, we have to do something about the fact that, practically, we deal with samples, not populations. In other word, as discussed in Chapter 4 to 11, we must make inference from the sample to the population. We use classical (non-Bayesian) regression methods.

A simple way to do so follows what was done for the population mean, μ; it applies to developing confidence intervals about the estimates of the parameters a and b, which are symbolized as $\hat{a}, \hat{b}$. The formula of the standard error of the estimate for the parameter b, $\hat{b}$, is (without derivations):

$$se(\hat{b}) = \sqrt{\sum_{i=1}^{n} e_i^2/(n-1)} / \sqrt{\sum_{i=1}^{n} (x_i - \bar{x})^2}.$$

Figure 3.8 depicts the two-tailed 95% confidence interval for b, $b = \hat{b} \pm 1.96se(\hat{b})$.

f(b)

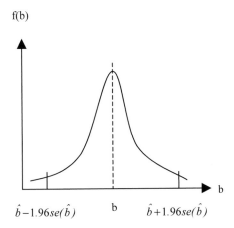

$$\hat{b}-1.96se(\hat{b}) \qquad b \qquad \hat{b}+1.96se(\hat{b})$$

Figure 3.8. Two-tailed Confidence Interval for the Population Parameter b, Given the Population Model $Y = a + bX$.

The *se* for $\hat{a}$ is:

$$se(\hat{a}) = \sqrt{\sum_{i=1}^{n} e_i^2 /(n-1)} \sqrt{1/n + \bar{x}^2 / \sum_{i=1}^{n} (x_i - \bar{x})} \, .$$

The two-tailed confidence interval for the population parameter a is $a = \hat{a} \pm 1.96se(\hat{a})$. More discussions and examples can be found in Chapters 4-11.

3.2.5. Modeling Supply and Demand Statistically

Simultaneity between supply and demand can be modeled through a system of (linear) equations that shows the dependencies between price and quantity of a good or service. We exemplify such simultaneity and, to make the discussion more realistic, also account for income Y on the demand side, and weather, T, on the supply side. To place the algebraic system relating the demand and supply equations (in which P is price) in a statistical context, the system must account for the errors (symbolized by u) in each equation. The necessary condition to obtain a unique solution is that the system of equations must be *identified* (it yields unique estimates of the parameters through a statistical method such as the least squares). We discuss the identification step for brevity and because it gives rise to three important situations. In general, establishing that the necessary condition holds is enough for the type of statistical analyses discussed. The chapters that follow extend these discussions and provide additional methods for obtaining estimates of the parameters of statistical models used in risk assessment and management.

1. If the system of equations is *exactly identified*, then there is a formula that yields a unique value for the parameter or parameters. Exact identification means that the number of excluded independent variables is equal to the number of dependent

variables minus one. For example, let the (static) demand (subscript d) and supply (subscript s) *system* be:

$$Q_d = \alpha_0 + \alpha_1 P + \alpha_2 Y + u_1$$
$$Q_s = \beta_0 + \beta_1 P + \beta_2 T + u_2$$

Then Q_d is exactly identified because it excludes Y and includes P and Q. Similarly, Q_s is exactly identified because it excludes the independent variable T, but includes P and Q.

2. If the system of equations is *over-identified*, then uniqueness is lost in that two or more equations can yield expressions for the same parameter.

3. If the system is *under identified*, then the number of independent variables that are excluded is lower than the number of dependent variables minus one. In this case, it not possible to obtain parameter estimates. Thus, the system:

$$Q_d = \alpha_0 + \alpha_1 P + u_1 \, ,$$
$$Q_s = \beta_0 + \beta_1 P + u_2 \, .$$

is *under-identified* and the expression in which the reduced-form parameters are expressed as a function of parameters of the original system of equations is probably meaningless.

The estimation of the parameters of this system (and more complicated ones) uses the reduced-form equations, discussed by Kennedy (1998) after the identification step is completed. There are several techniques for obtaining estimates of the reduced form equations. A simple method is the two-stage least squares. Estimation by the two-stage least squares consists of the following stages:

Stage 1. From the reduced-form model, obtain estimates of the parameters using the ordinary least squares. For an exactly-identified model, this means the equation:

$$\hat{P} = \hat{\pi}_3 + \hat{\pi}_4 Y + \hat{\pi}_5 T \, .$$

Stage 2 The estimated values of the dependent variables are substituted for the original values in the reduced form equations and the parameters are then estimated using the OLS. These estimates include the appropriate standard errors of the estimates and all other statistics generally associated with estimation. What we need is numerical values for $\hat{\pi}_i$, the estimated values. The next example uses statistical estimation to produce estimates of the parameters of a system of two simultaneous equations. It is designed to exemplify how the simple concepts of estimation can be extended to deal with regressions that include more than one independent variable, as well as two dependent variables.

Example 3.15. I created a data file consisting of a sample of one hundred observations on the simultaneous equation model that follows, and used the two-stage least squares procedure (TSLS routine) in SYSTAT 8.0 to obtain the estimates. The simultaneous system consists of two linear equations, with X_i symbolizing the dependent and independent variables, as follows:

$$X_2 = f(X_3, X_4, X_5) = \alpha_1 + \beta_3 X_3 + \beta_4 X_4 + \beta_5 X_5,$$

and:

$$X_4 = g(X_6, X_7) = \alpha_2 + \beta_6 X_6 + \beta_4 X_4 + \beta_7 X_7.$$

Using the TSLS procedure, the form of the model is stated in terms of *instruments*. These are: constant, X_5, X_6, X_7 and X_8. The estimated values of the parameters of the first equation, their associated standard errors and the *p-values* for a two-tailed test of the hypothesis, are as follows:

Estimated Constant = *25.059*, se = *5.156*, p-value = *0.000*.

$\hat{\beta}_3 = -1.347$, se = *0.76*, p-value = *0.080*.

$\hat{\beta}_4 = 1.813$, se = *0.822*, p-value = *0.030*.

$\hat{\beta}_5 = -1.678$, se = *1.368*, p-value = *0.223*.

Only two of these estimated parameters are statistically significant at $\alpha = 0.05$, namely the constant and β_4. The role of the instruments, $X_4 = g(X_6, X_7)$, is to account for the simultaneity and yield the appropriate estimates for the parameters of the variables X_3, X_4, X_5. If we use the regression: $X_2 = f(X_3, X_4, X_5) = \alpha_1 + \beta_3 X_3 + \beta_4 X_4 + \beta_5 X_5$, and not account for simultaneity, we obtain:

Constant = *26.654*, p-value = *0.000*.

$\hat{\beta}_3 = -0.008$, p-value = *0.957*.

$\hat{\beta}_4 = 0.19$, p-value = *0.202*.

$\hat{\beta}_5 = -1.109$, p-value = *0.056*.

The statistical significance of the results has now changed, thus leading to different and incorrect conclusions.

The relationship between the price and quantity of private goods and services is first developed mathematically and then the parameters of the model are estimated using statistical methods. As shown, adding income and temperature provides for a more realistic model. The reason why income can be relevant is as follows. The willingness-to-pay (the demand function) for wage-risk tradeoffs provides a theoretically plausible measure of value for those at risk. Compensation commensurate with the added risk is measured by a premium which is added to the basic wage, through open negotiations and agreement.

Example 3.16. A chemical factory employs *600* people who will hold the same job for *40* years. Initially, each has an occupational fatality (hazard) rate of *0.001* per year and a non-occupational (background) hazard rate from all other sources of *0.02* per year. A proposed change in the factory will increase the occupational fatality hazard rate to *0.002* per year, but would raise average productivity and salaries by *$2,000* per work-year. Each worker would be willing to accept the increase in annual occupational hazard rate in return for an extra *$2,000* per year of compensation if and only if, he survives for at least another *45* years to enjoy the increased wealth. Under these assumptions, is the proposed change acceptable? Initially, each worker has an expected remaining lifetime of *1/(0.001+0.02) = 47.65* years. The proposed change would decrease expected remaining life to *45* years, which might make it just barely acceptable to each worker based on his ex ante expectations. However, the median survival time would only be *0.69/0.022 = 31.36* years, due to the skewness of the exponential survivor function, which has a median equal to *0.69* times the mean. The majority of workers will therefore achieve increments in lifetime wealth and enjoyment that are insufficient in retrospect to make the increased risk worthwhile; this is predictable in advance. There is a conflict between the ex ante risk projections of each individual and the statistically predictable pattern of population risk. Each individual expects to live longer than each of them actually will. Whether public decisions should be based on ex ante majority preferences for implementation of the proposed change or on statistically predictable ex post majority preference for consequences is a policy question to be resolved by public policy-makers.

Many risky situations involve activities that are not priced by the operation of the markets for good and services. Nonetheless, if a CBA is performed or considered, those values have to be accounted for, even if their accounting is imperfect. In general, if the analyses are reproducible and based on sound economic theory, the information about intangible values can be important in social balancing of cost and benefits. Moreover, such valuation is often legally mandated (Ricci, 1995).

3.3. EVALUATING NON-MARKET PRICED GOODS AND SERVICES

The previous discussions provide the basic framework for the theoretical and empirical analysis of economic relations that are relevant to risk assessment and management. In many cases, the market signals (prices and wages are some of the key signals) are available. But, other situations such as those to which the precautionary principle may apply, this data can be unavailable or inappropriate. Typically, the market may not provide the information needed or, if it does, it may be through the *sale* price of nearby land or improved property. For instance, for assessing comparable worth, that information may have to be supplemented and assessed further to deal with unusually large variability and robustness of the results. On this reasoning, economists have developed several methods to deal with these evaluations.

Most of the literature on these methods is based on the evaluation of environmental resources. Nevertheless, the methods apply to site selection and evaluation, as well as to other areas of concern to risk assessment and management. Perhaps needless to say, environmental legislation and many environmental cases, some decided by the US Supreme Court, have been concerned with their accuracy, including uncertain (*hypothetical*, in the words of Justice O'Connor) causation relating event to environmental injury, the magnitude and severity of that injury (Ricci, 1995).

In this section, we describe some of the salient points of the better-known methods to establish defensible monetary value for non-market goods and services. These methods include:

- Contingent Valuation Method (CVM)
- Contingent Ranking Method, (CRM)
- Travel Cost Method
- Hedonic Price Method
- Production Function Method

Contingent Valuation Method. The Contingent Valuation Method directly elicits answers from respondents through questionnaires (Bockstael et al., 2000; Carson, 2000). This information is used to determine each interviewee's willingness-to-pay (or the willingness-to-accept compensation) for a change in risks and her ability for paying for the change. Thus, because the CVM attempts specifically to account for the respondents' *ability* to pay, it avoids the pitfall of just determining someone's *willingness* to pay. The method has been applied to assess the damage caused by the tanker Exxon Valdez, which was reportedly *estimated* to range between three to five billion US dollars. CVM creates a quasi-market condition to determine the value of a change in risk or damage, under the assumption that the respondents can determine and provide their true preferences.

The aggregate willingness-to-pay can be modeled using, for instance, a linear statistical model, such as the population model with k SES socio-economic variables that describe the characteristics of those over whom the CV is calculated (Dasgupta and Pearce, 1978; Bishop and Heberlein, 1979; Bradford, 1970). The model is:

$$CV_{ij} = \sum_{i=1}^{n} (a_{jk} * SES_{ik}) + u_{ij}.$$

In this model, CV_{ij} is the contingent valuation by the ith respondent (n is the number of individuals responding to the questionnaire) of the jth change in the public good availability to that respondent. SES_{ik} is the kth variable for the ith respondent; a_{jk} is the parameter associated with the kth variable; u_{ij} is the random error. The random error accounts for variables that are not included in the model as well as for measurement errors (Kennedy, 1998). The analysis consists of developing a survey that produces a sample of data, which is used to estimate the parameters of the model, namely a_{jk}. (*SES* means socio-economic variables)

The aggregation over the respondents with the SES characteristics modeled can be used to calculate the consumer surplus associated with changes in value. An application of CVM consists of binary questions (with yes or no answers) for two alternatives, one being the base line (or status quo) and the other the alternative to that option. For instance, each respondent (characterized by the appropriate set of SES variables) is informed that the alternative to the status quo can raise her tax rate. Repetitively using this procedure identifies the levels of money that different individuals would or would not be willing to pay. This maps their willingness to pay.

The statistical analysis of these responses to the questionnaire (assuming the appropriate design, questions and adequate sample size) determines the aggregate willingness-to-pay. The CVM has both advantages and disadvantages. These are summarized in Table 3.6 (Kahneman and Knetsch, 1994; Mitchell and Carson, 1989).

Table 3.6. Advantages and Disadvantages of the Contingent Valuation Method

Advantages	Disadvantages
Broad applicability because it is based on direct (expressed) responses.	Choice of open-ended questions or referendum affects the responses. Respondents may have a "warm" feeling for the environmental issue
Biases can be reduced by the proper formulation of the questions and their administration by trained personnel. The questions can be *yes* or *no* or be open-ended.	The successive administration of the questionnaire may force answers. Embedding occurs (a service is given the inappropriate WTP value if it is included in another service)
The elicited responses are accurate, provided the sample is randomized.	A respondent behavior may differ from her expressed attitudes. The information base of the respondents can be limited
Possibly more accurate than some other empirical methods.	The respondents can play strategic games. Symbolic effect may take place: the issue is contrived or hypothetical.

The US Department of Interior has studied CVM and found that it is a practical and theoretically sound method for determining the value of natural resources not priced by the market mechanism. However, this department would not use CVM if a natural resource can be restored for less than its lost value or the *diminution-in-value* method with estimates based on existing and reasonable market data.

Contingent Ranking Method. The Contingent Ranking Method (CRM) consists of the direct elicitation of attitudes that are determined through individual response to pictures or photographs of the natural resource being assessed. CRM produces ranks of preferential responses. Rankings are correlated with the cost of an environmental resource, such as visibility, and then developed as a function of the value of that resource. An outcome, for example, can be some amount of visibility improvement from reducing particulate matter causing haze, which is emitted from stationary sources of energy, such as a power plant. Unlike the CVM, the CRM does not attempt to develop a quasi-market for the attributes being studied. The illustrations must provide evidence that is as proximate as possible to the outcome of an impact, haze on visibility, and the cost of reducing that impact. The theory of the CRM is a random utility model that is used to determine changes in value as a function of changes in utility and, from these, the consumer surplus associated with those changes. The basic assumption is that the specification of a random utility model is representative of individual utility. The advantages and disadvantages of the CRM are similar to those of the CVM.

Travel Cost Method. The Travel Cost Method, TCM, consists of determining, through surveys administered to visitors, or potential visitors, the cost of travel from several zones nearby the areas of interest to those visitors. The TCM is used to estimate the parameters of the individual demand function for the trade-offs made by those visiting a park or other area. Each trade off is between travel (which takes time and thus detracts from the total time available for leisure) and the amount of leisure time that remains available to the visitor. TCM establishes a measure of consumer surplus (Brown and Rosen, 1982). The nearer the respondent lives to the site, the lower her cost of travel and, therefore, the higher the number of visits to that site. Those who live further away will have higher travel costs, spend more time on traveling and will visit less frequently because of the trade-off between travel time and leisure time. Demand for those areas is a function of the unit cost of traveling to them. The consumer surplus measures the value to the respondent (Mendelshon and Brown, 1983). At equilibrium, the marginal cost of traveling to a site equals the marginal value of the visit to the site and the respondents have the same behavioral characteristics (Kneese and Sweeny, 1985). This method may be difficult to apply to travel involving short distances and is generally single purpose focused (e.g., recreational travel). Although the method is well suited for determining the value of a recreational area to visitors, it can overestimate value if travel has other purposes or travel cost is not an accurate proxy for an entrance fee to the area being visited. An alternative to TCM is to use a random utility model that focuses on the time spent by visitors to the recreational (or other) areas (Lipton et al., 1995).

Hedonic Price Method. The Hedonic Price Method (HPM) attempts to place a monetary value on public goods and services through the price of other observed goods (Ronan, 1973). Values are captured by the changes in the market value of housing or other factors for which there is a market value due to an environmental factor, such as noise (Maler, 1974). The HPM presupposes that each individual acts as a utility *maximizer*. The public component of the price structure is measured through statistical models, such as the regression model, in which their specification includes the non-market with the market factors. The model consists of setting up a regression model in which price depends on the determinants of price. A simple example is the rent model where the price is the market value of land, while socioeconomic factors, pollution and geographic location are the variables that measure the factors that affect the price of land. For example, noting that the specification of the HPM may not be linear, a hedonic price model may be:

$$PL_i = f(SES, PC, D_i^m, ...; b_k) + u_i,$$

Where the *PL* is the price of the ith parcel of land, *SES* is a set of socio-economic variables, *PC* is a measure of pollution, *D* is the distance from the area of study to the ith land parcel, *m* is a model's parameter, b_k is the (vector) of parameters to be estimated, and *u* is the error term. The model, justified by the theory of rent, can capture the effect of pollution on the value of land (Ricci, Perron and Emmet, 1977). The HPM applies to all respondents; random fluctuations are accounted by the error term (Ricci, Glaser and Laessig 1979; Scottet al., 1998). The principal assumption is that the price of land reflects the optimal and rational choice of the purchaser. The benefit to society from the site (e.g., a park and lake) is obtained through the aggregate marginal benefit calculation. The product for which the hedonic price is being sought reflects the utility of the attributes that contribute to forming and explaining it.

Example 3.17. Theory leads us to expect that the price of land depends on the quality of the land parcel, *L*, the quality of the neighborhood, *A*, and a measure of air pollution, *AP*. The functional relation is:

PRICE = g(L, A, AP).

This model is used with data from land parcels of similar characteristics, the homogeneity assumption, *PRICE* measures hedonic value. The first (partial) derivative of the function, $\partial(PRICE)/\partial(AP)$, is the marginal price of air pollution.[6] This evaluation presupposes that the individual acts as a utility maximizer, the price function is differentiable and the market of land values is at equilibrium. Then, the willingness-to-pay for a change in air pollution can be determined by the marginal price for that parcel of land.

Production Function Method. The Production Function Method, PFM, measures the amount of goods and services produced by the environment, and then places a value on that output. This is done by establishing a mathematical relation between that output and the utility derived from it. Therefore, the PFM can include specific environmental goods or services to measure the total output produced by the specific environment being studied, for example a marshland. Consider a marshland. It serves a number of ecological, hydrological (e.g., aquifer recharge, retarding food waters), water filtration; provides fish habitats and other ecological and biological services. If these services can be priced through known human analogs, e.g., a rapid sand filter as an analog for water purification, the value of that marsh can be approximated by a set of such analogies, if they are realistic. The cost of physical inputs to produce the environmental output can be studied through a cost function, well-defined in microeconomics and engineering.

3.3.1. Comment

The practical application of the methods reviewed in this section requires samples and statistical techniques. The essence of the statistical techniques has been briefly outlined in earlier examples in this chapter. Chapters 4-11 add the details and examples that can be used with the economic modeling discussed in this chapter, as well as with epidemiology and toxicology. The idea to keep in mind is that the basic statistical methods discussed in this chapter can be used in a wide variety of risk-related studies, almost regardless of the measurements taken on the variables of a model. This should have been made evident by the simple regression model: $Y = a+bX$. Thus, a model can be a damage function; another can be a demand function; yet another can be a value function relating land prices to its factors.

Theory, sample and the parameters' estimates determine the actual model; the estimates of parameters of the model, with confidence intervals, provide a basis for inference from the known sample to the unknown population.

---

[6] We take partial derivatives because there is more than one independent variable in the model.

3.4. DEATH OR MORBIDITY AVERTED IN RCBA

The balancing of risks, costs, and benefits can require placing values on statistical lives saved, deaths averted, illness averted and similar outcomes. Adopting a monetary value on a human life raises difficult ethical and philosophical issues. These are not part of this textbook. What is generally true is that, implicitly or explicitly, such evaluations occur whenever society justifies social expenditures, sets priorities and limits budget allocations. However, there is controversy about the propriety of assigning a value to human life. In risk management, that controversy is further complicated when those at risk do not benefit (or benefit indirectly) but suffer from exposure to hazards over which they have little or no control. The nature of the hazards themselves can heighten the fear of exposure to toxic agents, particularly when science cannot provide reliable answers. Nonetheless, because those evaluations do take place routinely, we summarize some of their salient aspects and leave the reader to think about the ethical questions inherent to such evaluations on their own.

Often, a reason for an evaluation is that lives either have been lost or can be lost due to a risky activity. One method of valuation used in setting a lower bound on such value is the *human capital approach*. It consists of setting the value to equal the loss, suitably discounted, including ancillary costs (Rice, 1966). An upper bound on this evaluation is suggested when society *should* spend as much as it possibly can to save lives. In this case, the approximate upper bound would be several millions to hundreds of millions of dollars per life saved. Both of these bounds can be questionable, as the reader can imagine. There is also much disparity between the evaluations of life-saving measures adopted by the public sector.[7] Considering health care, for example, the median cost per life saved is *$19,000* (sample size of *310*). In transportation, the cost rises to *$56,000* for the reduction of fatal injury (sample size of *87*); in health and safety, the cost is *$68,000* (the sample size is *16*, with costs approximately up to *$1.4* million for *toxin* control). The highest expenditures per statistical life saved occur in environmental health: namely, approximately *4.2* million dollars per statistical life saved (from a sample of *124*). Based on the median cost per statistical life saved for five US regulatory agencies, the Federal Aviation Administration incurred a median cost of about *$23,000*, the Consumer Products Safety Commission *$68,000*, the National Highway Traffic Safety Administration *$78,000*, the Occupational Health and Safety Administration *$88,000*, and the Environmental Protection Agency *$7,600,000*.[8] The approximate ranges of costs are given in Teng et al., (1995).[9]

Between the least costly solutions measured by the cost per life saved, the change in safety standards for concrete constructions (from 1971 to 1988) was insignificant

---

[7] Adapted from Table I, TO Teng, ME Adams, JS Pliskin, DG Safran, JE Siegel, MC Weinstein, JD Graham, *Five-Hundred Life-Saving Interventions and Their Cost-Effectiveness*, Unpublished Manuscript, 1995.

[8] Ibid, adapted from Table III.

[9] Ibid., adapted from Table IV.

relative to the regulation of chloroform emissions from private wells at about fifty pulp mills (*$99,000,000*).[10]

A method to determine the cost per life saved is as follows.[11] The computations adopted to determine the incremental cost per life saved calculate it as the difference between the discounted net cost of the option chosen to save a life and the discounted net cost of the status quo. This quantity is divided by the difference between the discounted net numbers of life-years with the option minus the discounted net life-years without the option, as follows:

$\Delta C/\Delta E = (C_L - C_B)/(E_L - E_B)$.

The variables in this equation represent: costs, C, discounting, r, and net life years, E, The specific model consists of the following equations:

$$C_L = \sum_{i=1}^{N_C} (G_{Li} - S_{Li})/(1+r)^i,$$

$$C_B = \sum_{i=1}^{N_C} (C_{Bi} - S_{Bi})/(1+r)_{C_{Bi}},$$

$$E_L = \sum_{i=1}^{N_C} (V_{Li} - D_{Li})/(1+r)^i,$$

$$E_B = \sum_{i=1}^{N_C} (V_{Li} - D_{Bi})/(1+r)^i.$$

In these equations, L is the life-saving option, B is the status quo option; C_1 is the net discounted cost of the option, C_B is the net discounted cost of the status quo, E_L is the discounted net life-years associated with the option and E_B is the net discounted life-years from the status quo. G_{Li} is the gross dollar cost of the action in the ith year and G_{Bi} is the gross cost of the status quo. S_{Li} and S_{Bi} symbolize the savings for the action and for the status quo. The life-years saved, for the option and the corresponding status quo, are V_{Li} and V_{Bi}. The corresponding life-years lost, for the option and for the status quo are D_{Li} and D_{Bi}, respectively. Finally, N_C is the number of periods (measured in years) and r is the discount rate.

Example 3.18. Tseng et al., (1995) calculate the cost per life-year saved associated with rear seat belts. $(C_l - C_B)$ equals $791,000 and $(E_L - E_B) = 0.8177$, from which their ratio is *$967,702* (the incremental cost per life saved, in 1986 dollars). This cost is adjusted to 1993 dollars by multiplication by the ratio of CPIs for the two years: the factor *(145.8/109.6)*. The operation yields *$1,287,326*. The discounted life-years per life is calculated from the expected age of death (*35* years), life expectancy given age equal *35* years (i.e., the *actuarial* remaining expected life-years for a person who is 35 years old and does not die at age *35*), which is *42* years. The discounted life-years per life (*r = 0.05*) is calculated to be 17.42. Thus, the incremental cost (\$US 1993) per life-year saved is: *$1,287,326/17.42 = $73,886*.

---

[10] Ibid., adapted from Tables V and VI, Teng et al., (1995).
[11] Ibid.

Inflation and other economic changes reflect 1993 prices, using the Consumer Price Index, CPI. The calculation consists of multiplying the ratio of the changes ($\Delta C/\Delta E$) by the consumer price index for 1993 ($CPI = 145.8$), divided by the CPI for the years (indexed by CPI_x) in which the data is assessed.

3.5. INTERGENERATIONAL EQUITY

Discounting over several generations must account for technological changes and social preferences that are uncertainty. It is unlikely that we can predict today what technology and social preferences and how political institutions may shape the future of distant generations. It follows that we can only approximate how those distant generations will benefit from today's expenditures on pollution and the distribution of future risks and costs that follows from those initial expenditures. Discounting, as discussed, is the trade-off between known (or determinable) preferences for consumption today relative to preferences of a later period that we can, however , assess today. Intergenerational tradeoffs are fundamentally uncertain because intergenerational preferences are unknown. We can however make plausible assumptions. For example, assuming optimism, future generations will probably be better off than today's generation because of changes in technology. Accordingly, the discount rate should be very small, near zero but positive. Portney and Weyant, at the Resources for the Future in Washington DC, suggest using a *1%* discount rate. Others suggest a varying (time-dependent) interest rate, relatively high in the first generation, becoming progressively lower to follow behavior and technological advances, as these become better understood.

Intergenerational equity has economic, philosophical and political aspects and implications. Focusing on the economic aspects, consider the following speculations. Decisions made today may have serious consequences to future generations. Yet, the political viability and economic soundness of actions taken in the present may be predicated on a low concern for future (meaning long-term) consequences because future hazards can appear to be conjectural. However, the combination of low social discount rate and national and international trust funds may be able to redistribute wealth according to the dynamically evolving understanding of risks over generational times. This combination may help shield future generations from actions based on scientific conjectures taken by the current generation. Transactions made well within a generation emphasize current science and market exchanges driven by prices and budgets. Future generations face the fallout of these transactions. There can be assets, if conservation practices have been instituted and enforced. They may also be deficits, if natural resources are exhausted earlier and environmental and health burdens are shifted to the future. The equity of considering future generations relative to the present has a single perspective. It measures welfare by growth in consumption and sees the environment as a collection of goods for consumption. The issue is that, as the economy grows and natural resources become depleted, the challenge shifts from the task of an equal distribution of wealth to one of equitably rationing shrinking resources, if technology does not fulfill earlier expectations. More specifically, non-renewable resources, such as timber, natural gas and coal are priced by the operation of today's market, including options and long-term contracts. As demand for these non-renewable resources grows, with a growing population and increasing demand for energy, these resources continue to be depleted. Scarcity causes price to increase and can result in geopolitical instability further affecting the potential for equitable distributions.

Without a strong and balanced conservation policy, future generations will pay proportionally higher prices for non-renewable resources and will value remaining assets higher. The preservation of non-renewable resources becomes an early important aspect of intergenerational equity, particularly if technological advances do not provide adequate substitutes for the depleted resources. Moreover, geopolitical factors can dramatically affect the otherwise orderly progression of technological improvements and change. Placing environmental resources on an economic scale emphasizes their ability to contribute to economic wealth. Considering that one of the basic assumption of economics is one of continued economic growth (apart from periodic recessions) and the fact that non-renewable resources are increasingly depleted with time, a market-driven belief is that built capital can substitute for natural capital. However, substitutability can be theoretically and ethically challenged because substitution may be physically either impossible or extremely difficult. If the aesthetic and ecological values of the natural environment cannot be replaced or enhanced, then the substitutability argument can be incomplete for policy analysis.

Responsible decisions by the current generation concerning the economic use of exhaustible or unique natural resources must be made with regard to equity as well as economic efficiency. Economic analysis for risk management includes subsidies and taxes because they both affect the distribution of costs and benefits through the price system. For this reason, we provide a short overview of the effect of taxes and subsidies on price and quantity of goods and services affected by them. Additional material can be found in macroeconomics and public finance textbooks. It should be evident that the RCBA of many projects of some magnitude should account for both taxes and subsidies because the effect of the project on local taxes and subsidies can substantially change the net benefit calculations and thus their ranking. The impact of taxes and subsidies can be assessed using the supply and demand framework described in this chapter.

3.6. SUBSIDIES

A financial subsidy is revenue added by the government or private entity to enable the output of an activity to be a lower cost than the actual costs incurred in producing that output. Subsidies might take the form of special tax allowances, rebates, direct payments, and can be indirect. For instance, when the cost of environmental impacts are not recovered from the party damaging or disrupting the environment and profiting from the disruption, that party is *indirectly* subsidized. Similarly, using the environment as a sink for some emissions that a firm would otherwise have to control enables higher, but unjust, profits. As another example, if environmental impacts are mitigated by public agencies or private firms that have not caused the impacts, the costs avoided by the actual polluter is a form of subsidy for that polluter.

The market does not correctly price a good or service when subsidies are not *internalized* in the price of that good or service. This distorts the signals of the market mechanism provided through prices (prices are relative measures of worth). When the loss of economic activity and opportunities forgone can be estimated using direct economic pricing or approaches such as hedonic pricing, evaluation is relatively simple. The evaluation of externalities can introduce subjective elements.[12]

The reasons are as follows. For some, the value of environmental attributes is simply the worth of the environment's services that it can provide for human consumption. For others, a pristine wilderness, a diverse ecosystem, a unique species or other ecological system simply does not have a market value. These are fundamentally priceless. Cultural differences, variation in priorities, economic changes and political causes also increase the difficulties of valuing the environment. Nonetheless, beliefs notwithstanding, under most versions of legal precaution, some valuation must take place. It follows that the effect of reducing environmental subsidies can provide incentives to producers to develop and use improved technologies to control environmental impacts.

These additional costs are generally passed to consumers through higher prices for the products and services they purchase. This is the result of the direct economic operation of the market. The downside is that some individuals can be excluded from being able to purchase the good or service at higher price, particularly if those goods do not have less expensive substitutes. The implications of *removing* subsidies, if those are directly paid to producers, depend on the response from producers and consumers as well as on technological innovations.

3.7. TAXATION

A tax is an added financial burden placed by the government to generate revenues from those on whom the tax falls. A tax increases the cost to the ultimate consumer of the good or service. We are all familiar with several types of taxes, ranging from individual income taxes to tariffs. An important aspect of using taxes is not only a way to fund governmental activities or some specific activity but also redistributes income. The *legal* incidence of taxation determines who pays the tax and the *economic* incidence of taxation determines the change in the distribution of some part of a private income. To describe the effect of taxes we can use the supply and demand model, which now depicts the aggregate demand and supply for the economy.

In Figure 3.9, the initial position of aggregate demand and supply for a taxable good or service is identified by the labels *supply* and *demand before tax*. It can be used to exemplify the impact of taxation on the price of goods and services.

---

[12] An externality is a factor that is not included in the analysis. The exclusion generally results in a bias that favors those that could be affected by including the externality.

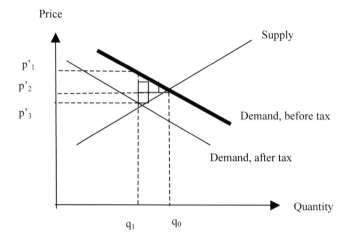

Figure 3.9. Impact of a Constant Tax on Demand and Supply

To simplify the analysis, we use a constant tax rate. If that (constant) tax rate is applied to the price of the good, then the demand function will shift to the left, assuming that the change in the tax rate does not change the behavior of consumers. The reason for the uniform shift is that all prices are equally affected by the single tax. Therefore, the demand curve shifts to the left, as shown on Figure 3.9.

The effects of taxation can be calculated using the methods described by Rosen (1992). In Figure 3.9, the equilibrium price before tax is p'_2 and the quantity supplied is q_0. This equilibrium price is changed to p'_3 by the tax.

The supplier will now produce q_1 units of output, but the consumer will pay p'_1 for that amount. These disequilibria take place because the effect of a tax is immediate. The adjustments achieved through the operation of the market affect consumers, resulting in a new equilibrium, the point p'_3, q_1, which can take some time to occur. The crosshatched triangle measures the total amount of consumer welfare lost due to the disequilibrium, as shown on Figure 3.10.

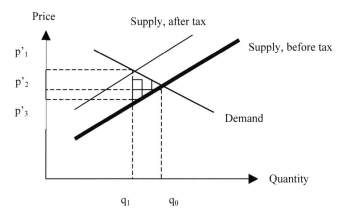

Figure 3.10. Impact of a Constant Tax on Supply

3.8. BUBBLES, TRADEABLE PERMITS, AND BANKING

Traditionally, environmental pollution standards apply throughout a country. These are imposed by law and are monitored according to specific protocols that are also legally enacted and enforced. This form of environmental protection has been called *command and control*. Typically, an environmental statute sets principles, legal tests and other forms of guidance that are then used by a regulatory agency, or an authority of the government, to set pollution standards. Standards are then set through those regulations (secondary legislation) uniformly on a country or region, unless there are specific regional requirements that result in special standards for that region. Although this is a simplification, as those familiar with special provisions of the law for state implementation plans know, it is a basis for command and control regulatory standard setting. An example is the US National Primary (protecting human health) and Secondary (protecting other welfare aspects such as visibility, vegetation and so on) Air Quality Standards for sulfur dioxide, particulate matter, carbon monoxide, ozone, nitrogen dioxide and lead that were promulgated, under authority of the Clean Air Act, by the US EPA as discussed.

These standards apply to specific *stationary* sources of pollution, such as fossil fuels burning power plants, are numerical and include suitable averaging times, (e.g., for SO_2, the US federal primary standard is *80* micrograms/cubic meter of air, annual arithmetic mean). The standards have been set at the level appropriate to protect sensitive individuals, not just the average member of the US population. As discussed in Chapter 2, Primary Standards were set without the CAA explicitly demanding that the US EPA should account for the costs of regulation before issuing the standard. Command and control regulatory framework is often associated with forcing new pollution technology, such as *the best available control technology* to achieve requisite level of emissions. Achieving the concentration of pollutants set forth in those standards has resulted in complex relations between states and the federal government. These, in turn, have created State Implementation Plans, New

Source Performance Standards, Prevention of Significant Deterioration regulations, Lowest Achievable Emission Rates, Reasonably Available Control Technology and so on (Tietenberg, 2000).

Legislators have come to realize that the carrot (profits) is often preferable to coercion. The US EPA has been working with economic incentives, either as alternatives or complementary policies to *command and control*, such as the offset program, since the 1970s. The *offset program* requires that a new stationary source operating in a region that had not met the national ambient air quality standards would have to purchase a sufficient amount of emission reductions to offset its contributions of pollution to the region. The *bubble* policy, developed by the US EPA, was an early approach, in the late 1970s, to regulating air emissions. An emission bubble is placed over several sources of air pollution, and the contributions from individual sources can be modulated to meet permitted, overall, emission levels that can be met by accounting for different efficiencies of the individual contributors to that overall bubble.

Emission trading and banking is a relatively recent addition to the portfolio of economic instruments used to reduce the emission of pollutants. Emission banking results in a pollution credit, which form when there is a gap between the environmental standard and the emissions of pollutants from a firm, if those emissions are below the regulatory standard. Firms can use these credits as if they were currency: they can be banked, transferred, sold or auctioned under the supervision of the environmental agencies. The environmental standards cannot be exceeded: the gain, relative to the more uniform command and control system, is that economic efficiency is allowed to work at the level of the firm, stack or other to minimize pollution and costs and thus generate aggregate societal benefits. The reason for allowing pollution trading through a government-regulated permit system is that those permitted to do so are driven by the market to reduce pollution, increase the efficiency of production and opt for local or regional flexible control strategies. Specifically, market-based instruments such as pollution permits, can accurately account for the different marginal costs of production, technological innovations and fuel use. The trading system is as follows: a company reduces its emissions below a legal level and thus accrues emission credits that can be used elsewhere or be traded. The economic incentives generated by the trading system contribute to minimizing pollution and increasing the efficiency of production.

Trading is generally based on the sale of pollution permits. This is an activity that, in the US, is controlled at the state level and can then be left to the market operation, provided that reporting and legally binding standards are met (Bryner, 1999). For example, the CAA, Title I, Section 110(a)(20(A) allows the states to use such economic instruments as *fees, marketable permits and the auction of emission rights* to provide incentives while maintaining competition and economic efficiency. The CAA's Title IV allows electric utilities specifically to trade sulfur dioxide emission allowances, under an emission cap, if they emit below that cap. If the cap is exceeded, the utility doing so loses the allowances and can be fined at levels that are several-fold larger that the marginal cost of compliance (Guerrero, 1997). The statute places monitoring and reporting requirements under the jurisdiction of the US EPA.

Under the CAA, economic instruments designed to provide incentives to reduce emissions can be combined. This has been done for acid rain resulting from emission from power generation burning fossil fuels, principally coal, at the national level. The cap on the total emissions was such that it would result in sulfur dioxide reductions of *10* million tons from the 1980 levels, in 2010. The allowances are based on one ton of sulfur emitted, with the permit system envisioned to operate to the year 2030 (Bryner, 1999). The trading of acid rain allowances began in 1993, it included auctions, spot and advance allowances. The proceeds from auctions are distributed among those from whom allowances were withheld. Trading and caps have also been combined at the state level. In California, the South Coast Air Quality Management District of Los Angeles administers the Regional Clean Air Incentives Market (RECLAIM), under the CAA. It developed a trading system for stationary sources emitting NO_x and SO_x. RECLAIM covers approximately four hundred stationary sources and it is open to sources that generate more than four tons of either of these pollutants per year (Guerrero, 1997).

Example 3.19. The combined action of chlorofluorocarbons and O_3 on stratospheric ozone depletion and the effect of alternative forms of regulation, such as command and control, a constant charge (equivalent to a tax) and marketable permits have been studied by Palmer, Mooz, Quinn and Wolf (1980). Among other results, they show the following results for a number of environmental policy options:

Environmental Policy Options	Reduction in Emission of CFCs (10^6 permit-pounds)			Total Cost of Compliance (1976 $1*10^6$)		
	1980	*1990*	*Cumulative reduction over the decade*	*1980*	*1990*	*Cumulative PDV (r = 0.11; 10 years)*
1. Emission Standards for CFCs	54.4	102.5	812.3	20.9	37.0	185.3
2A) Constant Tax ($0.5/pound)	54.8	96.9	816.9	12.3	21.8	107.8
2B) Marketable Permits (from $0.25 in 1980 to $0.71 in 1990)	36.6	119.4	806.1	5.2	35.0	94.7

These results suggest that the three potential policy choices are not equivalent: marketable permits are the least costly for obtaining roughly the same level of cumulative emissions of CFCs over the decade.

Tietenberg (2000) provides additional discussions and references to later events, such as the Montreal Protocol of 1988 and the fact that the US has adopted the marketable permit policy under this Protocol. The advantages of economic instruments include their tradability in the open (but regulated) market and signaling of the equilibrium between price and quantity. The market provides some anonymity to its entrants, if it is sufficiently large. Tradability of permits, which reflects the short life of some pollutants, enhances local pollution control because the operation of a cap and trade system can address local variability of pollution, such as is the case for tropospheric ozone. The way tradability reduces pollution is through the efficiency of the market place, which is directed to the source of the emissions of pollutants, rather than being technology forcing. This shift in focus reduces administrative costs and helps to achieve the least cost of control for the minimum emission rates. The disadvantages include increased monitoring of pollution and the trading and banking of permits. The accurate measurement of pollution emissions is also important, because of the allowances, as is knowledge of the physical and chemical characteristics of the pollutants, particularly when they do not mix uniformly. In these situations, market incentives may not be able to reduce local *hot spots* of pollution. This is where restricted trading, or trading weighted by specific ratios of chemicals, can be used correctly to represent the market's operation.

Some have suggested that using economic instruments to control emissions of air pollutants can reduce innovations, unlike the legal commands for such technologies as the *best available*, *best practicable* and so on. The reason for this concern is that some participants or entrants to the emission trading system can opt to purchase credits, rather than fund research and development. Pollution control through economic instruments does not do away with the *control* part of pollution regulations because enforcement is essential to the successful operation of the market for these incentives. Others are concerned with increasing costs of reporting and oversight costs. For example, the bubble program has been reported to have high administrative costs (approximately US *$10,000* per transaction) because it adds that system to the already existing environmental regulatory system (Bryner, 1999). Another adverse view to using economic instruments to reduce the emission of air pollutants is the instability of political choices and the potential for writing stringent laws that are then not enforced by political means, as can occur when environmental funds are withheld or reallocated. This problem is less likely to affect command and control regulations, once they are issued and put in place. It has also been argued that economic incentives conflict, legally and morally, with the *polluter pays* and other principles underlying many international environmental laws and regulations (Bryner, 1999). In some instances, efficiency of production at the *least cost-least emission* levels may advantage some firms over others, driving out of the market the least efficient firms even though these may still able to meet environmental standards. This outcome can create local employment shortages or other shortages such as, if the firm is generating electricity, energy shortages. The final issue that affects market-based emission controls is that using economic instruments laws can violate some socially driven laws. For instance, in 1997, the California Air Resources Board suspended the trading of VOC's emissions permits because a federal civil *right* law appeared to be violated on the grounds of an *inequitable* distribution of the air pollution exposure, resulting from emission trading (Bryner, 1999).

3.9. CONCLUSION

Making decisions about hazardous situations requires assessing not only the risks, but also the economic costs and benefits of hazardous situations and the way choices of technology or other means to minimize pollution can be ranked. The three elements, the *R*, *C* and *B* in RCBA can become understood at different rates of knowledge, points in time and with different uncertainties. In this chapter, we have addressed the economic aspects of RCBA from traditional economic analysis based on supply and demand. We have simple used supply and demand analysis to illustrate some of the policy implications of taxes and subsidies on price and quantity. From the demand function, after estimation of its parameters' values through statistical methods, we can calculate an individual's willingness-to-pay for private and public good or service. In the aggregate over all individuals, we can generate the aggregate demand curve and therefore obtain estimates of the magnitude of the societal consumer surplus inherent to risky choices. The supply of goods or services completes the information by describing the marginal cost functions of one or more producers. The value of the producer's surplus can be calculated from his supply curve. At (static) equilibrium, the quantity supplied and demanded is at equilibrium with price, which is a necessary condition for assessing if the allocation of resources under study is optimal.

Most RCBAs requires considering several issues. These include:

- The theoretical soundness of the framework underlying the causal aspects of the decision, measures of uncertainty, scales of measurement for the factors included in the analysis, the state of the information and time
- Eliminating overlaps in the set of actions available to meet the objectives of the RCA (with actions that are excluded being listed and discussed)
- Assessing the practical possibility of implementing each action and the political implications of an otherwise economically optimal action

Specifically, a RCBA-based decision should account for the:

- Distribution of benefits, their measures (dollars and possibly other measures of value) and the technical assumptions (for example, *efficiency* which means that marginal costs must equal marginal benefits) made in determining those benefits
- Policy implications relevant to each option
- Assessment of the changes in the aggregate economic welfare (measured by the consumer surplus) associated with each option
- Uncertainty and variability about the components of the costs and benefits
- Identification and inclusion of stakeholders, accounting for their attitudes and preferences towards each option and the full rationale for selecting the preferred option
- State and quality of the information, at the time of the analysis
- Resilience and robustness of each management action as time evolves and new information becomes available

The assessment of benefits is predicated on individual preferences that are aggregated across all affected, provided that: a) preferences are consistent; b) there are no externalities left to be included in the analysis or, at least, that those are fully disclosed and discussed; and c) preferences are disclosed. If there is an explicit accounting for these conditions, the process of reaching a choice may be fair. In practice, the decision maker's preferences are assumed to approximate those of the individuals for whom the choice is being made, using arbitration and negotiation.

The public sector's objectives are often a hybrid of economic, social, and political factors that can be sharply contrasted with the profit maximization objective of private decision-making by economic entities. The fact that an agency acts in the public interest has the effect of requiring representations of its objectives in a manner that can plausibly account for vague descriptions and values, in addition to having to deal with pervasive uncertainties and conjectures. Risks and economic costs can be transferred to third parties who do not benefit directly from an activity, causing uncertain economic externalities. Because it takes time to resolve the many uncertainties affecting an RCBA, there may be a lag between a policy choice and the reduction or elimination of a hazard. The potential for competing hazards also means that the decision-maker might focus on those hazards that appear to be the largest, leaving a number of less understood and less obvious risks out of the assessment. Any realistic and defensible objective of a risk management is contingent on the state of information used and available to the risk assessor. This suggests that not only the dynamics of the decision require careful modeling, but also the decision process should allow for the partial or even full change from the established path, as new information becomes available. For public decisions, the concept of resilient and robust decisions must guide the risk manager. Simply put, resilience suggests that the public decision-maker should prefer to have options that can be retracted at the least cost as new information becomes available. Robustness means that a choice is unaffected by bias, relative to those remaining in the set of options.

QUESTIONS

1. Suppose that the interest rate is *10%*, for a public project that generates the following stream of annual net benefits: *$12,000*, *$14,000*, *$10,000*, *$13,000*, *$22,000* and *$30,000*. Calculate the PDV of this net income. Then, using the same stream of net benefits, calculate the PDV using *4%*. Draw some policy conclusions and justify the eventual choice of discount rate for that public project. Does multiplying all net benefits by 10^3 make any difference to your conclusions?

2. In no more than 600 words, discuss the discount rate in terms of individual preferences for postponing consumption today for a later period. What assumptions would have to be made to justify a constant interest rate of the period of time?

3. Consider the cost/benefit ratio, namely *(Total Benefits, in €)/(Total Costs, in €)*. One alternative is discounted at an interest rate of *3%*; another is discounted at *10%*. Develop your own example using at least three alternatives to achieve the same objective. Discuss the effect of the different discount rates on the B/C ratio. Rank specific options by the appropriate ratio. Then, calculate the difference between the discounted benefits and the discounted costs. Do the ranking based on the difference *(B-C)*. Do you reach the same conclusions? What critical assumption governs the use of the *B/C* ratio to rank alternatives? Discuss in no more than 700 words.

4. Consider and describe the variability of B and C, as two random variables. How would you express that variability in calculating the difference $(B-C)$? Do you always reach the same conclusion about the rankings that you would have reached when using deterministic numbers for B and C? Why and why not?

5. Consider the B/C ratio to rank alternatives. How would you express the variability of the ratio? Do you think that the ranking reached with B and C as deterministic numbers (without variability) is the same as that reached when accounting for it? Why and why not?

CHAPTER 4.

EXPOSURE-RESPONSE MODELS FOR RISK ASSESSMENT

Central objective: develop and exemplify statistical models relating exposure to adverse response. The models include multiple regression, Poisson regression, logistic regression, Cox proportional hazards model, and time series analysis.

Damage functions are the essential element of risk assessment and management because they link risk factors (explanatory or independent variables) with adverse response (the dependent variable) to determine the magnitude and significance of environmental, occupational and indoor exposures. Dose-response models, discussed in Chapter 5, complete the discussion of damage functions. In risk management, we work with these models to study the effect of policies to reduce risks. Thus, they are important in formulating environmental and health policy, in setting environmental health standards, guidelines, and in litigation. An exposure-response model, in an air pollution study, can be stated as:

Lung cancer death rates (a continuous, non-negative variable) = f[*background rate of lung cancer, average daily concentrations of airborne arsenic* (a continuous non-negative variable), *particulate matter 10 microns in diameter* (a continuous non-negative variable), ..., *smoking* (measured as a yes or no), ...].

In this model, average daily concentrations of airborne arsenic and particulate matter represent exposure, adjusted for the effect of smoking and other risk factors. Response is the daily mortality rates from lung cancer, given those adjusted exposures. We begin with the classical multiple regression, in which the dependent variable is continuous. We will also exemplify models in which:

- The dependent variable is binary; e.g., the logistic regression model where either there is or there is not an adverse response: an individual is diseased and is recorded as *1*, another is not and is recorded a *0*
- The dependent variable is continuous; e.g., the classical regression model in which response is mortality rate (e.g., *3/1,000* or *2.1/1,000* per year)
- The dependent variable describes rare events; e.g., the Poisson regression model in which the number of responses are rare (e.g., the counts of outcomes can be *0, 0, ...*, *1, 0, ... 2, ...* per day)
- Models used in survival analysis to obtain the probability of surviving exposure, such as the Cox Proportional Hazards Model

- Time series models, in which observations are a function of time (e.g., yesterday's concentration is *1.5* ppb, today it is *1.2* ppb, and so on), focusing on the Autoregressive, Integrated, Moving Average (ARIMA) model

Because a researcher can use different models to relate exposures to response, it is often necessary to have to choose between models or even *mine* the data. Thus, we also discuss:

- How to choose between alternative statistical models
- Some of the advantages and disadvantages of data mining

The contents of this chapter relate as follows:

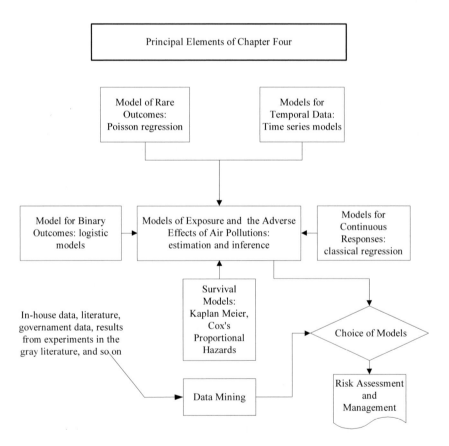

The simplified process of statistical model building is as follows. Begin with a review of the literature. If possible, consult with some experts and determine the appropriate form of the relationship. After these steps, data is collected. The last steps include formulating the model, using data to estimate the parameters of the model, discussing the findings and modifying the model according to what the data indicate, reanalyzing it, drawing conclusions, and identifying unresolved issues.

4.1. INTRODUCTION

We begin our discussions by considering a situation in which we seek to develop a defensible relation between exposure to air pollution and lung cancer mortality. Building exposure-response models requires combining past information from the literature and other sources through mathematical, statistical and probabilistic analyses. These models can account for different forms of exposure, including transient, cumulative, average or other. We can write a simple form of such an exposure-response model as:

$DR = a+bX+cW,...$,

in which DR is shorthand for daily lung cancer mortality *rates*, X represents the daily average concentrations of airborne arsenic, W is the daily concentration of PM_{10} and so on. If the choice of linear function is appropriate, the next task consists of collecting a sample from which we estimate the parameters a, b and c and inferentially relate them to the population at risk. The models discussed in this chapter are different forms of regressions.

Example 4.1. Suppose that the literature suggests that two air pollution variables, average daily concentrations of sulfur dioxide and average daily concentrations of particulate matter of 2.5 microns in diameter, are related to lung cancer mortality. She would let:

Lung cancer death rates = DR = f(exposure sulfur dioxide, particulate matter less than 2.5 microns in diameter, smoking).

This model can be *specified* as:

$DR = a+bSO_2+cPM_{2.5}+d*SMOKE$.

In this *multiple regression* model, the dependent variable is continuous, smoking is an independent variable that identifies whether either an individual smokes or does not; it does not measure the number of cigarettes smoked per day or the concentrations of the carcinogens in the smoke stream.

At this point, some may be confused by having to deal with an unknown population model. The confusion may be reduced as follows. An exposure-response model can be developed from the literature: from studies involving animals or from epidemiological studies, some of which are experimental in the sense that volunteers may agree to being exposed. This information helps to determine the initial form of the model that the risk assessor is building.

Practically, an *initial* choice of the specification (or form) of this exposure model is linear. Because of the empirical nature of modeling, this specification can be justified by Occam's principle of parsimony[1], discussed in section 4.4, by the literature she has reviewed or through a combination of both.

Example 4.2. Continuing with Example 4.1, the linear *specification* (or form) of the exposure-response function, *f(.)* the example is:

$DR = a+b*SO_2+c*PM_{2.5}+d*SMOKE.$

The coefficients of the model (*a*, *b*, ..., are unknown *population* parameters) will be estimated from sample data on *DR*, SO_2, $PM_{2.5}$ and *SMOKE*, thus requiring a change in notation, e.g., *â*. This equation is a *multivariate* linear regression. The term multivariate means that there is more than one variable in the equation.

The process, depicted in Figure 4.1, involves a series of steps that assess the modeling process by testing adequacy of fit of the model to the data, developing confidence intervals, studying and correcting potential violations of standard assumptions leading to the final model.

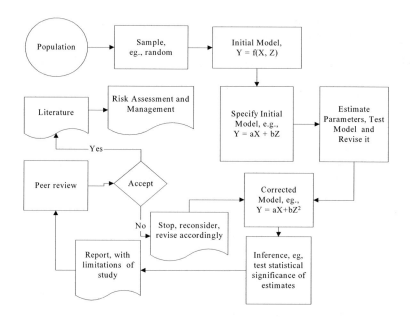

Figure 4.1. Model Building, Estimation and Inference for Exposure-Response Models

---

[1] Also spelled as Ockham, who was an English philosopher (1285? to 1349?).

The result of statistical model building is a function relating mortality or morbidity rates to exposure to environmental agents and other risk factors. This function is based on several considerations, which include:

1. Understanding the *theoretical* basis for the relationship or relationships that determine the set of independent variables (e.g., E for exposure, X for age and so on). This requires developing a literature review for guidance as well as using scientific knowledge specific to the problem at hand,
2. Formulating (specifying) the relationship between dependent and independent variables by setting up the mathematical form of the model linking the dependent variable to theoretically or empirically relevant independent variables. For example:

$$MR = \alpha + \beta X + \gamma E.$$

Here, α, β, and γ are the unknown population parameters of (say) *Age* and *Sex*. Note the often-tacit assumptions that the structure of the relationship established does not change (is invariant) with respect to time and space,[2]

3. Developing a data set (the sample), and
4. Estimating the values of the model's parameters through statistical methods. The values of the parameters and their statistical characteristics are the numerical results of interest.

Models are first an abstraction and then a synthesis of reality, depicted in Figure 4.2.

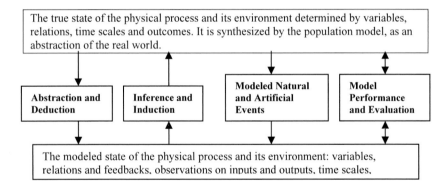

Figure 4.2. Abstraction for Building Statistical Exposure-Response Models

Figure 4.2 points to the fact that developing a risk model consists of formally describing the structure of the relationship between one or more independent variables, Y_i, and one or more independent variables, X_k. Specifically, for exposure-response models, their *physical* context means accounting for the biological, epidemiological, toxicological or other foundations of the relations.[3]

---

[2] The *causal order* of the relationship is assumed to be known as are the variables that are under the control of the researcher and those that are not.

[3] As discussed in Chapter three, relations of relevance to risk assessment and management can involve simultaneities.

Although this process is straightforward, it can often result in incomplete models and conflicting conclusions. Tables 4.1 and 4.2, which contain descriptions of some of the literature on exposure to air pollution and adverse health response, depict the fact that there are different plausible relationships between exposure to air pollution and adverse response. Given these results, increasing the confidence in the results obtained from estimation requires that:

1. The estimated coefficients of the models have the sign that theory demands, provided that theory provides that information a priori (by being based on sound scientific knowledge),
2. The theoretically-relevant independent variables (e.g., age and sex) be included in the relationship,
3. The interval of time (if time is included in the model) separating the values of dependent or independent variables (using lagged values of exposure, for instance: E_{t-1}, rather than E_t) must be empirically consistent with biological or other theoretical considerations.[4]

The interpretation of the parameter estimates is important for risk assessment and management. If a parameter is positive, then it measures the adverse effect on the dependent variable mortality rate (*a* is a positive number) per unit increase in the concentration of the pollutant. If it is negative (*a* is a negative number), the estimate measures the decrease in the value of the dependent variable per unit increase in the concentration. The coefficient can also measure the lack of impact, if it is zero. If the variables in the model are added (*additive* model), each variable influences the independent variable (response) independently from any of the other variables included in the model. The next example describes a data set that can be used to estimate the parameters of a regression equation.

Example 4.3. The risk assessor wants to determine the relationship between exposure to air pollution and lung cancer mortality. She develops a sample of data from *80* major cities from government databases, which provide summaries for a period of one year. This type of data is *cross-sectional* and the corresponding exposure-response model is also cross-sectional. The sample consists of a yearly average of observations taken for the same periods of time for different cities (same time periods for all the cities in the hypothetical example):

City	Average daily lung cancer mortality rates (per 100,000)	Average daily PM$_{2.5}$ ($\mu g/m^3$)	Average daily SO$_2$ (ppm)	...
San Miguel	3.5	2.4	5.2	...
Civita'	4.1	3.2	3.4	...
....	...	...	...	...

This sample is said to be *cross-sectional* because the measurements are assumed to be unaffected by time. The study is not experimental because the research has no control on the data. Statistical estimation methods, exemplified in Chapter 3, yield numerical values of the coefficients of the exposure-response model, for instance: $DR = 120+4.5*SO_2+0.02*PM_{2.5}+$ ... .

---

[4] There are doubts about the potential of complex statistical models correctly to obtain forecasts, because relatively simple models can be as good or even superior, to complex models.

4.1.1. Discussion

The important and practical question answered by statistical models used in risk assessment is as follows (Selvin, 2001):

"Are the estimated coefficients of a statistical model sufficiently large, relative to their measure of variability (e.g., the standard error of the estimate), to support the assertion that their variables systematically influence response?"

For the purpose of this chapter, statistical estimation applied to the mathematical model (the y-intercept is suppressed to simplify the discussion):

$$DR = \alpha(SO_x) + \beta(PM_{2.5}) + \dots ,$$

which consists of adding a random term, u, to the exposure-response model for the population.[5] This random component accounts for errors made in measuring only DR:

$$DR = \alpha(SO_x) + \beta(PM_{2.5}) + \dots + u.$$

The random term, the *error*, also accounts for omitted independent variables that contribute in minor ways to the overall *variability* of the model. As a further simplification, we can assume there are no errors made in measuring the *independent* variables; this assumption can be relaxed at the expense of some more complicated statistical work, discussed in the references.[6] Because each death rate has an associated random error, each death is the realization of a random process. Each u has its own distribution function: each u is identically distributed and independent (i.i.d.) and with constant variance.

The mechanics of estimation is an extension of the discussions developed in Chapter 3. It consists of calculations in which the objective is to minimize the error made in fitting the model to the data. The error is squared because other metrics, such as the absolute value, make the analysis more difficult and because negative differences become (once squared) positive.[7] The squared error is:

$$(Error)^2 = [DR - (\alpha SO_x + \beta PM_{2.5} + \dots)]^2 = u^2.$$

To repeat, the *error* is the difference between the *actual* observation and the *fitted* value for the population if that information were available. This was shown in Chapter 3. This method of analysis is the already discussed OLS.

---

[5] More complicated errors can be used, but are outside the scope of this work.
[6] This limitation can be resolved using errors-in-variables models and estimating the parameters with modified generalized least squares method or the two-stage least squares (Judge, Griffiths, Carter-Hill and Lee, 1980).
[7] Methods such as robust regression use the absolute value of the error (Green, 1997).

Those estimation formulae (for which we gave the solution in Chapter 3 for the simpler model $Y = a+bX$) are applied to the set of measurements on *DR*, SO_X, $PM_{2.5}$ and other variables to yield the numerical estimates of the parameters.[8] The population's parameters (e.g., α, β) are unknown, but from a sample we obtain the values *a*, *b* and such other parameters as the standard errors of the coefficients, the coefficient of correlation and so on. Non-linear models require additional complications, which not discussed in this textbook, but available in the references. The sample model conveys exactly the same idea that is contained in the population model. With a sample, the exposure-response for the population is re-written as:

$$DR = a(SO_X)+b(PM_{2.5})+, ..., +e,$$

in which *e* is the *measured* error associated with the sample data and model. This is the only practical information available because the error associated with the population (i. e., *u*) is unknown. If the results of estimation were: $a = 3.545$, $b = 0.001$, ..., the sample-based model is:

$$DR = 3.545(SO_X)+0.001(PM_{2.5})+... .$$

Notice that the error associated with the sample has now disappeared from the equation because the parameters of the model have been estimated. The variability of each parameter is summarized by the confidence interval associated with each estimated value. That is, each of the estimated parameter values (*3.545, 0.001* and so on) should include confidence intervals and other statistical information. For *b*, the estimated value of β, that report might be the 95% confidence interval (the lower and upper confidence limits). If the distribution is normal,[9] these two limits for the population parameter β should be written as: $b\pm1.96*se(b) = 0.001\pm1.96(0.0.0001)$.

4.1.2. Some Exposure-Response Models and Results for Air Pollution

Table 4.1 is a summary of some prominent studies of the relationship between air pollution and human health. The methods for analysis used in these studies are those discussed in this chapter (abbreviations are given at the end of this chapter).

Table 4.1. Sample of Air Pollution Exposure-Response Models Discussed in this Chapter

Year	Author	Area	Period	Pollutant	Statistical model	Aim of Study
1994	Schwartz	Birmingham, USA	1986-1989	PM_{10}	Poisson Regression	Examine the association between PM_{10} and ozone and

---

[8] Optimization provides the set of equations that meet this objective. Specifically, the first derivatives of the expression to be minimized are set to zero and then the resulting set of simultaneous equations is solved for α, β, and so on for all of the parameters of the model, yielding the estimation formulae. Thus, if $Y = \alpha X_1+\beta X_2$, then $\partial Y/\partial(\alpha, \beta) = 0$ yields the *normal* equations used to estimate the values of these parameters.

[9] *1.96* is the *z*-value, from the *Z*-distribution, which is a normal distribution with zero mean and unit variance, for a two-tailed test of statistical significance set at *0.05*.

						hospital admissions for respiratory disease in the elderly
1995	Ostro et al.	Los Angeles, USA	1992, 13 weeks.	PM_{10}	Logistic Regression	Study the association between air pollution and asthma exacerbations among asthmatic children
1995	Lippmann et al.	London, UK	1965-1972	SO_2	Simple Regression	Study the effect of air pollution on daily mortality with separation of the effects of temperature
1995	Moolgavkar et al.	Philadelphia, USA	1973-1988	SO_2	Poisson Regression	Examine the association between air pollution and daily mortality
1995	Xu et al.	Beijing, China	1988	SO_2	Linear & Logistic Regression	Examine the acute effects of air pollution on preterm delivery < 37 weeks

More information about some of these studies is summarized in Table 4.2. It provides an overview of the statistical results obtained by modeling the relations between exposures to the air pollutants described in Table 4.1 and adverse health outcomes.

Table 4.2. Example of Findings Using Exposure-Response Models

Concent. Range	Other Air Pollutants	Findings	Additional Independent Variables	Adverse Effect(s) & Reference
TSP: 40-100 $\mu g/m^3$	TSP was also associated with development of AOD & chronic bronchitis	Exposure to concentration of PM_{10} that exceeded 100 $\mu g/m^3$, a statistically significant but small positive associations with: (a) AOD (RR=1.17, 95% CI=1.02-1.33) (b) chronic productive cough (RR=1.21, 95% CI=1.02-1.44) & (c) asthma (RR=1.30, 95% CI=1.097-1.73)	Seasonal fluctuations were controlled by seasonal & non-seasonal regression equations	Increased development of AOD, chronic bronchitis & asthma. Abbey et al., (1993)
PM_{10}: New Haven: 19-67 $\mu g/m^3$ mean 41 $\mu g/m^3$. SO_2: New Haven: 23-159 $\mu g/m^3$ mean 78 $\mu g/m^3$	O_3 was a marginal predictor of hospital admission	50 $\mu g/m^3$ increase in PM_{10}, with a significant association of respiratory hospital admissions in New Haven (RR=1.06, 95% CI=1.13 - 1.00) & Tacoma (RR=1.1, 95% CI=1.17 - 1.03). For a 50 $\mu g/m^3$ increase in SO_2, significant respiratory hospital admissions, New Haven, (RR=1.03, 95% CI=1.05 - 1.02) & Tacoma (RR=1.06, 95% CI=1.12 -	Regression variables controlled for humidity, temperature, dew point temperature & seasonal patterns	Increased respiratory hospital admission or emergency room visits. Schwartz (1995)

		1.01)		
Summer: mean 17μg/m³ max. 51μg/m³ Winter: mean 211μg/m³ max. 478 μg/m³	TSP was also found to be significant and independent predictor of medical diagnoses	SO₂ was found to be significant, independent predictor of internal medicine visits in both summer & winter when adjustment was made for surgical visits	Weather, day of the week, & season were adjusted for	Increased hospital visits for respiratory disease. Xu et al., (1995)

An important aspect of statistical modeling involves the assessment of the results from estimation. Specifically, statistical assumptions must be studied and explained after estimation. That is, obtaining parameter estimates is just the beginning of a complete statistical analysis. For example, the assumptions of the ordinary least squares are that:

1. The errors are normally distributed,
2. The errors have constant variance,
3. The errors must not be correlated,
4. The independent variables must be orthogonal (there is no *multicollinearity* between them)[10],
5. The error and the independent variable or variables must not be correlated.

There are several diagnostic tests that can be used to assess whether a model's results meet these assumptions. One simple way to make the assessment is to use the residuals from the regression and plot them against the value of an independent variable.

Example 4.4. The regression model is: *Median Housing Values = mvalue = 30.98-0.123*Age.* The relationship between *mval* and *age* does not show constant variability, the constant variance assumption. As the *age* of the house increases, the variability of the data for this variable also increases. The plot of the residuals (errors) against *age* confirms it:

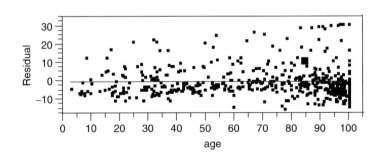

---

[10] Multicollinearity means that two or more independent variables are correlated: it is a *sampling* issue. The effect is that it may become difficult to disentangle their individual effect on the dependent variable. This can be important to the risk assessor and risk manager attempting to find causal factors and isolate their individual contribution.

> This plot suggests that the errors appear not to be normally distributed (possibly violating assumption *1*) and that the variance is increasing as age increases. Statistical tests can analytically assess these and other violations of the assumptions of the OLS method.

As this example suggests, statistical results cannot be accepted without detailed diagnostic tests that can lead to correcting the initial model (Ricci and Wyzga, 1983). For example, a linear variable (say, X) is changed to a quadratic (X^2), if the plot of residuals for this variable indicates that such change is appropriate. Other diagnostic tests can suggest that some independent variables can be multiplied to detect their interactions. For example, we can multiply W and X: the form of this interaction is stated as $\beta_i(W*X)$.

Because art and science combine in building exposure-response models, the associated difficulties have lead to the development of protocols to be followed before a study can be used in science policy. In the next section we discuss one of the better-known protocols (the Air Pollution and Health European Approach, APHEA) for assessing the relations between exposure to air pollution and adverse human health responses. This protocol serves as a summary of these discussions and provides a nexus between statistical models, causation and the relationship between environmental health and exposure to air pollution.

4.1.3. The APHEA (Air Pollution and Health, a European Approach)

The European Union has supported research in the health effects of air pollution. On that basis, a number of European universities and research centers have developed the APHEA Protocol, which was used to study the short-term adverse health effects of urban air pollution and health for 15 European cities. The specific objectives of the Protocol that are relevant to this chapter are to (Katsouyanni et al., 1996):

"provide quantitative estimates of the health effects, ... further develop and standardize the methodology for detection of short term health effects in the analysis of epidemiological time series data ... select and develop meta-analytic approaches ..."

Katsouyanni et al., (1996) used daily counts of all deaths, cause-specific deaths and hospital admissions. The air pollution variables were SO_2, *TSP*, PM_{10}, *British Smoke*, NO_2 and O_3, consisting of *24*-hour and other values. For instance, O_3 data consisted of hourly maximum and maximum eight-hour daily concentrations. Temperature, in degrees Celsius, and relative humidity were used to *control for the potential confounding effects of weather*. (Katsouyanni et al., 1996). The Protocol established a process for the *admissibility* of air pollution data regarding urban or other areas. It also develops a process for determining when data can be sufficiently complete. An example of *completeness* is the protocol's requirement that *at least 75% of the one-hour values on that particular day had to be available* for the *24*-hour and one-hour data for NO_2 and SO_2. On the other hand, *completeness* for O_3 meant satisfying the criterion that data would be those from *9* am to *5* pm due to the characteristic diurnal peaks of this chemical. The problem of *missing* data is addressed by including monitoring data from nearby stations that had a more complete record for the same day. The statistical exposure-response models include Poisson regression and time series analysis.

The APHEA protocol used a criterion of parsimony to reduce the number of variables with model diagnosis, which was based on the analysis of the errors and several formal statistical tests. The protocol requires that confounding due to seasonal diseases and other cyclic factors should be accounted in developing an unbiased estimate of the effect of the air pollutants on human health. Specifically, if a cycle of adverse health responses coincides with a cycle of high exposures and high temperature and relative humidity, the increased mortality and morbidity may be spuriously ascribed solely to air pollution, when they are due to other factors. Conversely, the association may spuriously be negative. The protocol also requires controlling for cyclic patterns. Fourier series and dichotomous variables ($1 = effect$, $0 = no$ $effect$) can account for long-term cycles. Short-term cycles are studied using polynomials in which time is the independent variable. These analyses also included lagged variables to model past effects effect of temperature and relative humidity on response (previous day exposure). Sporadic diseases, such as influenza, day-of-the-week (e.g., higher reported disease rates on Monday) and holiday effects are studied by adding the appropriate independent variables.

4.2. THE MAXIMUM LIKELIHOOD ESTIMATOR (MLE)

As an extension of previous discussions, estimation may consist of adopting the maximum likelihood method, which was introduced in Chapter 2, rather than the OLS. As always, we seek formulae to estimate the parameter or parameters of an exposure-response model, given the data on the variables of the model. Consider a random variable X with distribution function $f(X; \theta)$, where X is the random variable, and θ is the single population parameter for that distribution. Each observation of this random variable has its own distribution. The key assumption is that each observation, x_i, is independent and identically distributed. The likelihood function, L, is formed by the product of each of these individual distributions:

$$L = f(x_1; \theta)*f(x_2; \theta)*....*f(x_n; \theta).$$

The maximum likelihood method maximizes this product, relative to the unknown parameter θ. The calculations can be simplified by taking the logarithm of L:

$$ln(L) = ln[f(x_1; \theta)] + ln[f(x_2; \theta)] +, ..., + ln[f(x_n; \theta)].$$

The necessary condition for the maximization is:

$$d[ln(L)]/d\theta = 0.$$

This yields the formulae for estimating the value of θ. The sufficient condition requires investigating the second derivatives of $ln(L)$. If these equations can be solved, we obtain formulae to estimate the coefficients of the model.

Example 4.5. Let the random variable X be normally distributed with density function:

$f(X) = (1/[\sigma\sqrt{(2\pi)}]\{exp-1/2[(x_i-\mu)/\sigma]^2\})$.

What are the ML estimators of σ and μ? Taking the natural logarithms obtains:

$Ln[f(X)] = \{1/[\sigma\sqrt{(2\pi)}]\}\{-1/2[(x_i-\mu)/\sigma]^2\}$.

The density function has two parameters to be estimated, namely μ and σ. Because the MLE is an optimization method, and we are using a continuous function, we use partial derivatives. Setting these two partial derivatives equal to zero does the maximization:[11]

$\partial[lnf(X)]/\partial\mu = 0$
and

$\partial[lnf(X)]/\partial\sigma = 0$.

We thus obtain $[x_i-\mu/\sigma^2]$ and $[-1/\sigma+(x_i-\mu)/\sigma^3]$. Equating these two solutions to zero and solving for the two parameters (the unknowns) produces the formulae for estimation. For example, we obtain the following familiar result for the estimator (the formula to estimate) of the population mean from a sample: $\mu = (1/n)\Sigma_i x_i$.

Example 4.6. Consider a binomially distributed random variable that takes two values, 0 or 1, to indicate either success or failure.[12] Let the size of a binomial experiment consist of n trials in which there are r successes (r = *the number of successes*). It follows that there are n-r failures. Suppose that in an experiment $r = 3$ and $(n-r) = 10$ and assume that the probability of success, pr, is known from past experience to be 0.15. The binomial probability is $C(n, r)pr^r(1-pr)^{n-r}$, in which $C(n, r) = n!/[r!(n-r)!]$. The probability of obtaining this specific result is $0.15^3(1-0.15)^7 = 0.00108$, omitting the term $C(n, r)$ for simplicity.[13] Let us change pr to equal 0.50. The new probability is $0.50^3(1-0.50)^7 = 0.000977$. The likelihood function can be calculated for selecting several values of pr to determine where the function reaches its maximum: the maximum (likelihood estimate) of pr is approximately $0.60^3(1-0.60)^7 = 0.00035$.

4.2.1. Likelihood Ratio

The *likelihood ratio* is the ratio of likelihoods to the maximum value taken by the likelihood function. Therefore, the likelihood ratio will be less than or equal to 1 (Clayton and Hills, 1993) The log-likelihood *(LL)* changes a ratio to a subtraction, $a/b = ln(a)-ln(b)$, to make the calculations easier. Let the proportion of success be π. Let F be the number of failures, n the sample size and the number of successes $(n-F)$. Note that the most likely value of π is F/n. The log-likelihood is:

---

[11] The second partial derivatives confirm that we are actually maximizing the expression.
[12] Outcomes characterized by either success or failure (e.g., coded as either 1 or 0) that are independent from one another are called *Bernoulli trials*.
[13] This omission is a simplification. The point of this example is to exemplify the search for the maximum likelihood, not the calculations of the probabilities.

$LL = [(F)ln(\pi)] + [(n–F)ln(1–\pi)].$

The log-likelihoods are negative; each model has likelihood. The difference between two log-likelihoods for two exposure-response models is approximately *chi*-square distributed (*-2* is added to achieve this approximation), that is:

$X^2 = [-2ln(LL_{model\ 0})] - [-2ln(LL_{model\ 1})]$

in which model *1* is the more general model and model *0* is developed from a subset of the independent variable included in model *1*. The number of degrees of freedom for the likelihood ratio is given by the *difference* between the number of parameters in the two models. This statistic, used to assess the fit of two models to the data (Selvin, 2001), is a way to meet Occam's principle of parsimony.

Example 4.7. Let $\pi = 0.50$ and calculate the log-likelihood as: $3ln0.5 + 7ln0.5 = -6.93$.

For another value of π, say 0.3, we calculate the log-likelihood to be: $3ln0.3 + 7ln0.3 = -12.04$.

The log-likelihood ratio is *1.00* at the value of the maximum likelihood and values less than *1* for all other values of the log-likelihood function. Thus, for $\pi = 0.50$ and $\pi = 0.3$, the likelihood ratio is $[-12.04 - (-6.93)] = -5.11$.

4.3. BASIC STATISTICAL FORMULAE (Also See Chapter 11)

Most of the work discussed in Chapters 4-10 involves statistical calculation based on the concepts discussed next. The basic statistical quantities are the mean and the variance of a random variable. The formulae for obtaining the *population* mean, μ, and its variance, *var(X)*, that is, σ^2, of the random variable X (using discrete values, population of size N, and each observation is symbolized by x_i) are:

$$\mu = 1/N \sum_{i=1}^{N} x_i ,$$

and:

$$\sigma^2 = Var(X) = 1/N \sum_{i=1}^{N} \left(x_i - \mu\right)^2 .$$

Using the same random variable X, if the population mean and variance are not known, they must be estimated from sample data. The estimators of the population mean and variance, from a sample of size n, randomly taken from a population of size N, are:

$$\bar{x} = (1/n) \sum_{i=1}^{n} x_i .$$

The estimated variance is:

$$s^2 = var_{estimated}(X) = [1/(n-1)] \sum_{i=1}^{n} (x_i - \bar{x})^2 .$$

The standard deviation is the square root of the variance. The expected value of the random variable X is:

$$E(X) = \sum_{i=1}^{N} pr_i x_i ,$$

in which pr_i is the probability associated with the x_i value of the random variable.

Example 4.8. The expectation is calculated by accounting for the probabilities of each value of the random variable. Let the random variable be X, its values are $x_1 = 2$, $x_2 = 4$, $x_3 = 1$ and $pr(x_1) = 0.33$, $pr(x_2) = 0.50$ and $pr(x_3) = 0.27$. The expected value, $E(X)$,[14] is:

$$E(X) = \sum_{1}^{3} pr_i(x_i) = (0.33)(2) + (0.50)(4) + (0.27)(1) = 0.66 + 2.0 + 0.27 = 2.93 .$$

A statistical quantity used in hypothesis testing and forming confidence intervals is the *standardized normal deviate*. Consider a random variable X that has a mean, variance and so on. Clearly, the area under the distribution of raw values (2, 1, 1.5 or any other value) will not equal 1.00, as is required to use probabilities. A simple transformation transforms (*normalizes*) the values of a random variable. This transformation generates a new random variable, Z, ($z \in Z$), such that the area under its density function, $f(Z)$, discussed and exemplified in Chapter 11, equals 1.00. The transformation that accomplishes this contraction of the X-axis to the Z-axis is:

$$z = (x_i - \mu)/\sigma ,$$

in which μ is the population mean and σ is the population standard deviation. A similar formula is used when the population parameters are unknown but the sample mean and standard deviation are known. Inference from the sample to the population can use confidence intervals. The upper and lower confidence limits represent length of the confidence interval (*CI*) about a population parameter such as the mean, standard deviation or other population parameter, symbolized by θ for generality, for a specific level of confidence $1-\alpha$ are:

$$(1-\alpha)CI = \theta_{estimated} \pm k[se(\theta_{s\,estimated})] .$$

For example, assuming a normal distribution for the random variable X, $k = 1.96$ for $1-\alpha = 0.95$, the confidence interval for the population mean,

$$0.95CI = \bar{x} \pm 1.96[se(\bar{x})] .$$

---

[14] It can also be symbolized by $<X>$.

The standard error of $\bar{x}$, $se(\bar{x})$, for a *known* population standard deviation, σ, is calculated as $\sigma/\sqrt{n}$. Other standard errors can be calculated for the variance median and so on. When more than one independent random variable is modeled, the covariance (abbreviated as *cov, Cov*) determines how these variables influence each other. Let the two random variables be X and Y. Their covariance, *cov(X, Y)*, is:

$$cov(X,Y) = \sum_{i,j}(x_i - \mu_x)(y_j - \mu_y)pr(x_i,y_j),$$

in which x_i and y_j are the observations of the random variables X and Y; μ_x and μ_y are the population means of these two random variables, and *pr(x_i, y_j)* is the joint probability of that pair of observations.

4.4. MODELING EMPIRICAL CAUSAL RELATIONS BETWEEN EXPOSURE AND RESPONSE

The statistical models discussed in this chapter are empirical. Empirical causation is inductive. Therefore, it has some important limitations. For example, the data used in empirical risk assessments is often non-experimental (recollect that we can use data from a variety of sources, most of which may be not random sampled). This data is observational: the risk assessor takes that data as given and cannot generally assess the sampling done by others. Nonetheless, non-experimental data can guide and often justify regulatory action to decrease exposure (Ricci and Wyzga, 1983; Health Effects Institute, 2000; US EPA 1997). As discussed, regression models such as the multiple regression, logistic and Poisson regressions are developed through literature and other informed judgments. From these, exposure is empirically and causally associated mathematically to a disease outcome, such as chronic obstructive pulmonary disease, asthma, death from bronchial cancer and so on through a statistical model. The choice of independent variables and the form of the model may follow the principle of parsimony put forth by William of Occam, known as Occam's razor. This principle states that *entia multiplicanda non sunt praeter necessitatem*, meaning that *entities should not be multiplied without need*.

Table 4.3. Criteria for Judging the Performance of an Estimator in Large Sample Theory

Property of the Estimator	Discussion of Assumptions
Bias	The estimator is centered on the true population parameter value. The ordinary least squares in unbiased estimator of central tendency for regression model.
Efficiency	An efficient estimator is centered on the true population parameter value and has minimum variability about that parameter (its variance is the smallest).
Consistency	As the sample size tends to infinity, the estimator tends to concentrate on the true population parameter value. That is, the expected value of the estimate equals the population parameter value as the sample size tends to infinity.

The direction (the sign of the estimated parameters) and strength of the statistical results (the *p*-value, under the null hypothesis) are determined by the estimates of the parameters of the model. These results may confirm theoretical (such as prior)

expectation. Estimation can be based on large sample approximations, using results from asymptotic theory of estimation. In the alternative, estimation can use *exact* methods that do not make such assumptions. In this textbook, we use both. In classical statistical estimation theory based on asymptotic properties of estimators, the concepts of bias, efficiency and consistency can be important (Kennedy, 1998). In practice, an estimator may not have all of these characteristics and thus the risk assessor must exercise scientific judgment in her final choice of estimator. The estimators used in this textbook are generally unbiased, efficient and consistent under specific assumptions about their large sample approximations: the formulae are valid if these approximations are true. For example, the ordinary least squares have these properties when their five fundamental assumptions are met. If not, a modified OLS estimator should be used. The MLE can also be unbiased, consistent and efficient. There often is a trade-off between bias, efficiency and consistency.

A practical problem with building statistical models is *over-fitting* (joining the dots). This is the situation where a model looses its meaning and becomes an exercise in fitting the data as closely as possible. In other words, a model can include so many terms to fit all of the data perfectly. This would not satisfy either Occam's principle of parsimony or correct statistical model building because over-fitting tends to increase the variance of the model. Moreover, such a jagged line would be difficult to understand and could not realistically predict. Additionally, in the case of multivariate models, the standard application of statistical tests can show that a relationship exists, even though such a relationship is spurious or biased toward the null hypothesis (Freedman, 1983). In other words, the positive relation is an artifact. Some probabilistic models in risk assessment are of low dimensionality: one variable is sufficient to describe the process leading to risk.

The multistage dose-response model (Chapter 5) exemplifies a model of low dimensionality because only *dose* relates to response. The type of statistical models discussed in this chapter have high dimensionality because they tend to include the risk factors, including exposure, associated with the risky outcome, say excess bladder cancer mortality from ingesting low concentrations of arsenic. Those models are *multivariate* because they attempt to account for economic, exposure, age, diet and other factors on the mortality from bladder cancer. Building such statistical models requires rigorous assessments of the empirical literature and the theoretical foundations linking the independent variables to the dependent variable. Suppose that the exposure-response model is $Y = f(X) = a+bX$. A measure of the way the model fits the data fit that accounts for overfitting is the adjusted coefficient of correlation.

As discussed in Chapter 10, the bivariate *coefficient of correlation* (or determination, if it is squared; it is bivariate because there are two random variables, X and Y) is a common statistics that measures how well the model fits the data. The correlation coefficient for two random variables X and Y is: $\rho = cov(X, Y)/var(X)var(Y) = cov(X,Y)/\sigma_X\sigma_Y$, is *estimated* from a sample, using:[15]

---

[15] Thus, we can form confidence intervals about this estimated value.

$$R = (1/n - 1) \sum_{i=1}^{n} \{[(x_i - \bar{x})/s_X][(y_i - \bar{y})/s_Y]\}.$$

In this formula, n is the sample size, $\bar{x}$ and $\bar{y}$ are the sample means, s_X and s_y are the sample standard deviations. When there is more than one independent variable, the coefficient of correlation is called the coefficient of multiple correlation. The adjusted coefficient of correlation, R, is weighted by the loss of degrees of freedom associated with adding independent variables to the model. The meaning of degrees-of-freedom is discussed in Chapter 8. Practically, the number of degrees of freedom is calculated from the number of independent variables, generally indexed by k. The number of degrees of freedom is therefore the number of independent variables minus *1*. If there are eight independent variables, then $k-1 = 8-1 = 7$. The adjusted R^2 is calculated as $[1-(1-R^2)(n-1/n-k)]$, in which n is the sample size and k is the number of independent variables used in the correlation.

Example 4.9. Suppose that a multiple regression model has *8* independent variables. Suppose that the sample size is *40* ($n = 40$); the sample is not given in the example. The unadjusted R^2 for the *8* variables in the model is *0.50*. The adjusted coefficient of correlation is *Adjusted R^2 = [1-(0.5)(39/32)] = 0.39*.

The discussions of Type *I* and Type *II* errors also apply to modeling exposure-response. Type *I* error is measured by the probability ($pr = \alpha$) of rejecting H_0 when it is true. Type *II* error is measured by the probability ($pr = \beta$) of not rejecting (accepting) H_0 when H_1 is true. Figure 4.3 depicts the relationship between Type *I* and Type *II* errors for the difference between two proportions under the null and alternative hypotheses.

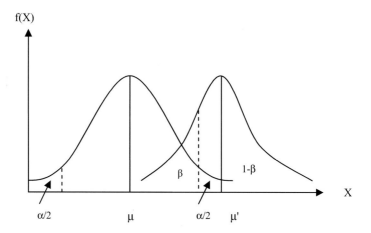

Figure 4.3. Type I and Type II Errors

The total area under each of the curves equals *1.00*; therefore the probabilities α and β are subtracted from *1.00*, as shown in Figure 4.3. The standard error of the estimate (directly related to the variance) identifies values on the x-axis. The next example shows the logic of applying the concepts of false positive and false negative to the logistic regression. Because sampling can be costly, it is critical to assess the power of the study and determine a prudent sample size. The next example uses the odds ratio, *OR* (also discussed in Chapter 9).

Example 4.10. Suppose that the null hypothesis is that the mean of the exposure variable is *50.00*, with a standard deviation of *5.00* and that the rate of adverse outcome at this mean is *0.20*. The null hypothesis states that there is no relationship between exposure and the event rate. Therefore, the event rate (*0.20*) is the same at all values of exposure.[16] Let us set statistical significance at *0.05*; the test is 2-tailed. The power of the tests is computed for the alternative hypothesis. Suppose that the alternative hypothesis is that average exposure values = *60.0* and the expected event rate is *0.50*. This corresponds to an odds ratio = *4.00* or a log-odds ratio = *0.14*. This effect is selected as the smallest that would be important to detect, in the sense that any smaller effect would not be of importance to the risk manager. Using Power and Precision, (2001) the regression must be based on a sample size of at least *120* individuals, with a power of the test = *0.80*: *80%* of studies would be expected to yield a statistically significant effect and thus lead to rejecting the null hypothesis (*OR* = *1.00*).

Power as a Function of Sample Size

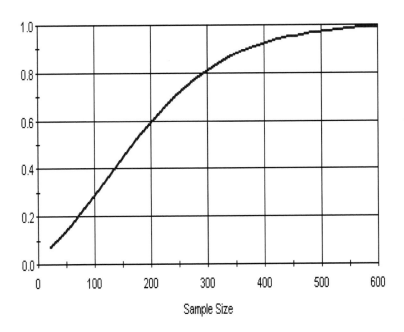

Figure 4.4. Power of the Test (probability *1-β*) as a Function of Sample Size (*α* = *0.05*, Two-Tails)

---

[16] Equivalently, the odds ratio is *1.00* and the log of the odds ratio is *0.0* because *ln(1.0)* = *0*.

This is the analysis of the *power of the test* (of significance). For $\alpha = 0.05$ and power of *0.80*, the sample size is *n = 294*.

4.5. LOGISTIC REGRESSION: The Dependent Variable is Binary

In many risk assessments, the dependent variable measures only whether the response (illness, for instance) is present or not. The logistic model prevents probabilities from exceeding *1.00* or being negative. The logistic model is depicted as a continuous sigmoid-shaped line in Figure 4.5.

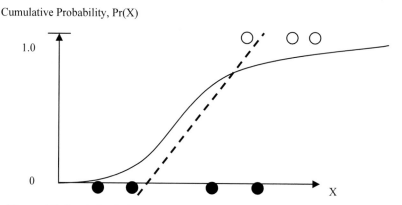

Figure 4.5. Simple Logistic Regression Model (dots are oversized points)

In this figure, the hollow circles identify the individual that responds while the black circles identify those individuals who do not respond: three individuals respond and four do not. The straight dashed line is a regression fit that does not account for the fact that the data for the dependent variable are binomial (yes or no), rather than continuous.

This type of regression model is one of best known for studying the effect of exposure to air, water or other forms of pollution; can be multivariate, including lagged variables as well as independent variables that only measure the presence or absence of a risk factor. The probability of success, the *response*, is *pr(y = 1)* and the probability of *no response* is *pr(y = 0)*. Because of the problem depicted in Figure 4.5, the dependent variable of the regression model must remain between 0.0 and 1.00. The logistic *transformation*, from which the logistic regression takes its name, accomplishes this task. The transformation is:

ln[pr/(1-pr)].

In this expression, *pr* is the probability of success and the ratio *[pr/(1-pr)]* is the odds of success. The reason for the transformation is that the linear probability model:

$pr(X = x) = a + bY$

can exceed the interval $(0, 1)$, as depicted by the dashed line in Figure 4.5. The *logit* is another transformation that is often useful, as will be exemplified in this section. It is defined as $ln[pr/(1-pr)]$.

The logistic model's parameters are estimated using the MLE. The *parametric* logistic model is consistent with the binomial distribution of the response variable.[17] The response random variable is Y; its realizations are the y_i observations, each of which occurs with probability pr_i. If the y_i are identically (binomially distributed) and independent, the likelihood function is:

$$L = \Pi_i\{n!/[y!(n-y)!]\}[(pr_i^y)(1-pr_i)(n_i-y_i)].$$

The likelihood function yields the formulae to estimate the model's parameters from a sample, as explained for the classical regression. The logistic regression can include nonlinear variables and interactions between the independent variables.

Example 4.11. A simple illustration of a logistic risk model is a 2*2 contingency table, where response equals *1* means that there is a response; exposure equals 1 means that there is response.

	Exposure = 1	**No Exposure = 0**	**Totals**
Response = 1	a	b	a+b
No response = 0	c	d	c+d
Totals	a+c	b+d	a+b+c+d

Exposure is not a series of concentrations. It is a binary variable: either there is exposure or there is no exposure. Response is also binary: either there is response or there is no response. The values a, b, c and d are numbers (or counts) representing the number of individuals in the study. Examples of such numbers would be $a = 3$, $b = 0$, $c = 10$ and $d = 5$. This model therefore does not include other exposures age, sex, smoking and other variables. It is a simple regression model. We use α, β and so on to identify the population parameters. We obtain the following relation (Kahn and Sempos, 1989):

$$pr(response) = \frac{exp[-(\alpha + \beta Exposure)]}{1 - exp[-(\alpha + \beta Exposure)]}.$$

The logit transformation of these two expressions is:

$$logit[pr(response)] = ln[pr(response)]/[1-pr(response)] = \alpha + \beta Exposure.$$

For a 2*2 cross-sectional or a case control study, the Odds Ratio, OR, is estimated as

---

[17] This distribution is discussed in Chapter 11. It is used to calculate the probability of exactly s successes, $pr(s)$, in n total outcomes, $n = $ (# successes + # failures); $pr(s) = \{n!/[s!/(n-s)!]\}(pr^s)(1-pr)^{n-s}$, with $s = 0, 1, \ldots, n$.

$OR = \{[pr(E=1|R=1)/pr(E=0|R=1)]/[pr(E=1|R=0)/pr(E=0/R=0)]\}$,

where E is exposure and R is response and the 0s and 1s represent absence or presence of an effect.

We can extend these results to more than one independent variable, which we now label X_1, X_2 and so on to X_k to be consistent with epidemiological modeling found in the literature. From the formula for the logit, $pr(response) = ln\{pr(response)/[1-pr(response)]\}$, we obtain the following expression:[18]

$[1-pr(response)] = 1-(1/\{1+exp[-(\alpha+\sum_k\beta_kX_k)]\})$.

Taking the logarithms of both sides of this expression, we obtain:

$ln[pr_i/(1-pr_i)] = \beta_0+\beta_1X_1+...+\beta_kX_k$.

This model is equivalent to:

$$pr(response) = exp[-\alpha + \sum_{k=1}^{K} \beta_k X_k]/\{1 + [exp-(\alpha + \sum_{k=1}^{K} \beta_k X_k)]\}.$$

The MLE yields numerical values for the coefficients α and β.[19]

Example 4.12. We use a dependent variable in which the values are binary (*yes* or *no*, coded as 0 or 1) and a sample of size 200. There are 28 no-responses (coded as 0) and 172 responses (coded as 1). The model is a linear function with the following variables and background response: *Response = f(education, age, IQ)*. The results of estimation using a linear model are as follows:

Parameter	Estimate	se	p-values
β_0	2.950	1.318	0.025
$\beta_{education}$	-0.098	0.071	0.170
β_{age}	0.046	0.040	0.253
β_{IQ}	-0.015	0.012	0.214

The term *se* is the standard error of each estimate. β_0 is the constant specifying background and is the only statistically significant estimate, if we take the level of significance to be $\alpha = 0.05$, because the *p*-value for $\beta_0 = 0.025$.

The odds of the response in those exposed (group 1) can be related to the odds of response to those unexposed (group 0) through the logistic regression, as follows:

---

[18] The logit function is symmetric about *logit* $= 0$, because $pr = 0.5$. That is, the logit transformation equals 0: $ln(0.5/0.5) = ln(1.0) = 0$. The transformation tends to $-\infty$ as pr tends to 0, and tends to ∞ as pr tends to 1.

[19] In case-control and cross-sectional epidemiological studies, the coefficient α cannot be reliably estimated. However, this coefficient is appropriate in a cohort study. The odds ratio is used in cross-sectional and case-control studies and the relative risk in cohort studies.

$[pr_1/(1-pr_1)]/[pr_0/(1-pr_0)] = pr(response) = \{exp[(\alpha+\sum\beta_k X_k)]\}/\{1+exp[(\alpha+\sum\beta_k X_k)]\}.$

The regression parameter α could be interpreted as the background odds of response at the individual level. However, such an interpretation may not make biological sense if there are individuals in the population at risk that have only some of the characteristics associated with the independent variables, summarized in X_i. Another interpretation of α is that it measures the aggregate background odds without any independent variables at all. In other words, α is the OR associated with the 2*2 table. A logit model with interactions between variables can be written as:

$logit [pr(response)] = \alpha+\beta_1 X_1+\beta_2 X_2+\beta_3(X_1 X_2).$

The coefficient β_3 accounts for the (multiplicative) interaction between X_1 and X_2.

4.5.1. Discussion

Consider the following logistic regression:

$logit [pr(response)] = \alpha+\beta_1 X_1+\beta_2 X_2+\beta_3 X_3+\beta_4 X_4,$

in which the estimates of the coefficients are known. That is, the estimation has taken place and the β_i have numerical values. A risk assessor can set $X_2 = 0$ and $X_3 = 10$. In other words, individual values of these variables are input into the model. The difference between two logits is:

$[logit\ pr(response|\alpha+\beta_1 X_1+\beta_2 X_2+\beta_3 X_3+\beta_4 X_4)\text{-}logit\ pr(response|X_2 = 0\ and\ X_3 = 10)].$

This difference is the change in the log-odds. When a variable is assigned a zero value in the logistic regression, that variable should be interpreted as *not* influencing the probability of response. Specifically, the change in the log-odds is measured by β_i. The ratio of the odds is:

$\{pr(response)/[1-pr(response)]\}/\{pr(no\ response)/[1-pr(no-response)]\}.$

The *risk odds ratio*, ROR, (Kleinbaum, 1982) is:

$ROR = [exp(\alpha+\sum\beta_1 X_1|X_1 = 1)]/[exp(\alpha+\sum\beta_1 X_1|X_1=0)].$

Note that $[exp(\beta_1\beta_2)] = [exp(\beta_1)exp(\beta_2)]$. Thus, the ROR is multiplicative. The goodness of fit of the logistic regression model to the data can be based on a statistic called the *deviance*, D, which is calculated as:

$D = -2[ln(L_c)\text{-}ln(L_f)].$

This difference has an approximate χ^2 distribution with $(n-k)$ degrees of freedom, k is the number of parameters, provided that n is sufficiently large. This distribution is explained in Chapter 11. The subscript c indicates the model fitted by the researcher;

the subscript f indicates the fully saturated (with all the independent variables included) model. Large D values indicate a poor fit. Specifically, the validity of the large sample approximation is a function of the total number of binary observations in the sample.

Example 4.13. A logistic regression model (Colford et al., 1999) was developed to model illnesses resulting from exposure to waste water-contaminated outflows. The effect of exposure is lagged up to fourteen days because the relation between exposure and response must be biologically realistic. For example, previous-days exposure results in gastrointestinal diseases but same-day exposure does not. The measurement of the outcome variable (measured by *sick leave use* as a proxy for *disease incidence*) was lagged at specific intervals of time, from *1* through to *7* days, and at *14* days. These lag exposure periods were chosen from the epidemiological and medical literature that disclosed the expected length of the incubation periods of four different groups of pathogens: bacteria, protozoa, helminths and viruses. The logistic regression model, for each lag period (t-k):

ln[odds (sick leave use)] $= b_0 + b_1(group\ exposed) + b_2(rainfall)_{t-k} + b_3[(group\ exposed\ or\ unexposed)(rainfall)_{t-k}]$.

In this model, b_0 is the intercept, b_i identifies the parameters of the model, group is a dichotomous variable (*1 = works inside the area of study*, *0 = works outside*) and rainfall is a continuous variable measuring either daily total rainfall or rainfall intensity. The length of the lag (t-k) means that the exposure affects response after k periods (namely, k days, $k = 1, 2, 3, 4, 5, 6,$ and *14*) have elapsed. The most relevant parameter is the interaction term b_3. It measures the joint impact of rainfall and work location, inside the area of study against outside. Under the statistical null hypothesis, the population interaction term (b_3) equals *0*. Hence, rainfall yields no additional risk of illness (as reflected in sick leave use) to workers working inside the area of study, compared to those working outside the area of study. The estimation of the population parameters is obtained by using the MLE. Note that the logistic model is parametric because the response variable follows the binomial distribution. The etiological basis of the logistic model suggests that the exposure-response model accounts for increased absenteeism among the exposed, relative to those who are unexposed. Attendance and absence records were obtained for *43* County Drainage and Maintenance Department workers and for *131* workers of the City of Sacramento Drainage Collection Department, from July 1, 1992 to June 30, 1993. For these two groups of workers, data were gathered regarding their sex, marital status, ethnicity and age. The *unadjusted* estimation (excluding the socio-economic information from the model) of the interaction coefficient, over the lagged exposure periods *1, 2, 3, 4, 5, 6, 7* and *14* days were not statistically significant (*95%* confidence interval). However, the MLE coefficients for lags *3, 4, 5* and *14* were greater than zero. The simpler logistic model was modified to include sex, ethnicity, marital status, as well the interaction of the variables work location and rain intensity as the independent variables. The parameter of the interaction term was positive at lags *3, 4, 5* and *14* days, but the results lacked statistical significance. These lags (*3, 4, 5* and *14*) identify situations where exposure is causally *associated* with response (measured by absenteeism). There was no evidence of increased sick leave use by either of these groups of workers inside the area of study at any other value of the lagged variables.

As discussed, the estimation of the parameters of a statistical model can create results that may not be accurate. A typical problem occurs when the data set is unbalanced. The next example describes the difference between assuming large sample characteristics for an estimator, having an exact (under the null hypothesis) method for estimation and assuming a large sample.

Example 4.14. I have used the computer software LogExact (Cytel, 2001). The data is:

Number Responding	Number at risk	Response (%)	Exposure (0, 1)	Sex (0, 1)
4	100	4	0	0
3	100	3	0	1
1	8	12.5	1	0
1	1	100	1	1

The results from using the MLE applied to the logistic regression and the exact estimation method demonstrate the point. The notable difference between using exact estimation methods or approximations based on large sample theory is the potential for obtaining different estimates. For example, note the borderline statistical significance of the coefficient of the *Exposure* variable, obtained using the MLE. Using exact methods, that result turns out to be strongly insignificant.

Estimation Method	$\beta_{estimated}$	Lower 95% CL	Upper 95% CL	p-value
Exposure (MLE)	1.60	0.094	3.29	0.064
Exposure (Exact)	1.57	-0.82	3.88	0.20
Sex (MLE)	-0.13	1.23	1.23	0.85
Sex (Exact)	0.13	1.45	1.45	1.00

The example suggests that the user of statistical results may want to investigate the results obtained from assuming large sample (asymptotic) estimation methods whenever sparse data sets are used or the estimation has to use small samples.

4.6. POISSON REGRESSION: The Dependent Variable Measures Counts of Rare Events

The Poisson regression is used when response, events such as death or diseases, is *rare*. The term *rare* refers to a suitably small period of time or small area. The counts of the rare responses distribute according to the Poisson probability distribution (called a probability mass distribution because it is not continuous). If the Poisson-distributed random variable is Y, its realizations y_i take values such as $0, 1, 2, \dots$, and so on with probability pr_i. Each y_i must be non-negative, discrete and be an integer. The Poisson probability mass is calculated from:

$$pr(Y = y) = exp(-\mu)\mu^y/y!,$$

with $\mu > 0$. The parameter μ is the expected value of the distribution. More information on the Poisson distribution can be found in Chapter 11.

Example 4.15. The Poisson probability mass function, for $y = 2$ and $\mu = 1$ yields the following result: $(1^2)[exp(-1)]/2! = (1)(0.367/2) \approx 0.18$. This is the probability of obtaining a count of 2, given a population mean that equals *1* count.

The Poisson regression relates the expected value of the response variable to one or more independent variables. The Poisson regression model can be written, taking the natural logarithm of the expected value of the random variable representing the responses, μ, as:

$ln(\mu) = ln[E(y)] = \sum_k \beta_k X_k.$

The summation is taken over the k independent variables, X_k. The reason for taking the natural logarithm of $E(Y)$ is that it must be greater than zero, a result of equating $exp[ln(\mu)]$ with $exp(\sum \beta X)$. This equality also states that the effects are multiplicative on response, as can be understood by noting that:

$[exp(a)][exp(b)] = [exp(a+b)].$

The Poisson model must account for the fact that the log of zero, $ln0$, is undefined. Thus, the zero values can be changed to a number greater than 0. This is done by inserting an arbitrary additive constant, c, to the observations take on the dependent variable:

$$E[ln(Y + c)] = \alpha + \sum_{i=1}^{k} \beta_i X_i ,$$

to account for data recorded as 0. The constant c equals either 0.5 or 1.0. In the Richmond air pollution study, discussed later in this chapter, Ricci and his colleagues (Colford et al., 1999) used $c = 1$.

Example 4.16. The dates of occurrence of *rare* outcomes, daily diagnoses of asthma in a very small area, are:

Date (dd/mm/yr)	Diagnoses (Counts)
1/1/95	0
1/2/95	3
3/1/95	0
4/1/95	7
....	....
30/6/95	0

These counts are outcomes of a Poisson-distributed dependent variable and are part of the set of data used in estimating the parameters of a Poisson regression.

Data measured for the independent variables, with the data for the dependent variable, are used with a computer software program to obtain estimates of the model's parameter.

Example 4.17. Let the Poisson regression model be: $ln[E(Y)] = \beta_0 + \beta_1 X_1 + ... + \beta_5 X_5$. After estimation from a sample not shown, the parameter estimates, obtained using the MLE in LogExact, are (omitting their confidence intervals for brevity):

$\hat{\beta}_0 = -4.35,\ \hat{\beta}_1 = 1.73,\ \hat{\beta}_2 = -0.22,\ \hat{\beta}_3 = 0.5,\ \hat{\beta}_4 = 0.9,$ and $\hat{\beta}_5 = -0.4.$

The estimation produces parameter estimates that are significant at $\alpha = 0.05$. The effect of a covariate on the response variable measured at its expected value, $E(Y = y_i)$, is given by the quantity $\mu\beta_k$, in which k is the subscript for the k-th independent variable. Assume that the expected value of the response variable is approximated by its sample mean and that such value is 0.5. Then, the incremental effect of X_2 on the dependent variable is calculated as: $\mu\beta_2 = (0.5)(-0.22) = -0.11$. The negative sign represents a *decremental* impact. Therefore, a unit change in X_2 will decrease Y by 0.11 units, all other effects held constant. The expected value of Y is calculated from specific input values for the independent variables taken from the sample, as:

$$exp[-4.35+1.73(2.5)-0.22(1.6)+0.5(0.3)+0.9(6.5)-0.4(1.1)] = 178.22.$$

The values in the parentheses are arbitrary input values for each independent variable. Multiplying 178.22 by one of the estimated coefficients yields the contribution of the coefficient to the overall effect. For instance, the quantity $[(-0.22)ln(178.22)] = -1.14$ measures the average effect of independent variable X_2 on the dependent variable $ln(Y)$. In addition, $exp(-0.22) = 0.80$ measures the marginal (measured by the first derivative of $exp\beta_2$ which is $exp\beta_2$) contribution of X_2 on Y.

Example 4.18. Consider the question: Does exposure to low (ambient) concentrations of hydrogen sulfide result in increased asthmatic and other responses in those exposed? With some colleagues, I have studied the residents of several Zip Codes, nearby or abutting a very large petrochemical refinery owned by Chevron-Texaco, in the Richmond area of the San Francisco Bay, in California. The residents in Zip Code 94801 live closer to the Refinery than those in the 94804 Zip Code. The working hypothesis is that the adverse effects of air pollution would be lower in the 94804 Zip Code than in the 94801 Zip Code. Air quality data was developed from four monitoring stations on and near the area of study (the area covered by the two Zip Codes). Three stations are operated by Chevron and one by the Bay Area Air Quality Management District, (BAAQMD), a public agency. The BAAQMD monitoring station located towards the far end of the study area in North Richmond exceeded *10* ppb of sulfur dioxide approximately *8* hours a year. The BAAQMD's 7th Street station, located in North Richmond exceeded *10* ppb of sulfur dioxide only *5* hours per year. Most of the time, the monitors registered either zero or one ppb. Additional analyses occasionally detected exposure levels in the North Richmond area reading between *7* and *13* ppb (hourly average) and between *30* to *60* ppb (hourly average) of sulfur dioxide. Thus, we were studying the *potential* effects of sulfur dioxide and hydrogen sulfide at ambient levels well below federal and state (California) ambient air pollutions standards. In our analyses we also included temperature (T) and exposure to other air pollution, such as to PM_{10}. We used daily diagnoses of a number of air pollution-related diseases from January 1, 1993 to September 30, 1995. The diseases included acute respiratory disease, chronic respiratory disease, and control diseases (such as gastrointestinal diseases) for all age groups in these Zip Codes. Sulfur dioxide and hydrogen sulfide were found to be statistically significant and with positive association on acute respiratory disease. As expected, the adverse effects were stronger in the 94801 Zip Code. SO_2 levels are significantly associated with increased acute respiratory disease for all age groups when air quality measurements from the 13th Street station are used. There are also significant effects of SO_2 when using data from the 13th Street station and H_2S data from the Gertrude station together. The Table below summarizes the results obtained for four Poisson regression models, two for all ages and two for children, for each Zip Code. These models are specified as:

Daily Diagnostic Counts = f(daily median exposure to H_2S, temperature).

The results for Acute Respiratory Disease, Gertrude Station Hydrogen Sulfide (Daily Median, DMD) and Temperature (T, Maximum Daily, and degrees-F) measurements are:

Population at Risk	Pollutant or Factor	Lag	Estimated β	se(β)	p-value
All ages, 94801, Acute Respiratory Disease, Gertrude H$_2$S, Zero Lag	H$_2$S DMD[1]	None	.026	.0167	.117
	T-max[2]	2	-.0205	.0023	.0001
	Intercept	NA	.91	.32	.0001
All ages, 94804, Acute Respiratory Disease, Gertrude H$_2$S, Zero Lag	H$_2$S DMD	None	.0022	.019	.91
	T-max	2	-.021	.0022	.0001
	Intercept	NA	2.39	.14	.0001
Children, 94801, Acute Respiratory Disease, Gertrude H$_2$S, Zero Lag	H$_2$S DMD	None	.0239	.024	.32
	T-max	2	-.0237	.0033	.0001
	Intercept	NA	1.785	.208	.0001

DMD is daily median concentration; *T*-max is maximum daily ambient temperature; NA is not applicable.

in which *Y* is the expected number of diagnoses, *H$_2$S* is daily average hydrogen sulfide concentration (measured in ppb) and *T$_{t-2}$* is the average temperature two days before the diagnosis. The results show a gradient from 94801 to 94804 Zip Codes, although it is statistically insignificant. Specifically, the size of the estimated parameters (the magnitude of effect) drops from *0.026* to *0.0022* (for all ages) and from *0.0239* to *0.0072* (for children). Such a relationship points to a zero lag effect, meaning that the effect of exposure is contemporaneous to response. A Poisson model is *ln(y) = 0.91+0.026(H$_2$S)-0.205T$_{t-2}$*.

As was discussed for other regression models, the practice of risk assessment often has to deal with sparse data sets. In those situations, estimation based on asymptotic properties can be inappropriate. It is therefore necessary to use exact methods.

Example 4.19. The data set is:

Responses (counts)	Exposure (1, 0)	Sex (1, 0)
0	1	0
0	1	0
1	1	1
0	1	1
1	1	0
0	1	0
2	1	1
0	0	1
8	0	0
0	0	1

The model is a Poisson regression in which *Counts = f(Exposure, Sex)*. The partial results from estimation are:

Estimation Method	Estimated β	Lower 95% CI	Upper 95% CI	p-value
Exposure (MLE)	-1.22	-2.34	-0.097	0.033

Exposure (Exact)	-1.22	-2.58	0.030	0.057
Sex (MLE)	2.11	0.059	4.16	0.044
Sex (Exact)	2.11	0.18	5.89	0.024

The MLE estimate of the parameter associated with exposure is statistically significant, but the exact estimation of that parameter indicates that there are estimates of borderline significance.

4.7. SURVIVAL ANALYSIS: Successes and Failures Over Time

Survival analysis is used in risk assessment to assess how the probabilities of successes and failures are distributed over time. The *failures* can be death, morbidity, or some other form of failure. Note that dealing with survival implies dealing with failures because the probability of survival is the complement of the probability of failure. That is, if the probability of surviving is *pr*, the probability of failure is *1-pr*. Several approaches can be used with survival models. Those discussed in this section include parametric, non-parametric and the Cox proportional hazards model. The parametric approach presupposes knowledge of the distribution of survival times. The non-parametric approach has no such restriction, but generally requires a larger number of observations than the parametric approach. Cox's model has both features: it is semi-parametric. The methods of survival analysis are generally independent of the unit of analysis: it can be a patient or a bolt or anything else that is consistent with some mechanism of failure. In this section, however, we deal with human health risks. In practice, the vertical axis (*y*-axis) of the survival function measures frequencies of response in a sample of individuals.

4.7.1. Survival Analysis

The survivorship (or survival) function, *S(t)*, gives the probability of not having failed at time *t*. In other words, it yields the probability that an individual at risk survives to an age greater than *t*:

S(t) = (number at risk that survive beyond time t)/(total number of individuals at risk),

S(t) equals *1* at the beginning of the study and decreases to *0*, when *t = T*. It is a cumulative distribution. Age need not be stated in years. In general, it is important to compare the survival rates for the same time for different diseases or outcomes or for the same disease with and without exposure. Thus, a comparison may be made of the 5-year survival rates for acute myelogenous leukemia from exposure and non-exposure to benzene. Figure 4.6 depicts the fact that at the beginning of a study all individuals survive. As time progresses, some of those individuals begin to experiences failures.

The curves are the best estimates, in the sense of the maximum likelihood, of survival. The right tail of the distribution (as stated, it is a cumulative distribution of the probability of survival) is developed from small numbers. Therefore, the tail of the distribution of survival can be accurate when the number of individuals surviving is small. One of the best-known methods for analyzing the survival of a group of individuals is through using the Kaplan-Meier estimator[20] (K-M), which is non-parametric. The K--M estimator yields estimates of the probability of survival at time t, (for $t_1 < t_2 < \ldots < t_j < \ldots < t_k$, these being the times-to-failure). It is:

$$\hat{S}(t) = \prod_{j:tj<t} (n_j - d_j)/n_j .$$

The hat over $S(t)$ symbolizes the fact that we are dealing with an estimator, Π is the multiplication sign, the colon is read *such that*, d_j is the count of failures at time t_j; n_j is the number of individuals at risk at t_j. The number of individuals at risk is the number of individuals present (neither failed nor censored) at time t.

The function $S(t)$ is not defined in the last period of time of survivorship. If needed, the count of censored individuals (individuals that left the study unexpectedly) in the interval (t_j+t_{j+1}) can be included by modifying the formula (Feinstein, 2002). The shape of $S(t)$ is not smooth, but it will be a staircase function, beginning at *1.0, t = 0*. The larger the sample, the smaller are the steps.

Example 4.20. The Kaplan--Meier estimator of $S(t)$, $\hat{S}(t_j) = \prod_{i<j}^{T}(1 - d_j/n_j)$, is applied as follows:

Time of failure, t_j	Number of failures at time t_j, d_j	Number of individuals at time t_j, n_j	$\hat{S}(t_j)$
6	3	21	0.86
...	...	...	0.81
10	1	15	0.75
...	...	...	...

The first entry is calculated as $[(21-3)/21] = 0.86$ and the other entry is calculated as $(0.807)[(15-1)/15] = 0.75$. The estimated variance of $S(t)$ can be calculated by Major Greenwood formula, although we omit these calculations for brevity's sake, which is:

$$\hat{V}(t) = [\hat{S}(t)]^2 \sum_{j:t_j<t} [d_j/[n_j(n_j - d_j)] .$$

The idea of the K-M estimator is that the conditional probability of failure at time t_t equals the observed conditional relative frequency of failure (d_j/n_j) at time t_j (Kalbfleisch and Prentice, 1980). This estimator of $S(t)$ has useful properties. For example, if the sample size is sufficiently large, the difference between the true and the estimated survivorship curves at a pre-specified time is asymptotically normally distributed with variance approximately equal to:

$$\hat{S}(t)[1 - \hat{S}(t)]/r(t) .$$

---

[20] Also called the *product-limit* estimator.

Simultaneous confidence bands for the whole function $S(t)$ can be derived by inverting the non-parametric Kolmogorov-Smirnov statistic for goodness of fit (Cox and Oakes, 1984). Moreover, in many risk studies, we find that some observations are censored. The example that follows describes a set of calculations that accounts for censoring.

Example 4.21. (Adapted from Clayton and Hills, 1993). The column with the l_j heading is the column that identifies censored observations.

t_j	n_j	d_j	l_j	Conditional frequency of death	Frequency of survival	Cumulative frequency of survival
0	50	2	-	0.040	0.960	0.960
1	48	1	-	0.0208	0.9792	0.939
2	47	2	-	0.0426	0.9574	0.897
3	45	1	1*	0.0222	0.9778	0.8875
...	...	...	...	...	...	...
60	17	1	1*	...	...	...

The frequency of failure at $t = 0$ is $2/50 = 0.040$ and the frequency of survival is $1 - 0.04 = 0.96$. At $t = 1$, the frequency of failure is $1/48$ and the frequency of survival is 0.9792.

The reason for comparing two or more survivorship functions is to determine the statistical significance of the difference between exposures and survival, as shown in Figure 4.6, developed with *Power and Precision*.

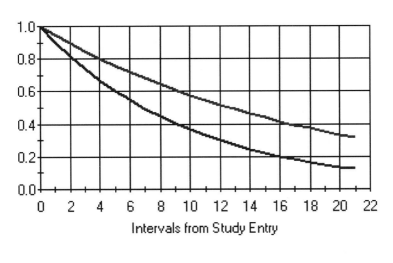

Cumulative survival

Intervals from Study Entry

———— Standard treatment   ———— New treatment

Figure 4.6. Comparison of Two Survival Curves

Methods such as Wilcoxon's can detect earlier differences in these survivorships, and yet other methods can detect later differences (Feinstein, 2002). The null hypothesis is that there is no statistical difference between the number observed to fail and the number expected to fail. The analysis is based on the expression $(d_{ij}-n_{ij}d_i/n_i)$, in which n_{ij} is the population at risk at time t_j (with j indexing a group), d_{ij} is the number of events at time t_j for the jth group, $n_i = \sum_j n_{ij}$ and $d_i = \sum_j d_{ij}$. The analysis is based on the formula:

$$\sum_i [w_i(d_{ij}-n_{ij}d_i/n_i)].$$

The term w_i is a weight that is equal to *1.00* for the *log-rank* test that detects differences when t_i is small or when the proportion failing is small over longer intervals of time. It uses the *chi*-square distribution with one degree of freedom. The log-rank method (Kahn and Sempos, 1989; Feinstein, 2002) can detect differences between survivorship functions when relatively long periods are analyzed. The *log-rank* test, for two groups (j and k, respectively) and *1 df* can assess if two or more distributions of failures over their entire period are equivalent. This is the null hypothesis. A formula for the log-rank test, for low risks in the interval of times considered, is (Kahn and Sempos, 1988):

$$\chi^2_1 \approx [\sum d_{ij}-\sum E(d_{ij})]^2/\sum E(d_{ij}) +[\sum d_{ik}-\sum E(d_{ik})]^2/\sum E(d_{ik}).$$

In this formula, $E(.)$ is the expectation of the term in the parentheses. The expected numbers for two groups (r, s) over t time periods, are:

$$E(d_{tr})=[O'_{tr}/(O'_{tr}+O'_{ts})](d_{tr}+d_{ts}),$$
$$E(d_{ts})=[O'_{ts}/(O'_{tr}+O'_{ts})](d_{tr}+d_{ts}),$$

where:

d_{tr} = number of failures in group r at time t,

d_{ts} = number of failures in group s at time t,

O'_{tr} = number of individuals at risk at time t in group r,

O'_{ts} = number of individuals at risk at time t in group s.

The last period does not contribute to the calculation of the expected and observed failures.

Example 4.22. In this example (adapted from Kahn and Sempos, 1988, p. 190) there are two groups that are experiencing deaths that are recorded at the beginning of each period, rather than at the end. These two groups are r and s; failures represent deaths:

t	# at Risk, O'_t in Groups			# of Failures, d_t in Groups			Proportion of O'_t in Groups		Expected Failures (H_0) in Groups	
	r	s	Tot.	r	s	Tot.	R	r	r	s
0	300	150	450	25	20	45	0.67	0.33	30.02	14.98

1	275	130	405	4	3	7	0.679	0.321	4.75	2.25
2	271	127	398	3	2	5	0.681	0.319	3.40	1.60
3	268	125	393	1	1	2	0.682	0.318	1.36	0.64
4	267	124	391	2	1	3	0.683	0.317	2.05	0.95
5	265	123	388	1	1	2	0.683	0.317	1.37	0.63
6	264	122	386	-	-	-	-	-	-	-
Tot.				36	28	64			42.95	21.05

Tot. means Total.

The calculations of statistical significance are based on the estimated *chi*-square:

$$\hat{\chi}_1^2 \approx \sum_{i=1}^{2}\left[\sum_{j=1}^{t}O_{ij} - \sum_{j=1}^{t}E(d_{ij})\right]^2 \bigg/ \sum_{j=1}^{t}E(d_{ij}).$$

In this formula:

$i = 1, 2$ (the two groups being compared),
$j = 1, 2, \ldots, t$ (the intervals being compared).

The estimated *chi*-square is: $\hat{\chi}_1^2 \approx [(36 - 42.95)^2 / 42.95] + [(28 - 21.05)^2 / 21.05] = 3.42$. Figure 4.7 depicts the theoretical *chi*-squared distribution for one degree-of-freedom. This distribution is discussed throughout Chapters 9-11.

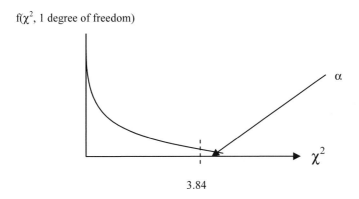

Figure 4.7. Chi-square Distribution for One Degree of Freedom

The calculated value of the *chi*-square test statistic in the preceding example equals *3.42*, which is less than (to the left of) the theoretical value of the chi-square distribution with one degree of freedom and $\alpha = 0.05$, which equals *3.84*. Therefore, the conclusion is that the null hypothesis cannot be rejected. The failure rates (deaths) in the two groups are not distinguishable from one another.

4.7.2. Discussion of Hazard Functions and Distribution Functions

Non-parametric maximum likelihood methods can be extended to i) censored data; ii) observations that include covariates for each individuals; and iii) observations taken at several stages in a causal sequence. Many aspects of risk analysis deal with failures over time: the random variable is time. When dealing with failures, it is often possible to use parametric methods to develop estimates of survival and, consequently, of failure. If the distribution function of failures is known, the hazard function of that distribution can be obtained by straightforward analysis.

Suppose that the distribution of the hazard rates is exponential. The reason for choosing the exponential distribution is that we know (from other sources) that adverse outcomes occur at a constant rate. This rate is generally symbolized by λ. This knowledge allows writing the general expression for the distribution as follows, in which the density function of the random variable t is $f(t)$: $f(t) = \lambda[exp(-\lambda t)]$, $\lambda > 0$. Then, $S(t) = exp(-\lambda t)$; $h(t) = \lambda$, which is constant. The shapes of these functions are depicted in Figure 4.8.

S(t)

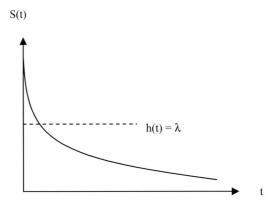

Figure 4.8. Shape of the Function $S(t) = exp(-\lambda t)$.

The relationship between failures and probabilities is summarized as follows. Suppose that the density function of a random variable T is $f(t)$, such that the values $\{t\}$ of the random variable are greater than or equal to 0. The reliability (or survivorship function) of a system characterized by the density function $f(t)$, where $F(t)$ is the cumulative distribution function, is:

$S(t) = 1-F(t)$.

The hazard function associated with the density function $f(t)$ is:

$h(t) = f(t)/[1-F(t)] = f(t)/S(t)$,

where $h(t)$ represents the conditional probability that a failure occurs in the (next, very small) period of time dt, given that the system has survived up to time t. For instance, the assumption of the multistage model is that a cancer develops at some time, t, under a constant dose rate (as discussed in Chapter 5). The *cumulative* hazard function, $H(t)$, is:

$N/k![\Pi_i^k(a_i+b_id)t^{k-1}]$.

The cumulative hazard function, $H(t)$, is an important concept in risk analysis. An estimator of this function is:

$H(t) = \{-ln[S(t)]\}$,

in which $S(t)$ is estimated using K-M's method, discussed earlier. Another estimator of $H(t)$ is:

$$\hat{H}(t) = \sum_{j:tj<t} (d_j/n_j),$$

in which d_j is the number of failures at time j, and n_j is the number of individuals within which the failures occur at time j. For instance, suppose that at time $j = 3$, $n_j = 22$ and $d_j = 2$; it follows that $d_j/n_j = 2/25 = 0.08$. If we plot the values taken by this ratio over all of the time periods of the study we obtain the distribution of the hazard function $h(t)$. Both estimators are asymptotically equivalent.

Example 4.23. Using the Kaplan--Meier estimator yields the cumulative hazard function $H(t)$, meaning that the $h(t)$ calculated at each period of time are added thus forming $H(t)$:

t_i	d_j	n_j	$S(t)_{KM}$	$H(t) = -ln[S(t)]$
1	2	21	0.91	0.10
2	2	19	0.81	0.21
...	...	...	...	...

The *extra risk* is calculated by noting that $H(t)$ is the addition of the background rate and the increased cumulative hazard due to exposure to dose rate d. It follows that:

$H(t; d) = H_{background} (t, d=0)+H_{exposure}(t, d)$,

because $\{1-exp[-H_{backround}(t; d=0)]\} = Pr(t;d=0)$. Moreover,

$Pr[(t, d)-Pr(t, d=0)] = exp[-H_{backround}(t, d=0)]\{1-[exp-H_{exposure}(t, d)]\}$.

It follows that the extra risk is: $[Pr(t, d)-Pr(t, d=0)]/[1-Pr(t, d = 0)] = 1-exp[-H_{exposure}(t, d)]$.

At low dose rates the extra risk is:

$1-exp[-H_{exposure}(t, d)]$,

which is approximately equal to $H_{exposure}(t, d)$, regardless of the type of exposure (Kodell, Gaylor and Chen, 1987). These authors also provide equations to handle the general case of intermittent exposures.[21]

---

[21] The hazard rate, *HR*, equals *(# failures)/(# person-years)*.

Example 4.24. Suppose that there are *5* failures in *10* people exposed for *10* intervals of time, each of which is *0.25* years in length. The failure rate is: *5/(0.25*10) = 2* failures/year. This rate, which can be greater than *1.00*, is a frequency-rate.

Example 4.25. Let the number of failures be symbolized by *F* and consider *N* intervals of time. The number of successes is *(N-F)*. Assuming independence, $pr^F(1-pr)^{(N-F)}$ is the likelihood function. Let λ be the constant hazard rate and suppose that the intervals of time, Δt, are sufficiently small. The conditional probability of success is *pr* and $\lambda = pr/\Delta t$. The likelihood can be re-written as:

$(\lambda \Delta t)^F (1-\lambda \Delta t)^{(N-F)}$.

The log-likelihood is $[F*ln(\lambda)+F*ln(\Delta t)]+(N-F)*ln(1-\lambda \Delta t)$. The small numbers inherent to letting Δt be small and *(N–F)* be large allows $ln(1-\lambda \Delta t) \approx -\lambda \Delta t$. Because $(1-\lambda \Delta t)^N = N*ln(1-\lambda \Delta t)$, using the approximation $ln(1-\lambda \Delta t) = -\lambda \Delta t$ and letting $N\Delta t = T$ yields the cumulative hazard rate, $-\lambda T$.

4.7.3. Comment

By definition, we know that *S(t) = 1-F(t)*. *S(t)* is:

$S(t) = pr(t \geq 0) = \int_0^\infty f(x)dx$.

It follows that:

$H(t) = f(t)/[1-F(t)] = f(t)/S(t)$.

Therefore, $h(t) = -(dS/dt)/S(t) = -d[ln(S(t))]/dt$. More specifically,

$H(t) = \int_0^t h(x)dx$.

It follows that:

$S(t) = exp[-H(t)]$,

and thus:

$H(t) = -[log S(t)]$.

The estimation of *S(t)* uses the maximum likelihood. The K-M estimator also has a Bayesian justification. It is the mean of the posterior distribution for *S(t)* when a degenerate *no information* prior is updated in light of the observed time series *{d(j)}*, for *j = 1, 2,..., t* (Cox and Oakes, 1984). The *chi*-square distribution can be used to develop point and simultaneous confidence bounds for *S(t)* and for the hazard function, *h(t)*. Using a parametric model means that the distribution function *is* known. The reason for their importance is that the formulae for the hazard function can be applied to such distributions as the Weibull, normal, log-logistic and so on. Because the Wiebull distribution and other distributions are used in modeling cancer and other chronic diseases to determine hazard and survivorship functions, it is important to have a sense of how they are derived.

Chapters 5 and 11 provide more details about cancer models and the best-known distribution functions used in risk assessment. For example, applying these to the exponential distribution generates the formulae given above for that distribution. The implication of these formulae suggests the following. First, the product of $h(t)$ and dt, $h(t)dt$, is the instantaneous probability of failure. For larger intervals, $h(t)\Delta t$ is the probability of failure in the interval Δt, given survival to that time period. More specifically, $h(t)$ is the instantaneous rate of failure that can represent mortality rates, illness rate and so on, if we are dealing with continuous density functions.

4.8. COX PROPORTIONAL (RELATIVE RISK) HAZARDS MODEL

Cox proportional hazards model (PH) describes a hazard function. The proportional hazards model makes a specific (parametric) hypothesis about the effect of exposure (or dose) on a measure of risk. It leaves unspecified background risk. Because of this construction, the proportional hazards model is semi-parametric. Recollect that the ratio $h_j(t)/h_0(t)$ defines the relative risk, RR, when this model uses incidence rates (new diseases/unit time) in the exposed and in the controls. On this basis, it follows that a relative risk model is: $h(t) = k[h_0(t)]$, where the constant k equals 1 for unexposed individuals and is greater than 1 for those exposed. In this model, the exposure or dose variable can be defined as $X = 0$ for the no exposure and $X = 1$ for the exposure. The general form of the proportional hazards model is (Kahn and Sempos, 1988):

$$h_j(t)/h_0(t) = exp(\textstyle\sum_i \beta_i X_{ij}).$$

In this equation the independent variable is indexed by i and the individual by j. In this PH model, t is age, $h_j(t)$ is the incidence rate at age t in those individuals characterized by the risk factors X_i and $h_0(t)$ is the background incidence. As with other regression models discussed in this chapter, we want to obtain estimates of the parameters of the independent variables, X_{ij}. The independent variables act *multiplicatively* on the risk because they are exponential. The exponential form of the PH model can be simplified by taking the natural logarithm of both sides of the model, as follows:

$$ln[h_j(t)/h_0(t)] = ln[exp(\textstyle\sum_i \beta_i X_{ij})] = (\textstyle\sum_i \beta_i X_{ij}).$$

Now, the independent variables act additively. The background hazard function, $[h_0(t)]$, is an arbitrary function of time. Its distribution does not need to be known. This non-parametric assumption can be changed. For example the background hazard function in Cox's model can be specified as a parametric function, for instance as a Weibull distribution. If both hazard functions, the background hazard and the hazard function related to exposure, are specified, the model is fully parametric.

The PH model describes the relation between background risk and the risk due to exposure, as a function of age. Age is the only independent variable. Figure 4.9 depicts the natural logarithm of the instantaneous hazard (the risk), accounting for age, under the PH assumption of proportional risk between background and exposure.

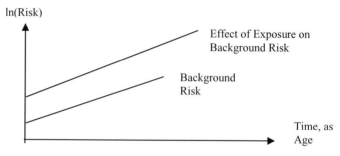

Figure 4.9. Natural Logarithm of Risk for Exposed and Unexposed Groups

Estimation yields numerical values of the parameter vector β from observed no-response times, presence or absence of response, and other independent variables. Unlike the logistic and Poisson regressions in which the stratification is generally limited to a few strata, the proportional hazards model uses each individual as a stratum. *Time* represents the *age* at which those who are at risk respond by failing. The background hazard function for an individual is calculated by setting means of the values of all covariates to zero. Once the parameters of the model are estimated through such methods as the MLE, they quantify the effect of each independent variable on the risk. The expression $exp(\beta_i X_{ij})$ is the relative risk due to the i-th variable affecting the j-th individual. The relative risk is the ratio of the incidence (new diseases) in the exposed to the incidence in the non-exposed, as further discussed in Chapters 8 and 9. The statistical null hypothesis is that the parameter values equal zero and thus $exp(\sum_i \beta_i X_{ij}) = 1$.

Example 4.26. Suppose that the results of estimation using the PH model in the context of different types of exposure are:

Variables	Estimated Parameter values	Measurement
Exposure 1	-0.95	1, if exposed; 0 otherwise
Age	0.015	Age, years
Exposure 2	0.082	1, if exposed, 0 otherwise
Exposure 3	0.0030	1, if exposed, 0 otherwise

The independent variables, Exposures 2 and 3, refer to different areas of exposure. From the estimates of the parameters of the model (*0.015* and so on) we can calculate the relative risk, RR, from age 35 to 50. It is calculated as *exp[(0.015)(50-35)] = 1.25*. The RR of exposure *2* for an individual aged between *35* and *50* is calculated as: *exp[(0.015)(15)+(0.082)(1-0)] = 1.36*, from Exposure 2 taking the value *1*. Non-exposure 2 is indicated by the same variable taking the value *0*. Suppose, to generalize this example, that there are two individuals, individual j and individual k. The RR for the j-th individual (relative to background) for three exposure variables, X_1, X_2 and X_3, is:

$$exp(\beta_1 X_{1j} + \beta_2 X_{2j} + \beta_3 X_{3j}).$$

A similar argument applies to the k-th individual: the three-variable PH model applies to that nth individual. Assume that the values of these 3 variables for the jth individual are *1, 0, 10* and for the nth individual, *0, 0, 10*. It follows that $RR_j = exp[\beta_1(1) + \beta_2(0) + \beta_3(10)]$ and $RR_k = exp[\beta_1(0) + \beta_2(0) + \beta_3(10)]$. These two individuals' risk ratio is $exp[\beta_1(1) + \beta_3(10)] / exp[\beta_3(10)] = exp[\beta_1(1)]$. Taking the natural logarithms of these expressions yields the logarithms of the relative risk and risk ratio.

4.9. TIME SERIES ANALYSIS

Time series analysis deals with data and relations between data that depend on time through regression methods. For instance, Table 4.4 contains a time series of yearly tax rates per pound of chlorofluorocarbons, CFC.

Table 4.4. Time Series of Yearly Data of Tax Rate (in US dollars per pound of chlorofluorocarbons emitted) for the United States

Year	1990	1991	1992	1993	1994	1995	...	2000
CFC-11 & CFC-12	1.37	1.37	1.67	2.65	2.65	3.10	...	5.35

US Congress, Joint Comm. on Taxation (2000).

Example 4.27. This example shows a time series of daily maxima SO_2 (SO2DMX) concentrations in which the time axis measures *24*-hours periods (*n = 149* days taken from the full time series of data of the air pollution and diagnoses of respiratory and other diseases (Colford, et al, 1999). The time series is:

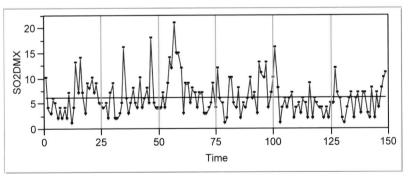

The mean, standard error of the estimate and sample size of these data are:

Mean of the daily maxima SO_2 concentrations	6.26
Standard error of the estimate	3.70
Sample size	149

Time series analysis is used in risk assessment to: i) describe the behavior of the data by assessing their variability; periodicities of different types (such as seasonality) and trends; ii) address the mechanism that underlies the data; and iii) make inference and prediction after the parameters of the time series model are estimated. The family of models that we will discuss in this section is the Autoregressive, Integrated, Moving Average (ARIMA) model, which is often used to study the temporal relation of exposure to air, water, soil pollution and adverse health and environmental outcomes. The statistical approach consists of developing a mathematical expression in which time is an explicit independent variable, obtaining a sample as a time series of data, and estimating the parameters of the time series. Estimation uses the ordinary least squares. The *error*, the measure by the difference between the observed value and its corresponding fitted value, again plays a central role in estimation and inference. In practical risk assessments, the prominent type of statistical time series analysis is called *time-domain* analysis, discussed next.

4.9.1. Time-Domain Analysis of Time Series, ARIMA

The development of an ARIMA model requires three sequential steps as well as some iterations:

1. *Identification and Model Selection.* This step determines the *order* of the ARIMA process.
2. *Estimation.* This is the result of using an estimator, such as the ordinary least squares, to obtain numerical values for the parameters of the AR and MA components of the ARIMA model, as well as confidence intervals and other statistics.
3. *Diagnostic Analyses.* This step assesses the adequacy of the empirical results and can be used to refine the initial specification of the ARIMA model by iterating the analyses.

The first step in time series analysis of a set of data, $\{x_1, x_2, ..., x_n\}$ generated by the random variable $X(t)$, is the *identification* of the time series model. After the time series of data is plotted, the identification step consists of applying mathematical formulae to the errors to determine if the underlying time series is random. It may be necessary to *difference* the original time series of data to eliminate trends because the estimation of the parameters of the models requires, for technical reasons, that the time series be *stationary* (Cressie, 1998). A trend occurs when the mean of the random variable changes over time. In some cases, *stationarity* is achieved by transforming the values of the random variable of interest (e.g., by taking the logarithms of the values of the data). Differencing consists of taking the differences between two successive values of the time series. For example, (x_2-x_1) is a differenced value of *order 1* because only one difference has been taken. The reason for differencing is to make the resulting time-series *stationary* with respect to the time series mean, variance and covariance.[22] Specifically, if a time series is

---

[22] The covariance of the observations measures the correlation between the observations of a random variable over time. That is, the observation y_t and y_{t-1} can be correlated and that correlation must be detected and accounted for in the estimation.

stationary its mean and variance are approximately the same through the time series. Higher order differencing can remove complicated periodicities (seasonalities or cycles); however, the new time series will have less data than the original, undifferentiated time series. Practical analyses are limited to taking first and second order differences.

Example 4.28. We have done the following operation on the data: $Bx_t = x_{t-1}$. The differenced time series (with first order differencing) is:

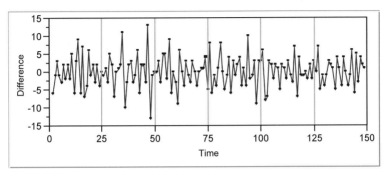

Taking the differences of second order, calculated as $(1-B)^2 x_t$, the differenced time series is:

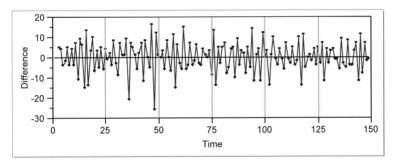

Differenced Mean of the daily maxima SO_2 concentrations	0.045
Standard error of the estimate	7.42
Sample size	147

The parameters of the ARIMA model are estimated through the methods of Box and Jenkins (1976), which require that the time series of data is stationary before estimation. The autoregressive, integrated, moving average, ARIMA model is a combination of two stochastic processes: the autoregressive (AR) and moving average (MA) processes. *Integration* is the name given to the differencing step, I(d), in which d is the order of the differences taken over the observations of the time series. According to Chatfield (1975) *integration means that the stationary model fit to differenced data must be summed to model non-stationary data*. The following notation indicates differencing of two values of the random variable X, in which x_t is a value taken by the random variable at time t and x_{t-1} is the immediately preceding value: $\nabla x_t = x_t - x_{t-1}$. The inverted triangle, *del*, can be applied to different values of the time series, for example to obtain a higher order we can use: $\nabla_{12} x_t = x_t - x_{t-12}$. These operations on the original data of the time series *pre-process* it before estimation. Therefore, a new time series is developed and then used to obtain estimates of parameter values as well as confidence intervals and so on. The ARIMA model can be separated out into two main components, the MA and the AR.

4.9.2. The Moving Average, MA, Process

A time series model that relates the present value of Y_t to past values of its errors e_t is the Moving Average (MA, of order n):

$$Y_t = \mu + a_0 e_t + a_1 e_{t-1} + \ldots + a_2 e_{t-n}$$

The term μ is the mean of the moving average process and is independent of time, $E(Y_t) = \mu$; e_t represents the effect of random and unknown disturbances about the trend of the time-series. It is a random variable with zero expected value, constant and finite variance, and zero covariance between the errors. The lags account for the effect of the error at time $t\text{-}k$. If the random variable is Y, the contemporaneous time series of data is represented by $\{Y_t\}$, the lag-*1* time series by $\{Y_{t-1}\}$ and lag-*2* by $\{Y_{t-2}\}$.

4.9.3. The Auto Regressive, AR, Process

An Auto Regressive process (AR, of order p) relates the current value Y to the p past values of Y, as follows:

$$Y_t = b_1 Y_{t-1} + b_2 Y_{t-2} + \ldots + b_p Y_{t-p} + d + e_t.$$

In this model, d is the trend (shown as not depending on time and thus being a constant), and e_t is the random disturbance with zero expected value, constant and finite variance and covariance properties. The implication of this model is that the effect of past values of the dependent variable determines its present value.

4.9.4. Simple ARMA Process

The AR and MA processes can be combined as:

$$Y_t = c + b_1 Y_{t-1} + e_t + a_0 E_t.$$

This model is the result determining (through diagnostic tests done at the identification stage of time series analysis) that the time series is ARMA model of order (*1, 0, 1*). Order means that there is a *one*-lag AR component, there is zero differencing, and there is one MA component that relates to the errors in the model.

In other words, the terms between the parentheses are a shorthand way to inform about the structure of the ARIMA selected for analysis. The next sub-subsection provides some of the details inherent to the estimation and diagnostic checks. To simplify the discussion, we repeat the time series of daily maxima SO_2 concentrations, of the Richmond, CA air pollution and daily diagnoses of respiratory and other diseases, Figure 4.10.

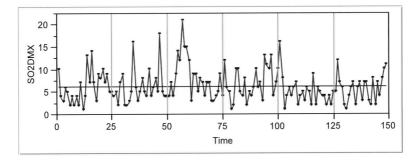

Figure 4.10. Time Series of Daily Maxima SO_2 Concentrations, Richmond, CA

This time series can be tested statistically for stationarity by considering its mean and covariance: $E[X(t)] = \mu$, covariance $[X(t), X(T-t)] = g(t)$. The covariance depends on the lagged observations. The practical statistic is the *autocorrelation* coefficient, $r(t) = g(t)/var(t)$, which is asymptotically unbiased: it is bounded between -1 and 1.

The autocorrelation coefficient measures the population's correlation between the values of the pair-wise numbers in the series:

$$\rho = E[(y_t-\mu_y)(y_{t+k}-\mu_y)]/\sigma^2,$$

in which $E(.)$ is the expected value of the difference of the quantity between parentheses, μ is expected value of the time series, and σ^2 is the variance of the time series. The sample autocorrelation coefficient for k lags (assuming a stationary process and equal means), is estimated by:

$$\hat{r}_k = \{\sum_{t=1}^{t-k} [(y_t - \bar{y})(y_{t+k} - \bar{y})]/[\sum_{t=1}^{T}(y_t - \bar{y})]^2\}.$$

In this expression, T is the total number of time periods considered. The autocorrelation coefficient is used to study the properties of the time-series through the *correlogram*, which consists of plots of estimates of $r(t)$ versus the number of lags, k. If the time-series is random, r is approximately normally distributed, $\sim N(0, 1/n)$, the sample size of the time series (the number of observations) is n. The plot of various $\hat{r}_k$ and their confidence limits, one for each lag, beginning with lag 1, as a function of the number of lags, empirically establishes whether the time-series is stationary, it is generated by AR or MA, or by a combination of both. An autocorrelation is a random variable and thus has a distribution and can be tested through standard statistical methods for inference. Inference requires dealing with the statistical variability of the estimation. The standard error of the autocorrelation for the *k*th lag can be used to test the significance of the results by choosing the appropriate level of confidence. Under the assumption that the time-series are generated by a random process and that the realizations of the process are identically and independently distributed, the approximate width of the *95%* confidence limits is $[\pm(2/n^{1/2})]$. The correlogram helps to determine the order of the AR process through the number of significant lags. For instance, if the first three lag values are significant, then the process is an AR(3). However, the order of the AR process is generally difficult to establish; the plot of the partial autocorrelations helps this task. The order of the MA component (symbolized by *q*) can be determined by inspection of the partial autocorrelation correlogram. Identification of the AR and MA components of the model is further aided by the partial autocorrelation function, PAF, which is bounded between *-1* and *1*. PAF measures the association between two values of a random variable, $(y_t\text{-}y_{t\text{-}k})$, when other lags up to *k-1* are constant. The formula for a two-lag (indicated by the numeral *2*) PAF is:

$$PAF(2) = \{ACF(2)\text{-}[ACF(1)]^2\}/\{1\text{-}[ACF(1)]^2\},$$

where ACF is the autocorrelation function, calculated at lag *1*. For a one-lag model, PAF(1) = ACF(1). For lags greater than two, the expressions for partial autocorrelation are cumbersome, but computer programs such as STATISTICA, SYSTAT, JMP, STATA and SPSS provide that information. A summary of the uses of the correlogram is provided in Table 4.5.

Table 4.5. Uses of the Correlogram in Time Series Analysis

Correlogram description	Additional comment	Remarks	Choice of Model
Rapidly decreasing values of $\hat{r}^2$ as the lag number (k = 1., 2, ..., k) increases	Significant and high initial values, followed by low but significant values, with overall tendency to zero as k increases.	Stationary time-series	Autoregressive
Alternating positive and negative values of $\hat{r}^2$ as k increases, tending to zero as k increases	Significant and high initial values fluctuating about zero, for small k values, dampening to zero as k increases.	Stationary time-series	Autoregressive
Very slow tendency of $\hat{r}^2$ to zero as k increases	Generally high and significant values over small and large k, tending to zero only as k becomes very large	The time-series is not stationary and the correlogram is not an appropriate measure. May difference the time-series to restore stationarity	Moving Average with a trend.

Cyclical, significant high and low values of $\hat{r}^2$, without tendency to zero	The cyclic behavior of $\hat{r}^2$ mimics the original data.	Cyclic variability can be removed to recalculate $\hat{r}^2$ and determine its pattern leading to the choice of model	Autoregressive

Example 4.29. The autocorrelation and partial autocorrelation functions for *15* days (the almost straight lines are the *95%* lower and upper confidence limits) of the original (not differenced) time series of SO_2 concentrations are depicted next. The autocorrelation function is:

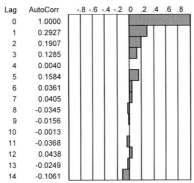

Lag	AutoCorr
0	1.0000
1	0.2927
2	0.1907
3	0.1285
4	0.0040
5	0.1584
6	0.0361
7	0.0405
8	-0.0345
9	-0.0156
10	-0.0013
11	-0.0368
12	0.0438
13	-0.0249
14	-0.1061

The partial autocorrelation function is:

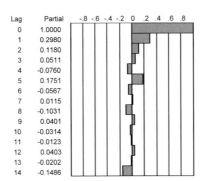

Lag	Partial
0	1.0000
1	0.2980
2	0.1180
3	0.0511
4	-0.0760
5	0.1751
6	-0.0567
7	0.0115
8	-0.1031
9	0.0401
10	-0.0314
11	-0.0123
12	0.0403
13	-0.0202
14	-0.1486

This and other information can specify the time series: an initial specification of the model is ARIMA (*1, 0, 1*) process.

The ACF and PAF can be used as summarized in Table 4.11.

Table 4.11. Assessment Using ACF and PAF

Process will be:	If the estimated PAF is:	Comment
AR(1)	PAF(1)**	All other PAF are insignificant.

AR(2)	PAF(1)** and PAF(2)**	Same
AR(k)	PAF(1)** to PAF(k)**	Same
MA	PAF exponentially tends to zero	None
Stationary	Insignificant after 2 or 3 lags	Mean and variance are constant
Non-stationary	Significant after several lags	None
Seasonal	Significant at specific seasons	All other PAF are insignificant

** Means statistically significant at the level of significance chosen for the analysis.

Example 4.30. Using the time series of daily maxima SO_2 concentrations, discussed in earlier examples, the ARIMA *(1,0,1)* model has these overall results:

Degrees of freedom	*145*
Sum of Squared Errors	*1861.54*
Variance Estimate	*12.84*
Standard Deviation	*3.58*
Akaike's Information Criterion	*383.76*
R-Square	*0.055*
Adjusted R-Square	*0.042*
-2LogLikelihood	*379.117*

This ARIMA *(1,0,1)* model, which can be used to forecast the future values of SO_2, for thirty time periods, with upper and lower *95%* confidence limits about these predictions, is depicted as:

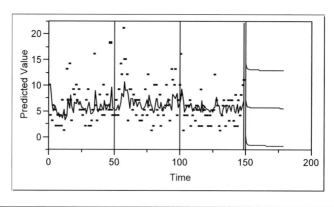

4.9.5. Discussion

Estimation yields numerical values for each parameter in terms of central tendency and dispersion. The parameters associated with the AR model are relatively simple to obtain through the ML estimator, under the assumption that the random disturbances are independent and normally distributed. Approximate non-symmetric confidence bounds can be obtained using the asymptotic properties of the ML estimator. For the MA model, an iterative method must be used to yield the ML estimates of the parameters and their variance, assuming that the errors are normally identically distributed and independent. The joint test (over *k* lags) is the Box--Pierce *Q*-statistic:

$$Q = (n - d) \sum_{k=1}^{t} \hat{r}_k^2,$$

which is distributed as a *chi*-square statistic with $(T\text{-}p\text{-}q)$ degrees of freedom. In this equation, T is the maximum number of lags, and d is the number of differences taken. Q can be used to test the joint significance that $\hat{r}^2$, under the null hypothesis, for several lags. In general, these tests of the significance are difficult unless the time series is stationary with respect to the mean and the variance (Ljung, 1986). The last step in the analysis of statistical time series is the *diagnosis* of results obtained from statistical estimation. This step consists of assessing: i) whether the estimated disturbances identify certain specifications of the variables left out of the analysis; and ii) whether certain variables need not be retained in the model.

The assessment of how well the model fits to the data uses the Akaike Information Criterion (AIC). The formula is:

$$AIC = -2n * ln(s_a^2 + 2k).$$

In this equation, k is the number of estimated parameters and s_a^2 is the estimated variance of the random error. The optimal order of an ARIMA is that for which the AIC is the minimum.

4.9.6. Frequency Domain Analysis

Recollect that the APHEA protocol uses trigonometric series to account for periodicities. These series are used in the air pollution and daily diagnoses of respiratory and other diseases and in economic analyses, such as in the evaluation of intangible economic values (Ricci et al., 1977). A complementary way to time-domain analysis of time series is through trigonometric series. The cyclical aspects of the data of a deterministic variable $y(t)$ can be characterized by a function in which the amplitude (A) and periodicity $(f*t/n)$ are described as follows:

$$y(t) = A sin[(f*t/n)2\pi + \theta],$$

where A is the amplitude of the sine wave, f is frequency measured by the number of cycles per unit time, θ (in radians) measures shifts along the time axis and t is an index. In this model, the quantity $2\pi f$ is number of radians per unit time. The term $(f*t/n)$ transforms discrete values to a continuous proportion of 2π radians. Letting t represent time, amplitude $= 1$, $n = 24$ and the shift, $\theta = 0$ we obtain:

t	ft/n	$Y(t) = 1*sin[(1*t/24)2\pi + 0]$
0	0	0
1	1/24	$sin[(1/24)\ 2\pi] = 0.2588...$
2	2/24	$sin[(2/24)\ 2\pi] = 0.50$
...	...	...
24		$sin[(24/24)\ 2\pi] = 0.0$

Given a frequency of $f = 1$, the wavelength of the cycle is *24* time periods. One cycle occupies the entire period. As frequency increases, in the interval $(0, 2\pi)$, there are more sine waves. Their number depends on frequency. The height of the wave is measured by the amplitude: the larger the amplitude the higher the peaks and the valleys in the interval. Trigonometric series can represent many functions. A trigonometric representation is (dropping the multiplication sign between f and t): $\Sigma_i A_i[sin\ (f_i t/n)2\pi + \gamma_i]$. The probabilistic representation uses $Y(t)$ as the random variable. The stochastic population error $\varepsilon(t)$ can be included additively, as done for other regression models:

$$Y(t) = \Sigma A_i[sin(w_i t2\pi + \gamma_i)] + \varepsilon(t).$$

The *sample* error is represented by $e(t)$, the term $\varepsilon(t)$ being used for the error associated with unobserved *population* model. This equation is cyclic. In statistical modeling, amplitude and phase are not known; therefore they have to be estimated. A transformation involving the sum of the product of trigonometric sine, cosine, and phase angles allows estimation through such methods as the ordinary least squares. Estimation, applied to the transformed time series:

$$Y(t) = \Sigma[\beta_i sin(w_i t2\pi) + \gamma_i cos(w_i t2\pi)] + \varepsilon(t),$$

yields estimates of the amplitude and phase, namely estimates of the coefficients β_i and γ_i, if the frequencies are known (Kennedy, 1998). A way to deal with unknown periodicities is to use a trigonometric series such as:

$$1/2\alpha + \Sigma_k[\beta_i cos(\omega_i t2\pi) + \gamma_i sin(\omega_i t2\pi)],$$

estimate its parameters and then convert them into amplitude and phase angle for k frequencies, $k = (n-1)/2$.

Example 4.31. Using the daily maxima concentrations of SO_2, the spectral density of the time series is:

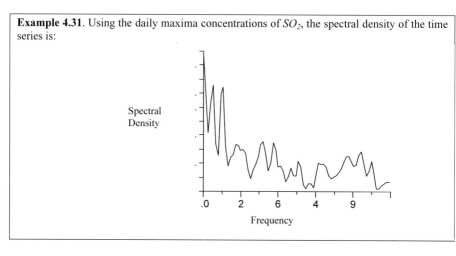

In simple cases, the spectrum (defined as the square of the amplitude plotted against frequency) appears as a spike and identifies an amplitude at a given phase angle and frequency. In other cases, there may be several spectral lines. The interpretation of the spectrum is as follows. When $f(\omega)$ is characterized by high values, large amplitudes are present in the time series. If $f(\omega)$ is horizontal to the ω-axis, the time series t is random. Random noise has constant average amplitude at all frequencies. Therefore, only the peaks are useful to determine periodicities above randomness. The peaks, in the plot of $f(\omega)$ against ω, identify the magnitude and the location of the periodicities.

4.10. MODEL SELECTION: Basic Considerations

Thus far, the discussion has indicated that the development of a statistical risk model is based on the literature and the context of the assessment and management of the risk we are assessing thorough statistical models. The steps taken by the risk assessor in that development are:

1. Model Development,
2. Data Development,
3. Estimation and Diagnoses,
4. Corrections to the Original Model,
5. Forecasts and Inference,
6. Report.

A question that requires some attention is: Where do models come from? An answer is that *models originate from theory and observations*. This answer to the question suggests that some of the practical considerations include assessing:

1. The general purpose of modeling (Is the purpose of modeling exploratory or causal?),
2. The specific purpose of the model and the data-gathering protocol (random sample or observational data taken from available data bases),
3. The choice of variables (linear, quadratic or other formulation),
4. The form of the model (linear or nonlinear),
5. Estimation (MLE or other),
6. Diagnostic tests and corrections (diagrams, statistical tests, and so on).

It would seem that measures of *best fit* are useful to assess the way alternative specifications or forms of a model fit the data from the sample. The combination of t-test on the significance of each coefficient of the model and the F-test of the joint significance of the coefficients provide several means to assess the performance of a statistical model, relative to the data from which its parameters are estimated. Diagnostic analyses based on plots of the errors (i.e., $y_i - \hat{y}_i$, the differences between the actual values and the fitted values) and values of the independent or dependent variables are also important in checking the results of estimation. The likelihood ratio has been traditionally used to assess models using the *chi*-square distribution, level of confidence and appropriate degrees of freedom, which was shown in log-form for two models as:

$LL = -2[ln(L_{Model\ 1})-ln(L_{Model\ 2})]$.

The larger the difference between the log-likelihoods, the more appropriate the model generating that difference. Several statistical tests, such as Durbin-Watson (used to study the correlations between these errors), can be used (Ricci and Wyzga, 1983; Kennedy, 1998) to strengthen the analysis of residuals.

Nonetheless, the problem of having to represent the *real word-theory-data-model-theory-real world* is not fully resolved. The reason is that any such resolution is partly within the physical domain and partly within probabilistic and statistical analysis. In risk assessment, the physical rationale can range from biological to metallurgical. Importantly, the risk assessor and manager should remember that *hypothesis testing is a poor basis for model selection* (Burnham and Anderson, 2000). Model selection principally deals with the form of the model, given that there can be several and different of models that can be fit to a set of data. Although Occam's razor can aid in the selection of models, parsimony is heuristic and practically limited to hypothesis generation because its prescription suggests that an investigation should be based on simple formulations (e.g., linear).

Example 4.32. To illustrate the problem of model selection, consider a statistical model in which there are *five* independent variables. The number of possible linear models that can arise is $2^5 = 32$ with the variables being present or absent. This number increases as the number of independent variables increases. For a *15*-variable model, the possible models are $2^{15} = 32,768$ models.

4.10.1. Akaike's Information Criterion, AIC

An alternative measure for choosing between models is Akaike's Information Criterion, AIC, (Burnham and Anderson, 2000) and its variants, used earlier in this chapter in the analysis of time series. It is a deceptively simple formula (it includes a conditional statement that was not obvious in the earlier formula for the AIC used in discussing time-series models):

$AIC = -2ln[L(\hat{\theta}|x)]+2K$.

In this equation, $\hat{\theta}$ is the parameter estimate of the population's parameter θ, x is the data for the independent variables and K is the number of parameters to be estimated. The AIC can be used to determine the relative importance of nested and non-nested models under the decision rule that the model with the smallest AIC is the preferable model. The AIC accounts for the trade offs made by increasing the number of independent variables and the penalty, measured by the increased variance, of doing so. The AIC formula for the linear regression model in which the errors are normally distributed, independent and have constant variance is $\{n[ln(\hat{\sigma}^2)]+2K\}$, in which n is the sample size and $\hat{\sigma}^2$ is the estimated variance of the errors associated with the OLS regression. As Burnham and Anderson (2000) remark, all of the parameters of the linear model must be included in the computation of the AIC. Specifically, this set must include the estimated variance, which is a parameter of the model.

The premise of the AIC is that an empirical model most closely approximates the unknown real-world model, even though that *real-world* model is unknown. Although this statement seems paradoxical, the need for knowing that real-world model is *integrated* out from the joint entropy of information that *does* contain the specification of the real model. This can be demystified by considering a known model as the truth and then working with the empirical model of choice. A similar rationale applies to *unknown* models in that integrating the unknown function out from an expression leaves the empirical function. The AIC can be calculated as a set of differences taken over the possible models being considered. Neither the AIC nor the Δ(AIC) are statistical tests, they are thus unlike the likelihood ratio test. The decision rule for the Δ(AIC) is that the smallest value of that difference identifies the best model. The AIC can be used to choose between models that are not nested. This is something that the likelihood ratio test cannot do. The criterion for choice under the AIC is unchanged: the model with the lowest AIC is selected. However, the same data set must be used in this assessment (Burnham and Anderson, 2000).

The AIC has been extended to account for models that can be affected by overdispersion, a problem that affects Poisson regression models. If it were found to occur, a modified form of the AIC should be used instead. Overdispersion can occur in counts when the variance of those counts exceeds the theoretical variance, which is the same as the expected value of the population, namely $var(n) = E(n)$. The modified AIC formula, in which θ is a (vector) of parameters, n is the sample size and K is the number of independent variables, is:

$$QAIC_c = -\{2ln[L(\hat{\theta})/\hat{c}]+2K+2K(K+1)\}/(n-K-1).$$

In this formula, $\hat{c} = 1$ if there is no dispersion (c is distributed as a *chi*-square variable with the appropriate number of degrees of freedom, Burnham and Anderson, 2000). Although overdispersion does not generally bias the estimates of the models' parameters, it can increase the variability of the estimates of the parameters. Burnham and Anderson, (2000) provide the details of this method and warnings associated with estimation. The estimates of c can be used with estimates of the (overdispersed) estimates of the variance to study how well the model performs. For example, if the specification of the model is appropriate, the method of quasi-likelihood can be used in obtaining estimates (Liang and McCullagh, 1993).

The AIC and its modifications are not based on hypothesis testing. The AIC is not a statistical test and therefore analyses based on statistical significance do not apply. In other words, we cannot consider AIC a statistic. As Burnham and Anderson (2000) remark, the choice of models under AIC or hypothesis testing generally yields the same ranking, if the sample is sufficiently large and if the models are simple. However, when these conditions do not arise the results can be quite different. Uncertainty in model selection methods based on the AIC relates to the variability between parameter estimates, but not to the uncertainty in the specification of the global model from which the choices of other models are made. Model selection suggests that a parsimonious model can be sufficient for risk management. This raises the issue of what to do when a ranking by the AIC may not include a variable that belongs to a model on mechanistic grounds. For example, the issue can be resolved by constraining the model to include the variable by forcing the coefficient of such variable to be greater than or equal to zero. This approach, well established in the statistical literature that deals with placing constraints on the parameters of a model before estimation, requires statistical methods that go beyond the scope of this chapter.

4.10.2. Cross-Validation

Cross-validation consists of splitting a sample for learning about the specification of a model and then testing that model on part of the sample. The approach selects several sub-samples from the data and uses these as *training* sets for estimating the parameters of the model (Hjorth, 1994). The remaining data set, called the *test* data set, is not used other than for assessing the forecasts obtained using the *training* data sub-samples. A similar approach, first introduced by Tukey as the *jack-knife*, consists of developing a sub-sample by taking all data points but one, then proceeding with estimation on that sample of size *n-1*. This jack-knife method is repeated several times to study the *stability* of the estimates; it is measured by changes in the variance of the estimates.

In model cross-validation, each training sub-sample is used to estimate the model parameters using methods such as the maximum likelihood. The accuracy of each model is determined by how well the models' estimates can predict the observed values in the corresponding subsets of data that were not used in the estimation. Cross-validation has been shown to approximate the ranking obtained using the AIC criterion under large sample approximations (Stone, 1977).

4.10.3. Direct or Reverse Regression?

Although exposure to an air or water pollutant such as benzene or arsenic can cause an adverse response, but *not* vice versa in those so exposed, many such relations may often not be as unidirectional as initially thought. In much empirical work, the *arrow of causation* is often ambiguous: $X \rightarrow Y$ can be just as plausible as $Y \rightarrow X$. Suppose that the question being addressed is the *nature* of the relationship between salary, education and sex. It is plausible to reason that salary depends on the education level and sex of the individual.

Specifically, as the level of education increases so should salary. It is also known that there is a (unjust) wage differential between women and men. Under these assumptions, an assessor might model the relation as a simple linear function, the *direct* regression:

Salary = $\alpha + \beta(Education) + \gamma(Sex) + error.$

The random variable is salary (education and sex are assumed to be measured with error, for simplicity) and ε is the random error of salary for the population. The objective of estimation is to obtain values for α, β and γ, from sample data. The parameter estimates from a sample:

$$\hat{Y} = \hat{\alpha} + \hat{\beta}(Education) + \hat{\gamma}(Sex) .$$

However, it may also be plausible to state that education depends on income and sex. This leads to the *reverse* regression:

$$Education = \alpha' + \beta'(Salary) + \gamma'(Sex) + error'.$$

In this reverse regression, the *prime* indicates that a completely new model is now considered. Which of these two models is correct? Estimation of a model by Kaldane and his co-authors yielded the following estimates:

Direct regression: $Salary|Education = 8.156 + 0.013(Education)$,

or:

Reverse regression: $Salary = 6.77 + 0.196(Education|Salary)$.

The noticeable difference between these results requires some resolution. Kaldane and co-workers suggest using Goldberg's test to resolve this ambiguity. The test involves an algebraic substitution between the direct and the reverse regressions.

4.10.4. Data Mining

There are different views of what data mining represents. Some are positive, some negative and some are neutral. The neutral view suggests that data mining is a means to make sense of the data particularly when theory has gaps and new insights are needed. Nonetheless, the warning given by Coase, reported by Peter Kennedy as *if you torture the data long enough, Nature will confess*, must be kept in mind. The concept of *data mining* suggests the assessor seeks to discover knowledge from empirical data in rather gross ways. This is not at all the case. The data being mined can have quite complicated structures and relations that go well beyond what can be discovered through flat database modeling. To do so, the assessor can specify an algorithm that defines knowledge as a threshold level. A metric must exist (and be unique) such that a pattern (e.g., a network, relation or other) can be related to that threshold. An expression that describes a pattern is then chosen as the simpler statement of all facts in the data, with a measure of uncertainty describing the threshold.

An aspect of data mining is opting for high coefficients of correlation and individually significant *t*-values associated with statistical tests of significance of the parameters of a regression model. It has been shown that relying on the *adjusted* R^2 can be misleading because these searches do not identify the correct specification of the model sought but, rather, explain local variability only. Kennedy reviews the research directed to understand the problem inherent in the search for high R^2 and indicates that randomly built models turn out to have both high R^2 and high t-values. The reason is that researchers who attempt to maximize these statistics commit a Type I error because of multiple sampling. A similar problem arises when independent analysts use the same data to investigate the relations between risk factors and increased risk. These sorts of searches are unsatisfactory and can be detrimental to risk management.

Data mining loses its pejorative connotation when there is a need to discover subsets of data that would not be known. This can occur when the databases are heterogeneous. The development and choice of the form of a model and the set of variables leading to the correct model that is relevant to the risk assessor can be aided by careful logical and statistical searches, particularly when the initial purpose of the search is *not* inferential. If the investigator limits her work to explaining, rather than inferring and forecasting, data mining does not present inordinate difficulties. Data mining methods can also be used for inferential purposes; in fact, they can account for uncertainty through probability measures, fuzzy measures and other representations of uncertainty. However, prior, domain-of-knowledge information is necessary in developing data mining work, particularly because, the problem inherent to data mining can be easily be forgotten, once the data have *spoken*.

The temptation to use exploratory data analysis as if it were sound theoretical modeling should be resisted, unless the methods are known to meet the standards of good modeling practice and be consistent with the relevant scientific knowledge. In this context, data can be studied using a number of statistical descriptions achieved by using methods such as stem-and-leaf diagrams, box-and-whisker diagrams, histograms and other methods that display the shape of the original data. It is also useful for splitting data into learning and testing sets for cross-validation. Non-parametric methods such as CART (discussed in Chapter 7) are also important in data mining.

Decision trees, such as those discussed in Chapter 3, can be useful in data mining because tree-pruning algorithms can reduce otherwise unmanageable trees to a manageable representation of events and their consequences. Several classes of methods from Artificial Intelligence can help to discover patterns and other structures in diverse data sets. Some of these methods do fall within the types of regressions that discussed in this chapter and can include such data reduction techniques as principal components, factor analysis, cluster analysis and so on (Selvin, 2001). The statistical basis for these methods presents additional information to the risk assessor. Other data mining methods include Bayesian Networks, also discussed in this book in Chapter 7.

Data mining also requires modeling, a formal language for expressing relations and criteria for searching the parameter space to obtain accurate estimates of those parameters. The search for the correct model, given the data, suggests using a family of models and then applying heuristic criteria to make the final choice. Artificial neural networks are an example of recent methods used in data mining. The concept is that stratified nodes with linear weighting of the inputs map to an output through a nonlinear function, such as a logistic model. The output from one stratum is the input into the next. Information propagates through the nodes, with errors being used to adjust the weight of the last output (Friedman, 1995).

Inductive logic programming, ILP, is another modern method of data mining that programs the logic of general relations to discover one or more specific relations, and can account for noisy data in a database (Dzeroski,1996). The construction of such programs requires developing logical arguments, predicates (meaning views or relations), rules and facts that are linked through a formal logic system that allows retractions (it is non-monotonic). ILP uses background information and learns about predicates (as relations), given other predicates (as views), from training on subsets of data. The process is deductive. A query to a database is deduced from that database according to the programming logic and restrictions that limit the extent of the query. The method has been applied to classify new samples of micro-invertebrates from samples taken in different layers of a river in different riverbed locations (Dzeroski, 1996). A combination of learning and programming logic were used to develop the symbolic model, which resulted in more than seventy rules for induction, and replicated the reasoning and results of human expert knowledge consistently.

4.11. CONCLUSION

In risk assessment, statistical models include independent variables from *a priori* knowledge of the risky situation being assessed. That knowledge may be biological or other, depending on what is studied. Statistical models can be descriptive, hypothesis generating or empirically causal. In the latter case, the results are causally associative. The important issue for the risk managers, whose decisions are supported by exposure-response models, is whether the results from modeling and estimation are sufficient to generate sound answers to the situations that managers confront. In this chapter, we have discussed several approaches to statistically modeling relations between different exposures, controlling for other factors associated with response. These models range from simple regression to time series analysis and require statistical estimation based on samples.

The Maximum Likelihood or other statistical methods provide the statistical framework for parameter estimation. For each statistical model, there are measures such as upper and lower confidence limits, goodness of fit and other statistics that can assess the performance of the model, given the data used in estimation. If parametric estimation is used, that estimation must be consistent with the distribution of the response variables. Non-parametric methods, which make mild distributional requirements, can be more data demanding than parametric methods.

We have also discussed how to choose between statistical models. This chapter contains both traditional statistical measures, such as the goodness-of-fit, as well as measures based on the entropy of information, such as AIC. Some of our findings are that:

1. All criteria for choosing between models require meeting the assumption that the fundamental structure of the models are understood and can be described mathematically.
2. The risk manager must be sure that model selection is sufficiently well-formulated that it can withstand peer review as well as legal and regulatory scrutiny.
3. The risk assessor and manager must be satisfied that the choice of a model based on statistical model selection criteria is not confused with statistical testing, such as tests that obtain under hypothesis testing.

Finally, many of the methods used for data mining can determine unusual patterns and can be quite useful to the risk assessor and the risk manager.

QUESTIONS

1. Consider the statement: *Time series models are descriptive; therefore, the researcher must use other models to develop causal associations.* Do you agree? Discuss your answer using statistical reasoning in no more than 600 words.
2. Why are time series useful to the environmental manager, relative to other regression models such as the multiple regressions, in assessing the risks of diseases characterized by long periods of latency or incubation? Answer with approximately 300 words.
3. Consider the assertion *Cox's Proportional Hazards Model (PH) can be used to assess survivorship functions.* Sketch an argument that supports this assertion and justify it. What would be an important use of the PH in risk assessment and how would such use benefit risk management? Answer with approximately 500 words.
4. Poisson and logistic regression models can use independent variables that depend on time. In other words: X can be replaced by X(t). What is the benefit of accounting for time explicitly? Use approximately 200 words.
5. Suppose that there is some literature to guide the selection of dependent (response) and independent variables (exposure and others) as well as the selection of the form (linear or non-linear) of the statistical model. You can use the information of Table 4.1 and 4.2. Could you determine whether that literature is sufficient to establish empirical causation for the risk manager so that she will provide funds for research that you think is necessary to determine empirical causation between exposure and response? Use about 600 words. (You may want to refer to Linhart H, W Zucchini, *Model Selection*, Wiley, New York, 1986, and Royall R, *Statistical Evidence, a likelihood paradigm*, Chapman & Hall, Boca Raton, 1999, provide extensive discussions of the principles and methods that can be used to answer this question.)
6. The K-M estimator can be used to test the survivorship function of a sample of individuals exposed to a mixture of air pollutants and a sample of individuals who are not exposed. Sketch these two functions assuming that exposure is deleterious to health while non-exposure is not. Could these two functions cross over? Explain why or why not using about 200 words.

7. Suppose that you can use a simple model of exposure and response but that you can specify a more complicated model based on the *physics* of your problem. The more complicated model includes more relevant variables than the simpler model. However, data to obtain estimates of the parameters of the more complex model are not available, nor will they be available in time to respond to an inquiry that a local group of concerned citizens has formed. You obtain some estimates of the parameters of your model using the simpler model. Do you think that the simpler model might be less accurate than the full model? Your manager asks that you go ahead and present your estimates in the public meeting because the estimates show no statistically significant results. *Should* you do it? If so, what caveats would you include in your work? Use about 600 words.

8. The time series analyses use autocorrelations and partial autocorrelations functions; their confidence intervals and other information having to do with goodness-of-fit. You fit an ARIMA (1,1,1) to the data and obtain a prediction. How did you use the ACF and PAF to justify your choice of ARIMA? Answer in 200 words.

9. Why should determining the size of the sample take place *before* the sample is taken? What aspects of this determination would you consider in developing your answer? Can you set up a simple set of relationships that show your reasoning? Answer with approximately 300 words.

SOME ABREVIATIONS USED IN AIR POLLUTION STUDIES

Abbreviations	Full Name
AOD	Airway obstructive disease
CI	Confidence interval
CO	Carbon monoxide
CO_2	Carbon dioxide
COPD	Chronic obstructive pulmonary disease
$FEV_{1.0}$	Forced expiratory volume in 1 second
FVC	Forced vital capacity
H_2S	Hydrogen sulfide
ICD	International classification of diseases
MAX.	Maximum
NO_2	Nitrogen dioxide
O_3	Ozone
OR	Odds ratio
PM_{10}	Particle with an aerodynamic diameter ≤ 10 mm
RR	Relative risk

S	Sulfur
SO_2	Sulfur dioxide
SO_3	Sulfite
SO_4	Sulfate
SDAs	Seven-day Adventist
TSP	Total suspended particulate matter

CHAPTER 5.

PROBABILISTIC DOSE-RESPONSE MODELS AND CONCEPTS OF TOXICOLOGY

Central objective: to describe probabilistic (stochastic) and toxicological dose-response models, and exemplify their practical uses in risk assessment

This chapter focuses on probabilistic dose-response models and their biological foundations, the second aspect of the essential element of damage functions (exposure-response being the first). Dose-response models use *stochastic* (probabilistic) processes that describe the initiation, progress and outcome of cancer, genetic, developmental and other diseases as a function of dose to specific biological targets, such as the DNA. Those models, which generally measure risks as hazard rates, relate the probability of an adverse outcome to doses in a mechanistic (that is, biologically motivated) way and can be used to assess risks at dose rates that are much below the experimental data. Thus, in risk management and public policy-making, dose-response models can determine acceptable, or tolerable, dose. We also discuss a class of statistical dose-response models, such as the probit or logit, used for assessing systemic adverse effects, such as liver damage, by describing the statistical distribution of individual thresholds for a particular health effect. However, these statistical models do not have the same fundamentally biological basis of stochastic dose-response models. The parameters of both types of models are estimated from experimental data. The models discussed in this chapter include:

- Single-hit and multi-hit models, in which a hit is an irreversible mutation caused by either ionizing radiation or a chemical that can probabilistically lead to cancer or other adverse health outcome
- The multistage (MS) family of models, including the *linearized* multistage model (LMS) used in regulatory risk assessment, in which probabilistic transitions are used to model the events turning non-cancerous cells into a cancer
- The Weibull and other probabilistic models
- The Moolgavkar-Venzon-Knudsen (MVK), stochastic cellular birth-death cancer model, which is a more theoretically complete cancer model than the multistage family of models
- Statistical threshold models, applied to toxicological outcomes
- Benchmark Dose or Concentration, BMD and BDC, methods from current US environmental regulatory work that use either statistical exposure-response or stochastic dose-response models

The contents of this chapter are depicted in the flowchart that follows:

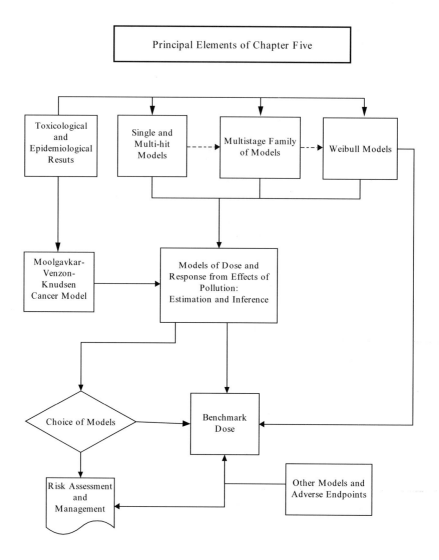

Because dose-response models often use the results from animal studies, this chapter also includes a discussion of some of the principles of toxicology that are relevant to estimation. More specifically, *exposure* to a pollutant may not measure the mass of a toxic or carcinogenic agent reaching a specific organ, such as the liver, a cell or the DNA. Dose, on the other hand, can account for biochemical or other changes within the human body that alters mere exposure. Thus, dose is the mass of the toxicant (or its metabolic byproducts) reaching a specific organ or tissue at the point where the actual adverse effect occurs (e.g., an irreparable mutation consisting of a base substitution in a particular gene).

5.1. INTRODUCTION

Dose is the mass of the chemical that is the likely cause of a specific response at either the organ or cell levels. Figure 5.1 depicts some of the processes that can result in adverse response.

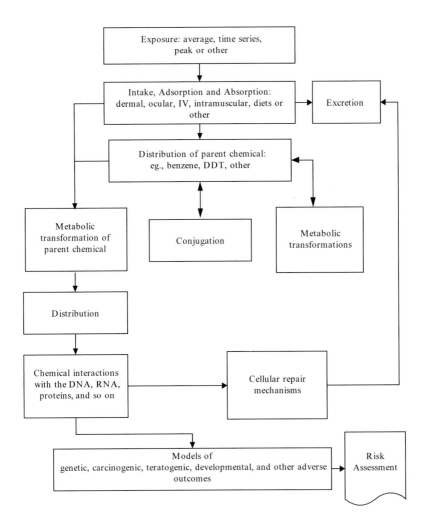

Figure 5.1. Principal Processes that Can Lead to a Toxicologically Adverse Outcome

The basis of dose-response model is biological realism. The choice of a dose-response model and the data used in a risk assessment are the result of judgmental choices. As may be expected, those choices will change as the scientific state-of-knowledge changes. Exposure (e.g., in mg of the toxic substance/m³ of air inhaled per hour, [mg/m³]) can also be used in dose-response models. However, dose to the target organ is the more accurate measure of the amount of a chemical because it can account for the transformations that exposure undergoes in the human body. The data sets used to estimate the parameters of the models discussed in this chapter consist of toxicological, animal or other experimental studies that do not involve humans. For dose-response, those studies have a dual purpose. The first is that they provide part of the biological rationale for developing dose-response models. Each study makes a different and often indispensable contribution to understanding the process of a disease and its modeling. The second purpose is that the data generated by a study are used to estimate the parameters of a dose-response model. These two purposes, with exposure assessment, are essential to risk assessment because they form the causal link between exposure, dose and response.[1] The relevance of toxicological studies to risk management is that they provide important scientific evidence as the legal basis of risk assessment. Figure 5.2 depicts the role of scientific knowledge in the development of dose-response models.

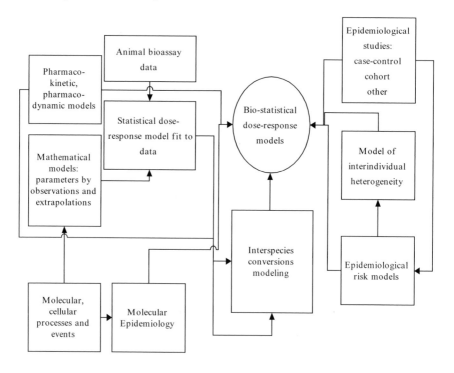

Figure 5.2. Scientific Knowledge in Dose-Response Modeling

---

[1] That is, *exposure → dose → response.*

US regulatory practice makes a distinction between cancer and non-cancer dose-response models. For example, for cancer risk assessment, exposure to chemical contaminants is often assessed using the (linearized) multistage (LMS) dose-response model (US EPA, 2003). This model is fit to animal data and is linear at low doses. The probability of carcinogenic response is the individual lifetime cumulative probability. The upper *95%* confidence limit of the estimated slope (the *linear* term of the LMS) is used to determine the tolerable dose, for a given regulatory level of tolerable risk (e.g., $1*10^{-6}$).

5.1.1. Cancer Risk Assessment

To illustrate a risk assessment process based on dose-response models, we consider the process for cancer risk assessment (US EPA, 1996). It consists of four steps:

1. Modeling the relationship between observed exposure and frequency of response in the sample through plausible exposure (or dose) response models,
2. Using inferential statistical methods and criteria for choice of model, such as the Akaike Information Criterion,
3. Extrapolating from the experimental results to the domain where observations are not available,
4. Using the linear exposure-response with *interpolation* to zero, when the procedure (steps 1, 2, and 3) is inapplicable, as a default.

In addition to these steps, partly because the state-of-scientific-knowledge about cause and effect is often limited, a number of regulatory agencies have attempted to make explicit those limitations. The International Agency for Research on Cancer, IARC, provided one of the first descriptive classification schemes to clarify the state-of-knowledge about the potential carcinogenicity of a chemical. Table 5.1 includes an example of a more recent carcinogenicity classification scheme (US EPA, 1994).

Table 5.1. Example of Criteria for Evidence of Carcinogenicity (US EPA, 1994)

Category	Nature of the Evidence
Category I: Might pose a carcinogenic hazard to humans under any conditions of exposure. The magnitude of the risk depends on dose-response relationship and extent of human exposure.	Evidence of carcinogenicity in either human or animal studies (strength of evidence varies …). No information available to raise doubts about the relevance to human model or results. No information available to raise doubts about the relevance of conditions of exposure (route, dose, timing, duration, etc.) under which carcinogenic effects were observed to conditions of exposure likely to be experienced by human populations exposed environmentally.
Category II: Might pose a carcinogenic hazard to humans, but only under limited conditions of exposure. Whether a risk exists in specific circumstances depends on whether those conditions exist. Dose-response and exposure assessment must be completed to identify	Evidence of carcinogenicity in either human or animal studies (strength of evidence varies …). Scientific information available to show that there are *limitations* in the conditions under which carcinogenicity might be expressed, owing to questions of relevance to humans of the animal models or results or relevance of the conditions of exposure (route, dose, timing, duration, etc.) under

conditions under which risk exists.	which carcinogenic effects were observed to conditions of exposure likely to be experienced by human populations exposed environmentally.
Category III: Notwithstanding the evidence of carcinogenicity in animals, not likely to pose a carcinogenic hazard to humans under any conditions.	Evidence of carcinogenicity in animal studies. Scientific information available to show that the animal models or results are not relevant to humans under any conditions.
Category IV: Evidence available to demonstrate lack of carcinogenicity or no evidence available.	No evidence of carcinogenicity or evidence of non-carcinogenicity (weight of negative evidence varies).

According to the US EPA (2003), there are five recommended standard explanations of carcinogenicity: carcinogenic to humans, likely to be carcinogenic to humans, suggestive evidence of carcinogenic potential, inadequate information to assess carcinogenic potential, and not likely to be carcinogenic to humans. Moreover, more than one conclusion can be reached for the risk of a specific agent. For instance, an explanation may state that an agent is likely to be carcinogenic by inhalation but not be likely to be carcinogenic by ingestion. Table 5.2 contains a summary of hazard assessment, in which *risk characterization* consists of *dose-response assessment* and *exposure assessment*; *risk characterization* relates to the guidance on dose-response modeling (US EPA, 1994).

Table 5.2. Steps in Hazard Assessment (US EPA, 1994)

Inputs	**Outputs**
Relevant biological information	Mode(s) of action
Physical and chemical information	Conditions of hazard a. Route b. Pattern
Tumor findings a. Human b. Animal	Human hazard potential a. Descriptor b. Narrative
Characterization of summary from the above information	Guidance on dose-response a. Biologically-based model b. Default model (linear, nonlinear, or both)

5.1.2. Cancer Slope Factors (CSF), Reference Doses (RfD), and Concentrations (RfC)

The cancer slope factor, CSF, is generally multiplied by the exposure estimate to generate an estimated risk. Thus, if CSF is zero, risk is also zero. In particular, if the calculations use maximum likelihood estimates (MLEs), rather than the confidence limits around those estimates, this should be clearly stated. The US EPA states that, according to Rhomberg, the EPA's method is also used by OSHA, but OSHA presents and focuses on the MLE dose-response curve instead of the UCL. Similarly, the CPSC also uses the same model as EPA, but modifies its constraints to obtain an MLE of the key parameters of the dose response function that is linear at low doses. Specifically (US EPA, 2004; also see sub-section 5.7.1 and 5.7.2) this agency states:

"When developing a non-cancer reference value (i.e., an RfD or RfC) for a chemical substance, EPA surveys the scientific literature and selects a critical study and a critical effect to serve as the point of departure for the assessment. The critical effect is defined as the first adverse effect, or its known precursor, that occurs in the most sensitive species as the dose rate of an agent increases …. Such a study, whether an occupational human study, a deliberately dosed animal study, or some other study, typically involves exposure at a range of doses. The highest exposure level at which there are no statistically or biologically significant increases in the frequency or severity of adverse effects between the exposed population and its appropriate control is … called the 'no-observed-adverse-effect level (NOAEL).' … The NOAEL is divided by appropriate uncertainty factors, UFs, (e.g., to account for intraspecies variation or study duration) to derive the final reference value. If NOAEL cannot be identified, then a 'lowest-observed-adverse-effect level' (LOAEL) … is identified instead. A LOAEL is the lowest exposure level at which there is biologically significant increases (with or without statistical significance) in frequency or severity of adverse effects between the exposed population and its appropriate control group. The NOAEL is generally presumed to lie between zero and the LOAEL, so an UF (generally 10 but sometimes 3 or 1) is applied to the LOAEL to derive a nominal NOAEL. Other factors are then applied to derive the reference value."

Furthermore, (US EPA, 2004):

"The Risk Assessment Forum RfD/RfC Technical Panel Report … defines an uncertainty factor as: one of several, generally 10-fold, default factors used in operationally deriving the RfD and the RfC from animal experimental data. The factors are intended to account for (1) variation in sensitivity among the members of the human population; (2) the uncertainty in extrapolating animal data to humans; (3) the uncertainty in extrapolating from data obtained in a study with less-than-lifetime exposure to lifetime exposure; (4) the uncertainty in extrapolating from a LOAEL rather than from a NOAEL; and (5) the uncertainty associated with extrapolation when the database is incomplete. … Some investigators evaluated the accuracy and limitations of allocating a value of 10 for each area of uncertainty …. For example, the subchronic-to-chronic UF, some conclude, is probably closer to a two-to threefold difference of uncertainty (95% of the time) and a UF of 10 should be considered as a loose upper-bound estimate of the overall uncertainty. Others … conclude that a 10-fold interspecies UF could account for all animals to human differences and likewise for interhuman variability.

There are opposing arguments that EPA's application of UFs may not be conservative enough … . EPA applies UFs in health assessments based on the available data and the scientific judgment of EPA risk assessors and internal and external peer reviewers. In cases where chemical-specific data are lacking, each UF is typically no greater than 10. For example, the majority of IRIS Toxicological Reviews provide justifications for the individual UFs selected for a particular chemical substance. … For several years, EPA has used a more qualitative approach to modify the usual 10-fold default values. For example, in deriving inhalation RfC values, the interspecies variability UF of 10 is used in the absence of data, where the distribution is assumed to be log-normally distributed. While EPA has not yet established guidance for the use of chemical-specific data for derived UFs, the reference concentration methodology guidance … provides opportunities for using data-derived interspecies UFs by subdividing the factor of 10 to allow for separate evaluations of toxicokinetics and toxicodynamics. The advantage to such subdivision is a default UF of *10* for interspecies variability that can now be reduced to 3 when animal data are dosimetrically adjusted to account for toxicokinetics …"

5.2. RISK ASSESSMENT THROUGH DOSE-RESPONSE MODELING

The *dose-response characterization* provides the dose-response information needed quantitatively to identify risk (U.S. EPA, 2000). Specifically, this is achieved by "including

- Presentation of the recommended estimates (slope factors, reference doses, reference concentrations),
- A summary of the data supporting these estimates,
- A summary of the modeling approaches used,
- The (point of departure) POD narrative,
- A summary of the key defaults invoked,
- Identification of susceptible populations or lifestages and quantification of their differential susceptibility, and
- A discussion of the strengths and limitations of the dose-response assessment, highlighting significant issues in developing risk estimates, alternative approaches considered equally plausible, and how these issues were resolved."

The US EPA (2003) states that quantitative human health risk estimates should include a descriptive weight of the scientific evidence to convey the qualitative uncertainty about the agent's adverse effect. The evaluation of the *mode of action* of a carcinogen uses a weight-of-evidence approach. Modes of action are associated with the chemicals in the database available from the US EPA. An example of a mode of action is mutagenicity (i.e., for benzo(a)pyrene and vinyl chloride). More specifically, a mutagenic mode of action for a carcinogen is developed from the scientific evidence indicating that one of its metabolites either is DNA-reactive or has the ability to bind to DNA.

The US EPA (1992) describes regulatory risk calculations based on dose-response models as follows:

"… when calculating risks using doses and 'slope factors', the risk is approximately linear with dose until relatively high individual risks (about 10^{-1}) are attained, after which the relationship is no longer even approximately linear. This result follows from the fact that, no matter how high the dose, the individual risk cannot exceed *1*, and the dose-risk curve approaches *1* asymptotically."

The model for assessing the probability of response, the risk, from a dose *d* of a noxious agent resulting in a specific adverse health outcome is $Pr(d) = f(d)$. This function can be linear, nonlinear or have a mix of linearity and nonlinearity. Let the linear (or approximately linear at low doses) risk model be:

$Risk(d) = R(d) = (slope)(dose) = b*d$.

In this model, risk is measured as a cumulative lifetime probability of the adverse effect; dose, *d*, has units of mg/kg/day. The *slope* of the model, *b*, has units of $[(\text{lifetime cancer probability}(/mg/kg/day))]^{-1}$.

Example 5.1. Let the tolerable risk level (set by regulation) be $1.00*10^{-5}$ and the estimated slope be given as *0.010* [mg/kg/day]$^{-1}$. The tolerable dose equals *0.001* [mg/kg/day].

More realistically, some dose-response models are (approximately) linear only at low-dose, low probability of response, and then become non-linear, as depicted in Figure 5.3. Because this model is a *probability* model, the vertical axis is constrained to be within *0* and *1*: that is, $0 \leq R(d) \leq 1$.

This constraint is imposed because extrapolations from the data would otherwise exceed the range (*0*, *1*) making the probability numbers either negative or greater than unity, which is inconsistent with the definition of probability. Moreover, this function can be supra-linear (be above a straight line from the origin), sub-linear (be below a straight line from the origin), or have a threshold.

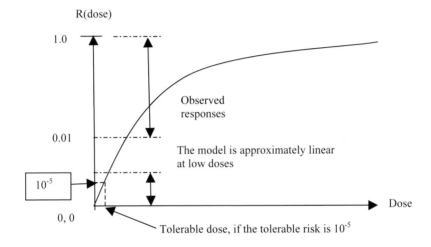

Figure 5.3. Dose-Response Model with Partial Linearity at Low Doses

In Figure 5.4, the tolerable lifetime probability of response is set to be 10^{-5}; accordingly, here is a dose and risk combination that can be tolerable. This model, if biologically accurate, is valid from *0* to some very large dose. Estimation of the *slopes* (not just the linear portion of the slope) uses the MLE applied to a sample. If a level of tolerable lifetime risk is given through regulation, the formula is: $d = R(d)/b$. Some environmental agencies have taken the upper *95%* confidence limit as a practical way to deal with uncertainty, at a specific value such as *1%* response, and then linearly interpolated a straight from that point to the (*0*, *0*) point. This is stated to provide a reasonable *upper bound* on the linear part of the dose-response model. Figure 5.4 depicts this approach.

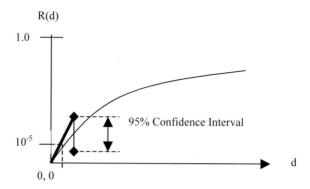

Figure 5.4. Hypothetical 95% Upper Confidence Limit and Interpolation to Zero

The interpolated line lies above the slope estimated by the MLE and, therefore, the risk level (which is stated as 10^{-4} in Figure 5.5) intersect the interpolated line before the MLE line. The issue that underlies each of these models is the effect of a choice of model on the acceptable level of exposure or dose, for a given level of acceptable risk. Figure 5.5, illustrates the issue. This figure illustrates the point that, for a constant risk level (10^{-4}), the tolerable (or acceptable) dose that corresponds to this can be very different.

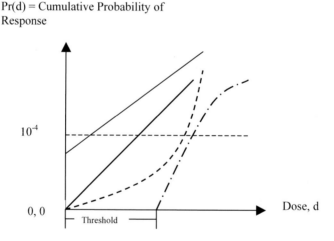

Figure 5.5. Differences in the Tolerable (Acceptable) Doses for the Same Tolerable Risk

Thus, the choice of model affects the level of clean-up that a risk manager may choose and thus the costs incurred at that level of risk protection. Therefore, the rationale for a choice can be critical. For example, some of the arguments that oppose the existence of a threshold for cancer use the single-hit model. That is, an irreversible mutation that is not repaired by a cellular process results in cancer, but it does so with an infinitesimally small probability at extremely low doses. The cell that has inherited a permanent damage eventually becomes cancerous through *clonal* replication. The event leading to a specific cancer is rare: that irreparable mutation is rare because of repair mechanisms. Many diseases of concern to the risk assessor can require more complicated cause and effect relations than the hit-based cancer theories. For example, Sivak, Goyer and Ricci (1987) reviewed the evidence of carcinogenicity of non-genotoxic carcinogens such as TCDD and dichloromethane. Their review suggests that a threshold model may be appropriate when carcinogens act as promoters, rather than initiators: managing health hazards requires considering the causal basis of those models.

Cancer models have been an important aspect of environmental and health risk assessment. The early rationale for developing risk models assumed that a single irreversible and irreparable interaction (a *hit*) between the carcinogen and the DNA could result in a cancer, with very low probability. These models result from studies of the effect of ionizing radiation, and were later extended to include chemical carcinogens. The probabilistic development of the single-hit and other hit-based models is as follows (ICRP, 1996; US EPA 1996). The Poisson distribution can be used to model rare events (rare hits) from a carcinogen to the DNA: dose does not affect response and the hits are independent. Let E represent exposure or dose. Given this distribution, the probability of survival is:

exp(-kE),

in which $k = 1/E_0$, with E_0 being the exposure that causes an average of a hit per cell at risk. Therefore, the probability of an adverse response is:

1-exp(-kE).

Under the Poisson assumptions, several cells have to be hit independently before irreparable damage occurs. The probability of failure is:

$\Pi_i[1-exp(-kE_i)]$.

It follows that the probability of survival is:

$\{1-\Pi_i[1-exp(-kE_i)]\}$.

The hit theory of carcinogenesis can account for more than one hit on the biological unit at risk. The Poisson distribution yields the following results. The probability of *0* hits reaching the unit at risk is *[exp(-kE)]*. The probability of *1* hit is:

(kE)exp(-kE).

The probability of *n-1* hits is:

$[(kE)^{n-1}/(n-1)!exp(-kE)]$.

It follows that the probability of surviving *n-1* hits is:

pr (surviving n-1 hits) = $exp(-kE)[1+kE/1!+(kE)^2/2!+...+(kE)^{n-1}/(n-1)!]$.

5.2.1. Toxicological Dose-Response Models

Some quantal (the proportion of individuals responding at a given dose level is the measurement for the response variable) toxicological dose-response models describe the heterogeneity of individual responses rather than the biological process leading to the disease. Each member of the population has a deterministic dose-threshold, which is a level of exposure below which no response can *biologically* occur. What is of interest is a model that accounts for the proportion of individuals exposed to a specific exposure, e_0, and respond. Intuitively and practically, different individuals exposed to the same exposure, e_0, will respond differently. The proportion of the population responding (or failing at e_0) is π_0. In practical work, raw exposure measurements are transformed by taking their natural logarithm: $d_i = ln(e_i)$.

Figure 5.6 depicts the cumulative distribution of the transformed variable for a generic toxicological dose-response model, without the data to which the model is fitted. In this figure, $F(d)$ is the cumulative distribution function of the transformed random variable D, $(d \in D)$. Under this transformation, d is approximately normally distributed with mean μ and variance σ^2. The Figure depicts the proportion of individuals, pr_i responding at dose d_i (specifically, at d_i the proportion will be equal to or less than pr_i).

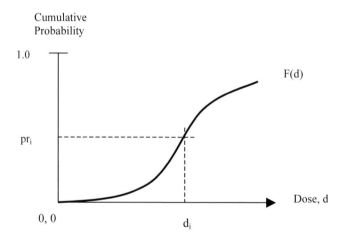

Figure 5.6. Typical S-Shaped Toxicological Dose-Response Model

A further transformation of the values of the random variable D, $(d \in D)$, is as follows. The steps of the transformation require setting:

$$D = Y = (x_i - \mu)/\sigma.$$

This change results in a new random variable, Y, which is also normally distributed, but now has mean zero and variance that equals 1. The proportion of the population responding at y_0 is now π_0, that is $\pi_0 = F(y_0)$. It follows that $y_0 = (d_0 - \mu)/\sigma$. We obtain the *probit* transformation when we modify the values, y_i, of the random variable Y, as:

5(y_i+5).

This change makes the values of Y usually positive (Johnston, 1991). The reasons for making these transformations have been supplanted by nonlinear estimators implemented in commercial software packages.

The statistical basis for a threshold model is experimental: the *s*-shaped function is fit to experimental data. The distribution of the individual sensitivity to disease is modeled as a function of response to *toxic* exposures. Thus, unlike the hit-based probabilistic dose response models, statistical models do not have a biological basis: they are purely experimental. Different toxicological health endpoints, for example diseases A and B, are described by different dose-response functions, each having the same general s-shape, but each being shifted along the dose axis. Each function is parameterized from different data sets. Figure 5.7 depicts these concepts, without showing the experimental data necessary for estimating the parameters of each model; the two *s*-shaped functions are hypothetical.

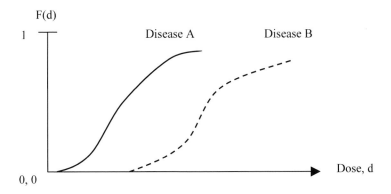

Figure 5.7. Dose-Response Functions for Two Hypothetical Different Diseases

5.3. MULTISTAGE (MS) DOSE-RESPONSE MODELS: Focus on Cancer

This section contains a discussion of the biological basis of cancer dose-response models and illustrates how it is expressed probabilistically. For example, the probability that zero dose equals zero response, *pr(d = 0) = 0*, can be used to suggest that no risk is acceptable. Because this suggestion is unrealistic, most jurisdictions have developed tolerable or acceptable risks. These are individual probabilities (or cumulative probabilities, depending on the context) that can range from $1.0*10^{-6}$ to some larger value, such as $1.0*10^{-4}$.[2] These lifetime tolerable excess or extra risk

---

[2] There are several reasons for tolerating a larger probability of death or injury in occupational settings. For example, workers are informed about the hazards they can face (e.g., by the government, by the enterprise for which they work, trade unions or other), may be compensated for the excess risk, face those hazards voluntarily and are protected by workers' compensation laws. The same cannot be said for members of the public.

values (over background) are determined by legislation or by secondary legislation such as regulations. These risk values are used to set clean-up levels or environmental and occupational health standards or guidelines.

5.3.1. A Regulatory Method

The multistage dose-response model is used for cancer risk assessment. For example, the US EPA has based its cancer risk assessment policies (the Guidelines for Cancer Risk Assessment) on the *linearized* multistage dose-response model, LMS. This model is biologically consistent with solid cancers, such as carcinomas, but it has also been used to model other cancers, such as leukemia. The LMS has been the mainstay of US EPA work on regulating exposure to the risk of cancer from airborne and other forms of environmental exposure (Ricci, 1985; US EPA, 2003). Let us first introduce two definitions (Ricci, 1985), in which dose, *d*, is stated in units of *mg/kg/day*. These are:

A. The *additional risk* at dose *d* is:

$[pr(d)-pr(d = 0)]$.

B. The *extra risk* at dose *d* is:

$[pr(d)-pr(d = 0)]/[1-pr(d = 0)]$.

The extra risk *linearized* multistage model, LMS, (developed by the US EPA, which we follow including their symbols), is a lifetime cumulative distribution of adverse response given dose:

$$[pr(d)-pr(d = 0)]/[1-pr(d = 0)] = 1-exp[-(\Sigma q_i d^i)], q_i \geq 0, i = 1, 2, ..., k.$$

In this model, d_i is dose and q_i is the parameter to be estimated from experimental results; the parameters are constrained to be greater than or equal to zero. This means that the estimation must be such that the estimated q_i cannot be negative. The LMS model is nonlinear at high doses. For most practical risk assessments, all that needs to be estimated is the value of q_i, which is done using the method of the maximum likelihood. At sufficiently low doses:

$$[pr(d)-pr(d = 0)]/[1-pr(d = 0)] = q_1 d^1.$$

This is the basis for the linear model, $R(d) = qd$, exemplified earlier. The approximate upper and lower 95% confidence limits on the linear parameter of the extra risk, $R(d) = q_1^* d$ are ($q_1^* d$) and R/q_1^*, respectively. The US EPA has defined the q_1^* term of the LMS as the 95% upper confidence limit on q_1. This term has taken on an important regulatory significance, as anyone can attest by reading the US EPA IRIS database (US EPA Integrated Risk Information System, National Center for Environmental Assessment)[3]. At low dose, the terms of the polynomial with

---

[3] http://www/epa.gov

exponent greater than *1* tend to zero. The model is: $Pr(d) = q_0 + q_1 d^1$, in which q_1 is the *potency* of the carcinogen. Consider the IRIS database on benzene (CAS Number 71-43-2). The cancer slope factor is $2.92*10^{-2}$ [mg/kg-day] and the unit risk factor, URF, is $8.5*10^{-6}$ [μg/m³]. Benzene is classified as a class *A* carcinogen, namely a known human carcinogen. Therefore, the inhalation risk for exposure to *1* μg/m³ of airborne benzene is:

$$R = 8.5*10^{-6}[\text{individual lifetime probability}/\mu g/m^3]*1.00[\mu g/m^3] = 8.5*10^{-6}.$$

In general, dose is a function of inhalation, concentration and duration of exposure. For example, standard assumptions are that the individual adult male weighs *70* kg, inhales *20* m³/day of air and his duration of exposure occurs for *70* years, *24* hours/day. These assumptions can be changed to reflect additional information.

Example 5.2. The estimated slope (the q_1 term of the linearized multistage model) for arsenic, a known human carcinogen, is *15.00* [(lifetime probability)/(mg/kg-day)]. This is the estimated value of the slope, based on epidemiological data and a dose-response model that is linear at low doses.

If exposure were for a shorter period of time than the expected animal lifetime, the shorter duration of exposure can be accounted for by multiplying the estimate of the linear parameter of the LMS by the factor:

[<animal life>/(duration of experiment)³],

in which <.> means expected value. This factor is developed from empirical knowledge about the function relating the incidence of cancer to age. For human exposures that are different from the expected lifetime exposure of a human (*70* years, by assumption), the calculation is simply the ratio of the length of exposure to the *70* years exposure. For instance, for *4* years of exposure, *6* hours per day, *165* days per year, the factor is: *4*[years]**6*[hours/24 hours]**165*[days/365days]/*70*[years] = *0.00065*.

The US EPA (1992) states:

"Consider a population of five persons, only one of whom is exposed. ... Assume a lifetime average daily dose of *100* mg/kg/day that corresponds to an individual risk of $4*10^{-1}$. Increasing the dose five-fold, to *500* mg/kg/day, would result in a higher individual risk for that individual, but due to nonlinearity of the dose-response curve ... . The average dose for the five persons in the population would then be *100* mg/kg/day. Multiplying the average risk of $4*10^{-1}$ by the population size of five results in an estimate of two cases, even though in actuality only one is exposed. Calculating average individual dose, estimating individual risk from it, and multiplying by the population size is a useful approximation if all members of the population are within the approximately linear range of the dose-response curve, this method should be not be used if some members of the population have calculated individual risks higher than about 10^{-1}, because the calculation overestimates the number of cases."

Example 5.3. A subset of experimental data from a lifetime experiment, in which rats were studied for evidence of cancer from exposure to benzene, is (Cox and Ricci, 1992):

Dose of benzene, mg/kg/day	0	25	50	100
Probability of response (number responding/number at risk)	0 (0/50)	0 (0/50)	0.02 (1/50)	0.14 (7/50)

These data are in a form that can be used to estimate the parameters of the LMS. The result (obtained by the MLE) yields the following LMS model: $Pr(d) = 1-exp(-0.15*10^{-6}d^3)$. This is a pure cubic model, in which the lower order exponentials (the linear and the quadratic terms) statistically turn out to be zero.

The linearized multistage model of dose-response can be formulated with a threshold by setting the probability of response at zero below (less than) a specific dose level, d_1, as follows:

$$Pr(d) = 1-exp[-q_1(d_0-d_1)^2-q_2(d_0-d_1)^3].$$

Pr(d) is used to indicate the fact that we are dealing with a cumulative distribution. Note that $pr(d) = pr(0)$ for $d \le d_1 = 0$.

Example 5.4. The following two multistage models are the result of fitting to the data of the skin paint experiment to study the carcinogenicity of benzo(a)pyrene of Lee and O'Neil (Crump, 1984):

Model 1. Threshold, $d < 6$, MS model:

$$Pr(d) = 1-exp[-2.3*10^{-4}(d-6)^2-4.1*10^{-7}(d-6)^3],$$

for $d \ge 6$, *0* otherwise.

Model 2. Low dose linearity MS model:

$$Pr(d) = 1-exp[-9.5*10^{-4}d-0.15*10^{-5}d^2-2.8*10^{-6}d^3].$$

The calculations of expected response in the range of the experimental data are:

Dose (μg/week)	Number of animals tested	Number of animals responding	Expected response, Model 1	Expected response, Model 2
6	300	0	0	0.0071
12	300	4	2.50	5.90
24	300	27	22.20	21.90
48	300	99	105.0	101.40

The EPA (2003) also provides the age-related adjustments:

1. For exposures before *2* years of age, the EPA uses a *10*-fold adjustment. It approximates the median tumor incidence ratio from juvenile or adult exposures in the repeated dosing studies. This adjustment accounts for pharmaco-kinetic and other differences between children and adults,
2. For exposures between *2* and *15* years of age, the EPA uses a *3*-fold adjustment. It includes an intermediate level of adjustment that is applied after *2* years of age,

when pharmacokinetic processes mostly resemble those of adults, through *15* years of age,

3. There is no adjustment for exposures after *15* years of age; the potency factor is that used for calculating cancer response in adults.

5.3.2. Discussion

The cancer slope factors are results obtained either from animal studies or from epidemiological studies of adult populations. The effect of early exposure is assumed the same, regardless of the age of the exposed individual. The EPA (2003) states that *studies* (that generate the data for dose response studies) *do provide perspective on the standard cancer risk assessment averaging ... and they contribute to concerns that alternative approaches for estimating risks from early childhood exposure should be considered.* The EPA concludes that existing animal results support the conclusion that there is greater susceptibility for the development of tumors as a result of exposures to chemicals acting through a mutagenic mode of action, when exposure occurs early in the life of the exposed. Thus, a risk assessment based on cancer slope factors, accounting for early-life exposure, appears feasible. For non-mutagenic modes of action, the available data suggest that new approaches need to be developed for early exposures as well. It follows that, as the EPA states, what we require is data from either human epidemiological studies on childhood exposures or studies with rodents involving early (postnatal) exposures. The EPA's study indicates that theoretical analyses and analyses of *stop studies* can contribute to decision-making and to scientifically informed policy. Theoretical analyses suggest that the differential sensitivity to dose depends in part on the mode of action (i.e., where, in the cancer process, the chemical acts) and that the lifetime average daily dose may underestimate or overestimate the cancer risk when exposures are time-dependent. *Stop-studies* begin exposure at the standard post-weaning age, but stop exposure after some varying periods (months), and use doses as high as or higher than the highest dose used in the two-year exposure. New methods (e.g., from genomics) can provide additional data and insights to a mutagenic mode of action.

5.3.3. Comment

The biological basis of the multistage model of disease is depicted in Figure 5.8.

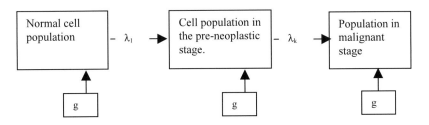

Figure 5.8. Cellular Biology of the Multistage Model

In this diagram, g is the cell growth rate and λ_i is a transition rate between these stages *without* the effect of an average dose d. In the model, the stages are traversed in the order shown, the transition rates are small, independent and constant, and cells grow at constant growth rates. The transition rates to account for an exogenous (meaning inhaled or ingested) carcinogen is obtained by setting $\lambda_i = (a_i + b_i d)$. The biological basis of this model has biological limitations (Cox, 2002). The principal ones include the following (Doll and Peto, 1981; Cox, 1995; Ricci, 1985):

- The effect of transient or peak doses are unaccounted for
- More than a single mutation may be required to trigger a transition
- Cell division rates can be dependent
- The dependency between exposure and dose may not be linear
- Cell death is unaccounted
- The size of the population of cells is not constant

Some of these limitations can however be relaxed (Ricci, 1985; Cox, 1995). The US EPA extended the work done by Doll and Peto in the early 1960s to develop the multistage model used in current cancer risk analysis. Its hazard function, including the effect of the dose of a chemical agent, d, and age t, is:

$$h(t) = (N/k!)[(a_1 + b_1 d)(a_2 + b_2 d)...(a_k + b_k d)t^{k-1}],$$

In which d is the dose rate of the chemical carcinogen and k is the number of stages a specific population of N cells must go through in the process of becoming malignant. The term $(a_i + b_i d)$ is the transition rate at which cells moves from stage $i\text{-}1$ to stage i through exposure to the chemical. Cox (1992) derives the multistage model from the cumulative hazard function, $H(d) = [q_0 + \Sigma(q_i d^i)]$, in which d is the lifetime average dose. The quantal multistage formula used in the earlier discussions follows from the relation $Pr(d) = 1 - exp[-H(d)]$. The approximate form of the MS model assumes that the cells at risk are rarely affected. This assumption is unnecessary in the exact formulation. Cox (1995) has obtained the exact formula for the (two-stage) multistage model. The relevance of the exact formula is that, when fitted to experimental data, it is the best model, if the multistage model is appropriate. Such knowledge is determined from molecular, cellular and physiological understanding of the specific disease, as well as from epidemiological data. The formula for the exact two-stage model (in which the cancer process consists of initiation and progression) is (Cox, 1995):

$$E(Pr) = 1 - exp\{-(\mu_1 LN)[1 - exp(-\mu_2 T)]\}.$$

In this equation, $E(pr)$ is the unconditional expected probability of cancer, μ_1 and μ_2 are mutation rates for normal cells and initiated cells, respectively; L is the number of periods for μ_1, N is the number of normal cells and T is the final time period in which μ_2 applies. With these symbols, the two-stage multistage model is: $Pr(response) = [1 - exp(-\mu_1 \mu_2 LNT)]$. This is a function that is convex-to-the-origin. In terms of the cumulative probability of cancer from a constant dose rate d, the exact formula (Cox, 1992), using linear transition probabilities, is:

$$Pr(d) = 1 - exp(\{[-(a_1 + b_1 d)]\}\{1 - exp[1 - exp(a_2 + b_2 d)]\}).$$

The exact formula for the MS dose-response model for risk assessment decreases the potentially large differences in the estimated risk, relative to the LMS, because:

1. The representation of the biology of the cancer for a specific chemical agent, through the proper representation of its effect on cell populations, is accurate,
2. The exact formulation may not have the same shape as the approximate solution used in regulatory risk assessment,
3. At low doses, the exact formulation will generally yield a smaller risk than the approximate formula of the MS model,
4. The shapes of the cumulative hazard functions will vary from convex to concave, depending on the experimental data to which the exact solution is fitted.

5.4. MOOLGALVKAR, VENZON AND KNUDSON (MVK) CANCER MODEL

The MVK model (Moolgalvkar, Dewanjii and Venzon, 1988,) is a two-stage stochastic model that accounts for cell growth, death and differentiation. Figure 5.9 depicts the two-stage MVK cellular process. It consists of a process in which two adverse and irreparable events must occur for normal cells to become malignant. The events may be mutations or other effects inherited by the cells. The cellular process consists of two stages, excluding the stage in which cells are normal) and the following transition rates, [cell/time]$^{-1}$, as follows:

1. Cellular death or differentiation, β_i, (a carcinogen decreases it),
2. Cellular division into a normal or pre-malignant cell, μ_i, (a carcinogen increases it),
3. Cellular division into two normal cells, α_i, the mitotic rate, (a carcinogen can increase it).

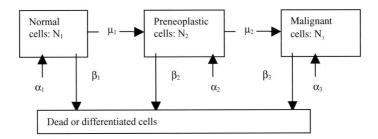

Figure 5.9. Cellular Biology of the Two-stage MVK Model

The parameters of the MVK model can be dose-dependent. As Figure 5.9 depicts, the MVK model can represent cell proliferation due to exposure to a chemical that aids such proliferation and can account for different cell division rates. The assumptions are: cancer is a two-stage process; cellular transformations are independent and that, once a cell becomes malignant, potentially cancerous cells proliferate independently of the normal cells resulting in a detectable cancer. The MKV is a stochastic model of cancer; its hazard function is:

$h(t) = \mu_2(t)[(\mu_1 N/\alpha_2)exp(\delta_2 t - 1)]/[1 - A + (B - 1)exp(\delta_2 t)].$

In this function, A and B are known functions of the parameters and the initial population of malignant cells is zero. The rate of transformation of normal cells to initiated cells is μ_1; the rate of transformation from initiated cells to malignant cells, at time t, is $\mu_2(t)$. The parameter α_2 is the cell division rate and β_2 is the cell death rate. The hazard function, when $\mu_2(t)$ is sufficiently small, can be simplified to the much simpler expression (Cox, 1992):

$h(t) = \mu_2(t)N_2(t).$

This model can describe and predict how a carcinogen can act. Typically, a carcinogen can increase the rates $\mu_1(t)$ and $\mu_2(t)$. However, unlike the multistage model, α_2 or β_2 can be affected leading to an increase in the number of cells in stage 2. Thus, an initiator will increase $\mu_1(t)$ while a promoter will only affect $\mu_2(t)$. If $\mu_1(t)$ is affected by a carcinogen, this effect will increase the hazard rate because it increases the size of N_2. If a carcinogen only acts on $\mu_2(t)$, that effect is transient and can cease after exposure (Cox, 1992). The MVK model is richer in detail than the MS because it includes more realistic cell growth functions. As all other dose-response models discussed in this chapter, the MVK model can also be coupled to a physiologically-based pharmacokinetic (PB-PK) model to obtain a more biologically realistic representation of the changes in the administered dose. Cox has also developed a modification to the MVK in which the hazard rate becomes a variable to avoid the condition that a tumor will eventually form with probability *1*. By using a random walk formulation for $h(t)$, Cox (2002) allows for the condition in which $h(t)$ = *0*. This result occurs when a pre-cancerous cell can become extinct and thus not become malignant. Hoogenven et al., (1999) have developed the exact solution of the model. The maximum likelihood estimates the parameters of the model and constructs upper and lower confidence limits. Cox (2002) gives the details, which are beyond the scope of this chapter.

5.5. OTHER DOSE-RESPONSE MODELS

5.5.1. The Weibull Model

The Weibull model is a probabilistic dose-response model that does not require assuming a constant hazard function (i.e., an exponential distribution of survival times). It describes the distribution of small, independent and extreme values of the gamma distribution, under asymptotic conditions. This distribution function can have several parameters. For the two-parameters form of this model, the Weibull density function of the random variable *time*, T with $t \in T$, is:

$f(t) = [(\alpha/\beta)(t/\beta)^{\alpha-1}][exp(-t/\beta)^\alpha],$

in which $\alpha, > 0$, is the shape parameter, $\beta, > 0$, is the scale parameter of the distribution, and $t \geq 0$. It follows that this two-parameters Weibull model can generate a family of distributions. For example, when $\alpha = 1$, the Weibull distribution becomes the negative exponential distribution, with constant rate of failure. The cumulative distribution of the Weibull density function of *dose* is:

$$F(d) = 1-exp[-(d/\beta)^{\alpha}],$$

where *d* is the dose. If $\alpha = 1$, then the distribution is exponential; if this parameter is different from unity, the Weibull model is nonlinear at low doses. The hazard function of the two-parameters Weibull distribution can account for increasing, constant or decreasing *hazard* functions. We can use the Weibull distribution in cancer modeling as follows. We let $Pr(d) = c+(1-c)[1-exp(-ad^k)]$,where Pr(d) is the lifetime cumulative probability of cancer when exposed to dose d, $0 \le c \le 1$, $a > 0$, $k > 0$, or $k \ge 1$ (*k* is the shape parameter of the distribution, the sharper the shape, the more potent the agent). When $k = 1$, the Weibull and the linear multistage model (with a single parameter, q_1, no threshold) predict the same estimate of risk at low doses. Risk increases at an increasing rate when $k > 1$. If $k < 1$, the risk decreases at a decreasing rate. This model is flexible because it can fit experimental data well, at least at relatively high dose levels, and can account for time-to-tumor and time-of-death due to cancer. As a final note, the epidemiological hazard function that shows declining hazard rate in the early years of life, a fairly constant hazard rate between early life and the onset of the decreasing efficiency of most organs and an increasing hazard rate in the later years of life, can be modeled by the *bi*-Weibull model. It accounts for two Weibull distributions, one for the *burn-in* period (the high risk early times of life) and for the *wear-out* period (aging and the high risks associated with age).

5.5.2. Hartley and Sielken Model

The Hartley and Sielken model (Ricci, 1985) accounts for the distribution of times-to-tumor. These times are independent and identically distributed. The model predicts the probability of tumor, by time *t* and at dose *d*, in which *time-to-tumor* and dose are the independent variables. It has the form:

$$Pr(t; d) = 1-exp\{[-\Sigma_i q_i d^i H(t)]\}, q_i \ge 0.$$

In this equation, $H(t) = \Sigma_j b_j t^j$, $b_j \ge 0$.

Example 5.5 (Modified from Crump, 1984). This example compares two different acceptable doses, for the same level of acceptable risk. Male rats were exposed to specific doses *p*-cresidine for *104* weeks. The results, for two levels of acceptable risk (10^{-5} and 10^{-6}) are as follows:

Cancer Risks	Doses Estimated with the Hartley-Sielken Model	Doses Estimated with the LMS Model
10^{-5}	0.0015	0.0014
10^{-6}	0.00046	0.000014
95% Lower CL on 10^{-5}	0.000097	0.0000050
95% Lower CL on 10^{-6}	0.000019	0.00000050

When $H(t) = 1$, the model can be used with quantal data. The example highlights the importance of the choice of dose-response model. Risk assessment provides the inputs for risk management. If the decision maker confronts vastly different estimates, such as those developed in this example, her final choice may be value-driven, rather than being driven by science. Some of the necessary information for a risk assessment includes:

- The rationale for the selection of the data about exposure and response
- The type of response or responses associated with exposure
- The rationale for any choice of a dose-response model
- The choice of the excess response (the benchmark response)
- The statistical analyses associated with estimation and their discussion

5.6. BENCHMARK DOSE AND CONCENTRATION (BMD, BMC)

The BMD (and its lower confidence limit, BMDL) and BMC approach to risk assessment is the result of the US EPA's effort to develop more scientific regulatory risk assessments. The BMD (or BMC, if a concentration is used) method is depicted in Figure 5.10

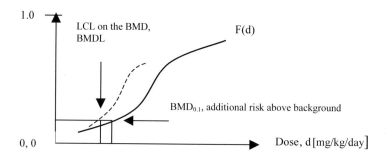

F(d), Cumulative Probability

Figure 5.10. Benchmark Dose (or Concentration) Method

In this figure, the term LCL is the lower confidence limit (BMDL) on the MLE estimate of the $BMD_{0.01}$. The dose-response model is monotonic. The US EPA suggests calculating the BMD that corresponds to an excess risk of *0.10* (US EPA, 2000): $BMD_{0.10}$. Crump (1984) has defined the benchmark dose, BMD, as the dose that produces a *specific* percent increase in the cumulative risk of disease, for a specific function relating exposure or dose to a toxic end-point.

The approach, which has resulted in a computer program available from the US EPA (2000), improves on the uses of toxicological end-points, such as the NOAEL and factors of safety, by introducing a process for the assessment of human health risks based on biologically plausible dose-response modeling, explicitly accounts for variability and for choices of model. The basis for calculating the BMD is as follows. For a dose function $f(d)$, BMD_α implies that:

$[f(d_\alpha)-f(d_0)]/[1-f(d_0)] = \alpha$

Consider the toxicological dose-response linear probit model in which d is dose:

$Pr(d) = \Phi(\beta_0+\beta_1d)$.

In this equation, $\Phi(.)$ is the unit normal.[4] The probit model describes individual responses that can occur at dose d, if that dose exceeds that individual's response threshold.

The individual thresholds are normally distributed. Stating the additional risk as:

$R(d) = Pr(d)-Pr(d = 0)$,

the BMD for the specific risk level α is:

$BMD_\alpha= \{\Phi^{-1}[\alpha+ \Phi(\beta_0)]-(\beta_0)\}/\beta_1$.

The *point of departure* for a risk analysis using the benchmark dose (or concentration) is the lower one-tail confidence limit of the benchmark dose, the BMDL. It is calculated, assuming a standardized normal distribution of doses, $z_{1-0.05} = 1.645$ (the 95th percentile of the standardized normal distribution), as:

$BMDL_{0.95} = BMD_{estimated}-\{1.645*[(standard\ error)*(BMD_{estimated})]\}$.

> **Example 5.6**. If the maximum likelihood estimated of the $BMD_{0.01}$ is 3.97, with an estimated standard error of 0.774, then: $BMDL_{0.95} = 3.97-(1.645*0.774) = 2.7$ [mg/kg-day].

The extra risk is equated to a specified α-value, for instance $[pr(d)-pr(d = 0)]/[1-pr(d = 0)] = 0.05$. The 95% confidence limit for the BMD is established by satisfying this constraint and the log-likelihood ratio: $2\{[lnL_{max}(d)/L(d)]\} = (1.645)^2$, in which 1.645 is the 95th percentile on the standardized normal distribution. The *chi*-square test statistic can be used to assess the goodness of fit of the models used to determine the BMD or BMC by comparing calculated percentiles to percentiles of the theoretical value of the chi-square distribution (Crump, 1984; US EPA, 1995; Krewski and Zhu, 1995).

5.6.1. Comment

The US EPA relies on *profile likelihoods* to calculate approximate confidence intervals, consistently with the use of the BMDL. The curvature of the profile likelihood is a graph that allows the investigation of the results from estimation.

---

[4] The standardization means that the random variable is normally distributed with mean that equals zero and variance that equals 1.

Figure 5.11 depicts these relationships using two diagrams, the LL is the axis perpendicular to the page, at *0, 0*.

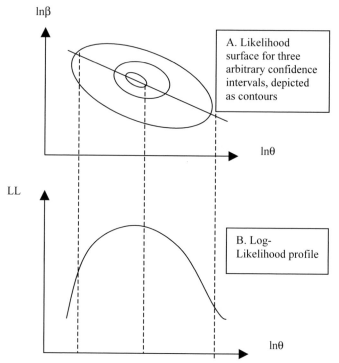

Figure 5.11. Typical Profile Likelihoods: contours view (A) and cross-section (B)

Panel A of Figure 5.11 depicts the *contours* of joint density function $f(\alpha, \theta)$. This three-dimensional joint density function is actually an elongated bell-shaped function. The $f(\alpha, \theta)$-axis is perpendicular to the page and is not shown. In this illustration, the values of $\ln\theta$ are fixed so that the values of $\ln\beta$ can be varied to find the maximum of the LL. Because the measure of interest to risk assessment is the BMDL, only half of the profile likelihood is relevant to this analysis: the left-most half of the graph showing the likelihood profile. This profile is an approximate set of confidence intervals about the maximum value of the log-likelihood. Consider an assessment that compares death rates. The null hypothesis is that there is no difference in the rates.

Consider, more specifically, two rates, λ_1 for the exposed and λ_2 for the non-exposed. Let the count of diseases in the exposed be D_1 and the count of diseases in the non-exposed be D_2. Let Y_1 and Y_2 be the person-years for those exposed and unexposed. The log-likelihood for the *unexposed* individuals (with hypothetical numbers) is (Clayton and Hills, 1994):

$D_1 ln\lambda_1 - Y_1 ln\lambda_1 = 20 ln\lambda_1 - 2,000\lambda_1.$

The most likely value for the rate is *20/2,000* per *1,000* person-years, namely *10*. For the *exposed* individuals, the log-likelihood is:

$D_2 ln\lambda_2 - Y_2 ln\lambda_1 = 40 ln\lambda_2 - 1,000\lambda_2.$

The most likely value for this rate is *40/1,000* per *1,000* person-years, namely *40*. The joint likelihood is:

$(20 ln\lambda_1 - 2000\lambda_1) + (40 ln\lambda_2 - 1000\lambda_2),$

with parameters λ_1 and λ_2. The maximum occurs where the two observed rates equal the two estimated parameters (Clayton and Hill, 1994). The parameters of the model just described can be changed to allow the investigation of the likelihood function. The log-likelihood, *LL*, is:

$$LL = (D_1 ln\lambda_1 - Y_1\lambda_1) + (D_2 ln\theta\lambda_1 - Y_2\theta\lambda_1) = (D_1 + D_2) ln\lambda_1 + Y_1\lambda_1 + D_2(ln\theta - Y_2\theta\lambda_1).$$

The reason for this change is that the new parameter θ allows the comparison of the ratio of the two rates.

The US EPA uses the AIC criterion as a means to choose between competing model dose-response models. The model with the lowest AIC is generally the most appropriate for risk assessment. However, in practice, some alternative models may have AIC values that are so near in magnitude that additional considerations must enter into the choice of model.

Example 5.7. Consider the following data DICHOTOMOUS (US EPA, 2001, http://www.epa.gov/ncea/pdfs/bmds):

Dose	Total	Percent Response
0	100	2.34
50	100	4.58
100	100	42.5
150	100	60
200	100	90.23

Using the multistage model: $Pr[response] = \beta_0 + (1-\beta_0)\{1 - exp[-(\beta_1 dose + \beta_2 dose^2)]\}$, in which, following the US EPA standard approach, β_i are restricted to be positive. The parameter estimates, using the BMDS software program (US EPA, 2001), are:

Parameter	Estimate	Standard Error
Background = β_0	0.0127	0.063
β_1	0.00	NA
β_2	4.67E-5	5.30E-6

NA indicates that this parameter has encountered a bound cause implied by the inequality constraint and thus has no standard error.

The AIC = *405.43*, the *chi*-square values = *10.45* with *df* = *3*, the *p*-value = *0.0151*.

We now need the BMD and BMDL estimates. We select a specified effect = 0.10, extra risk model and confidence level = *0.95*. We obtain BMD = *47.52* and BMDL = *43.042*.

As an alternative application of the BMD method, let us consider the cumulative Weibull probability dose-response model:

pr[response] = background+(1-background)[1-exp(-βd^γ)].

In this model, the dependent variable is percent response, the independent variable is dose and the power *γ* parameter is restricted to be ≥ *1*.

Example 5.8. Let us use the Weibull model: *Pr[response] = background+(1-background)[1-exp(-βd^γ)]*. The parameter estimates, using the data set DYCHOTOMOUS, are:

Parameter	Estimate	Standard Error
Background	0.0185	0.0117
β	4.054E-06	5.14E-06
γ	2.49	0.25

The *AIC* = *409.494*, *chi*-square = *7.38*, *df* = *2*, *p*-value = *0.0250*. The BMD and BMDL estimations, based on a specified effect = *0.10*, extra risk model and confidence level = *0.95*, are: *BMD* = *59.24* and *BMDL* = *49.45*. The plot of the Weibull model is:

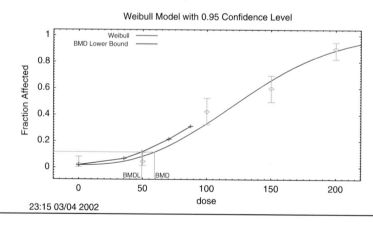

Example 5.9. Using the oral.cyl data set in Cytel (2001), we also use the two-stage multistage model for the excess risk at relatively low dose.

In this example, the *BMD* and *BMDL* shown are those for the extra risk = *0.01*. The next diagram depicts the results for both the multiplicative and the additive risk models, using a two-stage multistage model (*n* = *118*):

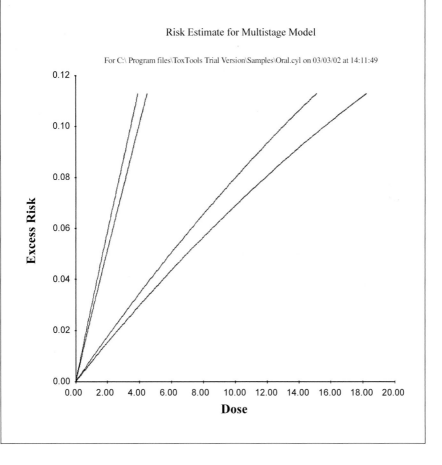

Risk Estimate for Multistage Model

For C:\ Program files\ToxTools Trial Version\Samples\Oral.cyl on 03/03/02 at 14:11:49

The *BMR* is measured as the *fractional increased added or extra risk relative to the predicted response in untreated (unexposed) animals* (US EPA, 2003). The BMR measures a percent increase in incidence for a specific BMD. More generally (US EPA, 2003):

BMR = [Pr(BMD; α, β, γ)-Pr(BMD = 0; α, β, γ)]/[1-Pr(BMD = 0; α, β, γ)],

in which BMD is a dose that results in a specific (e.g., *0.10*) increased incidence in the disease, above the incidence due to background in the controls.

5.6.2. Benchmark Response, BMR

The benchmark risk, BMR, for dichotomous responses, is:

$BMR = [Pr(BMD)-Pr(d = 0)]/[1-Pr(d = 0)]$.

In these models, $Pr(d = 0)$ is the proportion responding in the control group and $Pr(BMD)$ is the proportion affected. From this definition we can calculate,

$Pr(BMD) = Pr(d = 0)+[1+Pr(d = 0)](BMR)$.

Example 5.10. Let the BMR be *0.10*. Assume the following experimental results (US EPA, 2000):

Dose Rate	Number Affected	Percent Affected	Number at Risk
0	1	0.02	50
21	15	0.31	49
60	20	0.44	45

The probability of response, BMR, is calculated as *[0.02+(1-0.02)*0.10] = 0.118*.

The application of the benchmark dose ranges across many health endpoints and different dose-response models, including cancer risk. Different dose-response models can be used to establish the BMD or BMC. For example, the Weibull model applies not only to modeling cancer end-points, but also to modeling developmental toxicity or other end-points. Investigation of the relationship between the BMD and the no observed adverse effect level with factors of safety has shown that for a variety of health end-points the $BMD_{0.05}$ is similar to the NOAEL (Faustman, Allen, Kavlock, and Kimmel, 1994). The BMD approach nonetheless is more appropriate than the NOAEL divided by safety factors because it can be biologically motivated through the appropriate choice of health-point specific dose-response models and avoids the inherent arbitrariness of using factor of safety.

5.6.3. Effective Dose 0.01, ED_{01}

A quantitative risk assessment (NRC, 2001) uses a subset of ecological studies that relate exposure to waterborne arsenic to excess cancer risks based on the concept of Effective Dose *0.01*, ED_{01}. The computations follow the BEIR IV formula (US NRC, 2001), which relate to lifetime risk using hazard rates:

$$R_0 = \sum_i h_i/h_i^* S_i (1 - q_i)$$

where:

R_0 = the lifetime cancer risk due to non-exposure,
h_i = hazard rate at *i*th age interval
h^*_i = total hazard rate at *i*th age interval,
q_i is the probability of survival in the *i*th age interval, given survival to the beginning of that interval. It follows that the probability of failure is $(1-q_i)$.

The cumulative probability of failure is:

$$S_i = \prod_{j=1}^{i-1} (1 - q_i).$$

The risk to lifetime exposure to arsenic is obtained by multiplying h_i by a function of dose, $g(d)$. It follows that:

$R_d\text{-}R_0 = 0.01.$

Solving for dose, at the level of excess response *0.01* yields ED_{01}. Smith et al., (1992) have suggested a simple way to calculate the ED_{01} from epidemiological data. Let the baseline risk be R_0 and $g(d)$ be the risk due to dose d. Then:

$R_0[g(d)\text{-}1] = 0.01$

can be used to calculate ED_{01} from either mortality or incidence numbers. The quantity R_0 is (US NRC, 2001):

$$R_0 = \sum_{i=1}^{T} h_i / h_i^* \prod_{i}^{I} \left[1 - (exp - 5h_i) \right],$$

in which h_i^* is the background hazard. The rationale for the coefficient *5* is that the 5-age groups are used in the calculations. Table 5.3 contains an example of the formulae.

Table 5.3. BEIR IV and US NRC Probabilistic Models (US NRC, 1999, 2001)

BEIR IV (1988)	US NRC (2001)
$\prod_{k=1}^{i-1} q_k = S(1,i)$; $S(1,1) = 1.0$; probability of surviving up to year i, given survival in all of the previous years	$S_i = \prod_{j=1}^{i-1} (1 - q_i)$; "the probability of surviving to the beginning of the ith age interval."
$\prod_{k=1}^{i-1} q_k (1 - q_i) = S(1,i)(1 - q_i)$; probability of surviving to year i-1, but dying in the ith year, year i.	Not stated.

In these equations, the probability of surviving cancer death is q_i (given all possible causes and survival to year *i-1*). The term h_i^* is mortality due to all causes and h_i is the cause-specific mortality. The probability of surviving the *i*th year is $exp(-h_i^*)$; the probability of death is $(1-q_i)$. The excess risk in the ith year, due to a specific cause, is e_i.

When the proportional hazards model is used, the cause-specific mortality rate is $h_i(1+e_i)$, with overall mortality rate (from all causes, including the one of interest) is $h_i^* + h_i e_i$. Therefore, given exposure to arsenic, the probability of surviving the ith year is $q_i[exp(h_i e_i)]$. The US NRC states that *the BEIR IV uses relative risk from the epidemiological studies (of miners exposed to radon).*

Example 5.11. The effective dose, ED_{01}, is developed from the data in Ferreccio et al., (2000), Chiou et al., (2001), Chen et al., (1985 and 1992) in US NRC 1999 and 2001, for lung, urinary and bladder cancers from ingestion of waterborne arsenic as follows. The data is:

Author, study type	Risk Ratios (95% UCL)	Model	Cancer	Sex
Ferreccio et a.,l (2000), N Chile, case-control	OR: 2.4 (1.9, 2.9)	Linear regression	Lung, 1930 to 1994, 151 cases, 419 controls	M, F
Ibid.	OR: 1.4 (1.3, 1.5)	Linear regression	Lung, 1958 to 1970,	M, F
Chiou et a.,l (2001), NE Taiwan, prospective cohort	RR: 1.05 (1.01, 1.09)	Multiplicative, ln(dose)	Urinary, >8000 individuals	M, F
Ibid.	RR: 1.44 (0.63, 2.24)	Additive, linear dose	Urinary	M, F
Ibid.	RR: 1.21 (0.89, 1.64)	Multiplicative, linear dose) (<400 ppb)	Urinary	M, F
Ibid.	RR: 1.25 (0.89, 1.64)	Multiplicative, ln(dose) (<400 ppb)	Urinary	M, F
Ibid.	RR: 1.47 (0.58, 2.36)	Additive, linear dose (< 400 ppb)	Urinary	M, F
Ibid.	RR: 1.54 (0.81, 2.91)	Multiplicative, linear dose (< 200 ppb)	Urinary	M, F
Ibid.	RR: 1.77 (0.21, 3.34)	Additive, linear dose (< 200 ppb)	Urinary	M, F
Chen et al (1985, 1992)	1.15 (1.10, 1.14)	Multiplicative, linear dose	Lung	M
Ibid.	1.15 (1.13, 1.18)	Multiplicative, ln(dose)	Lung	M
Ibid.	1.26 (1.25, 1.27)	Additive, linear dose	Lung	M
Ibid.	1.16 (1.14, 1.18)	Multiplicative, linear dose	Lung	F
Ibid.	1.21 (1.18, 1.24)	Multiplicative, ln(dose)	Lung	F
Ibid.	1.46 (1.44, 1.49)	Additive, linear dose	Lung	F
Chen et al (1985, 1992), SW Taiwan	1.22 (1.19, 1.24)	Multiplicative, linear dose	Bladder	M
Ibid.	1.29 (1.26, 1.33)	Additive, linear dose	Bladder	M
Ibid.	1.98 (1.92, 2.14)	Multiplicative, ln(dose)	Bladder	M
Ibid.	1.25 (1.23, 1.28)	Multiplicative, linear dose	Bladder	F
Ibid.	1.34 (1.31, 1.38)	Multiplicative, ln(dose)	Bladder	F
Ibid.	2.57 (2.42, 2.73)	Additive, linear dose	Bladder	F

The results for the Chen et al., (1985) study (SMR is the standardized mortality rate, *M* is males and *F* is females) are as follows:

Author, study type	SMR	Cancer	Sex
Chen et al., 1985, Taiwan (no exposure data)	11.0 (9.3, 11.7) 20.1 (17.0, 23)	Bladder	M F
Ibid.	7.7 (5.4, 10.1) 11.2 (8.4, 14.0)	Kidney	M F
Ibid.	3.2 (2.9, 3.5) 4.1 (3.6, 4.7)	Lung	M F
Ibid.	1.7 (1.5, 1.9) 2.3 (1.9, 2.7)	Liver	M F

The Ferrecio et al., and the Chiou et al., are *the only studies that quantified exposure levels well enough to support a quantitative dose-response analysis* (US NRC, 2001) and thus support setting a maximum concentration level, MCL, for waterborne arsenic at *10* mg/liter. Individual water ingestion varies considerably between the Taiwanese and Americans.

The US NRC (2001) conducted a sensitivity analysis varying the amount of water from *1.0* liter, L, to *3.0* liters per day. For data from SW Taiwan males, using the BEIR IV formulae for ED_{01}, the ED_{01} ranged from *65* [μg/L] of arsenic to *246* [μg/L]; the lower *95%* confidence limits varied from *41* to *173* [μg/L] of arsenic. Further analyses, using the Poisson multiplicative regression and Bayesian methods, showed that the ED_{01} and lower *95%* confidence limits were much more stable (*142* to *145* for the ED_{01} and from*125* to *129* [μg/L] of arsenic for the *95%* LCL).

5.7. ANIMAL STUDIES IN RISK ASSESSMENT: NO OBSERVED ADVERSE EFFECT LEVELS, HORMESIS, AND OTHER CONCEPTS

Animal studies are important in risk assessment because they allow the study of the initiation, progression and culmination of a disease process *in vivo*. Some of the reasons for this importance include several factors. These include controlled exposure, diet, and other factors unlikely to be attained with many epidemiological studies. Moreover, animal studies can provide faster results than prospective epidemiological studies; provide detailed description of pathology and blood chemistry. The animal species and strains are characterized in terms of their background incidence of cancer, immunology, and other factors. Tests animals, such as rats and mice, are more homogeneous than humans (although *intraspecies* differences are noticeable). More than twenty years ago, the Office of Technology Assessment (1981) reported that the National Cancer Institute-National Toxicological Program (NCI-NTP) assessment of the carcinogenicity of *190* chemicals found that *98* chemicals had tested positive for cancer in either one or in both species and that *44* were positive for cancer in both species.

The NCI-NTP experiments generally involve both sexes of a rat strain and a mice strain, three dose levels and an unexposed group (the control). There are *50* animals in tests for each dose group and the bioassay lasts the expected animal lifetime (*24* months). However, animals are often exposed through routes, such as *gavage* (forced feeding to bypass upper respiratory tract physiological defenses), which are different from actual human exposure. Moreover, the biochemical and physiological make-up of experimental animals is sufficiently different from the human counterparts that scaling is required to obtain results applicable to humans. Recognizing these and other issues, the International Agency for Research on Cancer (IARC) has proposed that *sufficient* evidence in an animal study and *inconclusive* evidence from an epidemiological study are sufficient to conclude that the chemical presents a cancer risk for humans.

The cancer risk values used by the National Institute of Occupational Safety and Health (NIOSH) are often the *best estimates* obtained from lifetime mice bioassay and are appropriate for protecting environmental health when epidemiology is unable to provide the relevant results. An important use of animal studies for risk assessment is that they provide specific experimental doses to limit exposure to carcinogens and other toxic agents to humans. Some of these doses are the Lowest Observed Adverse Effect Level and the No Observed Adverse Effect Level.

5.7.1. Lowest Observed Adverse Effect Level (LOAEL), and the No Observed Adverse Effect Level (NOAEL)

The no-observed-adverse-effect-level dose, NOAEL, is the dose, d_0, at which there is no response in the sample of animals tested. Establishing these doses requires suitable animal test results. In general, the lowest observed adverse effect level, NOAEL, dose is more appropriate for use than the Lowest Observed Effect Level, which can be a fairly high dose (US NRC, 1986). The US EPA's general regulatory risk assessment uses the NOAEL. The EPA also uses uncertainty factors for the NOAEL or the LOAEL, if the NOAEL is not available. The US EPA (1989) defined the LOAEL as:

"In dose-response experiments, the lowest exposure level at which there are statistical or biologically significant increases in the frequency or severity of adverse effects between the exposed population and its appropriate control group."

The same agency defined the NOAEL as:

"In dose-response experiments, an exposure level at which there are no statistically or biologically significant increases in the frequency or severity of adverse effects between the exposed population and its appropriate control; some effects may be produced at this level, but they are not considered to be adverse, nor precursors to specific adverse effects. In an experiment with more than one NOAEL, the regulatory focus is primarily on the highest one, leading to the common usage of the term NOAEL to mean the *highest* exposure level without adverse effect."

Thus the LOAEL and the NOAEL are determined from animal bioassays and are used in risk assessment, often with factors of safety, to set up an acceptable or tolerable exposure (or dose) for humans. American regulatory law and other jurisdictions, such as Australia, have adopted safety factors in setting acceptable exposure values for many toxicological endpoints. The US EPA uses the NOAEL to establish the regulatory reference dose (RfD) for toxic effects, but not for carcinogens, toxicants to the reproductive system and genotoxic agents. The LOAEL and the NOAEL can be adjusted to account for differences in the exposure patterns experienced by humans at risk, relative to the length and the frequency of exposure of the animal experiment (Hallenbeck and Cunningham, 1988; Gratt, 1996).

Example 5.12. The probability that the *95%* confidence limit is the correct number on that dose (the NOAEL, d_0), for a sample of size *n* in which *no* animal is affected, is calculated as (US NRC, 1986): $(1-pr)^N = (1-0.95)^n$. Suppose that the animal study has a sample size of *50* animals and that, after the experiment is completed, exposure does not affect any animal. The calculation, $(1-0.058)^{50} = 0.05$, means that there is a *5%* probability that the true response at d_0 is *0.058* (approximately *6%* response), and *not* zero response.

5.7.2. Safety and Uncertainty Factors in Risk Assessment

Some aspects of risk assessment are non-probabilistic and use factors of safety. Consider the acceptable daily intake (ADI) for a substance that can be toxic or otherwise deleterious to human health. The World Health Organization, WHO, has developed factors of safety to account for different species and strains, interspecies conversion from animal to human, differences in susceptibility between humans and different exposure pathways. The US NRC (1994) has stated that:

"Known determinants of susceptibility vary by factors of hundreds or thousands at the cellular level. However, many of these risk factors ... tend to confer excess risk of approximately a factor of 10 on predisposed people, compared with 'normal' ones."

There was disagreement about the appropriate value of this and other safety factors among the writers of the NRC document. The consensus reached was that *ten* is the *reasonable best estimate of the high end of the distribution*, thus indicating that factors of safety are essentially values on a specific distribution. Specifically, the ADI is calculated from experimental results in animals by dividing the NOAEL dose by the product of factors of a number of factors of safety (Barnes and Dourson, 1988):

$ADI = NOAEL/\Pi_i(SF_i),$

in which the SF_i is the *i*th safety factor. The rationale for using safety factors is that they account for: (i) individual heterogeneity among the members of a population at risk, (ii) animal to human conversion and (iii) for using short-term tests. Each factor of safety generally equals *10*. Several factors of safety are used at the same time; each of these is assigned an individual value; their multiplication yields the total factor of safety. Additionally, a modifying factor (MF), ranging from greater than unity to *10* (with the default value *1*), can be used to account for additional uncertainties. Bogen (1994) has suggested that the denominator of the NOAEL should be:

$exp[\Sigma_i(lnSF_i)^2]^{1/i}.$

Example 5.13. Suppose that each safety factor equals *10*, one to account for uncertainty in the interspecies scaling and the other for the uncertainty in the short-term to long-term differences of studies. If the factors are multiplied the denominator of the NOAEL is *100*. If the expression $exp[\Sigma_i(lnSF_i)^2]^{1/i}$ is used instead, that denominator is $exp[(ln10)^2+(ln10)^2]^{1/2}$. That is:

$$exp(2.3^2+2.3^2)^{1/2} = exp(5.29+5.29)^{1/2} = exp(3.25) = 25.790.$$

The *ADI* is calculated to be the level of exposure that does not cause an adverse effect in humans, although it has been noted that the concept of the *ADI* does not apply to lipophilic substances. Establishing safety factors can preferably rely on statistical measures. For example, a specific percentile of the distribution of a toxic response or susceptibility can be the appropriate statistical measure, as discussed for the *BMD*. This value is a policy value, selected to be small or large, depending on the fraction of the individuals that can tolerably be at risk. As an example, taking the logistic distribution with known population mean, μ, standard deviation, σ, and the (logarithms) of a quantity of interest such as the lethal concentrations, *LC*, then:

$$ln(LC)_{0.01} = \mu - 1.62\sigma,$$

yields a probability that 1% or less of those exposed to that concentration or more will die (Aldberg and Slob, 1993).

An issue with this suggestion is that the population mean and standard deviation are unknown and samples are used. This can result in either overestimates or underestimates of the true value of the confidence limit when the sample sizes are small and the asymptotic approximations do not hold. Monte Carlo simulations have shown that this overestimation can be large (Alberg and Slob, 1993).

The US EPA (2004) states that:

"[i]t has been recognized that limited data are currently available for the a priori identification of susceptible populations and life-stages for many chemicals and risk assessments. In these situations, it is important that risk assessments clearly identify and summarize data needs and uncertainties, in addition to the available data. Typically, when data are limited, default practices are used: a) Non-cancer effects: An intraspecies uncertainty factor, UF, is used to account for variations in susceptibility within the human population .... This UF typically has a value of 10-fold, but can be increased or reduced when sufficient data are available. One can apply the same UF to carcinogens using a non-linear dose response model. For example, the IRIS chloroform oral carcinogenesis assessment considers the non-cancer assessment to be protective against cancer risk, and the same intraspecies UF is applied in the chloroform oral cancer and non-cancer assessments ... . A database UF may also be applied for deficiencies in the available data or when existing data suggest that additional data may yield a lower reference value ... . This UF is most often used when developmental or two generation reproduction studies are not available, but it may be applied in other situations to account for the lack of data for potentially susceptible populations or life-stages. b) Cancer effects: An evaluation should be made as to whether low-dose linear extrapolation is sufficient to protect susceptible populations ... . For example, available data indicate that early life exposure to mutagenic carcinogens may lead to greater incidence of cancer than the same lifetime average daily dose received later in life."

In the past (US EPA, 2004), the situation was that:

"when reference values were originally being generated, some combinations of UFs reached as high as 10,000. As we gained greater and greater experience and obtained more usable data, the trend has shown a decrease in the combination of UFs to such high levels. The more recent RfDs and RfCs are more in line with the recommendation from the Technical Panel: no derived reference value exceeds 3,000."

5.8. ALLOMETRIC FORMULAE: Converting Dose from Animals to Humans

A practical way to deal with the conversion of exposure to dose is through *allometric* formulae. These *scale* the intake of a chemical for an individual with a specific body weight accounting for organ size, differences in blood flow, respiratory and metabolic rates from one species to another. These rates empirically scale at a fractional power of an animal body-weight. The basis of such formulae is experience with chemotherapeutic drugs but *it does not have either a toxicological or pharmaco-kinetic basis and should not be applied to every chemical* (US NRC, 1986). Specifically, the basis for the allometric formulae is the relation between heat output and body mass, following a relationship empirically obtained in 1932:

[lnH(output, in kcal/day)] = a*[ln(wet body mass, in kilograms)],

where H is heat output and the parameter $a = 0.67$, is empirically determined from data on mouse, rat, guinea pig, cat, macaque, rabbit, goat, sheep, chimpanzee, steer, cow, and elephant. McMahon (1973) used approximations of the human body, namely arms, torso and legs to obtain a relationship of mass to the radius of the (cylindrical approximation) of these body parts. This empirical relationship (obtained by regression modeling) results in the estimate of $a = 0.75$, which is assumed to account for metabolic rate proportional to the radius of large muscles, because muscles contractions and power are related. The allometric formulae for risk assessment are generally stated as:

Dose = a*(body weight)^k,

in which a and k are determined from estimation. Thus, the human dose conversion formula, noting that the square brackets identify the units of each variable, is (dose is in [milligrams/kilogram], weight is in [kilogram]):

$$Dose_{Human} \text{ [milligrams/kilogram]} = Dose_{rat}\left(\frac{BodyWeight_{human}}{BodyWeight_{rat}}\right)^{1/3}.$$

Example 5.14. Suppose that the average human weight is *70* kg and the average animal weight is *0.50* kg. Suppose that the animal inhales *0.05* m^3/day of air with an average concentration of *10* micrograms/m^3 of a toxic agent. Human exposure from inhalation is:

$E_{human} = 0.05(m^3/day)10$[micrograms/m^3]$(70/0.50)^{1/3}$[kg/kg] = *13.70*[micrograms/day].

The term $(70/0.50)^{2/3}$ is the allometric term.

The assumption of homogeneity of individual animals, within a strain, can be questioned because individual animals can be genetically different. Controlled exposure between animals in the same dose group may be variable, animals may gain or lose weight at different rates, and undetected infections may occur between closely housed animals; other factors affect risk results. For example, some mouse strains, such as the $B_6C_3F_1$ mouse used in the NCI/NTP cancer bioassay, develop large numbers of liver cancers (hepatomas) naturally. Mouse liver tumors are commonly found at the end of a bioassay. For example, out of *278* chemicals evaluated by the NCI/NTP, *141* were carcinogenic of which about half resulted in mouse liver tumors. A plausible explanation for this high incidence is that the relatively *high* doses used in those bioassays may *promote* tumorigenesis through *toxic* effects on the cell. That is, the toxic lesion initiates the tumorigenic process. On the other hand, some rat strains used in those tests have relatively *low spontaneous tumor rates*. In general, a basic causal mechanism of rodent carcinogenesis consists of mutagenic lesions to the DNA.

For example, in a study of *73* chemicals that were part of the NCI/NTP long-term cancer bioassays, *44* chemicals were positive for carcinogenicity. Of these *44, 24* tested negative for mutagenicity in the *Salmonella* test (Reynolds, et al., 1987), noting that the Sprague-Dawley rat strain, used to test the carcinogenicity of dioxin, has naturally occurring liver cancer. Their cells play an important role in understanding cancer processes. For instance, the liver-focus induction system is now used to predict the *potential* for eventual tumorigenic properties of a chemical, although there is considerable debate regarding the relationship between such results and tumorigenesis. It is increasingly becoming apparent through molecular techniques that oncogenic sequences, transformations and genetic alterations can induce tumors. Other genetic events, such as gene amplification, can occur during cellular progression. These changes can be cumulative and selectively enhance tumorigenic activity within the cell. Reynolds et al., (1987) noted several that oncogenes occurred in *80% of spontaneously occurring hepatocellular carcinomas of the $B_6C_3F_1$ mouse and to a lesser degree (30%) in benign liver neoplasia*. These authors also state that:

"[t]he regulation of human occupational or environmental exposure to a chemical, based on long-term carcinogenesis studies in rodents, should take into account whether the chemical in question is mutagenic, cytotoxic, or has a receptor-mediated mechanism of promotion."

Genetic mutations are under extensive study. For instance, it has been determined that the instability of certain genetic activity relates to a mutation of a gene (p53) that has tumor suppressing capability (Harris and Hollstein, 1993). Ashby and Tennant (1991) have developed a database of rodent studies (conducted by the National Toxicological Program) which relate gene mutations and certain chromosomal aberrations to rodent tumors at multiple sites and type, within the test animal. These authors note that electrophilic chemicals can be carcinogenic in humans and cause the same types of lesions to the DNA. This can explain the high number of cancers at high maximum tolerated dose, MTD, reported in animal studies, but would not exclude a different mechanism of response at low doses.[5] If so, the usefulness of the MTD becomes questionable, as will be discussed shortly.

The NCI/NTP protocol for bioassays uses *50* animals per exposure group (one group is exposed to the MTD), two strains and the two sexes. There still is considerable debate on the usefulness of certain animal bioassay studies for cancer risk assessment in making environmental policy. Following the US EPA (2004):

"Carcinogenic effects are typically evaluated using a long-term rodent carcinogenicity bioassay, ... using at least 50 animals per sex per dose group in each of three treatment groups and in a concurrent control group, usually for 18 to 24 months, depending on the rodent species tested. One group is exposed to a high dose, often the maximum tolerated dose, or MTD ... Two groups are given lower doses of the test compound. The high dose in long-term studies is generally selected to provide the maximum ability to detect treatment-related carcinogenic effects while not compromising the outcome of the study through excessive toxicity or inducing inappropriate toxicokinetics (e.g., overwhelming absorption or detoxification mechanisms)."

The purpose of the two or more lower doses is to provide some information on the shape of the dose-response curve, including possible subtle, precursor, and/or early events that can provide a more meaningful dose-response relationship relative to anticipated human exposures. ... Because of biological variability and the relative low sensitivity of most bioassay designs, positive evidence at the high dose is considered a rebuttable presumption of carcinogenicity. These findings are not altered by negative data at lower exposures unless there are supporting studies that provide a biological basis to conclude that the mode of action responsible for the high-dose effects is not relevant at lower doses and therefore for lower levels of exposure experienced by humans."

Risk assessors are generally concerned with risks at exposures well below the experimental data. The US EPA (2004) uses either of two methods to assess human health risks; it states that:

"For mutagenic chemicals and those for which insufficient data exist to determine a mode of action, a point of departure (POD) at the low end of the exposures for which we have data is selected and a straight line is drawn from (the lower confidence limit on) that dose to zero. A nonlinear approach is selected when there are sufficient data to ascertain the mode of action and conclude that it is not linear at low doses and the agent does not demonstrate mutagenic or other activity consistent with linearity at low doses. The use of low-dose linear assumptions for mutagenic carcinogens has been the peer reviewed, common practice for decades for many federal regulatory agencies including EPA. Modifications for non-mutagenic carcinogens are currently under review."

For non-cancer effects, the US EPA (2004) uses the absence of an adverse effect (the NOAEL) in a relevant animal study for determining a negligible hazard due to exposure. Specifically:

"Adjustments are made to the NOAEL to account for limitations in knowledge, i.e., uncertainty, or limitations inherent in the test system, e.g., the use of relatively small numbers of genetically relatively homogeneous animals. Similarly, mode of action, physiological differences, and other factors have been used to improve the accuracy of estimation of the RfD

---

[5] The MTD is the dose that used to set up lower doses for testing animals in cancer bioassays. Lower doses are expressed as percentages of the MTD.

or RfC for non-cancer endpoints. The Agency has also been using and is continuing to refine the BMD methodology ... for estimating non-cancer risks. In this procedure, the POD is derived from the dose-response curve, and further adjustments analogous to those for the NOAEL may be necessary to account for similar limitations of the data. The EPA's RfD/RfC review ... delineates several options ... used as the Agency developed more accurate methods for evaluating these parameters from responses in animals. It also provides a series of examples, as well as lists of recommendations that address the predictivity of results from animal studies."

The MTD has been controversial US EPA (2004) because testing animals at the MTD *may cause damage which may lead to cellular proliferation, increased mitosis, and eventually carcinogenicity.* Ames and Gold (1990) have found that biochemical and physiological effects at high doses may lead to toxicity-induced carcinogenicity not likely to occur at low doses. An earlier study by Haseman (1985) found that *more than two-thirds of the carcinogenic effects detected in NTP feeding studies conducted would have been missed if the highest dose had been restricted to doses below the MTD, if the chemicals were, in fact, animal carcinogens.* Bickis and Krewski (1989) report that the upper confidence limit on the linear term (the term q_1^* in the linearized multistage cancer dose-response model), *in 263 data sets, was highly correlated with the maximum dose tested.* Thus, the US EPA (2004) includes a caveat:

"The draft final cancer guidelines (USEPA, 2003c) cautions the assessor in the use of results from the MTD and recommends that this issue be addressed on a case-by-case basis ... . The guidelines state that the results of such studies would not be considered suitable for dose-response extrapolation if it is determined that the mode of action underlying the tumorigenic response at high doses is not operative at lower doses."

Importantly, the US EPA (2004) reports that:

"the recent chloroform risk assessment was predicated on the recognition of a high-dose effect leading to cell proliferation and then carcinogenicity. The MCL was therefore developed based on an assumption of nonlinearity for carcinogenicity."

The US EPA (2004) note that conditions can exist to justify the use of MTDs, such as *failure to reach a sufficient dose reduces the sensitivity of the studies.* Because animal tests are conducted with a relatively small number of animals, dosing at the MTD is necessary to provide the best chance that cancer will be seen in test animals, if a chemical is carcinogenic. Further, the EPA recognizes that *overt toxicity or inappropriate toxicokinetics due to excessively high doses may result in tumor effects that are secondary to the toxicity rather than directly attributable to the agent stressing the importance of establishing that the MTD has not been exceeded.* The basis of such determination include whether *an adequate high dose would generally be one that produces some toxic effects without unduly affecting mortality from effects other than cancer or producing significant adverse effects on the nutrition and health of the test animal.* Thus, this Agency believes *that effects seen at the MTD may be appropriate for use in risk assessment when the data have been critically evaluated.*

Abelson (1992) stated that *the differences in metabolism of mice, rats, monkeys, and humans raise doubts about the relevance of the mouse experiments* in risk assessment. The issues in the debate range from deciding whether or not the increase in cell division (mitogenesis) resulting from exposure to chemicals limits the rate of cancer induction, to the experimentally demonstrated differences between the metabolism of humans and that of rodents. Using butadiene as an example, Abelson notes that a large quantity of the inhaled mass of butadiene is exhaled after inhalation, and that the by-products of the metabolic processes involving butadiene are tumorigenic. Some of the by-products are retained in much greater quantities in mice than in rats or humans.

Typically, mice have a lower level of an enzyme that metabolizes the mono-epoxide to a harmless substance. When $B_6C_3F_1$ mice in a long-term bioassay were exposed to *625 and 1,250 ppm* of butadiene, the liver and lung cancer mortality rates were so great that the experiment *was stopped at 60 and 61 weeks*. Part of the reason for such potency is that the mice have endogenous murine leukemia virus that enhances *the incidence of malignant (thymus) lymphomas when mice are exposed to butadiene.*

Abelson (1992) contrasted the epidemiological result from an occupational cohort exposed to high levels of butadiene, when the Threshold Limit Value (TLV) was *1,000 ppm during exposures that took place in the early 1940s. At those levels of exposure, the overall mortality from cancer was only 75% of the rate in the ordinary public.* On the other hand, Ames and Gold (1990) state that:

"Our analysis so far indicates that, in general, potency values for a given chemical do not vary much between males and females or, with a few significant exceptions, between rats and mice."

5.8.1. Hormesis

Hormesis (Calabrese, McCarthy, and Kenyon, 1987) has been invoked as a possible explanation for finding a *U-shaped* dose-response (or exposure-response) function between exposure and response, Figure 5.12. The *U*-shaped dose-response function has a portion that shows that exposures at low doses are apparently beneficial. The descending limb of the parabola indicates that more exposure results in lesser response, up to a point (where the first derivative of the parabola is zero). After this point, the relation between exposure and response consists of the ascending limb of the parabola, which restores the expected relation between increasing exposure and increasing adverse response. A biological mechanism described by such parabola suggests the health-protective effect of exposure, perhaps by supplying some essential mass of the chemical and, thereafter, the adverse effect of too much dosage of the agent (Davis and Svendsgaard, 1990).

A biological mechanisms that can justify a *U*-shaped function include: detoxification coupled to stimulus to overproduce gluthathione (Pisciotto and Graziano, 1980); enhanced homeostatic control; inhibitory and compensatory mechanisms mediated by the joint effect of two or more chemicals, or even by a single chemical. There is empirical evidence of *U*-shaped exposure-response functions. An early summary is provided by Davis and Svengaards (1990) who report (on the basis of studies dated from 1979 to 1988) that the *U*-shaped dose-response curve was found in ten rat studies, fourteen human studies, one mice study, and one study with monkeys (out of twenty-nine studies which included crab, minnows and pigeons). We show the prototypical biphasic dose-response for carcinogenic and other toxic outcomes in Figure 5.12 (note that we use two vertical axis in this figure).

Figure 5.12. Hormetic Responses Mechanisms for Cancer and for Toxic Outcomes

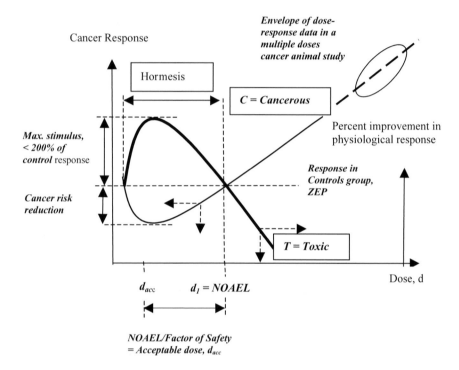

A hormetic dose-response (or biphasic) dose-response is a non-monotonic, unlike the threshold models from toxicological responses and the linearized multistage cancer risk model. The fact that we can empirically observe a *J*-or *U*-shaped dose-response has particular relevance to policy principles or procedures because a non-threshold model is more protective than using a regulatory agency default dose-response assumptions. Because hormesis is measured at exposure and doses below those causing chronic effects, then it must be used quantitatively. This is consistent with low-dose interpolation, which is quantitative, and thus fits well with the generalizability of default arguments understood as choice out of the set of choices. It is also consistent with rational decision-making under risk, rather than uncertainty.

Calabrese and Baldwin give more recent discussions and arguments about the role of hormesis in risk assessment. The finding of relevance to precautionary actions is that

(Calabrese and Baldwin, 2003) *hormetic response is more common than the threshold model in toxicology.* If so, then it should be a core component of risk assessment. The biphasic response consists of a toxic effect followed by a correction induce by a homeostatic response (Stebbing, 1982). Some (Chapman (2001) have suggested that hormetic dose-response should be used for detailed risk assessments. We think that this is incorrect because there is sufficient evidence that a hormetic dose-response is as legally valid as a linear, non-threshold model under both under rules of evidence for scientific findings that are tested under the law of (scientific) evidence cases *Frye, Daubert* and *Khumo* (Ricci and Gray, 1998, 1999). If a hormetic dose-response is not included in screening risk assessments, then the results from those assessments can, and seemingly will, be underprotective.

Rational decision-making and legal arguments point to their use because of the benefits that would result to society from modifying the coarse factor-of-safety, BMD, and ED_{01} methods. Calabrese and Baldwin (2003) have studied this issue and found that hormetic mechanisms are in fact more prevalent than threshold mechanisms. The key conclusion, based on an analysis of approximately 1,800 doses below the NOAEL, from more than 650 dose-response studies, is that:

"While the threshold model predicts a 1:1 ratio of responses 'greater than' to 'less than' the control response (i.e., a random distribution), a 2.5:1 ratio (i.e., 1171:464) was observed, reflecting 31% more response above the control value than expected (p-value < 0.0001). The mean response (calculated as % control response) of doses below the NOAEL was 115% ± 1.5 standard error of the mean (SEM)."

Calabrese and Baldwin (2001, 2003) have based their assessments of hormetic mechanisms on a selection of past studies, and a protocol also accounts for false negative and false negative probabilities of error.

5.9. SUMMARY OF TOXICOLOGY IN RISK ASSESSMENT AND MANAGEMENT

Toxicology provides several important inputs to risk assessment and dose-response modeling and, consequently, to risk management. In this section, we summarize some of those practical aspects. Shaw and Chadwick (1998), Birkett (2002), Klassen (1996, 2001) give more details. Several types of studies that are designed to provide information that ranges from the rapid assessment of the immediate cause of harm, such as death, to obtaining physiological and pathological data about the disease process. Tests can be conducted *in vitro*. For example, the *in vitro* (reverse) mutagenicity assays conducted on the *Salmonella* (with or without liver homogenate which is used to mimic the metabolic functions of the liver) can test for base pair or frameshift mutations, depending on the strain of Salmonella used (McCann, Horn and Kaldor, 1984).

The lymphocyte chromosomal aberration test is used to test damage to chromosomes from exposure to toxic agents. In this test, the lymphocytes (an exposed and a control group) are cultured. The exposed group is exposed to different doses of the chemical to study how exposure affects cell functions, during specific events that occur in the cell cycle (Shaw and Chadwick, 1998). *In vitro* testing can be used in batteries:

several tests can be used sequentially to identify different adverse endpoints. Batteries of *in vitro* and *in vivo* tests can determine the carcinogenic, developmental toxicity or other toxic responses in humans, as well as elucidate the function and magnitude of toxicological parameters. Figure 5.13 depicts the basic toxicological inputs into risk assessment.

TOXICOLOGICAL ASPECTS OF RISK MODELING

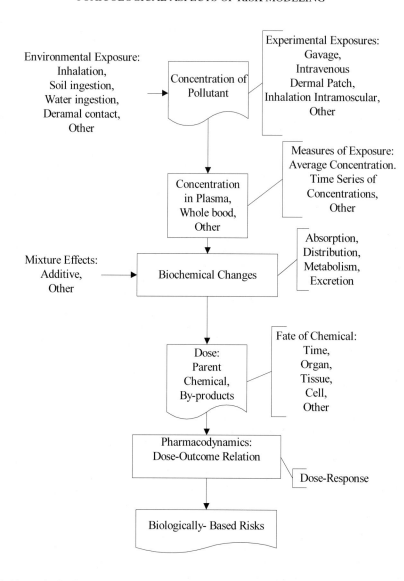

Figure 5.13. Toxicological Inputs and Outputs Commonly Encountered in Risk Assessment

Practically, human health risk assessment is based on the combination of *acute*, *subchronic* and *chronic* toxicity tests. These are *in vivo* tests, performed under close supervision of animals and ethical approvals. The results obtained from these tests are used to obtain estimates of human risk using interspecies conversion methods that range from allometric formulae to pharmacokinetic-pharmaco-dynamic methods or through factors of safety.

Acute toxicity tests are based on single doses to determine an acute adverse effect, such as the lethal or other less adverse outcome. For example, an experiment may consist of several groups of animal, each with *25* animals who are administered a dose of a chemical. One group can be the control group: this is the group that is not exposed to the pollutant being tested on the other groups of animals. Animals are observed for two weeks to determine distribution of deaths and to calculate the median *lethal dose 50 (LD_{50})*, the dose at which *50%* of the animals die. For example, for mice the oral LD_{50} for chlordane is *335* milligrams/kilogram body weight but the dermal LD_{50} is *840* milligrams/kilogram body weight (Hodgson and Levy, 1987). The LD_{50} for TCDD is *5* milligrams/kilogram body weight, phenobarbital has a LD_{50} between *50* and *500* milligrams/kilogram body weight and the LD_{50} for ethanol is between *5,000* and 15,000 milligrams/kilogram body weight (Klaassen, 2001). Acute toxicity tests are not designed to study the mechanism of the effect, but can be used to develop LOAELs and NOAELs. The LD_{50} is affected by body weight, animal species, route of administration of the agent, animal health and so on.

Subacute tests are used to study adverse outcomes that can occur at doses below the LD_{50}. Those tests are often conducted for three months under a daily (multiple dosing) regimen of exposure administered to the animals under test. The results can provide some of the details needed to understand and formulate bio-mathematically the potential mechanism linking exposure to response. The subjects of the study are examined for specific biological responses during the period (e.g., ninety days) of testing, for different types of lesions, after the experiment is terminated.

Chronic toxicity tests are designed to last the expected lifetime of the animals under test (e. g., two years for rats and mice). These studies have the highest likelihood of helping to determine the mechanism of action of the chemical tested. In this type of study, the groups of animals, other than the control group, are dosed daily for the duration of the experiment. Biological responses are measured during the experiment and pathological examinations occur at the end of the expected lifetime. This type of experiment has been extensively used in the United States for studying the carcinogenic characteristics of a chemical and provides the data used to estimate the parameters of dose-response models. Sub-chronic tests can also be relevant to risk assessment, but I do not discuss these for the sake of brevity.

Several factors control the absorption and distribution of chemical pollutants in the body. For example, some chemicals do not penetrate the skin well. Other can, if they are lipophilic. A chemical can be ingested or inhaled. When ingestion occurs, the chemical can penetrate human organs through the digestive tract. Chemicals can be actively transported through intestinal walls, a situation enhanced by the morphology of the intestinal walls' protrusions (*villi*) that greatly increase the surface area of the intestine. Factors such as intestinal flora, pH and others mediate the penetration of chemical pollutants from the gastro-intestinal tract, GI, into the blood vessels that further distribute them throughout the body, through the entero-hepatic circulation. Gases and small particles enter the body through inhalation. The lungs are the principal organs involved (they have an area that is much larger than the skin surface area, which is about *1.5* to *2.0* square meters). Gases enter the lung through the alveoli (an entry governed by ventilation rate). When particulate matter enters the lungs, its absorption is governed by diffusion, the electrical charge of the particles and other factors. In general, if the particles are large (> *5* micrometers, *µm*, in diameter), deposition governs their fate: they are limited to the nasal-pharynges. If the diameter of the particles is somewhat smaller, the particles can travel to the trachea and bronchi. Defense mechanisms, such as the ciliae, can effectively clear them efficiently but not entirely. Even smaller particles behave like gas and penetrate through the alveoli's walls into the blood. Their fate depends on solubility in the blood.

One of the principal biochemical processes that controls the distribution of a chemical toxicant within the body is mediated by plasma (the water component of the blood), interstitial, and intracellular water. The process that governs chemical distribution is the complex formed by a chemical pollutant and proteins in the plasma. The principal proteins include albumin and lipoproteins. As a rule of thumb, water-soluble toxicants preferentially bind to albumin while lipophilic toxicants preferentially bind to lipoproteins. The ligand-protein complex is partly governed by the number of binding sites on the protein molecule, the binding *capacity* of the protein. In general, there are several and different binding sites on a protein molecule. Binding *affinity* of the protein (a function of solubility of the toxic agent in plasma) is the inverse of the disassociation constant, $(k_{diss})^{-1}$. For toxicant-protein binding, high affinity is approximately $\leq 10^{8}$ moles/liter^{-1} and low affinity is approximately $\geq 10^{4}$ [moles/liter^{-1}] (Klaassen, 2001). Ligand-protein binding is a percent of the toxic chemical bound to the protein.

It is possible that, at low toxicant concentration in the plasma, a greater percentage of that toxic chemical binds to the protein if the chemical has high affinity for the protein. This can have important consequences if toxicants competitively bind. A more toxic chemical can become unbound and cause more damage. Binding affinity can therefore have significant and different implications for impaired individuals (susceptible population) and therefore on the eventual apportionment of the risk. The toxic ligand-protein reaction is:

$$Toxic_{free} + binding \xleftrightarrow{\ k1,k2\ } Toxic_{bound} ,$$

where the association constant is k_1, and the disassociation constant is k_2. This reversible reaction means that a toxic agent can bind and can also disassociate from the protein at some other site from where the initial contact between them began. Most ligand-protein complexes form through covalent and non-covalent bonds such as ionic, hydrogen and other bonds.

Example 5.15. The literature shows that DDT is about *99.9%* bound, with *35%* bound to albumin and the rest to lipoproteins. On the other hand, only *25%* of nicotine is bound to a protein of which about *95%* is bound to albumin. As another example, the chemical carcinogen benzo(a)pyrene binds to lipoproteins, but that binding can depend on the density of the lipoprotein.

5.9.1. Toxicological Concepts

This subsection contains a brief introduction to some of the toxicological concepts used in developing dose-response models. The toxicological parameters that are necessary to understand the fate of a toxic chemical in the body are *clearance* and *volume distribution*.

Clearance principally occurs in the liver (hepatic clearance, involving metabolic reactions) and the kidneys (renal clearance, which consists of filtration, secretion and re-absorption). Clearance is the flow rate of a chemical given a total blood flow rate (the units associated with the formulae are shown in the square brackets [.]), as shown below:[1]

Clearance = (flow rate$_{liver}$ [liters/hour])/(total flow rate$_{liver}$ [liters/hour]).

Clearance is the proportion of the quantity of the chemical eliminated from a physiological compartment. More specifically, M is mass, T is time and L is length:

Clearance[M/T]/[T/L^3] = (rate of elimination$_{compartment}$[M/T])/(input concentration$_{compartment}$[L^3/T]).

The fraction of a toxic agent that is metabolized (through enzymes) in the liver is:

(Liver Clearance)/(Liver Clearance+Kidneys Clearance).

Clearance rates can be added and subtracted. The total clearance can also be calculated as:

Clearance = (%dose)/(Area Under the Curve).

The area under the blood-concentration curve (AUC) is generally stated as:

AUC = (Dose/Volume)(Volume/Clearance).*

More specifically, using allometric formulae, clearance is the rate at which the chemical is eliminated from the compartment, k_c*(body weight)$^{0.7}$* and the rate of formation is given by k_f*(body weight)$^{0.7}$* (US NRC 1986). The AUC is used in determining the biological availability of a chemical after it enters the human blood.

---

[1] This is called *first pass* clearance.

Example 5.16. Let the total blood flow rate in the liver be *50* [liters/hour] and let *25* [liters/hour] be cleared. The ratio *25*[liter/hour]/*50*[liter/hour] is the clearance for that organ and for the chemical of interest. Specifically, this is the *first-pass clearance*.

The *distribution volume* is defined by the ratio of the mass of the chemical in the body to the concentration of that chemical in the plasma. It is calculated as:

Distribution Volume $[L^{-3}]$ *= (Mass* $[M])/(Concentration$ $[M/L^3])$.

Example 5.17. Suppose that there are *10* micrograms of the chemical in the body such that the concentration in plasma is *0.01* micrograms/liter. The distribution volume is calculated as:

10[micrograms]/*0.01*[micrograms/liter] = *1000*[liter^{-1}].

The distribution volume can predict a chemical's concentration by relating concentration to time, taking advantage of the fact that:

Distribution Volume = (dose)/(concentration in plasma).

Example 5.18. Suppose that a chemical's distribution volume is *10* liters and that the average individual weighs *70* kilograms. Then the dose required to cause a toxic effect is:

10/70[liter/kilograms]**1.00*[micrograms/liter] = *0.14* [micrograms/kilogram].

5.9.2. Modeling Changes in Concentrations

The changes in the concentrations of a chemical, *C*, can be modeled by the differential equation:

-dC/dt = kC,

in which the coefficient *k* is the rate of loss, a constant. The *half-life* of a chemical's concentration in a physiological compartment obtains from the equation (which is the solution of the differential equation immediately above):

C(t) = C(t=0)[exp(-kt)]*,

in which *C(t=0)* is the concentration at time zero (a given, in this equation) and *k* is the elimination rate (a constant, in this equation) which has units of *1/time* [1/T]. In other words, an elimination rate constant that equal *0.5* means that half of the chemical's mass is lost in a specific period of time (e. g., half the mass is lost per hour). This form of elimination is *first order*. The argument of the function, time, can be measured in the appropriate units of toxicological relevance: days, hours and so on. From the relation that halves the concentration, *0.5*C(t=0) = C(t=0)*[exp(-kt)]*, and taking the natural logarithms yields*: ln(0.50) = -kt$_{1/2}$*. It follows that:

k[1/T]= *0.693/t$_{1/2}$*,

where *ln(2.0) = 0.693*; so that:

$t_{1/2} = 0.693(Volume\ Distribution/Clearance)$,

and:

$k = (Clearance)/(Volume\ Distribution)$.

From these equations, the distribution volume is *Distribution Volume*$[L^3]$ = $(t_{1/2}*Clearance)/0.693$.

When multiple doses are used, the methods are an extension of the single dose study, but account for dosing over time and for the steady-state concentration. The *area under the curve*, AUC, depends on the time interval used in dosing and clearance. Figure 5.14 depicts the time series of the concentrations.

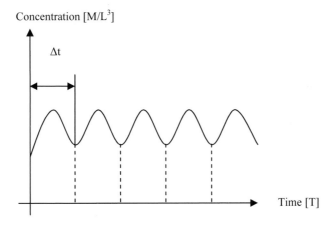

Figure 5.14. Concentration-Time for Consecutive Exposures

The rate of elimination of a chemical is:

Elimination$[M/T]$ = *Clearance*$[L^3/T]$*Concentration in Blood*$[M/L^3]$.

In general, the intake of a chemical produces a steady-state relationship between the intake and the mass of the chemical in an organ or tissue. That is, because steady-state means that the rate of intake equals the rate of elimination, then:

Elimination$[M/T]$ = *(Clearance*$[L^3/T])$*(Chemical Concentration at steady-state*$[M/L^3])$.

The AUC can be determined from the relationship between concentration and time, as depicted in Figure 5.15.

Concentration [M/L³]

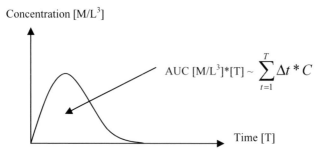

$$AUC\ [M/L^3]*[T] \sim \sum_{t=1}^{T} \Delta t * C$$

Time [T]

Figure 5.15. Area Under the Curve, AUC

The term Δt is a small interval of time; the summation is taken over all intervals, from *1* to *T*. The area under the curve can be used to calculate clearance as follows:

Clearance = (Dose[M])/(AUC[M/L³T]).

More specifically, clearance by organs such as liver or the kidneys relates to concentration at steady-state dose and to the AUC, depending on the way the substance is input into the organ. The simplest relationship that exists between blood and plasma concentrations of a chemical is their ratio, which is generally constant. The concentration in the plasma is simpler to obtain than that in whole blood. An important concept in dealing with exposure to chemical agents is biovailability. It depends on the way through which the chemical enters the body. If the chemical is injected intravenously it is assumed to be *100%* bioavailable. It follows that for other routes of uptake (e.g., oral) the ratio formed by the AUC-oral to AUC-intravenous describes bioavailability. In a single intravenous dose study, clearance is determined from the AUC. If administration occurs through ingestion, this formula accounts for the fraction of the chemical that becomes bioavailable.

5.9.3 Discussion

Excretion from the liver occurs through the bile. The liver can excrete both polar and non-polar chemical compounds. Urine, which is excreted by the kidneys, can contain compounds with high molecular weight. For humans, the literature suggests that the threshold between these two forms of excretion the molecular weights ranges between *500* and *700*. The threshold for rats it is about half that for humans. Excretion from the kidneys is a more complex process than that of the liver.

The importance of the relations developed to describe clearance by the liver and the kidneys is two-fold. First, it allows modeling the fate, transport and disposition of a chemical to which the body becomes exposed and the effect of bioavailability of the route of exposure. Second, it permits modeling different responses, such as when the liver or the kidneys are diseased and when enzymes are induced or inhibited by alternative agents. For instance, if the chemical has therapeutic functions, the study of its behavior in the body is critical to maintaining a concentration level that is neither toxic nor ineffective. For intravenously administered chemicals, systemic clearance controls the fate of the chemical; for orally administered drugs other factors make predictions more variable.

The metabolic reactions that occur in the liver are exemplified by considering the enzyme system cytochrome P450 (a mixed function oxidase system found in the endoplasmic reticulum, mostly in the liver, but also in other tissues). This is called the *Phase I* metabolism.[2] Other cytochromes also exist, e.g., the P420. Phase I metabolism (mediated by the P450) reactions generally tend to decrease toxicity. This mechanism generally adds oxygen to the substrate through hydroxylation. A typical example is the addition of *-OH* to the molecule of benzene. The P450 can also result in several metabolites, some of which may be more prevalent than others. For example, aniline results in phenol and in ammonia via the P450 (which adds *O*) metabolism.

In some instances, Phase I reactions can create toxic products from an otherwise non-toxic chemical. The P450 activity results in products that are more polar than the original chemical and thus can be metabolized and excreted by later conjugation, a series of reactions that characterize *Phase II* metabolism. These reactions include conjugation with glucoronic, sulphuric or amino acids and glutathione (glutamic acid, cysteine and glycine). Phase II reactions generally result in non-toxic end products that are excreted. The chemical group characterizing a toxic agent, for instance the hydroxyl and the carboxyl groups, triggers glucoronic acid conjugation mediated by glucuronyl transferases. Chemicals with a sulfate group are catalyzed by the sulphotransferases in the cell's cytoplasm yielding a non-toxic sulfate byproduct. The reactions involving an amino acid form a peptide bond between the amino group of the amino acid and the carboxyl group of a chemical toxicant. These conjugates are eventually excreted. Finally, glutathione forms conjugates with different chemicals, which are often further metabolized before excretion through the urine. For example, cholorobenzene conjugates to phenylglutathine (and *HCl*) via the glutathione transferase.

5.9.4 Molecular Studies in Epidemiology

The basis of molecular epidemiology is that some diseases, including some cancers, are related to chromosomal abnormalities linked to the activation of certain loci on a specific chromosome in all cells and gene transfers (Perera, and Weinstein, 1982; Ambrosone and Kadlubar, 1997; Khoury and Wagener, 1995; Schork et al, 1998). Molecular epidemiology combines quantitative methods with biochemical, molecular and other data, in which laboratory techniques are used, jointly to assess the changes in tissues, cells, and fluids induced by a carcinogen or other toxic agent. There is a large number of laboratory methods and techniques used in detecting an equally large number of different biochemical endpoints. The detection of a molecular marker generally involves the map of a chromosome, a rank-order by maximum likelihood of the markers, a representation of recombination frequencies, and a method for detecting those recombinations. For example, in the instance of a crossover between two chromosomes (involving an exchange of segments of two

---

[2] The *P450* term signifies that the spectrum of *CO* has a wavelength of *450* nanometers.

chromosomes within a cell) the detection of the marker (in this case a DNA marker) involves taking blood samples. DNA extraction, digesting of the DNA via a restriction enzyme, separation of the chromosome's strands by electrophoresis and denaturation, blotting on a nylon membrane, radioactive labeling of the DNA by a specific probe, exposure to a film sensitive to X-rays. The result is a photomicrograph of the map of (banded) patterns for each chromosome (OTA, 1986).

Molecular epidemiology is relevant to study both toxic and carcinogenic effects. The answers that these methods can provide include: i) the precise biological unit at risk (from a protein to a cell, rather than gross anatomical descriptions) of the effect; ii) an accurate measurement of the magnitude of the effect; iii) accurate descriptions of critical biochemical processes leading to a pre-disease stage; and iv) data used in stochastic and non-stochastic dose-response models. Thus, molecular epidemiology, by using the statistical techniques of epidemiology with molecular information, can help to determine some of the early steps that precede cancer or other adverse health outcomes. Molecular markers, such as DNA-adducts, mutations and abnormal cell populations can describe how a carcinogen can act with the DNA or a protein, whether the carcinogen switches on a particular gene and so on.[3] Understanding this causal chain is important to formulating and estimating the parameters of dose-response models.

Perera (1996) has developed a summary of biomarkers for cancer. Among these, PAH-DNA adducts formed in blood, the placenta and lungs, are biochemical markers of exposure to chemical carcinogens. For instance, genetic damage can result from inhaling PAHs in air or through cigarette smoke (Perera, 1996). Mutation in the p53 tumor suppressor gene, in the breast, liver and other organs, is a marker for increased risk of cancer from exposure to other carcinogens. Molecular epidemiology has also identified markers of inherited susceptibility to cancer; for example, the *H-ras-VTR* variant of the *H-ras* gene for breast cancer; a variation in the cytochrome p4501A1 gene marks for an increased risk of lung cancer from PAHs (Perera, 1996). Markers of acquired susceptibility to cancer have also been found in blood.

5.10. MIXTURES OF CHEMICALS

It should be apparent that we are exposed not just to single chemicals but to mixtures of chemicals. An important question is: What is the effect of chemicals acting together? In other words, do chemicals act synergistically (the effect of two or more chemicals is greater than their added effect) or antagonistically (the effect of the chemicals is less than the additive effect) or additively? Although the answer to this question is still being sought (ATSDR, 2001), nonetheless, it is relevant to discuss the salient aspects of how mixtures can be assessed to orient risk assessors that may want to study the issue further. Mixtures are combinations of chemicals in which the components of the mixture are the individual chemicals themselves. In practical risk

---

[3] Perera (1996) defines an adduct as a *complex* between *a chemical carcinogen* and either *DNA or a protein*.

assessments, the way to deal with mixtures is to add the concentrations of the components of the mixture. Individual response to the components of the mixture is characterized by the sum of the doses, adjusted for the potencies of each individual component of that mixture. The assumption is that each component causes the same adverse health endpoint, has the same mode of action, but differs in the potency. In other words, the slope of the dose-response curves of each component determines the overall potency. This is the basis of most practical risk assessments discussed.

Following ATSDR (2001), suppose that a mixture has two components, *1* and *2*. Then, for Y_i being the probit response for the *i*-th component, let the dose-response models be:

$Y_1 = b*ln(x) + a_1$,

and:

$Y_2 = b*ln(x) + a_2$

The potency, *p*, of component *1* relative to component *2* is:

$ln(p) = (a_1 - a_2)/b$.

It follows that:

$Y_2 = b*ln(px) + a_1$.

It also follows that for exposure x_1 and x_2:

$Y = a_1 + b*ln[x_1 + (px_2)] + a_1$.

When the components of a mixture are independent and have different modes of action, then responses can be added, rather than their doses. Following ATSDR (2001), suppose that a mixture has two components, *A* and *B*. The expected response *P* to the two components of the mixture, at doses that individually result in responses P_1 and P_2 is:

$P = P_1$ if the correlation = *1* and $P_1 > P_2$

or:

$P = P_2$ if the correlation = *1* and $P_2 > P_1$.

Alternatively, $P = P_1 + P_2$ if the correlation = *-1*. If the responses are statistically independent, then $P = P_1 + P_2(1 - P_1)$. An interaction model (ATSDR, 2001) is:

$Y = a_1 + bln(x_1 + px_2) + k(px_1x_2)^{0.5}$.

This equation accounts for a multiplicative interaction between the two components of the mixture; the strength and direction of the interaction is measured by the parameter k. Specifically, if $k = 0$, then the components are additive in the dose, x_i. If $k > 0$, the chemicals act synergistically; if $k < 0$ the chemicals are antagonistic.

5.10.1. Toxicity Equivalency Factor Approach (US EPA, 2004)

The US EPA (2004) states that:

"Several (commentators have) stated that the use of Toxicity Equivalency Factors (TEFs) is inappropriate because TEFs are based on the unproven assumption of additive toxic effects, compounded by the uncertainties inherent in the generation of TEFs, their application, and the heterogeneity of effects for which they are being used. However, some ... acknowledge the value of TEFs in certain circumstances. In 1986, EPA published the Guidelines for the Health Risk Assessment of Chemical Mixtures ... (that) ... provide the bases for evaluating human risk from exposure to combinations of two or more chemical substances, regardless of source or of spatial or temporal proximity. They were extensively peer reviewed, submitted for public comments, and reviewed again by EPA's Science Advisory Board. The guidelines have been supplemented by a technical support document ... (that) ... refer the assessor to several different methods for evaluating mixtures data of varying types, epidemiologic or toxicological data on the environmental mixture ... to which exposures occur; test data on a tested mixture judged to be sufficiently similar to the environmental mixture; data from a group of similar mixtures; and data on the mixture components.

EFs are used in evaluating mixture toxicity from data on components. Their use is based on an assumption of additivity among similarly acting components of a mixture ... the Agency provided procedures for developing the relative potency factor (RPF) method and described TEFs as a special case of RPFs. The RPF method is component-based and relies on two types of information: toxicological dose-response data for at least one component of the mixture being addressed (referred to as the index compound or IC) and scientific judgment as to the toxicity of the other individual compounds in the mixture and of the mixture as a whole. ... Both RPFs and the special case of TEFs are intended to serve as interim approaches for addressing any mixture, pending the development of new data on the mixture's toxicity. ...

The RPF method (including TEFs) is based on dose addition and assumes that the chemicals in a mixture share a common toxic mode of action that is relevant to the health endpoints being assessed ... Operationally, this means that mixture components tested in the same bioassay should have dose response curves of similar shape between the toxicity thresholds and the maximum response. The components are assumed to be true toxicological representations of each other, although their relative toxic potencies may differ. When one uses the RPF method, it is necessary to identify the constraints of its application. ... For example, a set of RPFs may be restricted to oral exposures and not be usable for exposures to the same mixture through the inhalation route; this was the recommendation for RPFs applied to PAHs. TEFs for chlorinated dibenzo-p-dioxins, dibenzo-p-furans, and coplanar polychlorinated biphenyls (PCBs) have no identified constraints."

5.11. CONCLUSION

The dose-response models and the toxicological experiments, formulae and models discussed in this chapter are used throughout the risk assessment and management methods that many countries have adopted. The preferable models use descriptions of the fundamental biological basis for the disease. An example of this modeling occurs when the biology of the disease is incorporated to predict hazard rates using stochastic dose-response models based on hazard functions.

The stochastic dose-response models discussed account for biological *hits* and the *stage*s through which a disease progresses before becoming manifest. Biology, ranging from molecular to toxicology knowledge, provides the basis for each model. The toxicological evidence, under the totality of evidence approach (US EPA, 1996), consists of an assessment of issues such as:

- Increased weight in animal *v.* decreased weight
- Evidence of human causality *v.* data are not available or do not show causality
- Evidence of animal effects relevant to humans *v.* data not available or not relevant
- Coherent inferences *v.* conflicting data
- Comparable metabolism and toxicokinetics between species *v.* metabolism and toxicokinetics not comparable
- Mode of action comparable across species *v.* mode of action not comparable across species

Most of these comparisons require models that are, as always, *caricatures of reality*. Nonetheless, the importance of an increasingly accurate caricature helps the risk manager deal with comparisons of risk, costs and benefits. Biological realism is essential to any risk assessment because risk assessments based on arbitrary factors of safety lack an epistemic basis. Although such assessments can be defensible, if they are adopted through a legal process that allows for full ventilation of the issues, they are inferior to stochastic models and statistical exposure-response. Even when a risk assessment based on safety factors is defensible on legal grounds, it inferior to a risk assessment that uses biologically realistic dose-response models and accounts for variability and uncertainty through the proper probabilistic and statistical methods.

The benchmark dose method is an example of biological realism and practical reasoning that includes statistical ways to deal with uncertainty. Such a method combines dose-response models with sound measures of risk that can justify precautionary regulations or guidelines. There are situations where stochastic models may not be available. In those situations, statistical distributions of thresholds can be particularly helpful in practical risk assessments. Thus, the formulae and methods of toxicology become important for at least three reasons.

The first is that toxicological studies can provide data for risk analysis much more extensively and rapidly than many epidemiological studies. The second reason is that toxicological formulae and models can properly account for interspecies comparisons and for changes from exposure to dose. The third is that toxicology provides the biological basis for understanding the processes leading from exposure to dose and then to adverse response. These reasons make a compelling case for using toxicological information and methods in risk assessment and management.

QUESTIONS

1. Consider the basis of the *hit* theory of cancer. Suppose that a normal cell needs two hits from the carcinogen before it is irreversibly damaged and becomes cancerous. Write the appropriate expression for the population of

cells survivorship function using the Poisson distribution and discuss its implications for risk management. Use approximately 400 words.

2. Discuss the differences, if any, between *hormesis* and thresholds. Limit your discussion to 400 words.

3. Do you think that probabilistic hit-related models are less biologically credible than models that use transition rates between stages and account for cell growth and death? Limit your answer to 600 words.

4. Suppose that conceptually different types of models fit the experimental data very well. On what basis would you *extrapolate* downward to a region in which the extra risk is about $1*10^{-5}$? What would an *interpolation* consist of?

5. Suppose that you are working for your country's environmental agency. The administrator of that agency asks you to use either the LOEL/NOAEL approach or the dose-response models to set a tolerable dose for a carcinogen. Which approach would you advocate? Describe your rationale using no more than 300 words.

6. What are the fundamental biological differences, if any, between the multistage and the MVK models and why are those differences important? Use about 400 words.

7. Why should the BMD approach *not* be applicable to human carcinogens? Answer this question in 300 words or less, discussing both sides of the views.

8. Let the distribution of individual thresholds be log-normal. The dose-response is $Pr(d) = a+\phi[b+c*ln(d)]$, in which ϕ is the standardized cumulative normal probability density function and d is dose. Note that:

$$\phi(x) = (2\pi)^{-1/2} \int exp(-t^2/2)dt,$$

the integral is evaluated between the limits $-\infty$, x. Discuss the meaning of a, b, and c.

9. Discuss the reasons why hormetic mechanism should be accounted for generally, as opposed to be accounted for in specific cases, in regulatory policy-making.

10. The issue is as follows. If we can generalize hormesis, like we generalize linear, non-threshold cancer dose response for regulatory risk assessment, what is the benefit to society for adopting this convention? Perhaps even more important, are there costs to society by having two major defaults, in both of which causation is empirically difficult to demonstrate.

11. Develop a formal way to represent the uncertainty in the choice of one default, relative to another. That is, assume that a linear and a non-linear model are biologically plausible, given the current state of knowledge about biology and other disciplines. Assume that two alternative data sets are available to estimate the parameters of each of these models. The issue that you address is: How do you portray, for the decision-maker, the alternatives that you face?

12. How can any regulatory choice of a causal argument be generalized? Take an exposure-response model of your choice. Suppose that its parameters have been estimated from data in a population of pigmies and that the

results are positive, meaning that the higher the exposure, the higher the risk (or mortality rate). Would it be proper to generalize the findings to the residents of Iceland? Why and why not?

CHAPTER 6.

MONTE CARLO, BOOTSTRAPS AND OTHER METHODS TO QUANTIFY OR PROPAGATE VARIABILITY AND UNCERTAINTY

Central objective: to discuss and exemplify practical methods used to propagate and combine the variability of the inputs of a formula or model into their output

This chapter introduces and exemplifies some of the methods that can be used to account for variability in formulae in which one or more inputs (the independent variables) are functionally related to the output (the dependent variable). In practice, many of the calculations made by risk assessors are deterministic and thus ignore the variability of the input variables. This simplification is often inconsistent with risk analysis and can create an illusion of accuracy that can later become indefensible. Even when the calculations are probabilistic, there are situations where the distributions of the input variables are unknown, the data set has missing values or it has many zeros. Fortunately, there are methods that can help a risk assessor correctly to account for these issues. Some of these methods include:

- Monte Carlo methods, which approximate the variability of an output, in situations when the distributions of the inputs of a formula or model are either known or can be assumed
- *Bootstrap* methods, which rely on the distribution of the data in the sample to characterize variability
- Markov chain Monte Carlo (MCMC) methods, which improve on traditional Monte Carlo methods by accounting for dependencies between inputs and for complexities in probabilistic calculations
- Aspects of the analysis of missing data and estimation of means and other statistical quantities, without having to use defaults such as inserting an instrument's limit of detection for the missing data
- Interval analysis and upper and lower probabilities to deal with forms of uncertainty inherent to partial information
- Entropy of information, which is used in models as discussed in Chapter 10

Recently, these methods have become part of standard regulatory work. For example, the US EPA has developed a policy for probabilistic analysis done in the context of risk assessment (EPA/630/R-97/001); MCMC methods are used in epidemiology and economics. It is thus important to have a grasp of the relevance of probabilistic analysis in the context of risk assessment and the implications for managing risks. The methods discussed in this chapter can answer such risk management questions as:

- What is the probability that a particular concentration of a chemical or other agent exceeds its regulatory standard or guideline value?
- What are the preferable ways to represent uncertainty when the risk assessor has limited information?

The connections between the methods discussed in this chapter and risk assessment are depicted as follows:

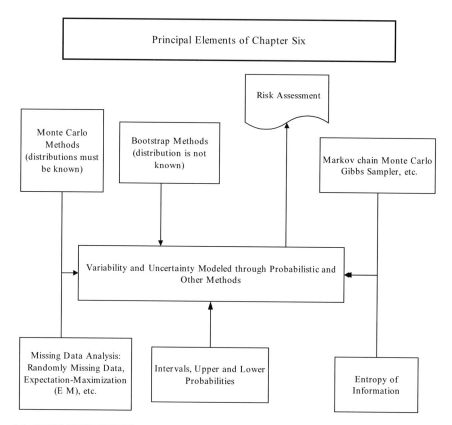

6.1. INTRODUCTION

Answers to these questions have both policy and science aspects. Consider the steady-state distribution of concentrations of arsenic in the ground water at a specific location. Suppose that the data is distributed log-normally.[1] This distribution can be used to limit exposure by a *policy* that sets a specific probability value, which is measured by the right tail of that distribution and consists of an area, a small probability, of exceeding a regulatory concentration of this contaminant. Because policy determines the *choice* of the probability of exceeding a particular concentration or the magnitude of an adverse effect, such policy has strong

---

[1] The density function applies to continuous random variables with the probability calculated as the product $f(x)dx$, where $f(x)$ is the density function of the random variable X.

foundations in risk-assessment. That choice, however, can be affected by untested assumptions. One of those assumptions is independence between input random variables. This assumption can affect the *tails* of the resulting output distribution, given the distributions of the inputs and the operations performed on them. That is, the tails can be either too fat or too thin, relative to what they should be if the assumption that the input variables are not correlated is correct. Some of the practical implications are as follows. The estimates of central tendency, such as the median or the mean, of the output distribution are either unaffected or mildly affected by correlations (measuring dependence). Importantly, the tails of the resulting distribution *are* affected. This result must be assessed because assuming independence can mask true variability. This is the variability of concern to precautionary risk management. If the assumption that two or more independent input random variables are uncorrelated is not true and the dependence, measured by the coefficient of correlation, is positive, then the *tails* of the output distribution will *underestimate* the true risk. If it is negative, although the true risk can be negligible, it may be overestimated. Furthermore, the form of the true but unknown dependency may not be linear. If so, an estimate of a *linear* coefficient of correlation, although mathematically computable from a sample, can be meaningless because it is not correct.[2] Figure 6.1 depicts the distributions of concentrations generated with and without correlations as the output of a hypothetical prediction of the distribution of a particular agent at a specific point and time.

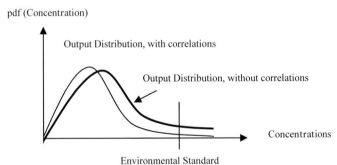

Figure 6.1. Hypothetical Results as Two Distributions of Concentrations at a Specific Location and Time with and without Correlations between the Input Variables

Making risk-based calculations suggests the following sequential activities:

1. Identifying the appropriate variables and specifying plausible (generally inductive and empirical) formulae or models,
2. Developing probabilistic conditioning, if relevant,
3. Propagating and combining the variability of each independent variable to determine the variability of the dependent variable or variables,
4. Accounting for missing or censored data,
5. Limiting the computational burden and time,
6. Discussing the accuracy of numerical approximations used in probabilistic analysis.

---

[2] Even when the estimated coefficient is zero, meaning no correlation, this *empirical* result does not mean that the random variables are *algebraically* independent.

A risk assessor who uses distributions to represent the variability of the inputs should account for some of the potential difficulties that can occur in doing this work. These difficulties include:

- Probability distributions that can be assumed and used without a proper justification
- The joint distributions of random variables may not be known, although the marginal distribution of each variable is known
- Statistical correlations between random variables may be excluded from the analysis
- Only the upper and lower bounds of an input variable may be known. In this situation, intervals, rather than the unknown distribution, can portray the uncertainty in that variable
- Missing or censored data
- Bayesian updating can give a false sense of security because it requires probabilities (conditional and unconditional) that may not be available

Finally, some uncertainties may not be represented by distributions because they are inherently qualitative or vague because they consist of statements such as *more likely than not, it is possible that* or other natural language statements that indicate uncertainty about a fact. Other representations of uncertainty include fuzzy numbers and fuzzy logic (Klir and Folger, 1986). Let us start with an example in which we combine the known variability of some inputs and the formula for those inputs is an addition.

Example 6.1. What is being sought is the variability of the output variable, Y, given the formula $Y = X+W+Z$. We assume that each input variable is normally (or log-normally) distributed. The variability of Y is described using the population standard deviation of the input variables, if these random variables are normally distributed. That is: $\sigma_y = \sigma_x + \sigma_w + \sigma_z$. Because these population standard deviations are generally unknown, we instead use their *estimated* sample variances and obtain the formula $s_Y^2 = s_X^2 + s_W^2 + s_Z^2$. The population or sample means of the input variables can also be added to yield the mean of Y: $\mu_Y = \mu_X + \mu_W + \mu_Z$. A multiplicative model can be made linear as follows. Let the new model be $Y = X*W*Z$. Taking the natural logarithms of both sides yields: $ln(Y) = ln(X)+ln(W)+ln(Z)$. Let the sample means of $X = 10$, $W = 5$ and $Z = 1$; the units of these variables are unstated. Then, $ln(10) = 2.302$, $ln(5) = 1.61$ and $ln(2) = 0.69$. Their sum is $2.302+1.61+0.69 = 4.602$. The geometric mean of the output, Y, is calculated as $exp(4.602) = 93.68$. The geometric standard deviation of the output Y is calculated as: $s_{geometric} = exp(\sqrt{\sigma^2})$, in which σ^2 is the variance of the *log*-transformed data. The three variances are *3.5*, *1.30* and *0.10*, respectively and sum to *4.90*. Therefore, the geometric standard deviation of Y is $exp(\sqrt{4.90}) = 9.14$.

This example shows that, in some instances, the central value and the variability of the output of an additive formula is calculated from the sample means and variances of the input variables. These calculations apply to variables with dimensions such as dollars, mass, length and so on. In this discussion, we are specifically dealing with the normal and log-normal distribution that can be characterized by their mean and variance alone. In some instances, even these sums can be inappropriate, as the next example shows.

Example 6.2. Suppose that the risk assessor combines the *95%* upper confidence limits (UCLs) of variables in a formula. These limits *are* random variables. Such combination, however, does not produce a *95% confidence interval*. Let X and Y be independent standardized normal distributions, then the sum of their *95% UCLs*, namely *1.96+1.96 = 3.92*, is *(3.92/1.4142) = 2.77* standard deviations above the mean of *0*, corresponding to a *99.4%* UCL.

In many situations, the random variables used in risk assessment have distributions that do not satisfy the additive property. In the next section, we consider the *superimposition* of cumulative distributions to combine variability and obtain an analytical solution to the density function that does not require simulation methods, such as Monte Carlo, discussed later.

6.2. COMBINATIONS OF DISTRIBUTIONS: Superimposition

The combination of continuous cumulative distributions through *superimposition* can be understood as follows. Suppose that two different mechanisms lead to the failure of a system and that each mechanism operates independently of the other. Let $F_1(.)$ be the cumulative distribution of the first mechanism and $F_2(.)$ be the cumulative distribution of the second mechanism. The superimposition of these two continuous distributions is their product, which is the cumulative distribution $F_{1,2}(.) = [F_1(.)F_2(.)]$. This is relevant when two or more groups of members of a population are heterogeneous. For two or more distributions, their superimposition is (Apostolakis, 1977):

$$F(x) = 1 - \prod_{i=1}^{n} [1 - F_i(x)].$$

The combination of continuous cumulative distributions through superimposition gives the distribution of the output random variable. The advantage of using the superimposition of distributions is that it produces a function, rather than an approximation via simulation. That is, we obtain:

1. A mathematical expression for the distribution sought,
2. The approximation errors made when a distribution is approximated numerically through a histogram, such as those generated by Monte Carlo simulations, are avoided.

To illustrate superimposition, we use the Weibull and the exponential distributions.

Example 6.3. Suppose that the random variable Y has a Weibull distribution, with two parameters known to be *2* and *1.5*. That is, the *scale* parameter is *2* and the *shape* parameter is *1.5*. The density function of this random variable (with $0 \leq y \leq \infty$) is:

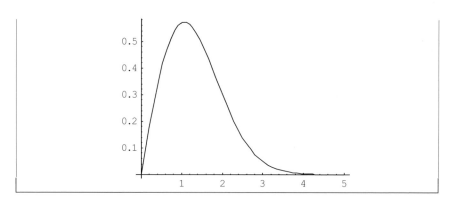

The cumulative distribution of this random variable is depicted in the next example.

Example 6.4. The cumulative distribution of Y is:

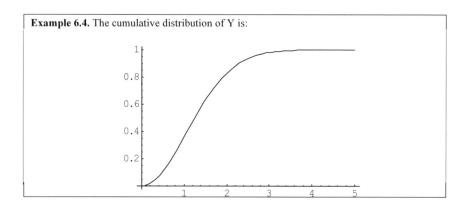

The (negative) exponential distribution is described in the next example.

Example 6.5. Suppose that the (negative) exponential distribution for the random variable *Z* is known to be $f(Z) = 0.5*exp(-0.5z)$, $z > 0$. Its density function is:

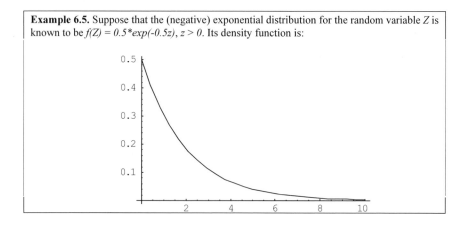

The next example develops the cumulative distribution of the random variable Z.

Example 6.6. The cumulative distribution function of the random variable Z is $f(Z) = 1-exp(-0.5z)$, which is depicted as:

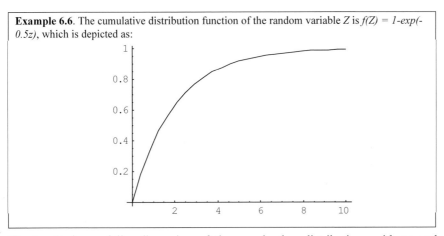

Chapter 11 has a fuller discussion of these and other distributions with several examples and reasons why it is often possible to choose a distribution function knowing the *physical* aspects of the process that generates the observation. Computer software such as JMP™, SYSTAT™, Statistica™ or other software, including mathematical software such as Mathematica or Mathcad, will be used in most practical work with superimposition. The formulae for the mean and variance of these and other distributions, as well as other statistics that can be relevant to a risk assessment, can be found in Andrews and Moss (1993), Granger-Morgan and Henrion (1990) and Grimmett and Stirzaker (1982). The superimposition of the exponential and Weibull distributions developed in the previous examples, using T (t is a realization of the random variable T) to describe time-to-failure, is:

$F(t) = 1-[1-exp(-0.444t^2)][1-(exp-0.5t)]$.

Suppose a system has two mechanisms of failure, one that is normally distributed and another that is an exponential distribution of failure over time: $F_1(t) = [1-N(\mu_T, \sigma_T)]$ and $F_2(t) = [1-exp(-\lambda t)]$. The superimposition of the *complements* of the distributions of time-to-failure yields the survivorship distribution for the entire system:[3]

$F_{1,2}(t) = [1-N[(\mu_T, \sigma_T)][1-exp(-\lambda t)]$.

This result also applies to a *parallel* system with two (or more) components with known (or assumed) distributions.

In some instances the risk assessor may want to indicate her degree of belief in a choice of distributions by weighing each of them by a judgmental probability. Consider heterogeneous failure rates of two groups in a single population and suppose that those failures are due to their genetic differences. Let $F_1(t)$ be the

---

[3] The symbol $N(.)$ identifies a normal distribution, with population mean μ and variance σ^2. The density and cumulative distribution functions of the normal distribution are discussed in Chapter 11.

cumulative distribution of failures in group *1* and $F_2(t)$ the cumulative distribution of failures for group *2*. The superimposition of these two cumulative distributions of failure times is their sum, which can be weighted by the probability of belonging to either of these two groups:

$$F_{1,2}(t) = pr_1F_1(t) + pr_2F_2(t),$$

where $pr_1 + pr_2 = 1.00$.

Example 6.7. Let the probability that a proportion π_1 of more sensitive individuals be pr_1, and the probability of other proportion, $(1-\pi_1)$, be pr_2. Assume that the distributions are exponential with failure rates λ_1 and λ_2, respectively. The superimposition of these two distributions is:

$$F_{1,2}(t) = \{pr_1[1-exp(-\lambda_1 t)]\} + \{pr_2[1-exp(-\lambda_2 t)]\}, \text{ with } \lambda_1 > \lambda_2.$$

For convenience we can agree that *variability* applies to situations in which the distribution of values of one or more random variables either is known or can be assumed. *Uncertainty* applies to situations where the distribution or distributions are either not known or cannot be assumed. The analytical technique for *adding* continuous distributions is the convolution integral of the distributions.[4] As the examples show, using superimposition can be demanding. In most situations, the risk assessor will not use integrals (for continuous distributions) or sum (for discrete distributions) and perform operations on their combinations. There are practical alternatives. One of them is a form of probabilistic simulation determined through Monte Carlo methods.

6.3. MONTE CARLO METHODS: Combining Variability through Probabilistic Simulations

The basic aspect of a Monte Carlo simulation can be exemplified, using Crystal Ball 4.01[TM], to obtain the distribution of the output variable for *given* or *known* distributions of the input variables.

Example 6.8. Let the formula be: $Y = A1 + A2 + A3$. Each of the input variables are log-normally distributed random variables: $A1\sim LN(3.5, \ 0.35)$, $A2\sim LN(1.01, \ 0.10)$ and $A3\sim LN(0.10, \ 0.01)$. The approximate distribution of the output random variable Y is obtained by Monte Carlo simulation (9,914 trials) as:

---

[4] The convolution can yield the probability density function of the sum of two (or more) random variables.

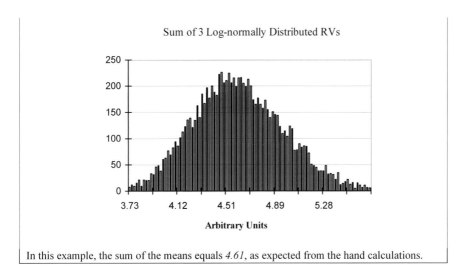

In this example, the sum of the means equals *4.61*, as expected from the hand calculations.

The Monte Carlo results approximate the distribution of the output, rather than resulting in a continuous *log*-normal distribution of that output, as would be obtained by the *convolution* of these three *log*-normal distributions. Figure 6.2 depicts the fit to the histogram generated by a Monte Carlo simulation using a *log*-normal distribution with mean *4.60* and standard deviation *0.36*.

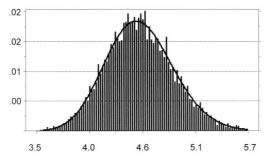

Figure 6.2. Fit of a Log-Normal Distribution, LN(4.60, 0.36) to a Monte Carlo Simulation

The rationale for replacing an integral with a simulation is that the evaluation of an integral can be difficult. The principal idea of the Monte Carlo method consists of replacing an integral with a stochastic simulation, based on sums of random numbers. Specifically, probabilistic simulation replaces integration to provide numerical *approximation*s of statistical quantities such as the mean, variance, other moments of the distribution and approximate probability distributions about these parameters. The Monte Carlo simulation estimates the value of an integral, used in calculating the expected value of a function of X, namely $h(x)$, with the expected value of a sum (Kalos and Whitlock, 1986):

$$\int_{-\infty}^{\infty} [h(x)f(x)]dx = E[N^{-1} \sum_{i=1}^{N} h(x_i)] .$$

The integral on the left-hand side is *estimated* by the expected value of the sum shown as the right-hand side. Statistically, this integral is an expectation (namely, it is the mean of a random variable in the previous equation). The simulation is depicted in Figure 6.3.

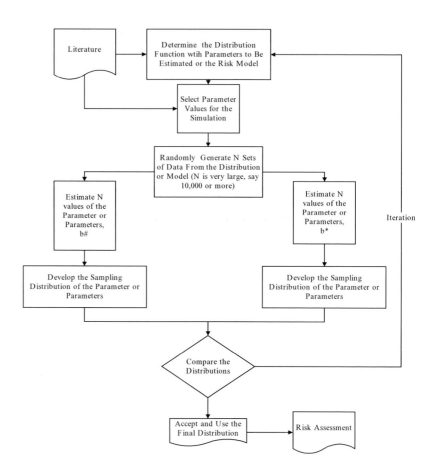

Figure 6.3. Flow-chart for a Simple Monte Carlo Simulation and End-use

The computational basis is a probabilistic simulation consisting of the following steps:

1. Draw values for the random variables, the x_i, from the known density function $f(x)$,
2. For each value of x_i calculate $h(x)$,

3. Compute the arithmetic mean of all the *h(x)*, as the expectation *E(S)*. This expectation is the *value* of the integral.

Example 6.9. Suppose that we wish to calculate the value of the integral $W = \int_0^l exp(-2x^2)\,dx$. This integral is related to the cumulative distribution of a normally-distributed random variable, $Z\sim N(0, 1)$. Therefore, $\Phi(2) = 0.977$ and thus $I = 0.598$. The approximation of the analytical value of *I* is due to the use of the numerical values in the table of values for the random variable *Z*.

The next step in the Monte Carlo simulation consists of developing the distribution of *S, f(S)*, using the Central Limit Theorem. The algorithm of a Monte Carlo simulation was described in Figure 6.3 (adapted from Kennedy, 1999). Increasing the sample size improves the accuracy of the estimation of the empirical distribution. Chebyshev's inequality can be used to determine the size of *N* for a simulation. This inequality, which does not require knowing the distribution function of a random variable, gives conservative confidence limits and can be used to back-calculate the number of trials needed in a Monte Carlo simulation. Each sample in the simulation, called a *trial*, is independent and identically distributed. Setting up the sum of those variables, developing the characteristic function of the sum and then inverting that function (with a Fourier transform), yields the probability density function of the sum of random variables. Similar reasoning, but a different transformation, is required for discrete random variables. When working with sums of random variables, the Monte Carlo estimator discussed is unbiased. The importance of Monte Carlo methods should now be apparent. First, integration made easier. Second, the risk assessor can study the distribution of a variety of parameters, such as the OLS estimates of variances, coefficient of correlation, central estimates of the regression model's coefficients and other statistics. Third, the Monte Carlo study can determine the bias of an estimator.

Example 6.10. Consider the formula *(A*B*C*D)/(E*F)*. Let *A* = 12, *B* = 50, *C* = 30, *D* = 0.75, *E* = 70 and *F* = 10. All variables have arbitrary units.

The deterministic result is:

X = *(12*50*30*0.75)/(70*10)* =19.286.

Let us define the following random variables and their distributions: *A~N(12, 3.5), B~U(0, 100), C~U(1.00, 45), D~T(0, 0.75, 1.00), E~T(5,70,120)* and *F~U(0, 20)*. Note that *A~N(12, 3.5)* means that *A* is normally distributed with *mean 12* and *standard deviation* = 3.5; *B~U(0, 100)* means that *B* is uniformly distributed with minimum value *0* and maximum value = *100*; *D~T(.)* means that *D* has triangular distribution with three values. The result of a Monte Carlo simulation with *10,000* trials yields the distribution of *X*:

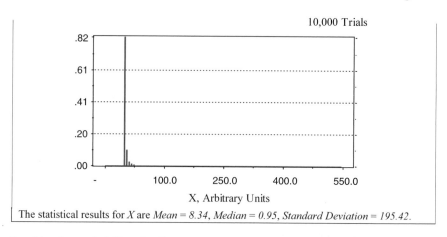

The statistical results for *X* are *Mean = 8.34, Median = 0.95, Standard Deviation = 195.42.*

Consider the probability distribution functions, *pdf(.)*, of random variables *A* and *B*: *pdf(A)*, *pdf(B)*. Combining them by the *convolution* of their density functions obtains the *pdf(Y)*, e.g., *pdf(Y)* = *pdf(A)⊗pdf(B)*, in which the symbol ⊗ is the convolution operation. What about their ratio? If the random variables are independent and their *pdfs* are either normal or log-normal, the result we seek is obtained by taking the ratio of the *pdfs*. However, this is not the case when the *pdfs* are triangular, uniform or take other forms. Specifically, the *back-calculation* of probability distributions generally requires an operation called *deconvolution* (Jansson, 1984; Bernabini et al., 1987).

Example 6.11. Taking *Z = X*Y*, we can calculate *Z* to be the product of two deterministic numbers. Suppose *X = 5* and *Y = 5, Z = 25*. Suppose that we now change the calculations of this trivial example to using random variables and their distributions. The question is: What is the distribution of the variable *Z*, for *Z = X*Y*?:

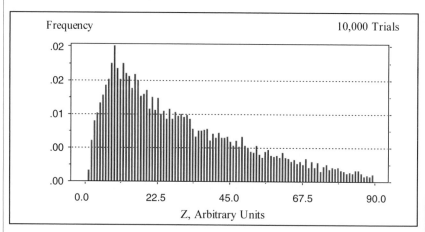

Let us use a Monte Carlo simulation to answer this question. Let X and Y be uniformly distributed: *X~U(1.00, 10.00)* and *Y~U(1.00, 10.00)*. The empirical distribution of *Z = X*Y*, obtained by Monte Carlo simulation with *10,000* trials.The statistics of Z are *Mean = 30.09,*

Median = 24.66 and *Standard Deviation = 21.21.* In this example, we know that *Z* is not uniform. Let us now consider the division of two random variables, such as *Y = Z/X*. The question is: Does *pdf(Y) = pdf(Z)/pdf(X)*? Assume that *Z* is log-normally distributed, *Z~LN(30, 20)* and we know that *X* is uniformly distributed, *X~U(1, 10)*. We obtain the following distribution of *Y*:

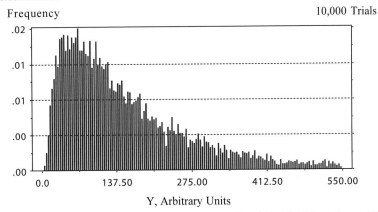

The statistics for *Y* are *Mean = 164.84, Median = 123.60* and *Standard Deviation = 143.34.* Yet, the risk assessor expected that the distribution of *Y* would have been uniform: *Y~U(1, 10)*, because that was known. The assessor should have deconvolved the distributions.

Example 6.12. The distribution of *Z* is Weibull (*1, 1.5*) and *X* to be uniformly distributed, *X~U(1, 10)*. The approximate distribution of *Y* is:

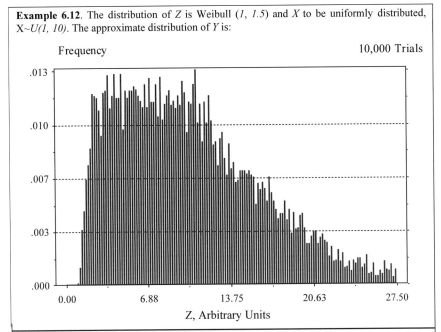

Example 6.13. Suppose that *X* and *Y* are statistically independent, identically distributed random variables. The value of each is equally likely to be *0.1* or *10*. Then both *X/Y* and *Y/X* are expected to exceed 1, the expected value of their ratio is *E(X/Y) = E(Y/X) = (0.01+1+1+100)/4 = 25.5*, even though *X* and *Y* are identically distributed. In words, if *X*

> represents the illness rate in a population exposed to *A* and not *B*, while *Y* denotes the corresponding rate in a population exposed to *B*, and not *A*, and if true response rates are driven by a risk factor not correlated with either exposure, then the expected value of the relative risk exceeds *25* for each factor.

Modeling a risky process through Monte Carlo methods requires knowing, or somehow being able to develop from theoretical or other knowledge, the distributions of the input variables. However, there are situations where risk assessments cannot rely on Monte Carlo methods. This occurs when:

- The distributions of the inputs are unknown
- Samples are small and therefore the asymptotic assumptions are inappropriate

In these situations, the risk assessor can use alternative statistical methods, such as the bootstrap methods, discussed next.

6.4. BOOTSTRAP METHODS

The rationale for the bootstrap is shown in Figure 6.4, modified from Efron and Tibshirani (1993), in which the subscript *est* means that the quantity with that subscript is *estimated*. The basis of this computational scheme is that the random sample of independent observations (*x**) *is* the empirical cumulative distribution sought.[5] The bootstrap method uses the empirical distribution function generated by the sample, which is taken to be the *estimate of the entire distribution* (Efron and Tibshirani, 1993).

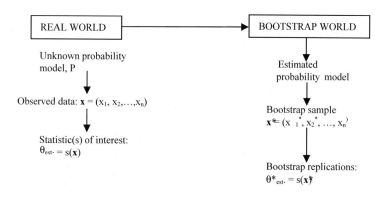

Figure 6.4. Basis of a Simple Bootstrap

The statistic sought is *s(x*)*; it can be the median; the coefficient of correlation or some other statistic. The bootstrap generates a relatively large number of independent samples, drawn at random (with replacement) from the single original sample, to

---

[5] A bold character indicates a vector or a matrix, depending on the context.

develop an empirical cumulative distribution function that approximates the unknown distribution of $s(x)$. The computation is summarized in Table 6.1, using a random sample generated from an unknown probability distribution $F(.)$, which is *ordered* as $\{x_1, x_2, ..., x_n\}$.

Table 6.1. Bootstrap Computational Steps

Step	**Computation**	**Comment**
1. The random sample is $\{x_1, x_2, ..., x_n\}$.	Order the observations from smallest to largest.	The set $\{x_1, x_2, ..., x_n\}$ contains the observed realizations of random variable X.
2. Define $F'(X = x_i)$.	This is the empirical cumulative distribution of the original sample.	X is the random variable.
3. Calculate $F'_n(X = x) = [\#(x_i \leq x)]/n$.	This sets up the calculations for the empirical cumulative distribution	$\#(x_i \leq x)$ means that the number of times the inequality holds is from 1 to n.
4. Let B → large number and calculate the $S*(\mathbf{x})$	Increase the number of samples to some suitable size N, and calculate the statistic	This generates the bootstrap empirical distribution from which the statistic is estimated.

The final step consists of graphing the discrete cumulative distribution for the random variable *X*. This is a histogram. Bootstrap algorithms can include Monte Carlo simulation to accelerate calculations and to achieve statistically efficient estimates. Each bootstrap result is an approximation of the true estimator. Note that the samples used in developing empirical distribution functions are assumed to be random (each value has an equal chance of being selected from the population). When this is not the case, more complicated methods must be used (Rao and Wu, 1988). An important use of bootstrap methods is developing the joint confidence intervals for multivariate models, such as exposure-response models (Hall 1987).

The formula for the standard error of a bootstrap simulation illustrates the type of computations the risk assessor may encounter in using this method. The sample mean equals $\sum_i x_i/n$; s_x^2 is the standard deviation of the sample. The estimated bootstrap standard error, for *B* bootstrap samples of size *n*, is:

$$se_B = \left\{ \sum_{B=1}^{B} \left[s(x^{*B}) - \left(\sum_{B=1}^{B} x^{*B}/B \right) \right]^2 / (B-1) \right\}^{1/2}.$$

In this equation, *B* is the number of bootstrap samples (Efron and Tibshirani, 1991). Specifically, x^{*1} is the first bootstrap sample, x^{*2} is the second, and so on through to x^{*B}. Efron and Tibshirani use x^{*B} to indicate a random bootstrap sample from the empirical distribution $F(x)$. Each bootstrap sample has its own statistic, such as the standard error $s(.^*)$. There are *B* values calculated by each $s(x^*)$, one for each bootstrap sample; $s.e._B$ is the standard error of the entire bootstrap replications.

Efron and Tibshirani (1991) have suggested that B might be in the range of *50* to *200* bootstrap samples for the distribution of the standard error of the mean. If a risk assessor were interested in estimates of the confidence interval, then the computational burden would increase by approximately *a factor of ten*. The two examples that follow illustrate the concepts.

Example 6.14. In an assessment of lead contamination data from two instruments (in ppm) we used a bootstrap with *1,000* replications to generate the medians and the upper and lower confidence intervals for the medians.

The median for instrument A is *202.5* (*95% CI* for the median = *64, 1,866*); the median for B is *185.5* (*95% CI* for the median = *62, 3,921*). Note that, in most practical situations, the confidence limits for the median are seldom estimated or reported.

In the example that follows we use a bootstrap method to estimate the parameters of a multiple regression model:

$$Y = c_1 + c_2 X + c_3 Z.$$

We also compare these results to those obtained by using the OLS estimation already discussed in chapter three.

Example 6.15. Let [*5 3.7 4.7 4.2 2.2 6.1 6.3 9.3 4.0 3.0 4.0 5.0 4.0 3.2 4.1 4.1 4.6 4.4 4.9 4.7*] be the vector of data for the dependent variable Y. Let [*20 30 40 30 34 38 41 42 55 35 30 29 30 41 26 41 45 35 23 30*] be the vector for the independent variable X and [*48 51 77 66 52 65 57 67 76 53 67 56 66 45 74 65 75 55 69 65*] be the vector for the independent variable Z. The units are arbitrary.

Suppose that the linear regression is $Y = c_1 + c_2 X + c_3 Z$. Using *1,000* bootstrap samples, the estimates of c_1 = *0.029*, c_2 = *-0.0047* and c_3 = *4.24*. The *5%* and *95%* bootstrap confidence limits are estimated as: c_1 = *(-0.037, 0.106)*, c_2 = *(-0.0156, 0.085)*, and c_3 = *(-3.02, 4.70)*. These results are statistically insignificant. By comparison with the bootstrap results, the regression using the OLS yields the following estimates: c_1 = *2.027*, c_2 = *0.012* and c_3 = *0.034*. These results are also statistically insignificant (e.g., at α = 0.05): the *p*-values of each of these three estimates are *0.405, 0.781* and *0.372*, respectively.

The bootstrap use of sample data as the best estimate of a population distribution is justified by a theorem proving that the empirical distribution (developed from the sample data) is a *sufficient* statistic, if the empirical frequencies are generated through random sampling (Efron and Tibshirani, 1991). Moreover, the bootstrap estimator of the empirical distribution is consistent: the sample distribution tends to the population distribution as the sample size tends to infinity. An immediate result is that sample estimates can be studied in detail, rather than being justified by (possibly inappropriate) asymptotic properties. Another advantage of the bootstrap is that all of the sample-to-population characteristics are empirical distributions.

An empirical distribution can be matched to theoretical distributions and then the amount of discrepancy can be shown to enhance understanding the results. This is useful in that the analyst can show to the decision-maker what the distribution looks like and can calculate upper and lower *coverage* intervals. There are different bootstrap methods. A reason for being careful with the type of the bootstrap estimator used is that, when a bootstrap method is used to develop upper and lower coverage levels, their accuracy is measured relative to theoretical confidence intervals. The rate at which the error between the numerical approximation and the theoretical value converge reflects the accuracy of the numerical stochastic approximation.

6.5. MARKOV CHAIN MONTE CARLO, MCMC

MCMC methods apply to Bayesian and classical statistical inference, such as by generating the posterior distribution or to implement complicated aspects of maximum likelihood estimation. The MCMC produces estimates of statistics such as the mean, variance and so on of *marginal* distributions of random variables, when the *joint* distribution is known.[6] This contribution is particularly useful because, although it may be possible to perform the required integrations, in practice the computational burden often is far too great or may be impossible. Recollect that Monte Carlo methods use an expected value calculated by a discrete sum to approximate an integral. Suppose that the expectation has to be applied to a problem that involves several dimensions. In these situations, the evaluation of the integral can become extremely difficult. Consider, following Gilks, Richardson and Spiegelhalter (1996), the Bayesian posterior distribution (θ is a parameter, either single or vector-valued):

$$pr(\theta \mid data) = \frac{pr(\theta)pr(data \mid \theta)}{\int pr(\theta)pr(data \mid \theta)d\theta},$$

in which the denominator is the total probability distribution, the normalizing condition already discussed in the context of Bayes' theorem using probabilities. The change introduced from the earlier discussion is that we are now dealing with functions because we are assessing several random variables. Although the calculation of expectations yields the mean, variance and higher order moments of the posterior distribution, it should come as no great surprise that the evaluation of those complicated integrals can be the source of analytical and numerical difficulties and yield the target distribution.

The basic MCMC method is as follows. Suppose that the sequence of random variables generated by the process described by the random variable X is $X_{t=0}$, $X_{t=1}$, ... such that as t increases, the sample is taken from the conditional distribution of $(X_{t+1} \mid X_t)$. This is a property of Markov chains: the previous realizations of the random variables do not matter to the estimation. All that matters is the immediate past, represented by X_t. This means that the evolution of the stochastic process described by the sequence of random variables that belong to X, namely $X_{t=0}$, $X_{t=1}$, ..., forgets the effect of the past; only the current realization represented by X_t affects X_{t+1}. Furthermore, as t increases, the chain will tend towards a stationary distribution. Therefore, the realizations of X will also be characterized by a stationary distribution.

---

[6] The probability distributions of random variable X and Y are, respectively $f(X)$ and $g(Y)$. For example $f(X)$ may be the log-normal distribution and $g(Y)$ may be an exponential distribution. Their joint distribution, $h(X, Y)$ means that we are dealing with $pr(X = x_i \, AND \, Y = y_i)$. These ideas are developed in Chapters 8 and 9.

Samples from this distribution have characteristics that can be used in estimation. The details can be found in Gilks, Richardson and Spiegelhalter (1996).

Recollect also that, in Monte Carlo simulation, as discussed, the *trials* are assumed to be independent and identically distributed. What if those samples are not independent, as Monte Carlo methods generally require? MCMC provides the answer in this situation, as discussed by Gilks, Richardson and Spiegelhalter (1996). First recollect that the normalization constant in Bayes's theorem can be unknown or very complicated. Thus, in practice, the quantity *pr(θ|Data)* is set to be proportional to *[pr(θ)pr(Data|θ)]*. That is, the proportionality sign replaces the equality sign in Bayes' theorem to avoid the problem of knowing the form of the normalization term (the denominator of Bayes' theorem). Yet, even this solution can be unacceptable when deal with several dimensions because the accuracy of estimation is adversely affected by omitting an important component of the formula. Let us define a vector-valued random variable, *X*, which can be sampled. Its posterior distribution, *π(x)*, is used in the expected value of the function:

$$E[f(x)] = [\int f(x)\pi(x)dx] / \int \pi(x)dx,$$

in which the small *x* indicates a realization of the random variable *X*. Markov chain Monte Carlo estimation yields the distribution of the end of the Markov chain. If the distribution is stationary we can develop its expectations. These include the mean, variance, mode, confidence limits and other statistical quantities. As Gilks, Richardson and Spiegelhalter note, MCMC can be used in classical statistics, in which case *π(x)* is the likelihood. The space of the realizations (a very large set) is the Cartesian product. Each of the realizations has a specific distribution. If the space of the realizations were characterized by independent random variables with the same distribution, then the estimation would be controlled by the strong law of large numbers, and the size of the sample would establish its accuracy. If the distribution of these random variables were stated as a product, then the feasibility of the calculations becomes manageable. The simulation of the Markov chain depends on this distribution and technical assumptions, including invariance, discussed by Norris (1997). As in Monte Carlo simulations, we sample from the uniform distribution *U(0, 1)*.

The Gibbs sampler is a method to develop the posterior distribution from the full conditional distribution using MCMC (Gilks, 1996). The essential aspects of this iterative method for sampling consist of partitioning a vector of values, such as parameters into subcomponents. For example, the vector $\theta = (\theta_1, \theta_2, ..., \theta_d)$ can be partitioned in a number of partitions, *d*, that can be sub-vectors or scalars. Sampling at random yields the conditional distribution:

$$f(\theta_j| \theta_1^t, \theta_2^t, ..., \theta_{j-1}^t, ..., \theta_d^{t-1}),$$

in which the superscript t identifies the iteration. The vector θ_j is updated on the most recent values of the vector θ from all other sub-vectors. The question answered through the simulation is: What is the conditional distribution of the vector θ? In the case of directed acyclic graphs, discussed in the context of Bayesian networks, the full conditional distribution for a node is the conditional distribution that includes the node and all other nodes in the graph. It shows the dependencies of the parents to the node, its descendants (or children) and any other co-parent to that node. It is *full* because it includes the prior and the likelihood distributions; each node has a full conditional distribution.

The Gibbs sampler algorithm is as follows (Spiegelhalter, Bets, Gilks, and Inskip, 1996):

- "Starting values must be provided for all unobserved nodes (parameters and missing data)
- full conditional distributions for each unobserved node must be constructed and methods to sample from them decided upon
- the output must be monitored to decide the length of the 'burn-in' and the total run length …
- summary statistics for quantities of interest must be calculated from the output, for inference about the true values of the unobserved nodes."

That is, the distribution $f(x)$ is generated by sampling from two conditional distributions: $h(x|y)$ and $g(y|x)$. The Gibbs sampler applies to different values of these two conditional distributions, beginning with an initial value of y, generating a sequence of values of y and x. The distribution of these x values (the sequence involves successive values of y and x, starting at y_0 and x_0) converges to $f(x)$, the distribution that is sought, as more values of the sequence are generated. The core of the idea is to use a *proposal* distribution (conditional distribution), which is equated to the full distribution of the value. Gibbs sampling technique can be used in modeling the nodes of a network connected according to graph theory.

As Cox (2002) discusses, MCMC methods can be used to choose between alternative model specifications, estimating Bayes' factors in Bayesian model averaging and in many other practical uses. For example, MCMC methods can be applied to making choices of model when using regression models such as Poisson, logistic and other statistical models. This section does not contain an example because specialized software is generally used for MCMC simulations. Those who are interested in this topic can refer to Gilks, Richardson and Spiegelhalter (1996) as well to the worldwide web for specific example and software.[7]

MCMC and Gibbs sampler have direct applications in missing data analysis and in a number of applications such as pattern analysis and signal detection. MCMC can be used in regression models as well, although the link may not always be obvious. The choice between alternative models can be studied through Bayes' factors, discussed by Raftery, in Gilk, Richardson and Spiegelhalter (1996). Another alternative to making formally defensible selections using a model decomposition based on MCMC is based on the Bayesian posterior distribution (Raftery, in Gilk, Richardson and Spiegelhalter, 1996). Latent variables can also be analyzed through Bayesian Networks and MCMC, an important aspect of formulating inductive causal arguments. The assessment of goodness-of-fit of a model to the data and the

---

[7] See omega.albany.edu:8008/cdocs/.

appropriated set of causal variables in the specification of the model are also handled through the Gibbs sampler (Gelman and Meng, in Gilk, Richardson and Spiegelhalter, 1996).

6.6. MISSING DATA

A topic of practical relevance to risk assessment and management is issues that relate to missing data. Missing data are not equivalent to censored data, because a missing number is unknown. By contrast, a censored value may be known for a period of time. Often and incorrectly, empty cells (blanks) are given a zero or some other value without disclosing the theoretical rationale for such action. For example, a common practice of filling-in missing data uses either the limit of detection, half the limit of detection or zeros to fill in for missing values. Cox (2002) and Cox and Ricci (2002) discuss censoring and how to deal with censored information in the context of statistical models used in risk assessment that were discussed in Chapter 4.

A useful classification for missing data is either being due to *observation unavailability* or to variable *non-response*. The former is the result of having to deal with populations and sampling. The latter is the result of human or other errors. More practically, data can be missing either at random or not. Recollect that data is drawn at random when the probability of observing each one of them is unaffected by the magnitude of the response or by the drawing mechanism. It follows that data is missing at random when the missing values of the response are unrelated to the values that would have been sampled (Little and Rubin, 1987). If data are observed at random and the unobserved data are missing at random, then the estimates of statistical quantities such as the mean that are obtained from the sample are unbiased. The main effect would be a reduction in the size of the sample resulting in more variable estimates than expected with the full sample and so on. Otherwise, the estimates can be biased, thus affecting managerial solutions.

The *imputation* of missing data in a regression uses estimates of sufficient statistics, such as the variance and covariance, to form functions of the parameters of the model. Estimation uses a special maximum likelihood algorithm, the SWEEP algorithm (Little and Rubin, 1990). Another aspect of data analysis related to missing data is that the patterns of missing data can be sparse (but still be random) or take specific patterns. For instance, one sample may have gaps in the data near the low values, another have gaps near the high values.

Multivariate samples, samples consisting of data on several variables, may have subsets (blocks) of missing data, the number of missing data may be higher for certain variables than for others or be clustered toward the tails of the joint distribution of the data. Suppose that the data sets were as depicted by the Table 6.2.

Table 6.2. Input Data Table

X_1	X_2	...	X_k
x_{11}	x_{21}	...	x_{k1}
x_{12}	x_{22}	...	x_{k2}
...	...	...	...

x_{1n}	x_{2n}	...	x_{kn}

Suppose that the missing data is x_{22}. The sample statistics for X_2 would be computed from a smaller sample size than the remaining variables. If the samples were randomly drawn, and the missing data points were also missing at random, the effect on the estimates would be loss of statistical efficiency but the estimates would be unbiased. Extending this reasoning, if some data for X_2 and X_1 were missing pair-wise, the missing pair could be excluded in such bivariate statistical analyses as the coefficient of correlation. The sample size would be reduced, again with a loss of efficiency. Aside from the practical fact that it is not scientifically sound to ignore missing data without full explanation, an important question is *Can the missing data be replaced and, if so, how?* In other words, are there statistical methods that can help *fill the gap?* An approach to fill the gaps is to substitute the mean of the sample for the missing observation (Little and Rubin, 1987). The rationale is that, for a random sample, the estimated mean is the most likely value of the sample. Using this method, however, biases the sample variance. The extent of the bias (an underestimation) is measured by:

Bias = (n-1)/(n_j-1).

In this formula, n is the sample size, and n_j is the sample size without the imputation. If there are two or more samples, the effect of imputing the mean value on the sample variance-covariance matrix must also be determined.

Another technique to account for missing data is to use the *conditional mean imputation* (Buck, 1960). The method uses the regression, a conditional form of estimation because of $(Y|X)$, through the steps that follow:

1. Several regressions (regressing X_j on X_i) are computed, each for each data set that is complete,
2. The data that are used to fill the gaps are those fitted by the regressions: the regression equation is used to impute the data (namely, the conditional mean) of the missing value of X_j.

If the variation of X_j is *explained* by X_i, the conditional mean imputation is a reasonable method for dealing with the missing data. Explanation is measured by the difference between each fitted value and the associated average value. Non-parametric maximum likelihood estimation methods can impute missing data (Maritz, 1981). Those methods can be used for i) censored data; ii) observations that include independent variables for each individuals; and iii) observations taken at several stages in a process.

Maximum likelihood methods, as discussed in Chapter 4, can construct confidence regions and compare alternative models through the likelihood ratio. The application of the maximum likelihood method to the problem of missing data involves the following steps:

1. Set up the probability model,
2. Develop the likelihood function, given that the distribution functions of the variables are known,

3.   Obtain estimates of the parameters of the function.

The joint density function formed by the product of the normal density functions for the complete data and for the incomplete data is used to obtain estimates of the probabilities associated with the missing values. The reason for using the maximum likelihood method is that its estimates are (statistically) efficient and consistent even when the missing data are not missing at random. The difficulties include the fact that the conditional likelihood can become difficult to compute and that inference requires large sample approximations, unless exact methods are used.

A simple way to handle missing data is to plot them against their probabilities and then extrapolate from the actual data towards the area of the missing values. Transformations of the raw data may bias the results of the analysis because the average of the data will be different from the average of the transformed data. Robust methods, such as robust regression, currently available in commercial statistical computer software, can help here. One of them is discussed in Chapter 7 of this book, in the context of Classification And Regression Trees, CART. MCMC methods can also be used either by themselves or with other statistical methods. One of those methods is the *expectation-maximization* (EM) method. It consists of assigning initial values for the parameter estimates and then calculating sufficient statistics (such as the sum of squares) from the experimental data, which includes data missing at random (Dempster et al., 1977). The likelihood of the data is based on the complete data set (which, however, consists of a matrix of the observations (subscript *obs*) and the matrix for the missing observations (identified by the subscript *mis*), $L(\theta|X_{obs}, X_{mis})$. The expectation step is based on the log-likelihood, which is conditioned on some initial estimate of θ. The maximization step maximizes this expression through a number of iterations aimed at producing larger values of the expected log likelihood (Little and Rubin, 1990). Calculating the maximum likelihood of the sufficient statistics is the *expectation* stage of the procedure. The expectation step consists of calculating the expectation of the data conditioned on the parameters and prior knowledge. The *maximization* step uses the sufficient statistics of the actual data; it is iterated to *maximize* the likelihood to obtain locally optimal parameters through convergence to a suitably small number (Little and Rubin, 1987).

The EM method requires that the data be missing at random, but the sample need not be random. The EM estimates functions of missing data such as the covariance and thus not points on the sample space. The idea is to find a local maximum for the likelihood function of the data conditioned on the parameters, the causal model and prior knowledge (Heckerman, 1996). The EM method can be applied to contingency tables and other statistical models, such as those discussed in Chapters 4 and 5.

Example 6.16. Suppose that the risk assessor wants to estimate means and correlation for the following random variables: population, life expectancy at birth, GNP, infant mortality, health expenditures and literacy. The estimates of the means of these variables, *omitting* the missing data, are:

POP_1983	LIFE_EXP	GNP_82	BABYMT82	HEALTH_84	LITERACY
18.995	63.571	4138.31	59.193	193.839	73.538

The expectation-maximization (EM) estimates of the means are:

POP_1983	LIFE_EXP	GNP_82	BABYMT82	HEALTH84	LITERACY
20.21	63.727	4240.037	57.937	208.932	73.563

EM estimated correlation matrix (lower triangular shown because the upper triangular is the same as the lower triangular; that is LITERACY and LITERACY have correlation 1.00, not shown):

	POP_1983	LIFE_EXP	GNP_82	BABYMT82	HEALTH84
POP_1983	1.000	--	--	--	--
LIFE_EXP	-0.025	1.000	--	--	--
GNP_82	-0.000	0.660	1.000	--	--
BABYMT82	0.078	-0.976	-0.656	1.000	--
HEALTH84	0.012	0.585	0.939	-0.590	1.000
LITERACY	-0.034	0.942	0.632	-0.929	0.559

Bartlett *chi*-square statistic = 324.719 with df = 15 and *p*-value = 0.000. If the missing data are omitted from the analysis, Pearson's correlation matrix (see Chapter 10) is:

	POP_1983	LIFE_EXP	GNP_82	BABYMT82	HEALTH84
POP_1983	1.000	--	--	--	--
LIFE_EXP	0.173	1.000	--	--	--
GNP_82	0.183	0.646	1.000	--	--
BABY MT82	0.117	-0.973	-0.644	1.000	--
HEALTH84	0.190	0.589	0.936	-0.600	1.000
LITERACY	0.205	0.936	0.610	-0.915	0.556

6.6.1. Limit of Detection

A typical situation that often occurs in practical risk assessment is when analytical chemistry methods are used to determine the amount of pollution present in a medium. Those methods rely on instruments that are characterized by *limits of detection*: the instrument is unable to provide measurement below a specific threshold. In risk assessment, the two aspects of missing data are related to the *limit of detection* and *limit of quantitation*. The former, the focus of this section, is the limit imposed on the resolution achievable by an instrument. Detection is related to data reported as *non-detects* or as *less than*. The data is unavailable below the limit of detection for the instrument because the *signal-to-noise* ratio does not, for a given instrument, permit an accurate detection of the signal from background noise. Using any specific instrument (or any other device) all observations will be above the detection limit.

Let:

D = the detection limit,
k = the number of values above the detection limit,
n = the total number of observed values.

Consider the data below the threshold (the missing data). Under the assumption that the missing data follow a normal distribution, a method to deal with missing data is due to Cohen (1959). He has developed the following formulae for the mean and variance in the presence of the data assumed to be left-censored (but still missing):

$$\bar{x}_C = \bar{x} - \lambda(\bar{x} - D),$$

and:

$$s_C^2 = s^2 + \lambda(\bar{x} - D)^2.$$

Example 6.17. The following sample for total arsenic, in parts per million [ppm], was measured with a hypothetical instrument with detection limit of *10* [ppm]:

13, 14, 14, 20, 21, 32.

The mean is *19* [ppm] and the standard deviation *7.21* [ppm]. The upper *95%* confidence limit is *26.57* [ppm] and the lower 95% confidence limit is *11.43* [ppm]. To illustrate Cohen's method, assume that the limit of detection is *15* [ppm]. In this case, $n = 6$, $k = 3$, and $D = 15$. It follows that:

$\bar{x} = (20 + 21 + 32)/3 = 24.33,$

and:

$s^2 = [(20-24.33)^2+(21-24.33)^2+(32-24.33)^2]/3 = 29.56.$

In addition, $\gamma = 29.56/(24.33-15)^2 = 0.34$, and $h = (6-3)/6 = 0.50$.

Using Cohen's tables, the (linearly) interpolated value of λ is *0.94*, from which the adjusted mean and variance are, respectively, *[24.33-0.94(24.33-15)]* = *15.56* [ppm], and *[29.56+0.94(24.33-15)²]* = *111.47* [ppm].

The statistical inefficiency of Cohen's method under small sample conditions has been discussed in the literature. Table 6.3, developed by J. Warren, US Environmental Protection Agency, depicts the differences between practical methods to deal with data not detected in experimental samples.

Table 6.3. Results from Methods to Impute Missing Values

	Detection Limit (ppm)	Sample Mean (ppm)	Sample Standard Deviation (ppm)	95% Upper Confidence Limit (ppm)
Sample	10	19.0	7.2	24.9
Missing Values Set to Equal Zero	15	12.2	14.0	23.7
Missing Values Set to between 0	15	15.9	10.1	24.3

and 15				
Missing Values Set to Detection Limit	15	19.7	6.6	25.1
Cohen's Method	15	15.6	10.6	24.2

In dealing with missing data, the risk assessor must identify two things. First, whether the method he uses is statistically sound, a risk assessment issue, and second whether the results are used in policymaking. Purpose is important because, if statistical results determine environmental clean-up levels or other management objectives, the level of statistical conservatism inherent to justifying the policy choice must be shown and discussed with the results obtained with other methods. In principle, the justification of data analysis *for* policymaking should be higher than scientific standard because of the social cost of making an incorrect decision.

6.7. INTERVALS

The risk assessor will find situations in which the uncertainty in the values of a variable cannot be described by a distribution function or an expert will be unable to provide such distribution. Intervals, two ordered numbers, can also characterize uncertainty.

Example 6.18. The interval *[2, 15]* associated with the hypothetical variable AGE is understood to mean that an individual age can be either *2* or *15* and that nothing else is known about that variable. By comparison, consider the statement *AGE is uniformly distributed between 2 and 15 years*. This involves:
1. The distribution is uniform and discrete, $U(a = 2, b = 15)$. It means that each discrete number within $U(2, 15)$ is equally likely. The uniform distribution $U(a, b)$ has a probability distribution $1/(b-a)$ for $a < x < b$, and zero otherwise. Its cumulative distribution is 0, for $x \leq a$; $(x-a)/(b-a)$ for $a < x < b$; and 1 for $x \geq b$.
2. The expected value and the variance of the uniform distribution, $U(a, b)$, are $(a+b)/2$ and $(b-a)^2/12$, respectively.

The uniform distribution can represent some lack of knowledge about the distribution of the random variable when expressing expert knowledge about the prior. This can be done with Bayes' theorem. A justification for the uniform distribution is the symmetry principle whereby, when the outcomes are homogeneous and *nature* cannot influence one outcome over another, then each event has an equal probability of occurring. A second justification for choosing the uniform distribution as the appropriate distribution to represent lack of knowledge is based on the idea of entropy of information. The algebra of interval numbers differs from the algebra of real numbers (numbers measured on the real axis). Table 6.4 depicts some of the operations used with intervals:

Table 6.4. Operations and Example of Results Obtained Using Intervals

Interval operation	Resulting Interval	Example	Comment
[a, b]+[c, d]	[a+c, b+d]	[23.54, 32.09]+[54.1, 98.5] = [77.64, 130.59]	None

[a, b]-[c, d]	[a-d, b-c]	[23.54, 32.09]-[54.1, 98.5] = [-74.96, -22.01]	None
[a, b]*[c, d]	Min[ac, ad, bc, bd], Max[ac, ad, bc, bd]	[23.54, 32.09]x[54.1, 98.5] = [1274.51, 316.87]	c, d > 0
[a, b]/[c, d]	min[a/c, a/d, b/c, b/d], Max[a/c, a/d, b/c, b/d]	[23.54, 32.09]/[54.1, 98.5] = [0.239, 0.593]	c, d > 0

Logical operations such as *AND* as well as *OR* applied to events *A* or *B*, result in *envelopes* of intervals:

(A AND B) = *envelope[max(0, A+B-1), min(A, B)]*,

(A OR B) = *envelope[max(A,B), min(1, A+B)]*.

Intervals represent the correct way to portray uncertainty about a quantity when the magnitude of that quantity *is* just an interval, rather than a distribution or range, for a given sample. Moreover, the dependence between the quantities represented by A and B do not have to be known because the resulting envelopes, although optimal, do not require that knowledge.[8]

The practical importance of not having to make the assumption of independence has two aspects. The first is that it simplifies probabilistic calculations: *pr(A AND B)* = *[pr(A)pr(B|A)]* simplifies to *[pr(A)pr(B)]*. The last expression states that the probability of event *A* does not affect the probability of event *B*: *A* and *B* are stochastically independent. The second is that we work with the marginal distributions of two (or more) random variables to *bound* the uncertainty of the dependent variable using operations (such as addition, multiplication or other) performed on the two marginal distributions, even if there are undisclosed dependencies.[9]

Example 6.19. Suppose that exposure to an agent is stated to be the result of being exposed between *20* and *45* years, at concentrations between *0.01* to *10* micrograms/liter of water drunk, given that between *1* and *3* liters of water are drunk per day. The amount of mass ingested may be calculated using the simple formula *Exposure* = *[(mass input)/(unit of exposure)]*(period of exposure)* yielding[10] the formula:

Exposure[Mass, micrograms of agent] = *A[µg of agent/liter of water]*B[liter of water ingested/day]*C[number of days]*.

---

[8] Based on Frechet's inequalities in real analysis, (Moore RA, *Methods and Applications of Interval Analysis*, SIAM, Philadelphia, PA (1979). Ferson S, Reliable Calculations of Probabilities: Accounting for small sample size and model uncertainty, in *Intelligent Systems: A semiotic perspective*, NIST, Oct. 1996; Ibid. What Monte Carlo Methods Cannot Do, *Human and Environmental Risk Assessment*, 2:990 (1996), stated these bounds as:
$max[0, pr(E)+pr(F)-1] \leq pr(E\ AND\ F) \leq min(pr(E), pr(F))$, and
$max[pr(E), pr(F)] \leq pr(E\ OR\ F) \leq min(1, pr(E)+pr(F))$.
[9] Williamson RC, T Downs, Probabilistic Arithmetic I: numerical methods for calculating convolutions and dependency bounds, *Int. J. Approx. Reasoning* 4:89 (1990).
[10] The exposure formula would require knowing the time series of concentrations, $\{C(t)\}$, and knowing the appropriate exposure statistic to be used.

Deterministic Analysis. The deterministic calculation yields: $E = A*B*C = 0.01*2*1,600 = 32$ micrograms of the agent. The minimum is $(0.01)(7,300)(1.00) = 734$ micrograms; the maximum is $(10)(16,425)(3) = 492,750$ micrograms.

Probabilistic Analysis. Suppose that the probability distributions of the random variables A, B and C are known (and have the same units as the deterministic case): *log-normal*$_A$ $(0.01, 0.001)$, *uniform*$_B$ $(1, 3)$ and *triangular*$_C$ $(200, 1,600, 3,600)$. The mean of the uniform distribution is 2 liters per day and the standard deviation is 0.57 liters per day; the mean of the triangular distribution is $1,600$ days with standard deviation 725.72. A Monte Carlo simulation (approximately $10,000$ trials, assuming *independence* between the input variables) of the expression yields the following empirical distribution, to which we fit a *log*-normal distribution:

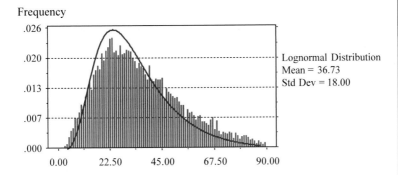

The mean of the *empirical* distribution is 36.20 micrograms of the agent, the median is 32.97 micrograms, the standard deviation is 18.30 micrograms and the standard error of the mean of the sampling distribution is 0.18. As a form of sensitivity analysis we change the standard deviation of the random variable A to be 0.01 (the same value of the mean), with nothing else changed. We obtain the following output distribution, to which we have fit an exponential distribution:

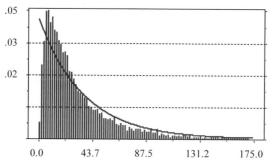

The mean of the empirical distribution is *36.26*, the median is *22.73*, the standard deviation equals *44.79* and the standard error of the mean equals *0.45*. What is noticeable is that the tails of the new distribution are quite different from the base-line calculation. *Interval Analysis.* Suppose that the uncertainty in *A*, *B* and *C* can only be represented by three intervals: *A* = *[0.00, 5]*, *B* = *[1, 3]* and *C* = *[200, 3,600]*. The uncertainty about the output variable, measured as an interval, is *[0.00, 54,000]*.

As always, for measures representing uncertainty, such as intervals and probability distributions, the algebra of intervals or random variables, respectively, should be used in inverting equations (Moore, 1979; Bernabini et al., 1987).

6.7.1. Discussion

If we have sufficient knowledge about a problem to be able to specify the distribution, then we know quite a bit about that problem. If that is not the case, it becomes useful to have other tools in our kit. Thus, a reason for including intervals and other measures of uncertainty is to guide in those situations in which the risk assessor may be unable or unwilling to provide either a single number or a distribution. This is the situation where, for whatever reason, an individual is *unwilling* to assert that the uniform distribution over that interval is the appropriate representation of her uncertainty.

Recent developments in probabilistic reasoning indicate that upper and lower probabilities appear to unify the frequency and Bayesian views of what is a probability. This advance would also account for intervals (Hampel, 2001). Upper and lower probabilities can be useful in practical risk assessments characterized by incomplete knowledge; the reader may refer to Howson and Urbach (1993) for discussions. The concepts are not difficult, but the reader may also access the references that provided by Howson and Urbach. Hampel (2001) clarifies most of the concepts used in current probabilistic reasoning.

6.8. ENTROPY OF INFORMATION

The entropy of information of the random variable X, ($X = x_1$, ..., x_n), is the expected value of the information contained in X. It is calculated as:

$$\Sigma_i\{-pr_i[log(pr_i)]\}.$$

When the probability equals 0, the expression $pr[log(pr)]$ is not defined and therefore requires a transformation from numbers in the interval $[0, 1]$ to $[0, \infty]$. A practical interpretation of uncertainty represented by entropy is that entropy reaches its maximum value when the uncertainty about the value of a discrete random variable uniformly distributed is at a maximum; it is zero when the random variable takes a specific value, ($X = x$), with certainty. A way to describe information is to consider a composition of *surprise* and *meaning*. The former is amenable to description through measures such as probabilities. The latter depends on context, cognitive knowledge and scientific theories. *Surprise* is a more intuitive and familiar concept than probability. The idea is to develop a transformation of probability to achieve a more practical representation of uncertainty.[11] This simple discussion develops the groundwork for causal models in Chapter 10. Information, $I(.)$, can be defined in terms of probabilities as:

$$I(E) = -k\,log_a[\,pr(E)],$$

where the logarithm is to the base a and the units are *bits* of information. This base is *2*, rather then *10* or *e*.

Example 6.20. Suppose that the probability of event *1* equals *0.25* and that of event *2* equals *0.077*. Then $I(E_1) \approx 2$ *bits* and $I(E_2) \approx 3.7$ *bits*, from which the joint entropy, $I(E_1$ and $E_2) = 5.7$ *bits*. The calculations are: $ln(0.25) = -1.386$, $ln(2) = 0.693$, $ln(0.077) = -2.565$, from which:

$log_2(0.25) = -1.386/0.693 = -2$, and

$log_2(0.077) = -2.565/0.693 = -3.01$.

As the probability number increases, the information content, measured by the value taken by $I(.)$, decreases. The larger $I(.)$ the greater the surprise. What has been accomplished is a mapping from the probability interval $[0, 1]$ to the information content measured on half of the real axis $[0, \infty]$ by the transformation $[-k\,log_2(pr)]$. The entropy of information is a summary measure of uncertainty is the expected value of the information measure

---

Example 6.21. Consider a random variable that takes values 0 or 1: $I(x = 0)$, $I(x = 1)$ and $log_2 pr(X = x) = log_2 pr(1/2) = 1$. Then: $H(x) = [-pr\,log_2(pr)] + [-(1-pr)log_2(1-pr)]$ which reaches its maximum value when $pr = 0.5$.

The entropy of information of two random variables is:

---

[11] Also see Cox (2002).

$$H(X,Y) = -[\sum_{j=1}^{n} \sum_{i=1}^{m} p_{ij} log_a (pr_{ij})].$$

6.9. CONCLUSION

Interval methods are a means to suggest that risk assessors have available alternative ways to represent uncertainty. Although many other measures of uncertainty exist, such as fuzzy numbers, intervals and their algebra are sufficiently simple to motivate further work in the practical representation of uncertainty. This chapter also introduces some ways to deal with missing data. Missing data is a fairly common occurrence in many practical risk assessments, as are censored data. Their analysis and the imputation of data where data is missing has become an important aspect of risk assessment, as are the limitations of imputing missing data. We note that some of the examples are simplified and that deconvolutions can be achieved using commercial software such as Crystall Ball and RAMAS.

QUESTIONS

1. Compare and contrast the Monte Carlo method with the bootstrap suggesting at least two important differences.
2. What might be the effect of deleting censored observation from the sample, and then using only the complete subset of observations on the accuracy of estimating the mean and the standard deviation of the sample?
3. Why should a risk assessor be concerned with propagating the uncertainty in the independent variables to describe the variability of the dependent variable in a formula?

CHAPTER 7.

CAUSAL MODELS: INFLUENCE DIAGRAMS, BAYESIAN NETWORKS, CLASSIFICATION AND REGRESSION TREES

Central objective: to introduce and exemplify recent, practical causal models based on graphical methods

The purpose of this chapter is to extend the discussions of empirical causation in Chapters 4 and 5 by including graph-based probabilistic and statistical methods. Most causal arguments made in environmental management combine qualitative and quantitative descriptions of causation, which is important in risk assessment and management because:

1. Causal explanations must be given to the stakeholders,
2. Laws and regulations are based on scientific and legal causation,
3. Scientific reasoning and explanations are or attempt to be causal,
4. Stakeholders need to know how and why risky events can generate adverse consequences,
5. Risk factors are used in apportioning liability to the sources of the hazard or hazards,
6. Risk reduction and minimization actions require causal knowledge to be fair and equitable.

This chapter contains discussions of four methods for obtaining assessments of relations between exposure and dose-response. These are:

1. Causal path analysis,
2. Influence diagrams,
3. Bayesian Networks (BN),
4. Classification and Regression Trees (CART).

These methods graphically describe how cause and effect are networked, and provide quantitative solution to empirical causation. The early qualitative principles of causation proposed by Koch, Bradford--Hill and Evans include:

1. Strength of the statistical association (e.g., a statistically significant relative risk of 2.00),
2. Consistency of the associations, such as replication by independent researchers with different databases,
3. Precedence of exposure relative to response, *Exposure* → *Response*, but not the other way around,
4. Direction of the relationship between exposure and response,
5. *Physical* (biologically-motivated) basis for constructing and justifying the causal relation,

6.   Specificity and sensitivity of the empirical results between exposure (or dose) and response.

These qualitative principles have increasingly become quantitative through the work of Shafer (1996), Galles and Pearl (1997), Robins and Wasserman (1998), Pearl (2000) and many others. Cox (2002) have reviewed and discussed them in detail. The linkage between correlation methods, such as regression, and the graph-based methods is completed in Chapter 10. The linkages between the methods discussed in this chapter are:

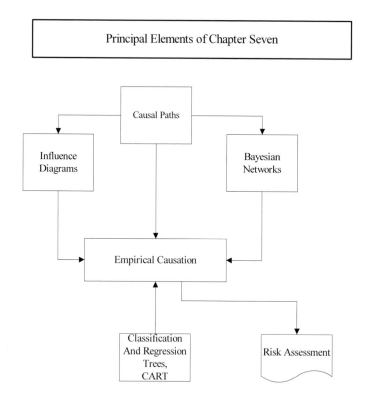

7.1. INTRODUCTION

In stepping from the qualitative to the quantitative, the criteria just listed are affected by confounding and several forms of bias that affect estimation and inference (Cox and Ricci, 2002). Cox (2002) has proposed several steps to overcome some of those problems, which we modify and reorder as follows:

1.   Identify a consistent, non-random association between exposure and response,
2.   Identify, explain and include in the causal relation the *physics* of the relation,

3. Eliminate, or at least explicitly account for, the effect of confounding factors,
4. Eliminate, or at least explicitly account for, sampling, information and modeling biases,
5. Test and confirm temporal precedence and conditional independence,
6. Study, explain and confirm the effect of policy interventions through changes in the value of the independent variables in the causal model on response.

Deterministic causation does not include uncertainty and variability. Practically, one or more (algebraic or differential) equations *completely* describe forward and backward predictions. Thus, the function $y = 3.12+0.06x$, can be understood as the statement that *we are given values for x, then we obtain values of y*. Inserting values for x yields values of y. If the dynamics of a system is modeled, the differential equation: $dc/dt = kc$ defines the rate of change of the quantity c, possibly a concentration, over time, t. Its solution is a continuous time-series of values for c, starting from some initial concentration and time point. Discrete methods (such as difference equations) are available. These methods are fundamental to modeling and describing chemical transformations, the uptake of pollutants and their pharmacological changes, in representing the spread of an infectious disease, all the way to modeling the relationships between health and economic factors. However, causation cannot always be described by *deterministic* algebraic or differential equations. The relationships sought may not theoretically be understood. In these situations, a deterministic model can be modified to account for variability and uncertainty by introducing one or more stochastic components. Alternatively, a purely stochastic model can represent a risky process. We begin with a graphical representation of causation that makes explicit the direction of the linkages between two or more variables. This representation is deterministic.

7.2. INFLUENCE DIAGRAMS

The causal relations of interest to the risk assessor are explicit through diagrams in which nodes represent variables and arrows (paths) link them. The development of this diagram initially is deterministic, as depicted in Figure 7.1.

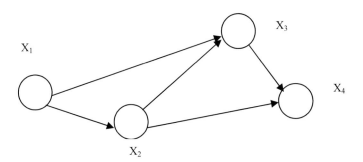

Figure 7.1. Causal Path Network Diagram

Figure 7.1 represents *a hypothetical* causal model consisting of four variables, X_1, X_2, X_3, and X_4; the arrows show the influences between them. An influence can be unidirectional, such as $X_1 \rightarrow X_2$, or can be more complicated. If the arrow has only one head, the variable at the end of the head of the arrow is influenced by the variable at the beginning of the arrow.

In Figure 7.1, X_1 influences X_2 and X_3, but these two do not influence X_1. If a bi-directional arrow, $X_1 \leftrightarrow X_2$ links two variables, then each variable influences the other. Figure 7.2 depicts a hypothetical influence diagram with the direction and strength of the linkages.

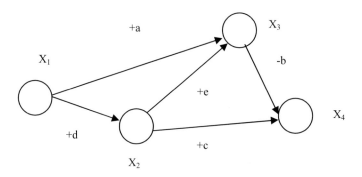

Figure 7.2. Influence Diagram with Linkages, Directions, and Strength of Associations

Theory, past experience and logic (prior scientific knowledge) suggest the variables and their linkages. The linkages can imply an additive (or more complicated) formal relation between them. Moreover, time is implicit in the network: multiple variables and linkages can account for changes or events. These diagrams can represent feedbacks, full and partial order of events, confounders, effect modifiers, direct and indirect cause and effect. If the diagram shown in Figure 7.1 included the strength and direction of the relations (e.g., X_1 positively affects X_3, with the strength of the association measured by a), it would be an *influence* diagram. The diagram depicted in Figure 7.2 has several important features. If the coefficient of a relation between two variables is positive, it implies proportionality: as X_1 increases so does X_2. The direction of the influence of X_1 on X_3 through the path $X_1 \rightarrow X_2 \rightarrow X_4$ is calculated by the multiplication of the signs of the coefficients, the product $(+)(+)$, which is positive. For the path $X_1 \rightarrow X_3 \rightarrow X_4$, the direction of the overall effect of X_1 and X_3 on X_4 is the product of their signs, $(+)(-)$, which is negative. These two chains show indirect effects (mediated by intervening variables). If a node-linkage-node path has zero value, then any path that contains it also has zero value.

In practice, causal paths are inductive and their analysis requires probabilistic and statistical methods. For example, the coefficients a, b and so on can be estimated through methods such as the Maximum Likelihood Estimator. An example of a DAG, with some of the key factors from the four key steps of a risk assessment *Emission $\rightarrow$ Exposure $\rightarrow$ Dose $\rightarrow$ Response*, is shown in Figure 7.3.

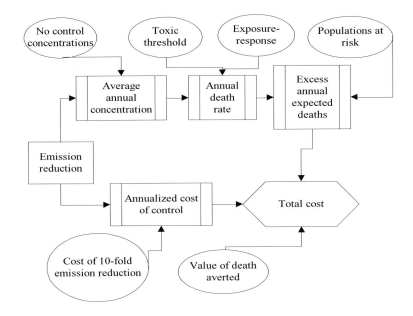

Figure 7.3. Influence Diagram of a Risk-Based Decision

The software Analytica2 (Lumina Decision Systems, 1999) provides an intuitive way to further develop these as a simulation. Table 7.1 depicts the variables used in the influence diagram.

Table 7.1. Variables in the Influence Diagram Depicted in Figure 7.3

Variable	Description	Distributions	Units
Concentration	Average annual exposure-uncontrolled concentration*(1-% emission reduction)	From Monte Carlo simulation	$\mu g/m^3$
Control Costs	10-fold reduction from no control,	Log-normal, $100*10^6$, $1.5*10^6$	$\mu g/m^3$
Emission Reduction	% mass emitted by source	Discrete, table of values	%
Exposure-Response Model	Linear, the estimate of the slope is the variable	Discrete distribution	death/yr
Excess Annual Deaths	Counts, above background	Log-normal	death/year
No control Concentration	Base-line concentration	Log-normal (10, 1.5)	$\mu g/m^3$
Population at Risk	Total number of individuals at risk, all ages, races, sex	Log-normal ($30*10^6$, $3*10^6$)	Counts
Threshold	Exposure below which response does not occur	Uniform (0, 10)	$\mu g/m^3$
Value of Death Averted	Subjective valuation	Log-normal ($2*10^6$, $1.5*10^6$)	$/death averted
Total Cost	(Control cost)+(value of death averted)*(excess deaths)	From Monte Carlo simulation	$/year

The exposure-response model is linear at low dose-rates; the objective of the analysis is to determine the least total social cost associated with technological choice of control level. In this diagram, the boxes in the flow chart identify specific operations. Ovals depict chance nodes, a rectangle represents a decision variable, a hexagon represents an objective; tables, and functions represent relations, and so on. Simulation of this influence diagram model yields the results shown in Table 7.2.

Table 7.2. Results from a Simulation of the Influence Diagram Depicted in Figure 7.3

Total Cost of Emission Control (constant $)							
Reduction (%)	0.00	0.3	0.5	0.7	0.8	0.9	0.95
Minimum Cost	0.00E+00	5.30E+06	1.03E+07	1.79E+07	2.39E+07	3.42E+07	4.45E+07
Median Cost	5.64E+04	1.97E+07	3.46E+07	5.79E+07	7.53E+07	1.04E+08	1.32E+08

7.2.1. Probabilistic Influence Diagrams

A probabilistic influence diagram can be defined (Hong and Apostolakis, 1993) as a *connected, acyclic, directed graph with two types of nodes or factors (decision and chance) and two types of arrows or arc (conditional and informational)*. Figure 7.4 depicts an influence diagram, based on Hong and Apostolakis (1993), which illustrates the levels of a decision and its outcomes; the initial and final boxes identify the beginning and end of the diagrams.

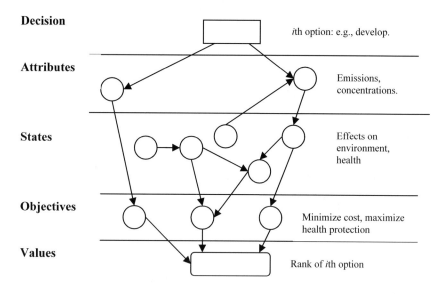

Figure 7.4. Influence Diagram with Decisions, Attributes, States, Objectives and Values

Influence diagrams can be coupled (by coupling one with another decision-maker's influence diagram depicting her network) to model the interactions that take place when two or more stakeholders are involved in a decision. Hong and Apostolakis (1993) have developed a case study consisting of two influence diagrams. These are coupled to represent two stakeholders, a private and a public decision-maker, and how their actions, conditioned on their knowledge of the process, lead to ranking of options. Influence diagrams clarify and quantify the outcomes of decisions and account for the conditioning of the probabilistic network.

A coupled network describes the decisions and the structure of a common decision for different stakeholders. It depicts the conditioning that an attribute (or other factor) of a stakeholder's decision exerts on another's option. The random variables will have dimensions such as dollars, counts of deaths, mass/volume and so on. Value nodes consist of utility or value functions that represent the stakeholders' different attitudes towards a risky choice. Influence diagrams can explain and describe stakeholders' attitudes (personal values) towards a risky choice.

Table 7.3 depicts the format and content of the information (adapted from Hong and Apostolakis, 1993).

Table 7.3. Risk Attitudes Held by the Two Stakeholders

Function	Attitude	Constraints	Node	Comment
1-0.5*[ln(Health)]	Risk Averse	10^{-4} to 10^{-6}	Workers' health	Regulatory constraint
1-[(Min. SES)/4]	Risk neutral	0, 1, 2, 3, 4.	Socio-economic (SES)	Statutory constraint

The relevance of influence diagrams can be further understood continuing with Hong and Apostolakis' (1993) two stakeholders' (DM_1 and DM_2) problem. The influence diagram is used to assess the optimal choice of pollution control technologies through game theory. The pay-off matrix, Table 7.4, depicts the outcome of their analyses. Two technologies are compared through their efficiency to remove contaminants. The cells with numerical values, (e.g., for DM_1: *0.77*, for DM_2: *0.50*) show the value to DM_1 and DM_2. The private decision-maker has a higher expected (or average) utility (*0.80 > 0.77*) in technology *B*, removing an air pollutant with *90%* efficiency, than its alternative. The public decision-maker has higher expected utility in technology *B* with *99%* removal efficiency because *0.90 > 0.68*:

Table 7.4. Range of Expected Utilities Held by a Stakeholder

Expected Utility of Private Decision-maker (DM₁)	Expected Utility of Public Decision-maker (DM₂)		
		90% removal	*99% removal*
	Technology A	(0.77, 0.50)	(0.76, 0.68)
	Technology B	(0.80, 0.73)	(0.71, 0.90)

In this study, the two stakeholders are not competing with one another and have full information about the outcomes of their potential choices. Furthermore, the pay-off matrix identifies the potential for equilibrium between these two stakeholders: DM_1 prefers *A* with *90%* removal, but DM_2 prefers *B* with *99%* removal. If the stakeholders can examine the decision space that affects them both, then the solution (*0.80, 0.73*) is optimal in the sense that the joint expected utilities are the highest. Therefore, technology *B* with *90%* removal efficiency is optimal: society benefits through agreement.

7.3. BAYESIAN NETWORKS

Practical risk assessment that involves causal explanations can benefit from causal graphs for ease of presentation and analysis. Representations by directed acyclic graph, DAG, include influence diagrams and Bayesian networks. Figure 7.5 depicts a directed acyclic graph.

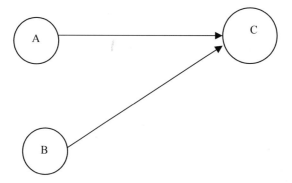

Figure 7.5. Directed Acyclic Graph

Acyclic means that there are no feedbacks (Pearl, 2000). Bayesian networks rely on directed acyclic graphs to represent the relations (nodes and arrows) between random variables. Arrows are *edges* or *arcs*. The initial structure of the graph is based on known or hypothesized relations, each relation consisting of a node, an arc and another node, two nodes, and two arcs and so on. The relations in a Bayesian Network (BN) build on ontological and deterministic reasoning (Pearl, 2000). In Figure 7.5, *A* and *B* are the parents of *C*; the terms *parent* and *descendant* describe the relationship between variables connected by a directed arc. Thus, *C* is a descendant of *A* and *B*. In this representation, *A* and *B* are *known* to cause *C*. Specifically, *A* can be concentrations of sulfur dioxide and *B* concentrations of particulate matter, then *C* can be the causal result from exposure, the *incidence of respiratory cancers*.

BNs use probabilities to represent the uncertainties inherent to such a causal statement. Specifically, BN uses data to compute relations in terms of such statistical quantities as conditional probabilities, marginal distributions and so on. An empirical deductive causal network *M* is defined by a set of dependent and independent variables forming the set *V*, each of the variables in *V* representing a distinct node. This representation includes intermediate variables that are explained by their predecessors and help to explain successors, unlike regression models. The arrows connect nodes with nodes such that each node is a function of the variables pointing to it.

The causal model represented by a BN can be succinctly stated as $M = (U, V, F)$ where U represents background variables, V the variables that are internal to the models and thus explicitly accounted for, and F represents functions. The probabilistic form of M, $[M, pr(U)]$, allows updating as new evidence becomes available over one or more variables (Pearl, 2000). Uncertainty in the specification of the causal BN can be written as $x_i = f(parents_i, u_i)$. Bayesian Networks describe probabilistic conditioning and thus the dependence-independence structure among the variables of the network. The simplification inherent to BN is that a joint distribution is the product of conditional distributions:

$$pr(Y_1, Y_2, ..., Y_k) = \prod_{i=1}^{k} pr\ (Y_i \mid Parents_i).$$

The common empirical situation confronted by risk assessors is that, although the relevant set of variables, V, is unavailable, the subset O of V is observable. Typically, $M = (Directed\ Acyclic\ Graph,\ \theta) = f(parents_i, u_i)$ requires finding $pr(M)$ and knowing the *DAG*. Because many models can be fit to what is observed, the criterion for choice of alternative models can be the likelihood ratio.

The causal structure of the BN should be known in advance: the variables and their relationships must be known, although the graphical representation can be used to determine the logical plausibility of alternative Bayesian Networks. To do so, there must be *meaningful directionality* between the variables in the network. Testing *whether a proposed set of causal relationships is consistent with the available temporal-probabilistic information* can also be obtained with Bayesian networks (Pearl, 2000). In BNs, a typical empirical causal statement is that *C should raise the probability of E* unless paradoxical results, such as Simpson's Paradox, discussed and exemplified in Chapter 9, negate this relationship.

7.3.1. Discussion

DAGs can be used for making inference, assessing the impact of policy interventions and predicting outcomes. These graphs can include different forms of scientific evidence and data in a single and computationally sound method. The *DAG*:

$$X \rightarrow Y \rightarrow Z$$

means that Z is conditionally independent of X given Y. Y is the (single) parent of Z. More specifically, the DAG $X \rightarrow Y \rightarrow Z$ has the following probabilistic meaning:

$$Pr(X, Y, Z) = pr(X = x)pr(Y = y|X = x)pr(Z = z|Y = y),$$

in which the joint distribution of X, Y and Z is $Pr(X, Y, Z)$. The importance of this representation is that the joint distribution of two or more random variables can always be factored into the product of unconditional, e.g., $pr(X)$, and conditional distributions, e.g., $pr(Y|X)$. The general statement is:

$$Pr(X_1, X_2, ..., X_n) = pr(X_1)pr(X_2|X_1)pr(X_3|X_1, X_2, X_3) ... pr(X_m|X_1, X_2,...,X_n).$$

Factoring (Spanos, 1999)[1] the conditional and unconditional probabilities reduces the complexity of the *joint* probability of the model:

$Pr(X_1, X_2, ..., X_n) = \prod_i [pr(X_i | Parents\ of\ X_i)]$.

These formulae use DAGs in which the nodes are variables (random or other); arrows depict conditional dependence, while the lack of an arrow indicates independence. As Cox (2002) shows, if the *DAG* $X \rightarrow Y \rightarrow Z$ is linear between the variables, the coefficient of correlation of X and Z is the product of the coefficients of correlations of the parents of Z:

$\rho_{ZX} = \rho_{ZY}\rho_{YX}$.

Once the direct effect of the parents of the descendant are given, the descendent is disconnected from its grand-parents, say A and B, (once the random variables have taken specific values, such as $A = a$ and $B = c$, with some probability). The model is now consistent and complete because the joint probability $pr(A, B, C, ..., Z)$ equals the product of all the relevant conditional probabilities. The intuitive rationale of Bayesian networks is that they can deal with statements such as *If I do A and B, then I will probably obtain C*, in which the outcome C is contingent on A and B. In summary, BN's are based on:

1. Empirical causal modeling,
2. The development of the graph of the causal model (the directed acyclic graph),
3. Calculating the local conditional distributions.

Example 7.1. Cox (2002) provides an example of using conditional probabilities in the context of DAGs. Suppose that the causal DAG is $X \rightarrow Y \rightarrow P$ and that these random variables can only take the values *0* or *1* (they are binary).

pr(Y|X)	X = 0	X = 1
Y = 0	1	0.50
Y = 1	0	0.50

and:

pr(P|X)	Y = 0	Y = 1
P = 0	1	0.20
P = 1	0	0.80

Then, *pr(P|X)* is calculated from the relationship *pr(P|Y)pr(Y|X)*, as shown below:

pr(P|X)	X = 0	X = 1
P = 0	1	0.60
P = 1	0	0.40

Supposing that a probability distribution for Y is κ, then *pr(P|X)* is calculated from *pr(P|X)*κ*, which is the conditional distribution for P.

---

[1] A theorem due to RA Fisher is used to factorize the likelihood function, $L(\theta, x)$. The idea is to express the likelihood function as the product of two functions, one of which is based on a *sufficient* statistic and the other only on the data and not on the parameter θ.

Example 7.2. Consider the following BN:

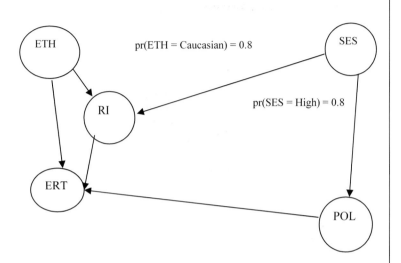

POL	ETH	RI	pr(ERT=yes|POL, ETH, RI)
low	Caucasian	yes	0.20
low	Caucasian	no	0.10
low	Not Caucasian	yes	0.20
low	Not Caucasian	no	0.05
high	Caucasian	yes	0.30
high	Caucasian	no	0.24
high	Not Caucasian	yes	0.40
high	Not Caucasian	no	0.10

SES	pr(POL = high|SES)
high	0.24
low	0.14

Note that *SES high = pr(RI = Yes|SES) = 0.04*. This network consists of an acyclic directed graph in which the random variables are: ethnicity (*ETH*), socioeconomic (*SES*), respiratory infection (*RI*), pollution exposure as concentration (*POL*), and emergency room treatment (*ERT*). They are causally related as follows (Ricci and Hughes, 2000, as a hypothetical example). The nodes in this BN are random variables and conditioning is shown in the tables associated with the graph. We used a simple representation to simplify an otherwise complex model and to complement the empirical analyses discussed in the context of regression models.

The probability that an individual is Caucasian (*ETH = Caucasian*) is not conditional and equals *0.80*. The probability that an individual will contract a respiratory infection (*RI*), given that that individual has *low* socio economic status (*SES = low*) is *0.14*, and so on. Consider the variables *SES*, *RI* and *POL*. The conditional probability distributions of *RI* and *POL* show that variables of each pairing (*SES, RI*), (*SES, POL*), (*RI, POL*) are correlated. The directed arcs between the node *SES* and nodes *RI* and *POL*, however, assert the conditional independence of *RI* and *POL*. This means that the probability of *RI* conditional on the states of both *SES* and *POL* is equal to the probability of *RI* conditional on the state of *SES* alone. Similarly, the probability of *POL*, conditional on the states for *SES* and *RI*, is equal to the probability of *POL* conditional on the state of *SES* alone:

$$pr(POL=high|SES=low, RI=yes) = pr(POL=high|SES=low)$$
$$pr(POL=high|SES=high, RI=yes) = pr(POL=high|SES=high)$$
$$pr(POL=high|SES=low, RI=no) = pr(POL=high|SES=low)$$
$$pr(POL=high|SES=low, RI=no) = pr(POL=high|SES=low)$$

Once the state of *SES* is known, the state of *POL* is irrelevant to the probability of *RI*, and vice versa, even though *RI* and *POL* are correlated. It follows from the concept of conditional independence that *SES* must control the probability of both *RI* and *POL* and thus the directed arcs have a direct causal interpretation (Ricci and Hughes, 2000).

Cox (2000) provides a generalized expression for these results. Moreover, although simpler calculations can also yield *pr(P|X)*, for instance if measurements for *P* are not used, conditioning will yield a better result. A result is that, if we have measurements on *X, Y* and *Z* for those at risk, that information can test if the model *X* → *Y* → *Z* is correct. The test is based on comparing *[pr(Z|Y)pr(Y|X)]*, with *pr(Z|X)*. When these conditional distributions differ, the assumed DAG model is incorrect.

A major advantage of a Bayesian network arises from the *causal* meaning of the directed arcs. If S^l represents a Bayesian network, S^2 another Bayesian network, and if *pr(S^l|D)* is larger than *pr(S^2|D)*, then the change in probability makes the causal link stronger. A further advantage of a Bayesian network is the concise and comprehensive representation of the relationship between the risk factors. However, the *closed-world* assumption, assuming knowledge of the distribution of each random variable, can be too demanding because of possible circularity. This assumption involves knowing all relevant factors and separating them into the causally relevant, background factors and irrelevant factors. An aspect of BN analysis that is useful in deciding between competing or alternative causal structures uses the principles of *stability* (meaning that there is an isomorphism between two structures) and *minimality* (meaning that the least complex of two structures, T_1 and T_2, given the same data, is preferred).

Minimality relates to the mathematical form and number of variables of the causal network. Stability relates to the probability of events in T_1 that make events in T_2 more probable. Stability refers to lack of extraneous, probabilistic conditional independences in a BN (Pearl, 2000). The stability principle states that it is most improbable that the two competing structures overlap. The idea is that a causal structure is minimal relative to a larger set of potential structure if no element of the minimal structure is preferred over those of any other structure in the class of structures considered (Pearl, 2000). This reasoning also applies to *latent* variables. Minimality and stability test the uniqueness of the network. Consider, for example, the photo of a chair; the objective is to discriminate between the following alternatives (Pearl, 2000):

T_1: the photo actually shows only one chair,

T_2: the photo actually shows either one chair or two chairs (the first chair completely hides the second).

Minimality suggests that T_1 is preferred to T_2 because it is simpler (Occam's principle) and consistent with the reality of the picture to argue for a single chair. Stability suggests that T_1 is much more likely than T_2 and therefore T_1 is preferable.

The combination of these two principles contributes to understanding empirical causation and account for uncertainty. It does not resolve, however, the fundamental question of whether or not the chair is the correct picture. Although BNs provide a means to assess complex causal patterns, a question is whether or not adding complexity, by adding additional variables and causal linkages to a model to make it more realistic, increases the statistical uncertainty in the outcome of the more complex model. Using a BN and results from entropy of information, the answer to the question is *no* (Cox 2002). Reducing a model (the *reduced* form) yields a smaller model in which some of the parameters of the reduced model are combinations of parameters of the original model. This operation does not destroy uniqueness, which is an important property in the analysis of causation for risk management.

Bayesian networks allow probabilistic updating through conditional probabilities to deal with confounders and apparent confounders Finally, there may be several *DAGs*. For a five variables *DAG* there are approximately *1,600* different Bayesian networks. Fortunately, the number of Bayesian networks that make physical and statistical sense is less than this. Pearl (2000) outlines practical algorithms for estimating Bayesian networks and covers aspects of Bayesian networks such as the identification of latent variables and the treatment of missing observations. Jensen (1996) and Pearl (1988, 2000) provide additional in-depth explanations.

The final aspect of this discussion concerns prediction and inference using the *DAG* methods discussed. The idea is to use the *DAG* for Bayesian inference through resolving the uncertainties from one node of the *DAG* to the other. This is preferred when the model's joint distribution contains a set of known and latent variables, when that joint multivariate distribution is too complicated to solve, or both. In some simple cases it is possible directly to use Bayes' theorem to develop the joint distribution so that the conditional probabilities involving known and unknown variables (Shenoy, 1992; Jensen, 1996). The issue is a situation in which only some evidence is available for some individuals, and we seek the posterior conditional distribution for a random variable (Cox, 2002).

Example 7.3. Let the *DAG* model be: $Y \leftarrow X \rightarrow Z$. We want to obtain $Pr(Z)$, which is the posterior distribution of that variable (note that in this example $Pr(.)$ is a distribution). Let $Pr(X|Y)$, $Pr(Z|X)$, $Pr(X)$ be observations on Y, symbolized by y^*, and on X, namely x. The observations are thus: y^* and x; but we do not have the full set of observation on Y, namely y, because that knowledge depends also on Z, which we do not have. Using arc-reversal methods (Cox, 2002) we obtain:

$Pr(X|y^*) = [Pr(y^*|X)Pr(X)]/Pr(y^*)$.

In this expression we have used known evidence. The second step, forward-chaining, consists of the expression which solves the problem:

$Pr(Z|y^*) = \sum_x Pr(Z|x)Pr(x|y^*).$

This is the probabilistic statement for the *DAG* model: $Y \rightarrow X \rightarrow Z$. As Cox (2002) explains, $Pr(Z|x)$ is calculated from the nodes in which $Pr(Z|X)$ are displayed and $Pr(x|y)$ is obtained from the arc-reversal calculations. Similarly, $Pr(y|X)$ is obtained from the nodes in which $Pr(Y|X)$ are calculated and $Pr(x)$ is obtained from the node $Pr(X)$, with:

$Pr(y) = \sum_j Pr(y^*|X = j)Pr(X = j).$

The importance of arc-reversal and forward chaining is that these go beyond the forward propagation of the uncertainty normally used in practical risk assessments. The information gain is that the evidence can update probabilities and be a lot simpler to obtain in practice. Cox (2002) provides further discussion including the application of Gibbs sampling to inference using *DAGs*.

To complete these discussions, other statistical methods exist to compare alternative *DAGs*. The statistical criteria, developed from a review by Cox (2002), range from the maximum likelihood to the distance criterion developed by Kullback and Leibler. For example, those methods include the maximum posterior Bayesian distribution method that determines the most probable model, given the data. Alternatively, the maximum description length method minimizes the amount of information. The method thus selects the model, from the set of candidate models, using the least amount of information (Cox, 2002; Madigan and Raftery, 1994).

7.4. CLASSIFICATION AND REGRESSION TREES (CART)

The essential idea of CART is to develop binary trees, beginning with the average value of the dependent variable, in which each tree has two branches for the independent variables. When the risk assessor has a wide range of potential variables, she may wish to develop a subset of independent variables justified by the properties of the sample. This idea is exemplified using the data developed by Harrison and Rubinfeld (1978) and extensively discussed by them. Their purpose was to assess the effect of several risk factors on median housing values, MHV, in Boston. Their full regression model, which is based on an observational study, is:

$ln(MHV) = \beta_1 + \beta_2(Average\ Number\ of\ Rooms) + \beta_3(\%\ Built\ before\ 1940) + \beta_4(Distance\ to\ Employment\ Centers) + \beta_5(Accessibility\ to\ Highways) + \beta_6(Property\ Tax\ Rate) + \beta_7(Student/Teacher\ Ratio) + \beta_8\ (\%Black-63)^2 + \beta_9[ln(\%Lower\ Status)] + \beta_{10}(\%\ Zoning\ Residential) + \beta_{11}(\%\ Industrial) + \beta_{12}(On\ the\ Charles\ River,\ yes = 1\ or\ No = 0) + \beta_{13}(NOx,\ in\ pphm)^b + \beta_{14}(Criminal\ rate) + u.$

In this model, *pphm* is parts per hundred million. Because the full model can appear to have too many independent variables, we can use CART to see if we can reduce its complexity by developing a simpler model from the entire data in the sample. Specifically, the theory of land rents provides the basis for the model. CART identifies a subset of variables, the most likely predictors of median housing values, through a binary selection from the full sample. CART results are depicted in Figure 7.6.

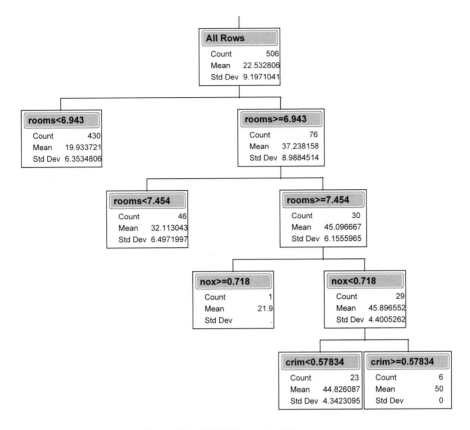

Figure 7.6. CART Regression Tree

The CART results show that the independent variable *ROOMS* is important because it appears on both sides of the initial branch and continues to be important as we drill down into the data. The variables NO_x and *CRIM* are of lesser importance. These results suggest the simpler regression model:

$MHV(*\$1000) = f(NO_x, CRIM, ROOM)$.

The information in each node is the median value of the houses, for a specific independent variable, such as ROOMS. We start out with *506* houses with median value *$22,532* and *sd* = *$9,197*. The first (left hand side) node contains 430 houses with approximately 7 rooms or less: we split on the independent variable ROOMS, in which there are *430* houses with less than *6.94* room. There are *76* houses with more than *6.94* rooms (the right hand split). Of those houses with more than approximately seven rooms, only one had a NO_x concentration greater than *0.718*. The predicted value of *MHV* is the average for this variable at the terminal nodes: that is $44,8261\pm\$4,342$ or *$50,000*, depending on the split based on *CRIM*, and so on. The advantage of using CART for risk assessment and management is that the interpretation of the results is straightforward, visually appealing, estimation is non-parametric and the method shows intermediate steps. For example, we might expect more expensive homes in areas that have lower pollution and less crime: this is what the tree depicts. The tree stops as shown; the stopping rules use the criteria discussed next.

7.4.1. Discussion

Consider the regression aspects of CART, the *R* in CART. As for the familiar estimation of the parameters of a regression model, obtained through the ordinary least squares, the criterion for estimation is the minimization of a quantity (the *resubstitution* estimate, Breiman, Friedman, Olshen and Stone, 1984). The quantity to be minimized (without using vectors) is:

$$R(d) = 1/N \sum_{i=1}^{N} [y_n - d(x_n)]^2 \, ,$$

in which *d(x)* is short-hand for a parameter estimate or predictor. *R(d)* can be formulated to include learning (*ls*) and testing (*ts*) samples. A learning sample is a subset of the information contained in the sample. Several learning samples are created at random by subdividing the learning sample (N_2). We thus obtain, for the testing samples, *ts*:

$$R^{ts}(d) = 1/N_2 \{ \sum_{(x_n, y_n) \in L_2} [y_n - d(x_n)]^2 \} \, .$$

The resubstitution estimate depends on the scale over which its components are measured and is normalized to yield the relative resubstitution estimate, *RE(d)* (Breiman, Friedman, Olshen and Stone, 1984). It has an estimator, *RE*(d)*, that is interpreted as a *measure of accuracy*, akin to the estimate of the proportion of the explained variance of a regression, if it is expressed as $[1-Re^{cv}(d)]$. The term *CV* is the cross-validation estimate of *R(d)*. This cross-validation estimate is derived from *v*-subsets of the learning samples, each of which should either be of the same size or very close to it (Breiman, Friedman, Olshen and Stone (1984). The optimal split at the node *t* has the maximum reduction in the error, *R(t)*; *R(t)* is the result of the summation of the *R(t)*, ($t \in \tilde{T}$, where the T-tilde is the set of terminal nodes):

$$R(t) = N^{-1} \sum_{x_n \in t} [y_n - \bar{y}(t)]^2 \, .$$

This yields $R(T) = \sum R(t)$, the summation being taken over the set of terminal nodes. The best split, *s**, is $\Delta R(s^*, t) = max[\Delta R(s, t)]$, $s \in S$. This optimal splitting rule is justified by the minimum of the weighted variance between the left, *L*, and right, *R* branches, $[pr_{LS}^2(t_L)+(1-pr_L)s^2(t_R)]$. The choice of the tree of an appropriate size, out of the possible sequence of trees, uses both learning and testing samples. The learning sample provides the sequence of pruned trees and cross-validation estimates. The T_k tree is the smallest tree consistent with the decision rule:

$$R^{CV}(T_k) \le R^{CV}(T_{K0})+se[R^{CV}(T_{k0})],$$

where $R^{CV}(T_{k0})$ is the minimum over all k of the expression $R^{CV}(T_k)$. Accuracy depends on the distribution of the data. A more robust estimation method is the regression based on the Least Absolute Deviations. This regression is particularly appropriate when the median is a better parameter than the mean to represent the tendency of asymmetrically distributed data. The (Bayesian) predictor $d_B(x) = g(Y|X = x)$, minimizes the absolute error, $R^*(d)$. The proof involves taking the derivatives of expressions stated in terms of absolute values and requires numerical solutions, unlike the minimization used to obtain the least squares and maximum likelihood estimators. Consistent with the previous discussion, the relative absolute error is:

$$R(d) = 1/[N\Sigma_i|Y_n\text{-}d(x_n)|].$$

The test sample estimate, cross-validation estimates, and their standard errors are similar to those just discussed. The splitting rule, unlike the least squares regression, minimizes the quantity $R(s, t) = R(t)\text{-}R(t_L)\text{-}R(t_R)$. In this function, the subscripts L and R mean *left* and *right*, respectively. The expression for the best split (based on sample medians, $v(t)$, rather than the expected values used in the earlier discussion) is:

$$\sum_{x_n \in t} |y_n - v(t)| - \sum_{x_n \in t_L} |y_n - v(t_L)| - \sum_{x_n \in t_R} |y_n - v(t_R)|.$$

Breiman, Friedman, Olshen and Stone (1984) give further details and discussions. The CART, as discussed, can describe complex data very well. It is preferable to use CART when the structure of the relationships is nonlinear and in studying conditioning. More specifically, following Cox (2002), suppose that the model being considered is:

M1: Fat consumption ← Socio-economic Status → Smoking → Cancer incidence.

Suppose that the variable *socio-economic status* is unobserved but it affects smoking and fat consumption. The data set consists of observations on fat consumption, smoking and cancer incidence. The alternative model:

M2: Fat consumption → Cancer incidence ← Smoking,

can be compared to the model with the latent variable socio-economic status through CART, as follows. For *M1*, if the classification obtained with CART begins with cancer, first splits on smoking and then on fat consumption, model *M1* is inconsistent with the data. The reason is that cancer is conditionally independent of fat consumption, given smoking. For *M2*, on the other hand, if the CART model starts with cancer, first splits on fat consumption and then on smoking, model *M2* is consistent with the data. Both fat consumption and smoking are cancer risk factors for the population from which the sample is drawn.

7.5. CONCLUSION

This chapter has developed several techniques used graphically to represent data-driven causation and has added to model building discussed in Chapters 4 and 5. These techniques use probabilistic analysis, such as those discussed in Chapter 6, applied to scientific causation in risk assessment and management. The combination of graphical and statistical modeling is particularly useful to risk assessors and to risk managers because of the intuitive appeal of representing relations through arcs and nodes. Causal path analysis, influence diagrams, Bayesian Networks (BN) and Classification and Regression Trees (CART) are practical methods for risk assessment because they are based on sound theory, account for deterministic and probabilistic reasoning and can deal with complex causal structures.

Scientific evidence in risk assessment and management consists of often-incomplete data sets and partially known casual models. As shown, the need to deal with causation for scientific, legal and managerial reasons can be answered with confidence using modern techniques of data analysis. However, there are *costs* when moving beyond the standard uses of modeling techniques. Those costs are the result of using techniques that may not be familiar to all stakeholders, the possible lack of theoretical guidance for the models and the lack of acceptance of new methods by the stakeholders. The usefulness of being able to use alternative methods for risk analysis should now be clear. Complex data sets may require using more than one method to make sense of the potential variables and forms of the relationships latent within the data. As always, the literature and past results must guide the analysis.

QUESTIONS

1. Why does a network provide useful insights in modeling the relationship between exposure to air pollution and adverse response? Use less than 400 words.

2. In *American Trucking Associations et al.* v. *Christine Todd Whitman* (U.S. Supreme Court 99-1426), Section 108 (National Ambient Air Quality Standards) and Section 109(a) of the Clean Air Act were litigated on vastly different values and understanding of the mandate of the Clean Air Act. One stakeholder, the American Trucking Associations' objective was to minimize the cost of environmental compliance. The US EPA, the other stakeholder, wanted to regulate truck with regulations that are, as mandated by these two Sections, health-protective and explicitly barred the consideration of costs in setting standards under these two Sections of the Clean Air Act. The US Supreme Court, which held for the US EPA, conflicted with American Trucking Associations's beliefs. The analysis of the interactions between these two stakeholders can use coupled influence diagrams. Please develop one. The coupling should reflect where, in the diagrams, one stakeholder's action or value affects the other. Use approximately 600 words.

3. Develop a simple *regression* model based on Bayesian Network of the air pollution example developed in this chapter from the work of Ricci and Hughes (2000).

4. Are there difference between the regression approach in CART and the regression models discussed in Chapter 4? Discuss and provide simple examples of the differences and similarities with less than 500 words.

5. Having to deal with empirical causation requires understanding the way in which nodes and linkages form. A question that may be considered is: What is the usefulness of developing nodes and linkages when the causal basis is

not scientifically well understood? How can limited information be used in developing causal diagrams such as influence diagrams for inference? Discuss these questions in less than 600 words.

6. Consider CART for regression. Sketch a CART model solution of exposure-response, using DEATH RATE, SO_2, $PM_{2.5}$, PM_{10}, AGE and RACE as the variables.

7. Sketch a Bayesian Network of exposure-response, using DEATH RATE, SO_2, $PM_{2.5}$, PM_{10}, AGE and RACE. Are there any differences between CART and BN? Please discuss those differences as well as similarities. Use approximately 500 words.

CHAPTER 8.

META-ANALYSIS, POOLING SAMPLE DATA, AND STATISTICAL DECISION RULES

Central objective: to develop and exemplify methods for aggregating the results of several risk studies and to summarize statistical decision rules for making statistical choices about the significance of estimated values

This chapter introduces and exemplifies some of the methods that risk assessors can use when they want quantitatively to analyze statistical *results* obtained by others and *published* in the general literature. Chapters 4-7 have dealt with the analysis of samples of data that link human exposure to environmental or occupational pollution to adverse health outcomes. Results contained in the literature, and thus outside the control of the risk assessor or user of that information, can justify a new study's choice of variables, model form and so on. In general, literature reviews are qualitative and consist of discussions and tables. This chapter discusses methods to synthesize results quantitatively to enhance their aggregate uses in replicable and predictable ways. This is relevant to justifying choices and conclusions that form the scientific basis of regulatory standards and guidelines because, in these processes, a very large amount of literature is qualitatively reviewed. The aggregation of summary results found in the literature can benefit from quantitatively combining them, particularly when those results are counterintuitive, and can therefore be ambiguous. Meta-analysis is a set of methods that take as inputs the results from studies found in the literature and statistically assesses the aggregate contribution of those studies' results. This is an important change and improvement from the traditional qualitative reviews of the literature.

Statistical studies, taken together or individually, must account for the variability of estimates and inference requiring the risk assessor to be conversant with hypothesis testing, inference, false positive and false negative errors. For example, a study may only report the probability of a Type I error; its results may rest on untested assumptions and other issues. An objective of this chapter is to discuss and exemplify the reasoning used in such tests through statistical decision rules involving hypotheses and confidence intervals.

In this chapter, we discuss and exemplify:

- Meta-analysis
- Pooling of data
- Statistical decision-making: hypothesis testing, confidence intervals and the power

of the test
- Analysis of variance (parametric and non-parametric) to test the equality of independent means for independent populations and tests applicable to dependent populations

The linkages between the major elements of this chapter and risk assessment are:

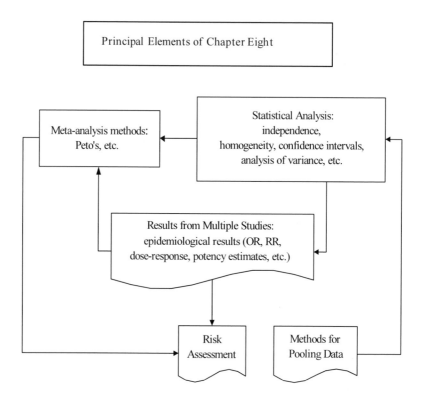

8.1. INTRODUCTION

Several statistical methods can combine (e.g., *pool*) data. Part of our discussion centers on meta-analysis. Another part focuses on methods to increase the sample size, by pooling sample data from two or more studies, to analyze them as a single, larger sample. Meta-analysis is the collection of statistical methods that deals with the analysis of results obtained by others and provides ways to combine and statistically interpret them in the aggregate. Each result, for instance the estimate of a regression parameter, becomes an individual and independent result and a single data of the meta sample. Thus, several of these form the *meta* sample: the totality of the data (as results) taken from the studies included in the meta-analysis. For example, if several studies report coefficients of correlation between exposure and response, these form another meta sample. Intuitively, their statistical aggregate contributes to understand the relationship between exposure and response in a way that qualitative reviews of the literature cannot do. The reason is that the aggregate of those results is analyzed statistically, rather than as a listing of results with qualitative judgments, yielding well-understood quantities such as the aggregate mean, variance, and confidence interval. Making decisions with data generally involves hypothesis testing and confidence interval analysis. We complete that discussion and include some of the practical methods used to test the differences between the means of several samples, using parametric and non-parametric analysis of variance (ANOVA) methods. The reason for discussing the ANOVA is that the (parametric) ANOVA is used in regression analysis to test the statistical significance of the overall regression (as opposed to the testing the statistical significance of each individual parameter of the regression model).

8.2. POOLING SEVERAL SAMPLES

Pooling two or more sample results in a larger overall sample size and therefore decreases the overall variability measured by the variance of the pooled data. The pooled estimate of a particular population's parameter is more accurate than would be obtained from each of the single samples alone. Two of the best-known statistics used in risk assessment are the mean and the variance. Their pooling is straightforward. For example, a useful pooling formula for the mean of two or more samples, yielding their total mean, is:

$$\bar{x}_{total} = \frac{\sum_i (n_i/s_i^2)\bar{x}_i}{\sum_i (n_i/s_i^2)},$$

where the number of samples is indexed by the letter i, $i = 1, 2, ..., k$. Each sample size is symbolized by n_l, with sample mean $\bar{x}_i$ and sample variance s_i^2. The aggregate sample mean is the weighted average of the two samples: the term n_i/s_i^2 is a weight that accounts for the variability of each individual sample. The statistical assumption is that the samples are independently drawn from the same population with mean, μ. An analogous approach pools the estimates of the population variance from two or more samples. For example, the pooled estimator of the population variance, under the assumption that the population variance is common to all samples, (shown for two samples) is:

$$s_{pooled}^2 = \left(\frac{n_x - 1}{n_x + n_y - 2}\right)s_x^2 + \left(\frac{n_y - 1}{n_x + n_y - 2}\right)s_y^2,$$

where n_x and n_y are the sample sizes of the two independent random samples, s_x and s_y indicates the two sample variances (Wolf, 1986). Testing a common hypothesis from several independent empirical results has been studied for some time (Pearson, 1933; Fisher, 1948). For example, suppose that a number of independent studies have generated estimates with some significant *t*-values. The individual *t*-values (the t_i) are pooled as follows:

$$Z_c = \frac{\sum_i t_i}{\sqrt{\sum_i \{(df)_i / [(df)_i - 2]\}}},$$

which is appropriate for $n \geq 10$ (Wolf, 1986). The term *df* is the number of degrees of freedom associated with each *t*-test; t_i is the *i*-th *t*-test. The testing of the hypothesis follows the *z*-distribution. The calculations are exemplified using the *z*-statistic because the aggregate sample size for the studies in the example is large.

Example 8.1. Assume that four studies are available. Let their $t_i = 2.05, 1.95, -2.1$ and 1.75. The sample sizes = $50, 60, 100$ and 20, respectively. The calculation is:

$Z_c = [2.05+1.95+(-2.1)+1.75]/[(50+60+100+20)/(48+58+98+18)]^{0.5} = 3.59$.

We use the one-tail test of the hypothesis because the direction of the finding is what matters, as opposed to the two-tailed test in which only the magnitude matters. That is, since we are pooling *t*-values from four studies, what can be relevant is whether the association is either positive or negative. The probability of obtaining the $Z_c = 3.59$ is greater or equal to 0.001 (using the tables of the *Z*-distribution found in most textbook of statistics). The result is statistically significant even though one of the studies has a negative *t*-value.

8.3. META-ANALYSIS

Meta-analysis consists of a portfolio of statistical methods used to assess empirical results from independent and essentially identical sampling studies generated by others. Meta-analysis captures the overall central tendency and variability of the results. Some of the early developments of meta-analysis are due to Glass (1977, 1983). Meta-analysis has been applied to develop the distribution of the estimated coefficients of exposure-response models (the estimates reported by many authors of a very large number of regression equations) from several independent epidemiological studies and to other risk-based studies. It can deal with fixed effect or random effect epidemiological models.[1] Meta-analysis is also an effective tool to assess the results of studies that have small samples. However, the issues that arise from relying on the mechanics of statistical analysis at the expense of causal reasoning at the physical level are still present.

Example 8.2. Suppose that risk measure used in a number of studies found in the literature is the odds ratio, OR, which (for a 2*2 contingency table) is calculated as: $OR = [(a*d)/(c*d)]$, where a, b, c and d are integer numbers ≥ 0. Recollect that the 2*2 contingency table, in which a, b, c and d are the number of individuals (the counts of individuals affected and not affected exposure and no exposure) is:

---

[1] A fixed effect mode assumes that all studies conducted belong to a single population. Therefore, under this assumption, if the sample size of each study were extremely large, the effect studied would be equal in all samples. If, on the other hand, different effects characterize a population, the random effect model is used (Borenstein and Rothstein, 1999).

	Exposed	**Not exposed**
Response	a	b
No response	c	d

If the $OR = 1.00$, the numerator and denominator are equal and therefore exposure does not affect response. The null hypothesis is that $OR = 1.00$. The estimated odds ratio of each study, $(OR)_i$, is estimated as $(ad/cb)_i$. The results of each study may be quite different. For instance, they can include from ORs that are less than one and ORs that are greater than one. Moreover, the statistical significance of those ORs can vary from study to study; see example 8.3.

Meta-analytic studies are observational because the researcher has no control over the data themselves: she takes them as she finds them. Importantly, she also has no control over the quality of the data and the statistical methods used by the original researchers. The risk assessor must: i) identify and use the relevant studies, ii) abstract the appropriate numerical information (e.g., coefficients of correlation, ORs or other) from them and iii) apply statistical methods to obtain aggregate results. Table 8.1 contains some of the steps for developing a meta-analysis, under the assumptions that the studies are independent.

Table 8.1. Steps for a Meta-analysis

Steps	**Description of Step**	**Comment or Question**
State the Research Issue or Issues Assessed through Meta-analysis	Defines the objectives of the meta-analysis	Is the result from the meta-analysis used for research or for policy?
Determine the Sources of the Original Studies to be Meta-analyzed	Literature searches that can be limited to published and peer-reviewed papers	These searches are often incomplete because of human limitations, databases limitations; errors of omission, researcher bias and so on. Include government reports? Include trade publications?
Establish and Justify Criteria for Inclusion and Exclusion of Available Studies	The included *and* the excluded studies must be reported and cited, with the criteria for exclusion clearly stated and attached to each excluded study	Sensitivity analysis can be used to study the effect of the exclusions on the meta-analytic results.
Establish the Statistical Protocol for Implementing the Meta-analysis	NA	Statistical methods and assumptions must be fully discussed and reported
Control Quality of the Studies, Weigh the Studies and Conduct the Meta-analysis	Independent analysts should verify the data base developed for meta-analysis and the analyses themselves	Discuss the results of this step

Note that a meta-analyst does not have access to the authors' original data and does not re-analyze them. Table 8.2 is a summary table of input data for a meta-analysis.

Table 8.2. Inputs for a Meta-analysis of Estimated ORs from the Literature

Author(s)	Number Exposed	Number Not exposed	Number of Cases	Number of Controls	Estimated ORs
AAA (19xx)	…	…	…	…	…
BBB (200x)	…	…	…	…	…
…	…	…	…	…	…

The example that follows shows the input data to the meta-analysis and a summary of results using hypothetical data.

Example 8.3. The example uses a data set available in the software program *Comprehensive Meta-Analysis* (Borenstein and Rothstein, 1999). It consists of a *hypothetical* literature review that includes the following results from thirty-three studies of survival in which the analysis compares immediate treatment with streptokinase to prevent mortality following myocardial infarction to delayed treatment (treatment reduces blood clots):

Study	Treated	Control	Effect	LCL 95%	UCL 95%	P-value
Australian 1	26/264	32/253	0.754464	0.435757	1.30627	0.313294
Australian 2	25/123	31/107	0.625411	0.341156	1.146512	0.127502
Austrian	37/352	65/376	0.562002	0.364501	0.866519	8.48E-03
Baroffio	0/29	6/30	6.39E-02	3.43E-03	1.191531	1.81E-02
Bassand	4/52	7/55	0.571429	0.156987	2.079987	0.391381
Cribier	1/21	1/23	1.1	6.44E-02	18.77436	0.947487
Dewar	4/21	7/21	0.470588	0.114025	1.942154	0.292407
Durand	3/35	4/29	0.585938	0.119998	2.861061	0.505242
European 1	20/83	15/84	1.460317	0.688741	3.09627	0.321937
European 2	69/373	94/357	0.635043	0.44669	0.902818	1.11E-02
European 3	18/156	30/159	0.56087	0.298199	1.054914	7.03E-02
Fletcher	1/12	4/11	0.159091	1.46E-02	1.731783	0.103522
Frankfurt 2	13/102	29/104	0.377761	0.183374	0.778208	7.01E-03
Frank	6/55	6/53	0.959184	0.288826	3.18542	0.945743
GISSI-1	628/5860	758/5852	0.806644	0.72076	0.90276	1.80E-04
Heikinheimo	22/219	17/207	1.248134	0.642814	2.423466	0.512012
ISAM	54/859	63/882	0.87205	0.598592	1.270432	0.475469
ISIS-2	791/8592	1029/8595	0.745551	0.676002	0.822255	3.82E-09
Italian	19/164	18/157	1.011877	0.509891	2.008067	0.973064
Kennedy	12/191	17/177	0.630956	0.292396	1.361532	0.237331
Klein	4/14	1/9	3.2	0.296053	34.58841	0.32179
Lasierra	1/13	3/11	0.222222	1.95E-02	2.533216	0.199674
N Ger Collab	63/249	51/234	1.21537	0.797093	1.85314	0.364449
NHLBI SMIT	7/53	3/54	2.586956	0.631588	10.59605	0.173952
Olson	1/28	2/24	0.407407	3.46E-02	4.795302	0.462833

Sainsous	3/49	6/49	0.467391	0.109974	1.986414	0.29402
Schreiber	1/19	3/19	0.296296	2.79E-02	3.141951	0.290426
UK-Collab	38/302	40/293	0.910417	0.565439	1.465868	0.699278
Valere	11/49	9/42	1.061404	0.391732	2.875891	0.906713
Vlay	1/13	2/12	0.416667	3.28E-02	5.29866	0.490279
White	2/107	12/112	0.15873	3.47E-02	0.727052	7.48E-03
Wisenberg	2/41	5/25	0.205128	3.65E-02	1.152611	5.29E-02
Witchitz	5/32	5/26	0.777778	0.198734	3.043957	0.717698

A summary of the meta-analysis of these thirty-three studies results is:

Meta-analysis Methods (Petitti, 1994)	OR; 95% Confidence Interval (Upper and Lower Confidence Limits)			Null hypothesis, Two-Tailed Test, 95% Confidence Interval	
Mantel--Haenszel:	Estimated OR	LCL	UCL	z-value	p-value
1. Fixed effects	0.765	0.717	0.816	-8.158	0.000
2. Random effects	0.762	0.682	0.851	-4.839	0.000
Peto, Fixed effects	0.765	0.717	0.816	-8.176	0.000

These results consist of three different tests. The Mantel--Haenszel test for fixed and random effect and Peto's test for fixed effects model only. The OR from the meta-analysis is less than 1.00. The meta-analysis of those thirty-three studies shows that, taken together, immediate treatment is more favorable to the patient than delaying treatment.

The width of the confidence interval may encompass the value of the null hypothesis: either $OR = 1.00$, or $ln(OR = 1) = 0$. The probability value (p-value) reflects this judgment and extends it to include even smaller probabilities. In this example, the p-value = 0.000 means that we are now dealing with probabilities lower than one in a thousand that the experimental result is due to chance.

Using *Comprehensive Meta-Analysis*, we have also developed two funnel plots to show the precision of the analysis, precision is calculated as *[1/(standard error)]* of the natural logarithm of the estimated OR.

Example 8.4. The funnel plot can determine the extent of the bias in the studies included in the meta-analysis. The size of each circle gives an idea of the variability of each study. The first funnel plot depicts the entirety of the evidence, which consists of thirty-three studies. The second uses a subset of studies in which the odds ratios are somewhat higher. These plots depict the tendency toward an aggregate OR that equals 1.

The selection of the studies to be meta-analyzed is judgmental and can support a particular result. This is a reason why it is important to develop a protocol for meta-analysis that includes the criteria for inclusion and exclusion of studies, as well as the complete literature searches that form the basis for the meta-analysis.

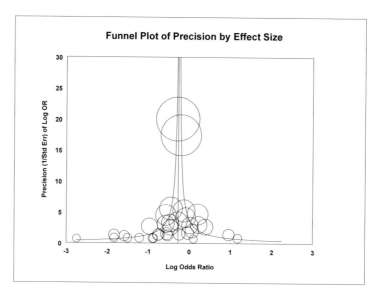

The second funnel plot uses a subset of fourteen of the thirty-three studies.

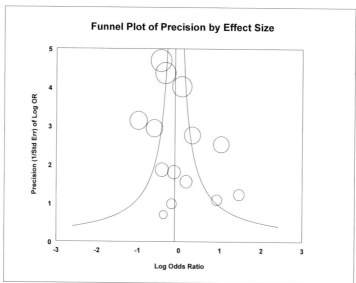

In this subset, the meta-analysis based on the M-H fixed effect model yields *OR = 0.8199 (95% CI = 0.676, 0.995).*

8.3.1. Discussion

Meta-analysis has raised several issues extensively discussed by many authors. These include the following (Wachter, 1988; Spitzer, 1991, Wachter and Straf, 1993, and Borenstein and Rothstein, 1999):

1. Difficulties to aggregate over studies that use different models and heterogeneous units of analysis,
2. Limitation to published studies (availability bias),
3. The impossibility of accounting for all of the appropriate studies: the "file-drawer" issue where those studies with insignificant results are unpublished and therefore are unavailable,
4. Inability to determine the use of a common set of data by different researchers, unless the sources are identified,
5. Inability to discriminate among studies that measure similar, but not identical, outcomes,
6. Reliance on the judgment of the investigator in the choices of studies to be included (personal bias),
7. Independence of the studies selected by the risk assessor,
8. The p-values of a meta-analysis can be inflated because the p-value can be higher merely because the sample size has increased.

Consider, for example, the *availability* bias, which is the result of using published studies. The bias arises because it is more often than not the case that studies with negative results are not published: the bias is introduced by having access to studies that only show statistically significant and positive findings. Thus, a meta-analysis can be biased by the selection criteria of an editor, its peer reviews or both. In this context, it can be relevant to determine the number of additional studies required to change the probability of positive findings. This number is called the *fail-safe* number, N_α. It is computed, for a specific level of statistical significance taken to be $\alpha = 0.01$ in the formula below, as:

$$N_{0.01} = \Sigma_i \left(\frac{z_i}{2.33} \right)^2 - N ,$$

where the z_i are the z-values reported in the individual studies and N is the number of independent studies used in the meta-analysis (Cooper, 1979). Another important issue in meta-analysis is whether the results used are homogeneous. A test to assess homogeneity is applied to the *OR* in Example 8.5.

Example 8.5. Each study (the ith study) of a set of studies yields an estimated OR_i, $OR_i = [(ad)/(cd)]_i$. The Mantel-Haenszel (M-H) formula for testing homogeneity is:

$OR_{MH} = \Sigma_i[(1/variance_i)(OR_i)]/\Sigma_i(1/variance_i)$.

This test can be applied to incidence density and cumulative incidence studies and therefore to relative risk studies (Kleinbaum, Kupper and Morgenstern, 1982). The confidence interval is: $95\%CI = exp\{[ln(OR)_{MH}]\pm[1.96var(OR_{MH})^{1/2}]\}$. A large sample justifies using the normal distribution and the value *1.96*. Let us define $Q = \Sigma_i[(1/variance_i)(lnOR_{MH}-OR_i)^2]$, where Q is distributed according to a *chi*-square distribution. Suppose that $Q = 20.00$, with *19* degrees of freedom. The result is statistically significant ($\alpha = 0.05$) and therefore the studies included in this meta-analysis are not homogeneous. The size of the (fixed) effect is not the same across the studies: the studies included in the meta-analysis are not homogeneous.

Shapiro (1994) has stated that *the meta-analysis of published non-experimental data should be abandoned.* Greenland (1994) states *Shapiro's condemnation of meta-analysis, while understandable, is premature at best, and potentially destructive.* Greenland's rationale for using meta-analysis is that a reasonable protocol can be set up to deal with some of the issues that affect the meta-analysis of non-experimental studies. For risk assessment and the use of risk-based results, there is little doubt that meta-analysis can provide valuable information. Risk assessment relies on aggregate information; therefore, analytical methods that permit formal and theoretically sound analysis are useful. Many non-experimental studies are good science, even though they may have not been randomized. Ignoring them omits relevant information.

8.4. BINOMIAL POPULATIONS

Outcomes are *binomial* because they are either mutually exclusive failures or successes (these are also called Bernoulli trials). A risk assessment can consist of two binomial populations that are portrayed in a table in which each cell contains the number of events for the two populations. For instance, a risky situation can be represented by a *2*2* table.[2] Table 8.3 contains the information used in formulating hypotheses and developing confidence intervals about proportions and differences in the probability of success (π_1) and the probability of failure (π_2). Note that π_i is the population parameter estimated from a random sample.

Table 8.3. 2*2 Table for Two Populations

	Population 1	Population 2	Row Totals
Successes	$y_{11} = a$	$y_{12} = b$	m_1
Failures	$y_{21} = c$	$y_{22} = d$	m_2
Column Totals	n_1	n_2	N

Table 8.4 depicts the observed numbers of success and failures.

Table 8.4. *Observed* Number of Successes and Failures in Two Populations

	Population 1	Population 2	Row Totals
Success	$x_{11} = a$	$x_{12} = b$	m_1
Failures	$x_{21} = c$	$x_{22} = d$	m_2
Column Totals	n_1	n_2	N

Example 8.6. Consider these two random samples taken from Population 1 and 2, respectively:

	Population 1	Population 2
Successes	$x_{11} = 3$	$x_{12} = 6$
Failures	$x_{21} = 10$	$x_{22} = 15$

---

[2] Chapter 9 continues the discussion of contingency tables and extends it to larger contingency tables, such an *n*k* table, which are used to stratify results to account for several (*n*) discrete responses and several (*k*) discrete exposures.

The null hypothesis is that $\rho = \pi_2/\pi_1 = 1$. The sampling results are:

1. *Observed* proportion for population, π_1, column 1 is *0.2308*.
2. *Observed* proportion for population π_2, column 2 is *0.4000*.
3. *Observed* ratio of proportions π_2/π_1 is *1.733*.

We use two statistical tests. One is *asymptotic* and the other is *exact* (StatExact5). The results are similar in that they are statistically insignificant at the *0.05* level of significance:

Tests	p-value, pr(t ≥ t$_{\alpha,\,df}$)	*95% CI for π_2/π_1*
Asymptotic	0.1484	(0.6490, 5.377)
Exact	0.2100	(0.6298, 7.591)

Suppose that a difference between two proportions, $\pi_{exposed} - \pi_{controls}$, is being assessed. The null hypothesis is stated as $\pi_{exposed} - \pi_{controls} = 0$. The alternative hypothesis is $\pi_{exposed} - \pi_{controls} \neq 0$.

Example 8.7. Let the two-sided null hypothesis be H_0: $\pi_2 - \pi_1 = 0$. The contingency table is as before:

	Population 1	**Population 2**
Successes	$x_{11} = 3$	$x_{12} = 6$
Failures	$x_{21} = 10$	$x_{22} = 15$

The *observed* difference of the proportions is $\pi_2 - \pi_1 = 0.1692$. The two statistical tests under asymptotic assumptions and an exact method (StatExact5) yield the following results:

Test	p-value, (t ≥ t$_{\alpha,\,df}$)	**0.95 CI for $\pi_2 - \pi_1$**
Asymptotic	0.1484	(-0.1558, 0.4305)
Exact	0.2100	(-0.1708, 0.4545)

Example 8.8. The risk assessor is planning a study in which he seeks to determine a difference between two proportions ($\pi_1 = 0.65$ and $\pi_2 = 0.50$). What are the sample sizes associated with a *95%* level of confidence ($\alpha = 0.05$) and several values of β (see sections 4.4 and 8.6)?

The results are depicted as (Borenstein, Rothstein, Cohen, 2001), in which power is measured on the *y*-axis:

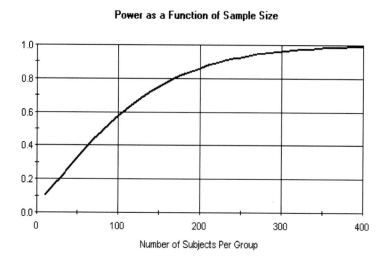

Power as a Function of Sample Size

From this curve, for example, *80%* of the studies will be statistically significant and show a size effect (confirming the alternative hypothesis) thus leading to the rejection of the null hypothesis. The *precision* of the estimate for a difference in proportions that equal *0.15 (from 0.65 - 0.50),* for $\alpha = 0.05$ and $n_1 = n_2 = 250$, is a confidence interval with the lower *95%* confidence level = *0.06* and upper *95%* confidence level = *0.24*. Power is *1.00* if both sample sizes equal *500*. Note that if the difference were reduced to detect a difference of *0.05*, rather than *0.15*, $\alpha = 0.05$ and $n_1 = n_2 = 250$, a study of that size would have *0.80* power.

This example is useful in developing a preliminary sample size for a study in which the researcher wants to investigate differences between proportions accounting for Type I and Type II errors. The example that follows further illustrates analysis and results.

Example 8.9. (Adapted from Kahn and Sempos, 1989). Suppose that H_0: $\pi_{expose} - \pi_{controls} = 0$ and that H_1: $\pi_{exposed} - \pi_{controls} = 0.01$. Let $\alpha = 0.05$, $\beta = 0.10$, $\pi_{exposed} = 0.030$ and $\pi_{controls} = 0.020$. Note that the variance of a difference of two random variables, $var(\pi_{exposed} - \pi_{controls})$ is the *sum* of the individual variances, $var(\pi_{exposed}) + var(\pi_{controls})$. The standard error of the difference in proportions, if H_0 is true, is:

$$se(\pi_{exposed} - \pi_{controls} = 0) = \{[(0.03)(0.97)/n] + [0.02(0.97)/n]\}^{1/2}.$$

Under the null hypothesis, the quantity $(1.96)[se(\pi_{exposed} - \pi_{controls})]$ controls the Type I error. The Type II error is controlled by the quantity $(1.28)[se(\pi_{exposed} - \pi_{controls} = 0.10)]$. It follows that the formula for determining the sample size, *n*, of this prospective study in which the alternative hypothesis is that the difference is *0.01*:

$$(1.96)\{[(0.03)(0.97)/n] + [0.02(0.97)/n]\}^{1/2} + (1.28)\{[(0.03)(0.97)/n] + [0.02(0.97)/n]\}^{1/2} = 0.01.$$

Solving this expression for *n* yields the sample sizes: approximately *5,700* individuals for the cases and *5,700* for the controls. The statement associated with this analysis is that *the*

> statistical study has the power of determining an effect of size 0.01 between the cases and the controls that equals 0.90. The probability of failing to reject the null hypothesis of a difference between the proportions that equals 0.010, when it is false, is $\beta = 0.10$.

In general, exact tests are computationally demanding. However, for 2*2 contingency tables, we can use Fisher's exact test. This test determines whether the observed *small* relative frequencies of a 2*2 table are rarer than would be expected from randomly occurring values. The relative frequencies are counts of binomial (yes, no) outcomes for exposure and response, treatment and no treatment and so on. The effect of small numbers in the observed counts can influence the magnitude of estimated *chi*-square statistic:

$$X^2_{df=1} = \sum_{i=1}^{4} [(Observed - Expected)^2 / Expected] .$$

If the number expected is small, the small number will tend to make X^2 be large simply as an artifact of the numbers. Fisher's exact text overcomes this problem. Table 5.5 depicts the 2*2 table.

Table 8.5. 2*2 Contingency Table Used in Fisher's Exact Test

	Exposure	No Exposure	Totals
Effect	x_{11}	x_{12}	$x_{11}+x_{12}$
No Effect	x_{21}	x_{22}	$x_{21}+x_{22}$
Totals	$x_{11}+x_{21}$	$x_{12}+x_{22}$	$x_{11}+x_{12}+x_{21}+x_{22} = n$

This test has one degree of freedom. Fisher's exact test estimates the following quantity:

$$pr(observed\ numbers) = [(x_{11}+x_{12})!(x_{21}+x_{22})!(x_{11}+x_{21})!(x_{12}+x_{22})!]/(x_{11}!x_{12}!x_{21}!x_{22}!n!).$$

The actual calculation consists of using this formula to calculate the probability of the observed numbers, and then calculate the probabilities of even smaller numbers, all the way to zero, by changing that small number, given that all other numbers are unchanged. These probabilities are added and the null hypothesis tested accordingly (Dawson-Saunders and Trapp, 1994). The next example clarifies the method.

Example 8.10. This example illustrates the calculations involved in Fisher's exact test. Consider the 2*2 contingency table in which only the cell of the relative frequencies are shown:

$y_{11} = 3$	$y_{12} = 6$
$y_{21} = 10$	$y_{22} = 15$

The complete analysis requires calculating the probabilities of more extreme events than those shown in the table. This is done by replacing the initial value in cell *1*, which is *3*, with more extreme values: *2, 1* and then *0*. The new values are used to calculate the probability of such table being due to chance, holding the row and column marginal totals constant. For example, for $y_{11} = 1$:

pr(observed numbers are due to chance) = $(7!25!13!21!)/(1!6!10!15!34!)$ =
$2.48 \times 10^{58}/1.009 \times 10^{60} = 0.0247$.

Probabilities add to *0.031* (using StatExact5, 2001). By contrast, the *chi*-square estimate obtained using Pearson's statistics with one degree of freedom equals *0.0009*. In this example, the null hypothesis is rejected under both calculations. The power of the test can be investigated as follows. Suppose that a proposed study has a Type I probability, $\alpha = 0.05$, the proportion for population *1*, π_1, is *0.50* and that for population 2, π_2, is *0.60*. A sample size of *500* per population has a power of *0.88*.

The important issue for risk assessors is that Fisher's exact test should be used when the traditional *chi*-square approximations do not hold. Chapter 9 provides additional results.

8.4.1. Sensitivity and Specificity of the Test

A form of statistical decision is hypothesis testing, which is developed before the sample is taken. In such testing, the risk assessor faces the Type I error: to conclude that the null hypothesis is false, when it is actually true. She also must confront the Type II error: to conclude that the null hypothesis is true, when it is false. We can gain additional insights on this issue using contingency tables. Let us consider the following 2*2 contingency table, Table 8.6.

Table 8.6. 2*2 Contingency Table for Sensitivity and Specificity of a Test

		Disease is Present:		
		Yes	*No*	**Totals**
	Positive	a = 10	b = 20	a+c = 30
Test is:	*Negative*	c = 20	d = 30	c+d = 50
	Totals	a+c = 30	b+d = 50	a+b+c+d = 80

In this discussion, π is a proportion. The *sensitivity* of the test, π *positive|disease is present*, is calculated as *[a/(a+c)]*. The sensitivity of the test measures the proportion of true positives. The *specificity* of the test, π *positive|(disease is absent)*, is calculated as *[b/(b+d)]*. The specificity of the test measures the proportion of false positives. Chapter 9 adds further discussions, with the same data.

8.5. COMPARISON OF DISTRIBUTIONS

Typically, under the null hypothesis we seek a test of whether the population means or variances estimated in a study are significantly different from one another. The selection and use of a statistical test involve:

- Scale of measurement for the data
- Assumption about the population distribution, unless the test is non-parametric
- Choice of test or tests
- Hypotheses and decision rule or rules
- Testing and assessing the results of statistical analyses
- Making statistical and other conclusions based on the results including comparisons

with past work and discussion of the assumptions and limitations of the study

To perform those tests, it is often necessary to assume knowledge of the distributions of the population for whom inference occurs through sampling: the *t*-test, *z*-test and the *F*-test are widely used. However, they may not be appropriate when the populations are normally distributed. Non-parametric tests can be used when it is unreasonable to make stringent assumptions about the distribution of one or more populations.

Common parametric tests are the *t*-test or the *z*-test, which have known distributions. However, these tests are applied when the populations are normally distributed (or approximately so) and when the data are measured on an interval scale. Otherwise, we can use non-parametric tests. Non-parametric statistical methods that follow the parametric testing protocol:

- Setting up the null and alternative hypotheses
- Developing one or more samples for the statistical testing
- Choosing a decision rule and level of statistical significance
- Comparing that numerical value to a table of critical values
- Making a decision on whether the null hypothesis should be rejected in favor of the alternative hypothesis

Consider the parametric comparison of data from two independent samples. The risk assessor may be interested in determining whether two samples were drawn either from the same population or from two populations (each of which has the same distribution).

8.5.1 Two-samples, Non-parametric Kolmogorov–Smirnov (K-S) Test

The purpose of this test is to determine if the samples arise from the same population. It is illustrative of the practical way in which non-parametric methods are used. It is one of several tests available to study and make inference from small and large samples when the scales of measurement are not interval-valued. Following Siegel (1956) and Sprent (1989), the K-S test first develops the empirical (sample-based) cumulative frequency distribution, for each sample, using the same class intervals. The mutually exclusive statistical hypotheses (null and alternative) are:

H_0: all samples are generated by the same distribution.
H_1: at least one sample is not.

The null hypothesis is rejected, at a specific level of significance, when the computed value of the K--S statistic is greater than the theoretical value. The cumulative function for one of the two samples, for example sample *1*, is:

$$S_{n1}(X) = \sum_{i \leq 1}^{n_1} x_i / \sum_{i=1}^{n_1} x_1 \, .$$

The cumulative distribution for the other sample is $S_{n2}(X)$. The largest absolute deviation between $[S_{n1}(X)-S_{n2}(X)]$ is the metric used in the K-S test. If the two distributions were similar, their largest deviation would be minimal, other than for some randomly occurring fluctuations. Accordingly, the (one-tail) K-S test uses the difference:

$$D = max[|S_{n1}(X)-S_{n2}(X)|],$$

where $max[.]$ means the maximum value of the bracketed expression. The terms S_{n1} = $[C(1)/n_1]$ and S_{n2} = $[C(2)/n_2]$, $C(.)$ are the cumulative total for each sample, as exemplified below. The two-tailed test requires taking the absolute value of the maximum difference in the expression for D. Values of D are available and can be found in Kanji (1999).

Example 8.11. Let us use two samples of concentrations of waterborne arsenic, in *ppm*. Let α = 0.05. In this example, $n_1 = n_2 = 7$. The data and the computations for the K-S test are as follows:

Sample 1	0.5	1.2	1.4	2.0	2.2	2.6	2.9
C(1)	0.5	1.7	3.1	5.1	7.3	9.9	12.8
Sample 2	1.5	1.8	2.2	2.8	4.3	5.0	5.5
C(2)	1.5	3.3	5.5	8.3	12.6	17.6	23.1
S_{n1}	0.039	0.13	0.24	0.40	0.57	0.77	1.00
S_{n2}	0.065	0.14	0.24	0.36	0.55	0.76	1.00
$|D|$	0.028	0.01	0.00	0.04	0.02	0.01	NA

The terms $C(1)$ and $C(2)$ are the cumulative totals for each sample; the S_{n1} and S_{n2} terms are calculated as $0.5/12.8 = 0.03906$, $1.7/12.8 = 0.1328$, ..., $23.1/23.1 = 1.00$. The maximum difference $(S_{n1}-S_{n2}) = 0.04$.

Using Kanji (1999) Table 16, the theoretical value of D equals 0.486. Because the computed value of D is less than the theoretical value, the null hypothesis cannot be rejected. Therefore, the conclusion is that both samples originate from the same distribution. D = $\{c[(n_1+n_2)/(n_1*n_2)]\}$, $c = 1.36$ for $\alpha = 0.05$, for two large samples.

The sampling distribution of D is known (Sprent, 1989) and tables exist for small and large samples. For large samples (either one- or two-tailed tests) D has an approximately chi-square distribution. This approximation works for small samples, but the results are more *conservative* (are wider). The critical value used in the tables provided by Siegel is the numerator of the largest difference between two cumulative distributions.

Example 8.12. The data used in this example are *scores*, or ratings. The researcher wants to test whether individuals exposed to pattern A are more affected than those exposed to pattern B. Impairment from exposure is measured, for a sample of *10* individuals, by the following *scores* from each of the two exposure patterns A and B:

A: 24, 29, 32, 33, 34, 36, 38, 40, 41, 42.
B: 40, 41, 41, 45, 46, 48, 48, 49, 52, 55.

For ease of computation, the data are arranged into classes of equal size; that is, into score intervals. The frequencies in each class are cumulative frequencies. The differences in the cumulative frequencies are computed for each class. The results are:

	Score Intervals								
	24-27	*28-31*	*32-35*	*36–39*	*40–43*	*44-47*	*48-51*	*52–55*	Total
n_A	1	1	3	2	3	0	0	0	10
n_B	0	0	0	0	3	2	3	2	10
n_{Acum}	1	2	5	7	10	10	10	10	55
n_{Bcum}	0	0	0	0	3	5	8	10	26
$D(n_{Acum}$ $-n_{Bcum})$	1	2	5	7	7	5	2	0	29

The test statistic is the largest class difference. In this case, the largest is *7*, which occurs in the classes *36-39* and *40-43*. Siegel (op. cit.) has provided two Tables of values for applying the K-S test. Because we are testing a directional hypothesis, the columns for a one–tailed test must be used.

The sample sizes are equal: $n_A = n_B = 10$, and the maximum difference is *7*. This difference is statistically significant at $\alpha = 0.01$. If the maximum difference had been 6, the difference would have been significant at $\alpha = 0.05$. Therefore, the two samples do not belong to the same population.

The statistical significance of the differences is:

Column 2 mean	46.5	t-Ratio	17.76
Column 1 mean	34.9	df	9
Mean Difference	11.6	Prob > |t|	<0.0001

8.6. STATISTICAL DECISION-MAKING

Hypothesis testing and confidence are forms of statistical decisions making which are important to determine the accuracy and precision of estimation and in statistical inference. Hypothesis testing precedes sampling; confidence interval analysis follows. This section extends the discussion to deal with risky situations in which samples are used for statistical inference. We focus on classical parametric statistical tests based on the z-distribution, the t-distribution, the F-distribution and the chi-square distribution.

Confidence interval analysis is based on results from sampling and is used in making inference from the sample to its population. Table 8.9 depicts the hypothesis testing system, modified from Dawson-Saunders and Trapp (1994). Specifically, the null hypothesis, H_0, is a statement about the population parameter. For example, $H_0: = 0$, states that the population mean is zero. Thux, we can have the mutually exclusive alternatives depicted on Table 8.7, which integrates past discussion.

Table 8.7. Hypothesis Testing Framework

		True States of Nature	
Conclusions from Testing the Hypothesis		*Difference (H₁ true)*	*No Difference (H₀ true)*
	Difference (Reject H₀)	Power, probability = (1- β)	Type I error, probability = α
	No Difference (Do not reject H₀)	Type II error, probability = β	Correct Decision

Recollect that the *t*-test for the null hypothesis that the population parameter $\mu = k$ (*k* can be zero or some other number that the risk assessor assumes to be true) is:

$$t = \frac{\bar{x} - k}{s / \sqrt{n}},$$

where: $s = (n-1)^{-1} \sum_i (x_i - \bar{x})^2$. If the sample is large, we use the *z*-test: $z = \frac{\bar{x} - k}{s / \sqrt{n}}$.

For $\alpha = 0.05$, the rejection numbers are: $|z| = 1.960$ for the two-tailed test, $z > 1.645$ or $z < 1.645$ for each of the one-tailed tests. The *t*-distribution for five degrees of freedom is shown in Figure 8.1. This distribution has *n-1* degrees of freedom and therefore can change shape; when *n* increases, the *t*-distribution approaches the normal distribution.

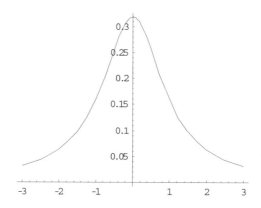

Figure 8.1. Plot of the t-distribution with df = 5

As discussed, the power (of the test) deals with detecting the magnitude of the effect under the alternative to the null hypothesis. Taking the relative risk as an example, an alternative hypothesis can be stated as $H_1: RR = 1.10$; the null hypothesis is H_0:

RR = 1.00. Power analysis is related to sample size, level of statistical significance, α, and power, *(1-β)*. The phrase *size of the effect* (effect size) is used in power analysis to indicate the magnitude of the response under the alternative hypothesis. The *power* of a statistical study is determined by keeping *constant* the magnitude of the effect and the probability of committing a Type I error (measured by α) and functionally relating power (measured by the probability *1-β*, a function of the sample size). Practically, we seek *a level of power to detect an effect* that is achievable within a budget and on time. The magnitude of the effect sought determines the time and labor needed to address the magnitude of the risk. The true magnitude is not known. That is why there will be several power curves for different effects sizes. Specifically, the classical statistical test of the hypothesis accounts for these errors as a function of the sample size. False positive and false negative errors are investigated, before a statistical study is started, for selecting a realistic sample size. The next example illustrate how the power of the test can be used in these analysis by developing a power curve based on $\alpha = 0.05$.

Example 8.13. A risk assessor is planning a study aimed at detecting a difference between two means, based on a random sample of data. Before starting her study, she wants to determine the *power* of the test, based on the *t*-distribution in which paired sample data are compared. She generates the power curve for $\alpha = 0.05$ (two-tailed test) under the null hypothesis that the mean *difference* of the two paired samples equals *0.00*.

Under the *alternative* hypothesis, the assumed difference equals *0.20*, (*sd = 0.500*). She wants to study the power of the tests under these conditions. The vertical axis of the diagram below depicts the power of the test, for different sample sizes:

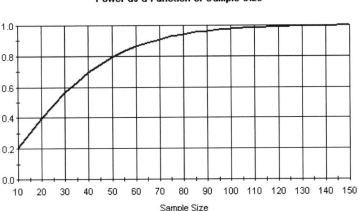

The results obtained in this example can be summarized as providing answers such as *If the effect size were 0.20, the power of a sample size of 50 would be 0.80, given that the probability of making a Type I error is 0.05*. The *precision* of the prospective study is shown below:

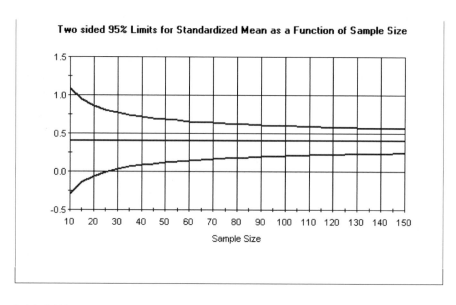

8.6.1. Difference of Two Population Means

Statistical decision-making for risk assessment studies can assess the significance of differences in the exposure, responses or other outcome. The null hypothesis for the difference between two population means is:

H_0: $\mu_1-\mu_2 = k$,

where μ_1 and μ_2 are two population means and k is an arbitrary real number. There are three alternative hypotheses:

H_1: $\mu_1-\mu_2 \neq 0$; $\mu_1-\mu_2 < 0$; or $\mu_1-\mu_2 > 0$.

Let us use parametric method; therefore, either the *t*- or the *z*-distributions can be used. The decision rules for the null hypotheses, the alternatives and the *z*-values and *t*-values for a specific level of confidence (*1-α*) are tabulated in Table 8.8.

Table 8.8. Decision Rules and Statistical Tests of the Null and Alternative Hypotheses for Differences in Population Means for Large and Small Samples (*z*-distribution for large samples, *t*-distribution for small samples, α = level of significance, *df* = degrees of freedom).

Large samples H_0: $\mu_1-\mu_2 = k$	H_1: $\mu_1-\mu_2 \neq 0$	$-z_{\alpha/2} \leq z \leq z_{\alpha/2}$
	H_1: $\mu_1-\mu_2 < 0$	$z < -z_{\alpha}$
	H_1: $\mu_1-\mu_2 > 0$	$z > -z_{\alpha}$
Small samples H_0: $\mu_1-\mu_2 = k$	H_1: $\mu_1-\mu_2 \neq 0$	$-t_{\alpha/2, df} \leq t \leq t_{\alpha/2, df}$
	H_1: $\mu_1-\mu_2 < 0$	$t < -t_{\alpha, df}$
	H_1: $\mu_1-\mu_2 > 0$	$t > t_{\alpha, df}$

Example 8.14. Suppose that the risk assessor wants to know if two new methods, A and B, to decrease occupational risks from routine events in the workplace differ in magnitude. The risk assessor sets $\alpha = 0.02$, using the two-tailed test, has a random sample of the number of incidents over a year. The hypotheses are: H_0: $\mu_A = \mu_B$, against H_1: $\mu_A \neq \mu_B$. The null hypothesis will not be rejected if the probability that $\mu_A = \mu_B$ is greater than *1%*, using the two-tailed test (that is, $\alpha/2 = 0.02/2 = 0.01$). There are 50 individuals that use the new method A ($n_A = 50$) and *40* individuals that use method B ($n_B = 40$). Let the sample averages and standard deviations be $\bar{x}_A = 14.9$ incidents, $se_A = 3.4$ incidents, and $\bar{x}_B = 16.8$ incidents, $se_B = 2.9$ incidents. The formula for the standard error of the difference between the means is:

$se_{A-B} = [(se_A^2/n_A) + (se_B^2/n_B)]^{1/2}$. Therefore, we obtain: $se_{A-B} = [(3.4^2/50) + (2.9^2/40)]^{1/2} = \sqrt{0.44} = 0.663$ incidents. When testing the difference between two sample means, the formula for the z-statistic is:

$$z = \frac{(\bar{x}_1 - \bar{x}_2)}{(s_1/n_1 + s_2/n_2)^{1/2}} .$$

Thus, $z = (14.9-16.8)/0.663 = -1.90/0.663 = -2.86$. Because we are using a two-tailed test, we use the absolute value, $|-2.86| = 2.86$. The decision rule compares the estimate of the z-value to its theoretical z-value, which is $z_{\alpha/2, [(n1+n2)-2]} = 2.33$. Using the table of z-values for the *98%* confidence level, the theoretical z-value is *2.33*.

The null hypothesis is rejected in favor of the alternative hypothesis if the estimated value of z is larger than its theoretical value. Since we are using the two tails of the normal distribution, the absolute value of *2.33* is compared with *2.86*. The test shows that there is a statistically significant difference between the means of methods A and B.

8.6.2. Decision Rules for the Population Variance

The variance σ^2 of a parametric distribution with population mean μ is estimated as:

$$\hat{\sigma}^2 = n^{-1}[s + n(\bar{X} - \mu)^2],$$

in which n is the sample size, s is the sample standard deviation, $\bar{X}$ is the sample mean and μ is the population mean. The variance has a *chi*-square distribution. Using X^2 to distinguish between a theoretical value of the *chi*-square distribution (χ^2) and its estimated value from a sample (X^2), the test statistic (the number of degrees of freedom equals *1*) is:

$$X^2 = (n-1)(s^2/\sigma_0^2),$$

where the subscript *0* identifies the value of the population variance under the null hypothesis. The term s^2 is the sum of the square deviations of the sample:

$$s^2 = \sum_i (x_i - \bar{x})^2,$$

for a sample of size *n*, drawn at random from a population that is normally distributed. Table 8.9 depicts the statistical decision rules that apply to the population variance, σ^2.

Table 8.9. Statistical Tests of the Null and Alternative Hypothesis for the Population Variance

Null Hypothesis	Alternatives	Decision Rules
$H_0: \sigma^2 = \sigma^2_0$	$H_1: \sigma^2 \neq \sigma^2_k$	$\chi^2_{1-\alpha, df} \leq X^2 \leq \chi^2_{1-\alpha/2, \, df = n-1}$
	$H_1: \sigma^2 < \sigma^2_k$	$X^2 < \chi^2_{1-\alpha, df}$
	$H_1: \sigma^2 > \sigma^2_k$	$X^2 > \chi^2_{\alpha, \, df = n-1}$

The role of degrees of freedom is also discussed in chapter 9. This number is important because it changes the shape of the *chi*-square distribution, thus affecting the areas in the tails of the distribution.

Example 8.15. The risk assessor wants to test the null hypothesis that a normal population has standard deviation that equals *0.01*. Without the details, let $X^2 = 25$. Suppose that the risk assessor is conducting a test with a sample size that equals *13*, to determine if the sample variance equals the population *variance*. Let the theoretical values of $\chi^2_{\alpha, df}$ be (for a two-tailed test with *df* = 12): $\chi^2_{0.975, 12} = 4.40$, and $\chi^2_{0.025, 12} = 23.34$. We conclude that, because the estimated $X^2 = 25$ is outside the acceptance region for the null hypothesis. We reject the null hypothesis.

8.6.3. Decision Rules for Two Equal Population Variances

The risk assessor may wish to determine if two different samples, with size n_1 and n_2, belong to two populations with different variances. Testing the equality of the variances has implications on the *robustness* of the analysis of data, which is associated with the assumption of constant variance (homogeneity). The concern is that, because the variance is directly related to estimates of the standard error, if the assumptions are not valid, the results from a statistical analysis can be erroneous or misleading. Table 8.10 depicts the statistical decisions for testing the null and alternative hypotheses.

Table 8.10. Statistical Tests of the Null and Alternative Hypotheses for the *Equality* of Two Population Variances

Null Hypothesis	Alternatives	Decision Rules
$H_0: \sigma_1^2 = \sigma_2^2$	$H_1: \sigma_1^2 \neq \sigma_2^2$,	$F_{1-\alpha/2, \, (n1-1, \, n2-1)} \leq$ Estimated F-ratio $\leq F_{\alpha/2, \, (n1-1, \, n2-1)}$
	$\sigma_1^2 < \sigma_2^2$,	Estimated F-ratio $< F_{1-\alpha, \, (n2-1, n1-1)}$
	$\sigma_1^2 > \sigma_2^2$.	Estimated F-ratio $> F_{\alpha, \, (n1-1, n2-1)}$

The ratio of the *population variances* is distributed according to the *F*-distribution, which has a number of degrees of freedom for the numerator, $df_1 = (n_1-1)$, and for the denominator, $df_2 = (n_2-1)$. Specifically, the ratio of two *chi*-square distributed random variables has an *F*-distribution. Figure 8.2 depicts the *F*-distribution with *14* *df* at the numerator and *10 df* at the denominator.

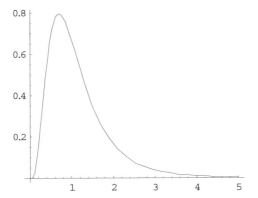

Figure 8.2. *F*-distribution with 14 df at the Numerator and 10 df at the Denominator.

The estimated *F*-statistic equals the ratio of the two *sample variances*, the estimated *F*-ratio, $\hat{F} = s_1^2 / s_2^2$, has *(n₁-1) df* for the numerator, and *(n₂-1) df* at the denominator. The null hypothesis is:

$$H_0 : \sigma_1^2 > \sigma_2^2 ,$$

and its alternative is:

$$H_1 : \sigma_1^2 \le \sigma_2^2 ,$$

whichever is larger. The one-sided *F*-test is used to determine if the variance of population *1* is larger than the variance of population *2*.

Example 8.16. The question being answered is: Are the population variances of two populations equal? Suppose that two samples have been taken. The null hypothesis is:

$$H_0: \sigma_1^2 > \sigma_2^2 ,$$

and the alternative hypothesis:

$$H_1: \sigma_1^2 \le \sigma_2^2 .$$

Let the sample average for the first population equal *195.3*, the sample variance equal *19.5* and $n_1 = 31$. The sample average from the second population equals *95.3*, the sample variance equals *6.00* and $n_2 = 31$. Let the level of significance, α, equal *0.02*. The decision rule for the null hypothesis is:

Reject H_0 if the estimated *F*-ratio $> F_{0.98, \ 30, \ 30}$, otherwise do not reject.

In this example, the numbers of degrees of freedom df_1 and df_2 both equal *30*. The *F*-ratio = $s_1^2 / s_2^2 = 19.5/6.0 = 3.25$. The theoretical value of the *F*-distribution (from the table of *F*-

values) equals 1.84. The null hypothesis cannot be accepted.

This test can be extended to correlated random variables (Kanji, 1999), thus relaxing one of the two assumptions need for the test of the hypothesis.

8.6.4. Decision Rules for the Equality of the Variances of More than Two Populations

We turn next to testing the equality of the variances of more than two populations. The null hypothesis is that the samples originated from normal populations characterized by the same variance. Rosenthal and Rubin (1982) have developed a test of homogeneity of two or more variances ($i = 1, 2, ..., K$), based on the *chi*-square distribution. The test statistic is:

$$X^2_{K-1} = \sum_{i=1}^{K} [1/var(d_i - \bar{d})^2]. .$$

The estimated variance of each study uses the difference between the *i*-th magnitude of the effect and the average magnitude, $\bar{d}$, is computed from all of the k studies included in the risk assessment. Under the null hypothesis, for a level of statistical significance (α), if the estimated value of *chi*-square is greater than the theoretical value, then the studies cannot be assumed to be homogeneous. They are heterogeneous. Wolf, (1990) provides additional information as well as useful tables of mathematical transformations for such statistics as t, F and others to obtain values for the magnitude of the effect under the alternative hypothesis to the null hypothesis.

8.6.4.1. Bartlett's Test

Bartlett's test is used to test differences between the variances of k ($j = 1, 2, ..., k$) normally distributed populations. The first step consists of calculating the individual sample variances, s^2_j, and then estimating the total variance (S^2):

$$S^2 = \sum_{j=1}^{k} (n_j - 1)s^2_j / \sum_{j=1}^{k} \left(n_j - 1 \right)$$

In this formula, n_j is the sample size of the *j*th study. Bartlett's test statistic is approximately distributed as a *chi*-square random variable with $k-1$ degrees of freedom under the null hypothesis that all variances are equal.

Bartlett's statistic (Walpole, 1974) is:

$$B = 2.3026 \frac{(N - k)log_{10}S^2 - \sum_{i=1}^{k}(n_i - 1)log_{10}s^2_j}{1 + [\frac{1}{3(k-1)}][\sum_{j=1}^{k}\frac{1}{n_j - 1} - \frac{1}{N-k}]}.$$

Example 8.17. In this example there are four separate studies. The sample sizes are $n_1 = 10$, $n_2 = 20$, $n_3 = 30$ and $n_4 = 40$, thus $K = 4$. Let the variances of these samples be 6, 3, 7 and 7, respectively. Then, $S^2 = [(10-1)(6)+(20-1)(3)+(30-1)(7)+(40-1)(7)]/(9+19+29+39) = 587/96 = 6.1$. The quantity $log_{10}(6.1) = 0.785$.

The denominator is $1+\{1/[3(4+1)]\}\{[(1/9+1/19+1/29+1/39)]-[1/(9+19+29+39)]\} = -0.776$. Therefore:

$B = (2.3026/0.776)[(9+19+29+39)(0.785)][(9)(0.778)+(19)(0.477)+(29)(0.845)+(39)(0.845)] = -5.822$.

The theoretical *chi*-square value for $\alpha = 0.05$ and 3 degrees of freedom equals 7.81 (from the table of *chi*-square values that can be found in statistical texts provided in the references). The results are statistically insignificant: the four variances are equal.

8.7. ANALYSIS OF VARIANCE (ANOVA): *Independent* Samples for K normally distributed populations

The analysis of variance (ANOVA) uses the F-test statistically to determine if k samples arise from K populations that have the same population mean.[3] More specifically, the ANOVA discussed in this section can test if the observed differences between means are random, is parametric, and should meet these assumptions:

1. K populations are normally distributed,
2. The populations' variances are homogeneous,
3. The samples are random.

The importance of the ANOVA in risk assessment is that it can determine the influence of several factors on the variability of the population means, although in this book we use a single factor. When the ANOVA is limited to one factor, therefore it is *one-way*. The null hypothesis, for the one-way ANOVA, for three or more population means, is:

H_0: $\mu_1 = \mu_2 = \mu_3 \ldots = \mu_k$.

The alternative hypothesis is that at least two of the means are different:

H_1: $\mu_1 \neq \mu_2$ or $\mu_1 \neq \mu_3 \ldots$, and so on.

Table 8.11 depicts the typical layout for the data in the ANOVA, in which we use standard notation for this form of statistical analysis.

---

[3] In this section and the section that follows, k and N indicate the number of factors (such as exposure levels) and the size of the population size, respectively. The context differentiates between sample size and population size.

Table 8.11. ANOVA Data Lay-out

	Sample from Population Number				
	1	2	...	k	Total
Measurement 1	x_{11}	x_{21}	x_{i1}	x_{kk}	--
Measurement 2	x_{12}	x_{22}	x_{j2}	x_{k2}	--
...	...	...	x_{ij}	x_{ik}	--
Measurement N	x_{1N}	x_{2N}	x_{iN}	x_{KN}	--
Total	$T_{1.}$	$T_{2.}$	...	$T_{.K}$	$T_{..}$
Means	$\bar{x}_{1.}$	$\bar{x}_{2.}$	...	$\bar{x}_{k.}$	$\bar{x}_{..}$

The mean of all population means is:

$$\mu_{Total} = (1/k) \sum_{i=1}^{k} \mu_i .$$

Each observation consists of three components:

$$x_{ij} = \mu + \alpha_i + \varepsilon_{ij}.$$

Specifically, the term α_i is the ith effect (the systematic component of the observation) with the constraint that:

$$\sum_{i=1}^{k} \alpha_i = 0 .$$

The term ε_{ij} is the random error of each observation. The estimated total variance, for *all* of the *nk* observations, is:

$$s^2 = \sum_{i=1}^{k} \sum_{j=1}^{n} (x_{ij} - \bar{x}_{..})^2 /(nk - 1) .$$

Table 8.12 contains the formulae for the one-way ANOVA (following the notation in Table 8.11).

Table 8.12. Formulae for the One-Way Analysis of Variance

Variability	Sum of Squares, SS	df	Mean Squares, MS	F-statistic
Between k groups	$SS_B = n\sum_i (\bar{x}_{i.} - \bar{x}_{..})^2$	$k - 1$	$MS_B = SS_B /(k - 1)$	None
Error	$SS_E = SS_T - SS_B$	$k(n - 1)$	$MS_E = SS_E /[k(n - 1)]$	$F = MS_B / MS_E$
Total	$SS_T = \sum_i \sum_j (x_{ij} - \bar{x}_{..})^2$	$nk - 1$	None	None

The ANOVA is used to study two forms of variability. The first is the variability *between factors*, which is measured by the difference between column mean(s) and the grand mean. The second is the variability *within factors*, which is measured by the difference between individual measurement(s) and the column mean(s). The ANOVA is *robust* to violations of the assumption of normally distributed populations. That is, small departures from this assumption do not invalidate the conclusions from an ANOVA. Under the null hypothesis, $MS_B/MS_E = 1$. More specifically, the decision rule is that if:

$$F\text{-}ratio > F_{\alpha, [k-1, k(n-1)]},$$

then the null hypothesis is rejected in favor of its alternative.

Example 8.18. The sample consists of fifteen individual adverse outcomes from three different levels of exposure. Exposure is a single *factor* with three levels, which identify the groups. The question is: Are the mean responses, on the average, equal? The sample is:

Column 1	Column 2
50	No exposure
31	No exposure
70	No exposure
60	No exposure
5	No exposure
30	Mid exposure
10	Mid exposure
20	Mid exposure
10	Mid exposure
11	Mid exposure
6	High exposure
11	High exposure
30	High exposure
20	High exposure
20	High exposure

The graph of the one-way ANOVA shows the three sample groups (high, mid and low exposures):

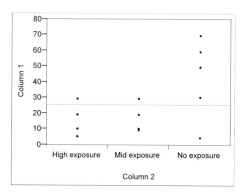

The results of the ANOVA are:

Source	df	Sum of Squares	Mean Square	F-Ratio	Probability > F
Column 2	2	2,326.80	1,163.40	4.22	0.041
Error	12	3,306.80	275.57	--	--
Total	14	5,633.60	--	--	--

Example 8.19. The risk assessor wants to plan a study in which there are five levels of exposure and five cities and wants to determine the appropriate sample size of his study. He opts to use the one-way ANOVA. The effect size he is interested in determining is 0.15. The columns of the data are exposure levels; the rows are the five cities. The level of significance has been set at *0.05* and power is set at *0.80*. The planning of the study (Borenstein M, Rothstein H, Cohen J, 2001) indicates that the study should include *108* cases per level of exposure for a total of *540* cases. The power curve (two-tailed test) for each level of exposure is:

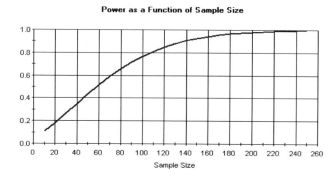

8.8. KRUSKAL-WALLIS TEST (K-W): Non-parametric Testing of k *Independent* Samples

The Kruskal-Wallis, K-W, test is also used for statistical comparisons of three or more independent populations from which random samples are taken. The observations must be statistically independent. Unlike the one-way ANOVA, the K-W test makes no distributional assumption about the populations. It can determine whether the *k* samples belong to *K* populations that have the same mean. The test is based on the ranks of the sample data; ties are resolved by taking the mid-ranks of the tied observations. Table 8.13 depicts the raw data and the ranks based on these data are an example of the inputs to the K-W test.

Table 8.13. Example of Raw Measurement for Three Samples, with Ranks

Sample 1		Sample 2		Sample 3	
Measurement	*Rank*	*Measurement*		*Measurement*	*Rank*
x_1	r_1	y_1	r_1	z_1	r_1
...		...		...	
x_n	r_m	y_m	r_m	z_m	r_1
Sum of Ranks	R_1	--		R_2	--	R_3

The null hypothesis is that all the random samples originate from the same distribution; the alternative is that at least two of them do not (Howell, 1995). Estimates of the K-W statistic are compared to the appropriate theoretical values of the *chi*-square distribution, for *k-1* degrees of freedom, and level of significance. A form of the test statistic (Sprent, 1989) is:

$$T = [(N-1)(S_t^2-C)]/(S_r^2-C).$$

If there are no ties in the ranks, we use:

$$T = 12S_t^2/[N(N-1)]-3(N-1) .$$

In these two expressions, $S_t^2 = \Sigma s_j^2/n_i$ and $S_r^2 = \Sigma_i\Sigma_j r_{ij}^2$, where r_{ij} is the rank for the *i*th observation of the *j*th sample. The term s_i is the summation of the ranks of the values of the *j*th sample. The following example, modified from Sprent (1989), shows the computations.

Example 8.20. The sample, from Sprent (1989), is:

Sample 1		Sample 2		Sample 3	
Measurement	*Rank*	*Measurement*	*Rank*	*Measurement*	*Rank*
93	6	29	1	126	10
98	7	39	2	142	11
216	15	60	3	156	12
249	20	78	4	228	17
301	25	82	5	245	18
319	26	112	8	246	19

731	34	125	9	370	28
910	36	170	13	419	29
-	-	192	14	433	30
-	-	224	10	454	31
-	-	263	21	478	32
-	-	275	22	503	33
-	-	276	23	-	-
-	-	286	24	-	-
-	-	369	27	-	-
-	-	756	35	-	-

The summation of the ranks for the first sample ($j = 1$), is $6+7+15+20+25+26+34+36 = 169$. The term n_j is the number of sample observations in the jth sample: for example, $n_1 = 8$. The coefficient $C = [0.25n_i(n_i+1)^2]$ is a correction for the mean. The simplified calculation for the T- statistic, when there are no ties in the samples, is:

$$T = \{[12(12865.7)]/[(36)(37)]\}-[(3)(37)] = 4.91,$$

which is the T-value to be compared with the theoretical *chi*-square value, $\chi^2_{\alpha=0.05,\ 2}$. This example has ($3-1$) degrees of freedom. The result leads to the acceptance of the null hypothesis because the critical value ($\chi^2_{\alpha=0.05,\ 2}$) = 5.99 is greater than the calculated T-value and to the conclusion that the three samples originate from the same population.

8.8.1. Discussion

This discussion follows Mehta and Patel (StaExact5, 2001). The null hypothesis is that each of the K populations has the same distribution. The alternative hypothesis is that the distributions differ. The individual raw data (u_{ij}) can be shown in a table (analogous to the one-way table discussed in the previous example and in the discussion of the one-way ANOVA), Table 8.14.

Table 8.14. Structure of the Data for the K--W Test

1	2	...	k
u_{11}	u_{12}	...	u_{1K}
u_{21}	u_{22}	...	u_{2K}
...	...	...	...
u_{2n}		...	u_{nK}

The samples need not be of equal size, as the previous example showed. The data in this table are generated by the random variable $U_{ij} = \mu+\tau_j+\varepsilon_{ij}$ where the overall population mean is μ, the *treatment* effect is τ_j and the associated error is ε_{ij} which is independent and identically distributed, with zero mean and unknown population distribution. The null hypothesis is:

H_0: $\tau_1 = \tau_2 = ... = \tau_k$.

The alternative hypothesis is that:

H_1: $\tau_i \neq \tau_j$ (for at least one pair of treatment effects).

The test statistic follows the *F*-distribution.

Example 8.21. Recollect the example used to develop the parametric, one-way ANOVA. Using that data set and the K--W test, we obtain the following results:

Estimated Chi-square	d.f.	Prob. > Chi-square
3.46	2	0.1770

The null hypothesis cannot be rejected. However, the computer software JMP5 warns that the sample is small and an exact method should be used.

8.9. K DEPENDENT SAMPLES

The previous discussion of *k*-samples has dealt with independence. In this section, we discuss dependent samples because these can also routinely arise in risk assessment. Specifically, the analysis considers *N* individuals (or *blocks*) that are treated and there are *K* possible treatments, or exposure levels. Assume that these correlate. Suppose a risk manager evaluates a program on the basis of the same number of attributes. Alternatively, responses may be evaluated over three or more levels of treatment. In these situations, the null hypothesis is that that there are no differences among the treatments (treatment is a general term that can include exposure or other factors) being considered. The alternative hypothesis is that there are at least some differences.

The two tests discussed below, Cochran's and Friedman's, are equivalent but Cochran's test applies to binomial outcomes only, Friedman's to ordinal and continuous outcomes.

8.9.1. Cochran's Test (Q) for K-related Samples with Binomial Responses

This test assesses the differences between *K* exposure levels on *N* individuals in which the effect is binomial. The observations are symbolized by u_{ij}, *i* is the row, *j* is the column. Thus, u_{ij} is either zero or one. The individuals are assigned to exposure at random. Table 8.15 shows the data used in estimating Cochran's *Q* test.

Table 8.15. Data for Cochran's Q Test

Block Number:	Treatment (or Exposure) Levels			
1	1	2	...	K
2	u_{11}	u_{12}	...	u_{1K}
...	...	...	...	...
N	u_{N1}	u_{N2}	...	u_{NK}

In Table 8.14 there are *K* levels of treatment and *N* independent rows, or blocks, of observations. A block is a unit on test. The treatment levels are ordered from low to high. The standard *chi*-square test is not directly applicable because the treatments are not independent. Cochran's Q test, in which there must be more than three levels of treatment, is:

$$Q = k(k-1) \sum_{j=1}^{n} (\sum_{j=1}^{n} C_j - \overline{C})/[(K * S) - \sum_{i=1}^{r} R_i^2] \; ; S = \sum_{j=1}^{n} C_j = \sum_{i=1}^{k} R_i \; .$$

In this formula, R_j is the row total ($i = 1, 2, ..., n$) and C_j is the column total. That is, j identifies the *treatment* and i identifies the block. The data in this table can be ranked, with ties being assigned their mid-rank value. The null hypothesis is that there are no differences between the K treatments; therefore order is irrelevant. The alternative hypothesis of at least one pair-wise difference means that order *is* relevant. The Q statistic is approximately χ^2-distributed with $K-1$ degrees of freedom. The decision rule is that if $Q > \chi^2_{\alpha, k-1}$, then the null hypothesis is rejected.

Example 8.22. Consider the following sample in which ten individuals are exposed to three (ordered) levels of exposure A, B and C. Each individual is a *block*; the response of each individual can only be either a success (1) or failure (0):

Individual	Levels of Exposure or Treatment		
	Level A	*Level B*	*Level C*
1	1	0	0
2	1	1	0
3	0	0	1
4	0	0	1
5	0	0	0
6	1	1	1
7	0	0	1
8	0	1	1
9	0	0	1
10	1	1	1

Notice that this is a relatively small sample, raising the potential for small sample problems. The question is whether the responses in the treatments are statistically different in terms of the effect of ordered levels of exposure. To test whether the results in the responses in the sample are different, we use StaExact5 asymptotic and exact estimation methods. Cochran's Q statistic $= 4.571$ has $K-1 = 2$ degrees of freedom. The asymptotic p-value (based on the *chi*-square distribution with 2 *df*) is: $pr(Q \geq 4.571) = 0.1017$. Using an exact method, the probability is 0.1235. These results are statistically insignificant, if $\alpha = 0.05$ is the criterion for accepting the null hypothesis. However, the exact probability that $Q = 4.571$ is 0.0412, which is statistically significant for $\alpha = 0.05$.

8.9.2. Friedman's T-test (T_F) for Ordinal or Continuous Observations

If the observations are either ordinal or continuous and correlated, Friedman's T_F test can be used statistically to determine the significance of K correlated treatments or exposures on N individuals. The observations are generated by the random variable $U_{ij} = \mu + \beta_j + \tau_j + \varepsilon_{ij}$ with population mean μ, β_j is the block effect, the treatment effect is τ_j, and the error ε_{ij} is i.i.d., with zero mean.[4] Each observation from the random variable U is u_{ij}. Individuals are assigned to a block at random.

The null hypothesis is:

$H_0: = \tau_1 = \tau_2 =, \ldots, = \tau_j,$

and the alternative hypothesis is:

$H_1: \tau_i \neq \tau_j,$ for at least one pair of treatments.

All of the raw observations, u_{ij}, are replaced by their ranks, r_{ij}. Friedman's T_F statistic is (Kanji, 1999):

$$T_F = 12/[NK(K+1)][\sum_{j=1}^{K} R_j^2 - 3N(K+1)],$$

where N is the number of rows, $i = 1, 2, \ldots; N; K$ is the number of columns, $j = 1, 2, \ldots, K$. The ranks are in order of increasing value. The sum of the ranks for each column is R_j, namely $\sum_{j=1}^{J} R_j$. If there are ties in the ranks, the test is modified (Kanji, 1999) as:

$$F_T = 12(K-1)[\sum_{j=1}^{K} R_j - N(K+1)/2]/(NK^3 - \sum_{i=1}^{N} f_i t_i^3).$$

In this formula, f_i identifies the tied observations and t_i the size of each group of tied observations. This test statistic is *chi*-square distributed, with number of degrees of freedom $K-1$. F_T is also approximately χ^2-distributed with $K-1$ degrees of freedom. The decision rule is that if $T_F > \chi^2_{\alpha, R-1}$, then the null hypothesis is rejected at the α-level of statistical significance.

Example 8.23. Consider the following data, obtained from three individuals who yield five response levels measured on a continuous scale. The question is whether the five sets of responses are statistically significant.

Individuals	Response Levels				
	I	*II*	*III*	*IV*	*V*
1	12	35	10	20	40
2	10	40	7	16	35
3	11	38	12	19	20

Using StatExact5, we obtain the following results. The asymptotic value for Friedman's statistic $=10.93$, $pr(\geq 10.93) = 0.0273$, with $df = 4$. The null hypothesis is not accepted. That is,

---

[4] $\Sigma_N \beta_j$ and $\Sigma_K \tau_j$ can be set to zero without loss of generality.

the responses are statistically significant and different from one another. The exact *p*-value is $pr(\geq 10.93) = 0.0028$, with the number of degrees of freedom unchanged. This result confirms the conclusion from large sample approximation, but we should note the order of magnitude difference between the two results (*0.0273* versus *0.0028*).

8.10. CONCLUSION

This chapter has developed some of the practical issues that often confront risk assessors and managers who must deal with literature reviews. Meta-analysis and pooling of data are important to risk assessment because they allow, in different ways, the use of several results as well as the combinations of data from one or more studies to develop a more precise understanding of the results. Although care is required in developing a meta-analysis, the insights gained can be important to make risk management decisions. Pooling data and assessing differences in results are also important in risk assessment because they provide more accurate information about the direction and magnitude of the potential risk. In particular, differences between means and other differences can be particularly important to risk analysts because inference from those studies' results can be used to assess differential or other type of health impacts.

Other sections of this chapter augment previous discussions of statistical decision making using tests of the hypotheses to describe these types of decisions. To do so we have used population means and variances and the associated sample means and variances as the basic statistics. This approach helps to address issues involving testing the equality of several means and variances, thus allowing inference and predictions from the samples to their populations. Addressing and possibly resolving these issues can be important because those answers allow statistical simplifying assumptions in modeling empirical causation. To give a sense of the potential differences between tests based on large sample theory and exact statistical methods, we have used different examples. Our intent is to suggest that risk assessors should be aware of the potential for differences in the results obtained by different approaches to the same test because the results may change from significant to insignificant. The implication of such changes on risk management can be profound. Many risk assessments use results that may not be randomized; as discussed in Chapter 4, the risk assessor must often deal with non-experimental data. In these situations, testing the homogeneity (e.g., of the variance) and other assumptions such as independence and identical distributions becomes important.

QUESTIONS

1. Suppose that the data set is: X = 1.7, 1.9, 2.0, 2.1, 2.2, 2.4; Y = 1.9, 2.0, 2.1, 2.4; Z = 1.7, 2.0, 2.0, 2.1, 2.2. Do not worry about the units of these data. How would you compare the population means of these three random variables? What is an appropriate decision rule for that comparison? Use about 200 words.

2. Discuss the key differences between meta-analysis and pooling of data from samples. Illustrate your discussion with two examples using hypothetical data for pooling and literature results from a risk assessment of arsenic in

the ground water, for the second. If you do not have access to a library or the Internet, use the air pollution literature results tabulated in Chapter 4. Use approximately 600 words,

3. Discuss and illustrate why the homogeneity of the variance is an important consideration in analyzing data. Use data from the literature or data in this chapter. Use about 300 words.

4. Discuss Bartlett's test and exemplify its application using $n_1 = 13$ and $s_1^2 = 2.5$, $n_2 = 7$ and $s_2^2 = 3.2$; and $n_3 = 9$ and $s_3^2 = 6.90$, using 200 words.

5. Discuss some of the reasons for using exact tests relative to tests based on asymptotic (large sample) approximations (no more than 300 words), using risk management as your point of departure for the discussion.

6. What is the difference between a regression model and the ANOVA, if any? Use any appropriate statistics textbooks, if needed, to answer this question. Use about 200 words.

7. Can meta-analysis account for the power of the meta-analysis study? What key aspects would be important to doing such power analysis and what implications would it have for the risk manager? Answer this question with no more than 700 words.

CHAPTER 9.

CONTINGENCY TABLES IN RISK ASSESSMENT AND MANAGEMENT

Central objective: to complete the discussions of contingency tables, their uses in associating risk factors to response and in rating agreement between individuals

This chapter extends earlier discussions by accounting for the effect of several levels of risk factors on responses. We use contingency tables to analyze associations between variables measured on scales of measurement such as nominal and ordinal scales. Specifically, the rows of a contingency table represent discrete values of the *dependent* variable and the columns discrete values of the *independent* variable. The general multiple rows and columns contingency table (with $R*C$, $r*c$ or $r*k$ identifying rows and columns) can be used to assess certain aspects of empirical causation or quantitatively to determine the degree of agreement between individuals on a particular issue. For example, a risk assessor can use counts of positive and negative result from high or low levels of exposure to estimate a relative risk or an odds ratio.

The statistical analyses of such tables use either the *chi*-square test or exact tests that are based on combinatorial methods. In this chapter, we exemplify:

- The construction of contingency (also called cross-classification) tables
- The statistical methods used for analyzing them with large sample (asymptotic) assumptions and exact methods

The *chi*-square distribution and associated tests can be used to:

- Test the goodness-of-fit of ordered data to a known distribution
- Test the independence of risk factors
- Test the agreement or concordance between two or more peer reviewers or decision-makers

The simplicity of contingency tables, and because they can include several classifications or strata, has made this analysis common in a variety of disciplines.

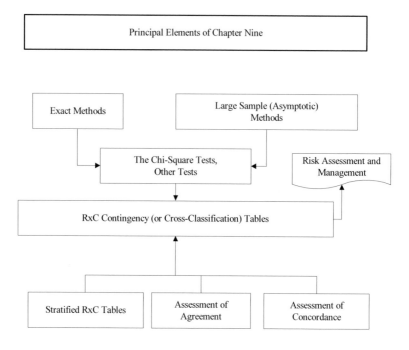

9.1. INTRODUCTION

Risk assessors can often develop simple and compelling representations of the relationships they seek to understand using discrete levels of responses and exposure, rather than continuous measurements on these variables. Such representation can be shown by the random variables (exposure and response) taking two values, as was shown in earlier chapters, using *1* and *0* as binary indicators for exposure or lack of exposure and response or lack of response. Table 9.1 depicts a contingency table for responses to two treatments, *A* and *B*, resulting in *k* responses that are mutually exclusive.

Table 9.1. 2*k Contingency Table

	Adverse Outcomes (1 to k)		
	1	**...**	**k**
Treatment A	O_{A1}	...	O_{Ak}
Treatment B	O_{B1}	...	O_{Bk}

The marginal totals for the rows, columns and the grand total, the sum of the counts in the marginal totals, are omitted for simplicity. Consider the usual 2*2 contingency table in which the entries are new cases, *incident* numbers, depicted in Table 9.2.

Table 9.2. 2*2 Contingency Table

	Disease	No Disease	Total
Exposed	a	b	a+b
Not Exposed	c	d	c+d
Total	a+c	a+d	a+b+c+d

For the exposed, the frequency of disease is: $a/(a+b)$; for the unexposed, the frequency of disease is: $c/(c+d)$. Because a, b, c, d represent incidence numbers, the relative risk, RR, is:

$$RR = [a/(a+b)]/[c/(c+d)].$$

The example that follows uses a 2*2 contingency table to illustrate how these tables are used.

Example 9.1. Consider the following contingency table:

Exposed to Air Pollution	*Number of Individuals without Cancer*	*Number of Individuals with Cancer*	*Total Number of Individuals at Risk*
Yes	$a = 150$	$b = 320$	$a+b = 470$
No	$c = 50$	$d = 570$	$c+d = 620$
Total	$a+c = 200$	$a+d = 890$	Total $= 1{,}090$

The probability of exposure to air pollution is $470/1{,}090 = 0.431$; the probability that respiratory cancers are not present is $200/1{,}090 = 0.183$. The probability that an individual who is exposed to air pollution and does not have cancer is $(470/1090)(200/1090) = 0.079$. The expected number is $\{[(470/1090)(200/1090)](1090)\} = 86.24$. It follows that there is a difference between the observed (150) and the expected (86.24). Consider the theoretical frequency of being exposed and having cancer. It is $[(470)(890)]/1{,}090 = 383.76$. In the 2*2 table, knowledge of the value in one cell and the marginal values is sufficient to allow the calculations of the values in the other three cells: $c_{12} = 470-150 = 320$, $c_{21} = 200-150 = 50$, $c_{22} = 620-150 = 470$.

The probability distribution used to answer the question of whether a causal association is statistically significant uses the χ^2 distribution. The *chi*-square density function (in which $v = df$) is:

$$f(x) = (x^{(v-2)/2})(e^{-x/2})[\Gamma(v/2)(2^{v/2})]^{-1},$$

if $x \geq 0$. The mean and variance of this distribution are v and $2v$, respectively. When the number of degrees of freedom is 1, $(df = 1)$, the *chi*-square distribution is the square of the normal distribution (Dougherty, 1990). As discussed earlier, the shape of the *chi*-square distribution depends on the number of degrees of freedom. The χ^2 density function, for five and one degrees of freedom, plots as depicted in Figures 9.1

and 9.2, in which the *y*-axis measures density and the *x*-axis measures the values of the *chi*-square distributed random variable.

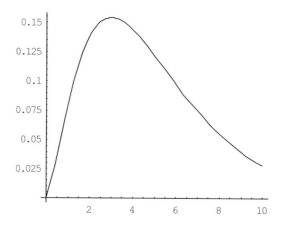

Figure 9.1. Plot of the χ^2 Density Function, 5 Degrees of Freedom

For one degree of freedom, the χ^2 density function plots as depicted in Figure 9.2.

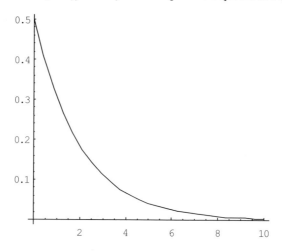

Figure 9.2. Plot of the χ^2 Density Function, 1 Degree of Freedom

The null hypothesis for the contingency table is that there is no association between exposure and response. The alternative hypothesis states that the association exists. The test statistic is *chi*-square distributed. The X^2 *statistic* (the derivation of the formula is omitted) is:

$$X^2 = N[(ad{-}bc)^2]/[(a+b)(c+d)(a+c)(b+d)].$$

Applied to the data in the example, we calculate:

$X^2 = (1090)[(150)(570)-(320)(50)]^2/[(150+320)(50+570)(150+50)(320+570)] = 101.50.$

The rejection of the null hypothesis of no association requires that the calculated test statistic be larger than the theoretical value, for a previously selected level of statistical significance and degrees of freedom.

Example 9.2. Consider the following contingency table, in which the values in the cells are counts of individuals affected from exposure to four ordered levels of air pollution. The individuals stratify into four age groups.

Age, A	**Severity of Air Pollution Impact, SI; I = none to IV = highest**				
	I	II	III	IV	Total$_A$
0-15	300	60	50	100	510
1-25	350	120	400	500	1,370
26-65	50	100	200	300	650
> 65	800	900	1200	1500	4,400
Total$_{SI}$	1,500	1,180	1,850	2,400	6,930

The distribution of the random variable *Severity of Impact* is calculated from the *Total$_{SI}$* and the distribution of the random variable *Age* from the *Total$_A$*. These distributions are called *marginal* distributions. Thus, the probability that an individual has *age > 65*, *pr(Age > 65)*, is $4,400/6,930 = 0.635$. The probability that an individual has *age > 65* and has been severely impacted, *pr(> 65 AND Severity of Impact IV)*, is calculated as $1,500/6,930 = 0.216$. The probability that the severity of the impact is highest, given that an individual has *age > 65*, *pr(SI = IV|Age > 65)*, is $1,500/4,400 = 0.341$. The probability that an individual older than 65 is severely impacted by air pollution, *pr(Age>65|SI = IV)*, is $1,500/2,400 = 0.625$. In other words, about 63 percent of those older than 65 years are impacted by air pollution, although only approximately 34 percent of those being most severely affected are older than 65 years.

9.2. THE *CHI*-SQUARE (X^2) TEST

This statistical test is widely used when dealing with observed and expected frequencies that fall into discrete categories, such as those shown in a contingency table. It is a common and important statistical test because it is:

- Non-parametric
- Suitable for data measured on the nominal, ordinal, interval and ratio scales
- Relatively easy to use

The *chi*-square test is based on the sum of the squared differences between observed frequencies (O_i) and expected frequencies (E_i), divided by the expected frequencies, as shown next:

$X^2_{df} = \Sigma(O_i-E_i)^2/E_i.$

The formula for the expected cell count in the *i*th cell is:

$E_i = (\Sigma all\ values\ in\ row_i)(\Sigma all\ values\ in\ column_i)/(\Sigma all\ rows,\ columns\ values).$

This test has a sampling distribution that approximates the *chi*-square distribution, as the sample size tends to infinity. The observations are statistically independent, by assumption. The expected frequencies are determined by their empirical distribution. For example, if the distribution were uniform (that is, if each outcome in the cell has an equal chance of occurring) then the expected frequency would be n/N. The number of degrees of freedom is a function of the number of cells minus the number of items provided by the sample and used in the test. The number of degrees of freedom is:

$df =$ (number of independent cells)-(number of constraints) $= (r-1)(k-1)$.

The number of parameters to be estimated determines the number of constraints. The marginal totals do not influence the calculations of the degrees of freedom because they are fixed by the experiment: we only use the *free* cells. In other words, under the null hypothesis, the total for the rows and columns are not free to vary. However, the observed values are free to vary in the sense that the combinations giving the marginal totals are variable. The basic ideas about the $r*c$ contingency table can be summarized as follows. We consider the information in Table 9.3.

Table 9.3. General r*c Contingency Table

		Exposure Levels				
		1	*2*	...	*c*	Total
Response Levels	*1*	n_{11}	n_{12}	...	n_{1c}	$n_{1.}$
	2	n_{21}	n_{22}	...	n_{2c}	$n_{2.}$
	...	...	...	...	...	...
	r	n_{r1}	n_{r2}	...	n_{rc}	$n_{r.}$
	Total	$n_{.1}$	$n_{.2}$	...	$n_{.r}$	n

Example 9.3. This application of the *chi*-square test determines whether the sample comes from a specific population distribution or not. The test consists of comparing the frequency of the individual observations in the sample to their expected frequency. Assume that we suspect that individuals are more likely to experience stress in situations where the severity of conditions is higher.

Highest	**Second Highest**	**Third Highest**	**Fourth Highest**	**Fifth Highest**	**Lowest**
12	13	8	8	5	2

The expectation is that there is no difference between the categories: the expectation is that stress is equally frequent. Since there were *48* events in all distributed across the six areas, the expectation is $48/6 = 8$ events in each area. Using the formula for the expected cell count in the *i*th cell: $E_i = (\Sigma \text{ all values in row}_i)(\Sigma \text{ all values in column}_i)/(\Sigma \text{ all rows, columns values})$. We calculate the difference between the two, square it and divide the result by the expected value. Then we sum the results across all the cells. The calculations are as follows:

	Highest	*Second Highest*	*Third Highest*	*Fourth Highest*	*Fifth Highest*	*Lowest*
Observed, O_i	12	13	8	8	5	2

Expected, E_i	8	8	8	8	8	8
$(O_i\text{-}E_i)$,	4	5	0	0	-3	-6
$(O_i\text{-}E_i)^2$	16	25	0	0	9	36

We obtain the value of $X^2_{df} = \Sigma_i(O_i\text{-}E_i)^2/E_i$ as follows $X^2_3 = 16/8+25/8+9/8+38/8 = 10.75$, with number of degrees of freedom equals 3. The theoretical *chi*-square values can be found in most statistical textbooks: the result obtained (*10.75*) is statistically significant, at the chosen level of significance.

The general formula for the X^2 statistic is:

$$X^2_{df} = \sum_i \sum_j (O_{ij} - E_{ij})^2 / E_{ij},$$

for ($i = 1, 2, ..., r$) and ($i = 1, 2, ..., c$), r being the row index and c the column index. The larger the value of this statistic, the more unlikely it is that the results are due to chance. Figure 9.3 depicts the statistical testing associated with this (or a similar) decision rule.

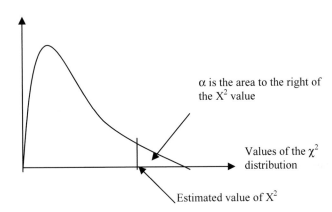

α is the area to the right of the X^2 value

Values of the χ^2 distribution

Estimated value of X^2

Figure 9.3. Hypothesis Testing Using the Chi-square Distribution and the X^2 Statistic

The level of significance is a probability, α, which is the *pr[(chance produces the association) ≥ (observed association), given that the null hypothesis of statistical independence is true]*. The decision rule is: $pr(X^2 \geq X^2_{df}) = \alpha$. The computational formula for the test can be written using row and column totals:

$$X^2 = \sum_{i,j} \frac{[(n_{ij} - n_{.i}n_{.j})n]}{(n_{i.}n_{.j}/n)}.$$

The null hypothesis, for a specific level of significance, is rejected when:

$$X^2 > \chi^2_{\alpha, df}.$$

Example 9.4. The risk assessor believes that type A individuals differ from type B in terms of their attitudes to budgeting. The research question is whether the pattern of attitudes is actually different between the two classes, A and B, of individual respondents. She takes two samples from two different populations. The two samples are independent and identically distributed. The null hypothesis is that there are no differences. The alternative hypothesis is that the frequency across categories is different. She obtains the following summaries from a survey (32, 14, 6, ..., 9). We obtain the expected frequency for each cell by multiplying its row with its column total, and dividing by the grand total. For example, take the B "in favor" (*12*): the expected frequency is its row total (*43*) multiplied by its column total (*44*) and divided by the grand total (*43*44/95 = 19.9*). The data and the expected frequencies are as follows:

Category	For	Against	Neutral	Total
A	32 (24.1)	14 (19.7)	6 (8.2)	52
B	12 (19.9)	22 (16.3)	9 (6.8)	43
Total	44	36	15	95

For each pair of observed and expected values, the computations consist of calculating the cell difference, squaring it; and then dividing the result by the expected value. The X^2 is:

$$X^2 = [(32\text{-}24.1)^2/24.1] + [(14\text{-}19.7)^2/19.7] + [(6\text{-}8.2)^2/8.2] + [(12\text{-}19.9)^2/19.9] + [(22\text{-}16.3)^2/16.3] + [(9\text{-}6.8)^2/6.8] = 10.67.$$

The number of degrees of freedom is $df = (r\text{-}1)(c\text{-}1)$. Therefore, the number of degrees of freedom is $(2\text{-}1)(3\text{-}1) = 2$. From values of the *chi*-square distribution, the result $X^2 = 10.67$ lies between *6.64* ($pr = 0.01$) and *10.83* ($pr = 0.001$). This means the probability of occurrence of a result like this by chance is less than *1%* and more than *0.1%*. If the significance level had been set at $\alpha = 0.05$, the risk assessor would have rejected the null hypothesis of no difference between the observed and expected numbers. She would conclude that there is a significant difference of attitudes between individuals A and individuals B.

Example 9.5. Consider the following 2*2 contingency table in which the data are binomial outcomes:

5	10
4	1

In the computations, Pearson's *chi*-Square statistic = *3.300*, or *1.684* with Yates' continuity correction, discussed next. Moreover, the *asymptotic* p-value (based on the *chi*-square distribution with *1 df*) for the two-sided test has probability equal to *0.0693*. The *exact* p-value for the two-sided test $pr(X^2 \geq 3.300) = 0.1273$.

When the *chi*-square test is applied to a 2*2 table, it has one degree of freedom. In this situation, the test does not approximate well the discrete sampling distribution of χ^2 and it is corrected by the continuity correction as follows:

$$X^2 = \Sigma(|O_i\text{-}E_i|\text{-}0.5)^2/E_i.$$

This correction, due to Yates, is used when the expected frequencies are between *5* and *10*, unless the cell frequencies are large. The extension of the 2*2 contingency table to the *r**c is straightforward.

Example 9.6. Suppose that a sample of *270* individuals *rates* three effects of air pollution on visibility. The null hypothesis is that air pollution and visibility are independent. The results are classified according to the following table; in it, low, middle and high amounts of air pollution are assumed to result in good, fair or bad visibility levels.

Visibility	Amounts of Air Pollution			
	Low	*Middle*	*High*	Total
Good	30	100	80	210
Fair	10	30	5	45
Bad	5	5	5	15
Total	45	135	90	270

The calculations for the estimated value of X^2 are shown in the table that follows (the relevant totals are stated in the last row of the table, the calculations are given in the first row only):

$pr_{observed}$	$pr_{theoretical}$	$(pr_o-pr_t)^2$	$(pr_o-pr_t)^2/pr_t$
30	(210)(45)/170 = 35.0	$(30–35)^2 = 25$	$(30–35)^2/35.0 = 0.71$
100	105.0	25	0.24
80	70.0	100	1.43
10	7.5	6.25	0.83
30	22.5	56.25	2.5
5	15.0	100	6.67
5	2.5	6.25	2.5
5	7.5	6.25	0.83
5	5.0	0	0
270	270	Not relevant	15.71

The estimated value of the X^2 is *15.71*, with number of degrees of freedom equals *(3-1)(3-1) =* *4*. This value is larger than the theoretical χ^2 value (*p*-value $\gg$ *0.05*); therefore, the null hypothesis cannot be accepted: the results are not random. In other words, air pollution is associated with visibility.

The next example compares asymptotic with exact values, when the contingency table has small values in some of its cells.

Example 9.7. Consider the following data:

	Low Exposure	*High Exposure*	Total
Affected	5	15	20
Unaffected	4	4	8
Total	9	19	28

The two-sided asymptotic *p*-value, based on the *chi*-square distribution with *1* degree of freedom, is *0.199* and the one-sided *p*-value is *0.099*. The exact probability values, calculated with StatExact, are *0.372* and *0.157*. Using the same contingency table, the asymptotic *p*-values for the likelihood ratio test are *0.209* and *0.104*. The exact *p*-values are *0.372* and *0.157*.

Example 9.8. Use the same 2*2 contingency table and let the outcomes be binary (binomial):

5	15	20
4	4	8
9	19	28

The *OR*, calculated from the ratio of the odds, is: $[\Pi_2/(1-\Pi_2)]/[\Pi_1/(1-\Pi_1)]$, where $\Pi_1 = 0.556$, $\Pi_2 = 0.790$, $(1-\Pi_1) = 0.44$ and $(1-\Pi_2) = 0.21$. Therefore $OR = (0.79/0.21)/(0.556)/(0.44) = 2.96$. The *95% asymptotic* confidence interval is $(0.54, 16.69)$. The *exact* confidence interval is $(0.38, 22.80)$ with the number of degrees of freedom calculated as $(2-1)(2-1) = 1.00$. Pearson's *chi*-square statistic $= 8.333$ $(6.750$ with Yates continuity correction). The asymptotic *p*-value based on the *chi*-square distribution with *1 df*, for the two-sided test, is $pr(chi\text{-}square > 8.333) = 0.0039$. The exact *p*-value for the two-sided test has $pr(chi\text{-}square \geq 8.333) = 0.0086$.

Example 9. 9. Let the sample be taken from two binomial populations:

	0	1
0	10	90
1	95	0

The *OR* is estimated as ∞, from the data in columns *1* and *2*, respectively, while the odds for population *1* is estimated as *0.095* and for population *2* is *1.00*. The confidence interval is not defined. Consider the table:

	0	1
0	10	90
1	95	5

The estimated *OR* is *171.0*, estimated from the odds for population *1* as *0.095* and for population *2* as *0.95*. The *95%* confidence interval is $(56.27, 519.70)$. The difference in proportions between these two populations, calculated as *0.85*, has the *95%* confidence interval as $(0.78, 0.92)$, with a *p*-value $= 0.0000$. Fisher's exact test has a *p*-value of *0.0000*.

The example that follows illustrates additional uses of contingency tables.

Example 9.10. The null hypothesis is that the *i*th *OR* equals the population's *OR*. Consider the following stratifications:

	0	1
0	3	7
1	9	4

The estimated *OR* for this table is *5.25*. The second table is:

	0	1
0	3	5
1	9	2

The estimated *OR* for this table is *7.5*. The last table is:

	0	**1**
0	5	7
1	9	4

Its estimated *OR* is *3.15*. The common *OR* is estimated to be *4.66*, with *95%* confidence interval (*1.64, 13.21*). Therefore, the estimated common *OR* is significantly greater than 1.00. A question that can be asked is: Are these three *ORs* homogenous? Using Breslow-Day's *D* test of homogeneity, *D* equals *0.43*; the *pr(D ≥ 0.43)* equals *0.81*. This test can be compared with the theoretical *chi*-square value with *df* = *2*. We conclude that these three odds ratios are homogenous.

9.2.1. Discussion

Contingency tables can describe associations or empirical causation. However, there are potentially serious issues when attempting to develop empirical causation.

Example 9.11. In the following table, response is related to exposure; the numbers in the cells are relative frequencies:

	Response, Yes	**Response, No**
Exposure, Yes	0.75	0.25
Exposure, No	0.40	0.60

This table depicts a relationship in which exposure does not *associate* with response. Suppose that the risk assessor attempts to control for age through stratification. The table is now rewritten to account for two age strata, as follows:

Young|Response	**Yes**	**No**
Yes	0.20	0.80
No	0.20	0.80

and:

Old|Response	**Yes**	**No**
Yes	0.80	0.20
No	0.80	0.20

These two tables depict the fact that those unexposed have the same frequency of responding as those who are exposed: therefore age is a confounding factor in the relationship between exposure and response because it affects the dependent and independent variable. Figure 9.4 depicts a form of confounding.

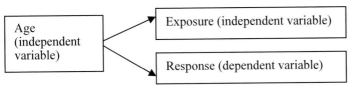

Figure 9.4. Age as a Confounding Factor for Response

Age, in this construct, should be correlated with response but not with exposure. In practice, this may not be the case. There are other issues that affect developing or using contingency tables. The first is that the stratifications can be incomplete and approximate: a *2*2* table is a rough approximation of a possibly more complex table. As in all forms of modeling, the construction of a contingency table requires a theoretical understanding of the mechanisms modeled. For example, the rows and the columns are associated for biological realism. If data for further stratifications is unavailable, additional data can be added, but perhaps at greater cost. In other situations, the counts that result from further stratifications are small numbers or zeros, requiring more complicated analyses. The risk assessor faces a difficult choice; add more data, running the risk of *oversampling*, or potentially incurring in Simpson's Paradox, in which the positive result in the aggregate is reversed when that group is stratified. Another issue, which we discussed in the context of regression models, is *multicollinearity*. It is the result of some independent variables being correlated (the variables are not orthogonal to each other). The result is that the risk assessor becomes unable to discern the impact of each separate independent variable. In the context of contingency tables, this issue arises when some strata are closely associated. The risk assessor faces the following dilemma. Developing a multifactorial (several strata) contingency table to account for the factors affecting response runs the risk of creating multicollinearity. Not doing so leads to the potential for simplistic models of cause and effect that are susceptible to confounding.

9.3. MEASURING AGREEMENT OR CONFORMANCE

Contingency tables are particularly useful in dealing with qualitative data, such as measuring agreement between examiners or raters. This area of analysis is relevant to risk management where rating environmental impacts or other events may be done in ways that can appear to be subjective. The assessment of agreement by reviewers is an important aspect of making decisions, as well as in situations when researchers assess the literature to select studies for inclusion in their work. If the factors over which the reviewers give their judgments are the same and the data are categorical, then it is possible to define a measure of agreement. *Agreement* is different from *conformance*. An *agreement* has no absolute baseline. *Conformance*, on the other hand, requires an absolute baseline. A well-known measure of agreement is Cohens's κ, read *kappa*. It is used for *nominal* data and requires that two, and only two, reviewers use the same number of factors in making their choices. The $2*2$ contingency table in Table 9.3 can be used to study if two individuals, A and B, agree.

Table 9.4. Contingency Table Used to Study Agreement

		A		
		For	*Against*	Total
B	*For*	a	b	f_1
	Against	c	d	f_2

	Total	n_1	n_2	n

The coefficient κ is bounded between *-1* and *1*. A measure of agreement is the proportion *(a+d)/n*. On this basis, the formula for Cohen's coefficient is:

κ= *[(Observed agreement)-(Agreement due to chance)]/(1-Agreement due to chance).*

A computational formula for κ is:

κ= *2(ad-bc)/[(b+c)n+2(ad-bd)].*

Example 9.12. Suppose that two raters, *A* and *B*, rate *170* individuals as follows:

Raters	Pass	Fail	Total
A	65	20	85
B	45	40	85
Total	110	60	170

There are *n =170* observations; κ= *0.2353*, the asymptotic standard error = *0.0712*, the lower and upper *95%* confidence limits on κ are *(0.0957, 0.3749)*. Asymptotic *p*-values for testing no association using the normal approximation one-sided test: *pr(statistic $\geq$ observed)* = *0.0007*. The two-sided value is calculated from *(2*one-sided)* = *0.0013*. If there were perfect agreement between *A* and *B*, the value of κ would be *1.00*; if agreement is random and the observed agreement the same, κ would equal *0*. When the observed agreement is less than the random agreement, the value of κ is negative. The formula for the proportion of agreement due to chance is:

Proportion of agreement due to chance = $p_A q_A + p_B q_B$.

In this example, p_A = *65/85 = 0.765*; q_A = *22/80 = 0.275*; p_B = *45/85 = 0.529* and q_B = *38/80* = *0.475*. The proportion of agreement due to chance is therefore *(0.765)(0.275)+(0.529)(0.475)* = *0.461*. The observed proportion in agreement equals *(58+38)/160 = 0.60*; therefore:

κ= *(0.60-0.461)/(1-0.461) = 0.139/0.539 = 0.258.*

The discrepancy between these results is due to rounding-off errors in the manual calculations.

Feinstein (2002) suggests that Cohen's measure can also be applied to more than two groups by accounting for the possible 2*2 pair-wise agreements in that table. This possibility is calculated using Cohen's weighted κ (Cytel, 2001). The reader may consult Fleiss (1981) for other methods.

Example 9.13. Suppose that two raters or judges, *A* and *B*, *order* their ratings on a particular issue as follows:

A	B				
	Sure	*Probable*	*Possible*	*Not Occurring*	Total
Sure	3	4	1	0	8
Probable	1	6	1	0	8
Possible	1	5	7	3	16
Not Occurring	2	1	2	1	6
Total	7	16	11	4	38

In this table of agreements, there are some partial agreements. Therefore, it is plausible to use a measure of agreement that accounts for partial agreements.

That measure is the *weighted κ*. We use StatExact5 for the calculations. These, for weighted *κ*, yield *κ = 0.2843, asymptotic standard error = 0.1585*, the lower and upper *95%* confidence limits are *-0.0265*, and *0.5950*. The asymptotic *p*-values for testing no association using the normal distribution results in a one-sided *pr(statistic ≥ observed) = 0.0366*. The exact test produces the same results.

9.3.1. Kendall's Coefficients of Concordance

These coefficients can measure the degree of association between raters, peer reviewers or judges on attributes or characteristics that are measured at least on an ordinal scale. A typical application is one in which the risk assessor attempts to determine whether the ranks assigned by several judges are consistent. Bivariate concordance occurs when the high values of a variable are associated with high values of another: the differences x_j-x_i and y_j-y_i have the same sign (Sprent, 1989). Discordance means the opposite. Bivariate concordance is measured by Kendall's τ, multivariate concordance by Kendall's W.

9.3.1.1 Kendall's τ.

Kendall's τ $(-1 \leq \tau \leq 1)$ measures *bivariate* agreement. We use a bivariate data set for X and Y, with x_i and y_i being the paired observations on these random variables. The observations are ranked in increasing order. Then we rank each couple of x_i, y_i and x_j, y_j. If their difference is positive, we assign a score of $+1$. If the difference is negative, we assign a score of -1. When the difference is zero, there is no score assignment. The scores are then summed to yield the number S. Kendall's τ is calculated as (Sprent, 1989):

$$\tau = (n_{concordant} - n_{discordant})/[0.5n(n-1)] ,$$

in which $n = (n_{concordant} - n_{discordant})$. The term $0.5n(n-1)$ is the number of all possible pairs. That is, if $n_{concordant} = 0.5n(n-1)$, then all possible pairs are concordant and the association is perfect. If $n_{discordant} = 0.5n(n-1)$, there is no concordance. The test statistic (for a sample greater than 10) is (Kanji, 1999):

$z = S/\{[n(n-1)(2n+5)]/18\}^{1/2},$

in which S is the sum of the scores and z is normally distributed.

Example 9.14. The data set (Smoke1, Cytel 2001) measures the years of smoking (*10*, *15* and *20* as the mid-points of three time intervals):

	10	**15**	**20**
Quit	22	9	8
Try	2	1	3
Did not	14	21	16

Kendall's $\tau = 0.13$, with *95%* confidence interval (*0.022, 0.24*).

Example 9.15. Suppose that the data ($n = 10$) and scores are as tabulated below and the level of significance is *0.05* (data and scores from Kanji, 1999):

X = x	7.1	8.3	10.7	9.4	12.6	11.1	10.3	13.3	9.6	12.4
Y = y	62	66	74	74	82	76	72	79	68	74
+scores	9	8	5	3	4	3	3	2	1	0
-scores	0	0	0	2	1	1	1	0	0	0

Using the formula for z we obtain $\{33/[10*9*(20+5)]/18\} = 33/125 = 0.26$. Therefore, the calculated value of z is 0.26, which is smaller than the theoretical value of $z_{10,\ 0.05} = 21$ (Table 27, Kanji, 1999). The null hypothesis that there is no correlation between X and Y cannot be rejected.

Example 9.16. Consider the following data set in which seven attributes (*A*, *B*, *C*, *D*, *E*, *F*, and *G*) are rated by two reviewers, *I* and *II*. These two individuals rank a research proposal (each attribute is ranked on a scale from *1* to *5*) as shown below:

	I	**II**
A	1	5
B	2	4
C	1	5
D	2	4
E	3	3
F	5	2
G	5	1

The sums of the ranks for judge *I* is *9.5*; for judge *II* is *11.5*. Kendall's coefficient of concordance is *0.082*, with a *p*-value of *0.450* (*1 df*).

Statistically, we can conclude that there is no significant disagreement between these two judges.

9.3.2. Kendall's W

Under the null hypothesis, the test can determine no consistency (that is, there is no concordance) among judges: their ranks are random. Concordance, in other words, means consistency of rating. Under the alternative hypothesis, there is concordance. The test is one-sided for a given level of statistical significance. Kendall's multivariate coefficient of concordance, W, is:

$$W = \frac{\sum_j [R_j - (\sum_j R_j / N)]^2}{[(0.0833)(k^2)(N^3 - N)]},$$

in which R_j is the summation of the ranks in the j-th row, k is the number of rows, and N is the total number of ranks.[1]

Example 9.17. We wish to assess the appropriateness of an evaluation protocol in the context of risk assessment of a catastrophic air pollution event generating acute respiratory responses. The protocol includes two sets of criteria; one for admission review, and the other for auditing the number of days of stay in a hospital. The admission criteria include:

1. Sudden onset of unconsciousness (coma or unresponsiveness),
2. Pulse rate (either below 50 or above 140 beats per minute),
3. Acute loss of sight or hearing,
4. Acute loss of ability to move major body part,
5. Severe electrolyte or blood gas abnormality,
6. Electrocardiogram abnormality,
7. Intermittent or continuous respirator use every 8 hours.

At least one of these criteria must be met for an admission to take place. Some criteria are quantitative: either the patient's pulse rate was outside the normal range of *50* to *140* beats per minute, or it was not. However, there is professional, qualitative judgment involved for the terms *sudden* and *severe*. One way to test the consistency of interpretation of several health care specialists is to ask them to rank a set of admissions in terms of a particular criterion. Specifically, three specialists (*A*, *B* and *C*) are given the same set of six admissions (given as case numbers) and each judge is independently asked to rank the cases with respect to whether they meet a single criterion from the protocol. The results are:

	Case 1	Case 2	Case 3	Case 4	Case 5	Case 6
Judge A	1	6	3	2	5	4
Judge B	1	5	6	4	2	3
Judge C	6	3	2	5	4	1
Sum, R_j	8	14	11	11	11	8

There are j columns ($j = 1, 2, \ldots, N = 6$) and k judges ($k = 1, 2, 3$). In the example, *Case 1* was ranked highest by judges *A* and *B*, but lowest by judge *C*. The sum of ranks, R_j, for each case is shown in the last row. We compute s, noting that: $\sum_j (R_j / N) = (8 + 14 + 11 + 11 + 11 + 8)/6 =$

---

[1] The reason for the subscript j, rather than i is that, in the example that follows, the raters are placed in rows, rather than columns.

10.5,

and $s = [(8-10.5)^2 + (14-10.5)^2 + ...] = 25.1$; we now compute:

$$W = (25.1)/[0.0833(3^2)(6^3-6)] = 25.5/157.5 = 0.16.$$

Note that, when checking the significance of the results, s rather than W, is used. Thus, s must be greater than the stated value, if the result is to be significant. The calculated value of s is 25.1, with $k = 3$ rows and $N = 6$ cases. The critical value for $k = 3$ and $N = 6$ is 103.9. Therefore, the k sets of ranks are independent. In other words, there is no significant concordance (agreement) between the judges. The calculations yield the (asymptotic) result, $W = 0.1619$, with $pr(W \geq 0.1619) = 0.79$.

If s is used, why calculate W? The reason is that W is bounded between *0* and *1* and can conveniently be used to compare results obtained from different trials, with different numbers of judges and ranked objects. There are other issues that need to be taken into account, including:

1. A different method of determination of significance should be used when N is large. A convenient method involves use of the *chi*-square statistic. In some cases, because chi-square is additive, the results from repeated experiments can be studied by adding the chi-square values (and adding the numbers of degrees of freedom) to obtain the required test statistic.
2. If the proportion of tied ranks is high, an adjustment to the formula to calculate W is needed. This is because tied ranks tend to lead to understatement of the value of W. Concordance, under the alternative hypothesis, only determines consistency between the persons or objects tested in the research design, and nothing more. In the example that follows, there are three raters (judges), and five attributes (clarity, technical content, technical merit, writing style and visual aids), of an issue.

Example 9.18. Consider the following data in which judges *A*, *B* and *C* have rated a risk assessment document for the following attributes: clarity, technical content, technical merit, completeness of the literature, visual materials. The rating scale is *1* to *5*; we let *1* be the *lowest* rating.

	Attributes				
Judge	*Clarity*	*Tech. Content*	*Tech. Merit*	*Style*	*Visuals*
A	3	2	4	3	2
B	1	4	3	4	2
C	5	2	2	2	1

The results show that $W = 0.26$, with $pr(W \geq 0.26) = 0.56$.

9.4. SENSITIVITY AND SPECIFICITY

In risky situations, as discussed in Chapter 8, it is important to have measures of specificity and sensitivity of a particular test, such as a medical test. Table 9.5 repeats the data used in Chapter 8; the data is hypothetical.

Table 9.5. Contingency Table to Determine the Sensitivity and Specificity of a Test

		Disease is Present		
		Yes	**No**	Totals
	Positive	a = 10	b = 20	a+c = 30
Test Is	**Negative**	c = 20	d = 30	c+d = 50
	Totals	a+c = 30	b+d = 50	a+b+c+d = 80

The *sensitivity* of the test, $\pi(positive|disease\ is\ present)$, is *[a/(a+c)]*. The sensitivity of the test measures the proportion of true positives. Thus, we obtain a sensitivity of 0.33. The *specificity* of the test, $\pi(positive|Disease\ is\ Absent)$, is *[b/(b+d)]*. Thus, we obtain a specificity of 0.40. The specificity of the test measures the proportion of false positives.

Example 9.19. Suppose that a difference between two proportions, $\pi_{exposed}$-$\pi_{controls}$, is being assessed. The null hypothesis is stated as $\pi_{exposed}$-$\pi_{controls}$ = 0. The alternative hypothesis is $\pi_{exposed}$-$\pi_{controls}$ ≠ 0. The value α (often set at 0.05) measures the probability that a particular null hypothesis is rejected as false, when it is true. The probability of rejecting the null hypothesis, when it is true, is β. In the example dealing with the effect of exposure to contaminated water in Sacramento, my colleagues and I dealt with the probability of making a Type II error by setting $\beta = 0.20$. Suppose that we had set the probability of making a Type I error at $\alpha = 0.05$. By making such a choice of probabilities, we imply that the Type I error is four times more important than the Type II error.

We can conclude our discussions on the versatility of contingency tables through an example with *matched* data. That is, we apply X^2 test can be applied to determine the statistical significance of responses when the *subjects are the same throughout the experiment* (that is, they serve as their own controls), and express their preferences for a particular outcome before the outcome is known and after the outcome is known.

Example 9.20. Suppose that the following table describe public preferences *before* an event has occurred and *after* that event has occurred:

Preferences before	Preferences after		
	For A	**For B**	Total
For A	a = 15	b = 10	a+b = 25
For B	c = 5	d = 20	c+d = 25
Total	a+c = 20	b+d = 30	50

The risk assessor is interested in determining if the *before* and *after* response rates are equal (in the sense of statistical testing under the null hypothesis). A formula for McNemar's version of the *chi*-square test is (Dawson-Saunders and Trapp, 1994):

$$X^2 = (|b-c|-1)^2/(b+c),$$

which is *chi*-square distributed with $df = (r-1)(c-1)$. Thus, $(|10-5|-1)^2/(10+5) = 1.07$. The number of degrees of freedom is *1*. The theoretical value of *chi*-square is *3.841* for ($\alpha = 0.05$). The response rates are similar: the preferences are equal. The *asymptotic* results are as follows: one-sided *p*-value: *pr(test statistic > observed)* = 0.0984 and two-sided *p*-value is *2*one-sided* *p-value = 0.1967.* The *exact* results (conditional on sum of discordant pairs) are as follows. The one-sided *p*-value is *pr(test statistic ≥ observed)* = *0.1509*, the *pr(test statistic = Observed)* = *0.0916*; the two-sided *p*-value is *2*one-Sided = 0.3018*.

9.5. CONCLUSION

Contingency (or cross-classification) tables can be used with a variety of scales of measurement, including continuous and nominal data, to develop relationships, assess rankings by interviewers and perform many assessments that are important to making risky decisions. As discussed in this chapter, we choose contingency tables to calculate odds ratios and relative risks, and adjust, using stratifications, for different aspects of those at risk. A typical adjustment uses different contingency tables for different age groups, sex or other suitable stratification. The examples that we have developed use asymptotic as well as exact methods for estimation. The importance of the latter in risk assessment becomes most evident when the tables have small numbers or are the data is sparse and includes zeros. In these situations, asymptotic theory is not helpful and thus exact methods become important.

The relationship between contingency tables and the methods for statistical analysis discussed in Chapters 4 and 5, as well as the hypothesis and confidence interval framework, is as follows. Specifically, stratifications are equivalent to using independent variables. Similarly, levels of exposure can be expressed as discrete values, interval or even as simple *high* or *low*. In addition to these aspects of risk assessment, contingency tables can formally assess the amount of agreement between judges, peers or other individuals. The usefulness of this aspect of contingency tables can be particularly important to risk managers who might poll several individuals on the merits of a particular option.

QUESTIONS

1. Consider the 2*2 cross tabulation in which a, b, c and are counts associated with two random variables:

	Accept	Reject
For	a = 15	b = 10
Against	c = 20	c = 50

   Label the random variables and complete the table. What does pr(For|Accept) measure? What does the probability pr(Accept|For) measure? Develop a formula for both conditional probability statements and apply it.
2. How would independence between random variables X and Y be determined? Develop a formula to determine if two random variables are independent. (Hint, if a random variable does not probabilistically affect the other they are independent. Therefore, $pr(Y = i|X = j) = pr(Y = i)$).
3. Consider the following table:

	Disease	No Disease	Total
Exposure	32	65	97
No exposure	54	7	61
Total	86	62	158

How would you calculate the agreement due to chance and the actual agreement? Apply Cohen's κ to fully analyze this table (we used this table earlier).

CHAPTER 10.

STATISTICAL ASSOCIATIONS AND CAUSATION FOR RISK ASSESSMENT AND MANAGEMENT

Central objective: to discuss and exemplify statistical correlations for measurements taken on different scales, the role and limitations of qualitative criteria of causal associations, and causal analysis based on entropy of information

In this chapter we review and exemplify several methods to develop practical associations, measured by correlations, for two or more variables. A well-known statistical measure of correlation is the coefficient of correlation, discussed in Chapter 4, to measure fit of a linear statistical model to the data. Correlations go beyond measuring goodness-of-fit; they quantify the strength and direction of potential *causal associations*. Although correlations do not imply *causation*, they provide an understanding of potential for causation. For example, correlations can support qualitative criteria of causation, such as Hill's. These can be useful in understanding aggregate knowledge about relations that can exist between exposure and response. In multiple regression models, for instance, we use pairwise coefficients of correlation between risk factors (the independent variables) to study some of the issues that can arise in obtaining estimates of the parameters of those models. An example of a generally unwanted correlation is *multicollinearity*; another is the *autocorrelation* between the errors in a time series of data. Both forms of correlation affect the estimation of parameters of statistical models in different ways. In using correlation, we deal with hypothesis testing, estimation, confidence intervals and inference. We conclude this chapter with an overview of empirical causation using the fundamental causal diagram, FCD, as well as with a brief comment about entropy of information.

Specifically, in this chapter, we discuss:

- Statistical associations as means to develop first approximations to causal associations
- Issues that may affect estimates of correlations
- Internal and external validity (consistency and generalization) criteria for empirical studies
- Causal analysis based on causal graphs and entropy of information

We develop several examples to illustrate how correlations can be useful in developing an aggregate understanding of the results from several studies. We also discuss several aspects of entropy of information and its uses in developing causal models for risk assessment. Although this form of causation is relatively advanced, we believe that it is important for risk assessors and managers.[1]

The principal elements of this chapter interconnect as depicted below.

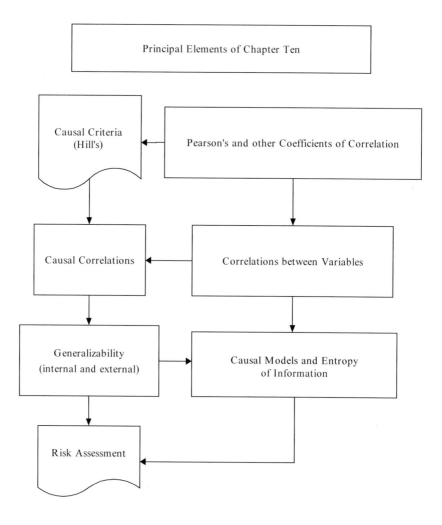

---

[1] LA Cox, jr. has provided the material for Section 6.10.

10.1. INTRODUCTION

We recollect the often-stated caveat: *correlation is not causation*. A reason for this caveat is that correlating two factors does not establish that X precedes Y, in the sense of the implication $X \to Y$. Rather, a correlation between X and Y implies that either $X \to Y$ or that $Y \to X$. Thus, we can calculate a coefficient of correlation between X and Y and between Y and X, regardless of the true, and unknown, causal relation between these variables. For example, nothing prevents us from correlating the number of gallons of beer drunk in Xiamen, China, with the size of the umbrellas used in Oxford, UK, and vice versa. Despite this correlation, it is unlikely that these two variables are causally associated. A correlation formally relates variable Y with variable X through one or more formulae. The formulae generally yield a single number, ρ, generally between -1 and 1, including 0. Descriptively, a coefficient of correlation between Y and X measures how the data relate to each other. Figure 10.1 depicts two scatter diagrams in which the data indicate a positive and a negative correlation, for an arbitrary set of data represented by the oversized circles, which represent points for X and Y, namely x_i and y_i.

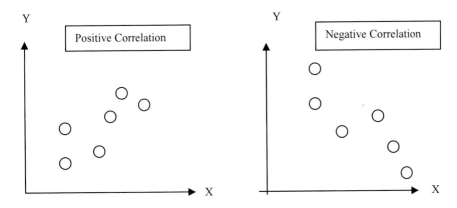

Figure 10.1. Scatters Depicting Linear Positive and Negative Correlations

High positive or negative correlation numbers means that the sample of points can easily be detected to occur close to a line that points either upwards or downwards. A low correlation number, which is more difficult to detect, means that the pattern is there but the correlation is weak. If there is perfect linear correlation between Y and X, then Y and X are perfectly matched ($\rho =$ either 1.0 or it equals -1.0), all the data lie on a straight line with either a positive or negative slope. Conversely, if Y and X are perfectly statistically uncorrelated, their coefficient of correlation is zero: the data are randomly dispersed. Thus, a measure of *correlation* is:

$Y = aX,$

where *a* is a *correlation* measure, *if such coefficient is normalized*, so that its value is dimensionless and lies in the closed interval $(-1 \leq a \leq 1)$.[2] Specifically, we substitute a straight line for the points in Figure 10.1. For several risk factors, the multivariate (linear) analogue of *a*, above, is:

$$Y_i = \Sigma_i \, a_i X_i,$$

where $(i = 1, 2, ..., k)$ and *provided that each a_i is normalized*, that is $(-1 \leq a_i \leq 1)$. Often, the *population* coefficient of correlation is symbolized by ρ. The estimated value of ρ, estimated from a random sample, is generally symbolized as either r, $\hat{\rho}$ or R.

A coefficient of correlation (Pearson's) for two normally distributed random variables X and Y is:

$$R = r = \sqrt{\frac{cov(X,Y)}{[var(X)var(Y)]}} \, .$$

The ratio of these statistics cancels out the units and bounds the value of the coefficient of correlation, r, to be between -1 and 1. A formula for estimating this coefficient of correlation for a sample of size n, with the data for X, shown as x_i, and the data for Y shown as y_i, is (Feinstein, 2002):

$$r = \frac{\sum_{i=1}^{n}(x_i - \bar{x})^2 (y_i - \bar{y})^2}{\sqrt{\sum_{i=1}^{n}(x_i - \bar{x})^2 \sum_{i=1}^{n}(y_i - \bar{y})^2}} \, .$$

When the coefficient of correlation is squared, r^2, it is called the coefficient of determination. In a linear regression with more than one independent variable there will be a single coefficient of *determination*, R^2, for the entire regression. However, there will also be several coefficients of correlation, depending on the coupling of the independent variables in multiple regression models.[3]

Example 10.1. The paired data for X and Y are:

X = 104, 122, 141, 168, 195, 215, 238, 246, 284, 307, 327, 423.
Y = 32, 41, 28, 56, 63, 49, 78, 91, 71, 60, 95, 110.

The coefficient of correlation for Y and X is positive and equals: $r = 0.87$.

---

[2] Normalization means that the correlation measure accounts for variability and has no units.
[3] As the number of random variables increase so does the number of coefficients of correlation. For instance, if a model has *10* variables, the number of pair-wise (or *bivariate*) coefficients of correlations that can be calculated between them is *45*. This is the result of the combination of two elements taken at the time, out of *10*.

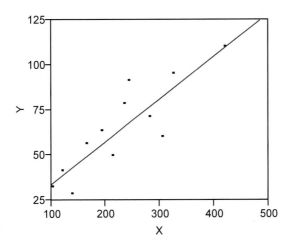

If we interchange Y and X, we obtain:

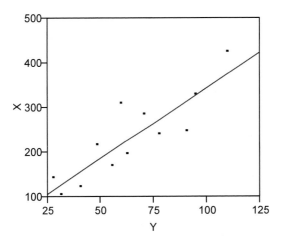

In this case $r = 0.87$: interchanging the variables does not change the positive association or the magnitude and direction of the coefficient of correlation.

In this example, the data for X and Y can be a sample or a population: all that we know is that x_i and y_i are paired. No inference or prediction is attempted: we are merely describing a potential relationship. If a sample of points does not approximately lie around a straight line, this fact does not mean that there is no correlation. It may be that the correlation relationship is non-linear (for instance, a parabola) or has cycles (e.g., a wave which can be represented by the relation $y = sin\theta$ or by a more complicated waveform). The liner approximation may be appropriate in some cases, but be an oversimplification in other cases. The assessment and the literature review justify the plausibility of a linear approximation, relative to the actual data. Specifically, the analysis of correlations must account for outliers, the number of modes in the distributions of the variables, correlations between the errors and other technical issues. In this context, as in the context of regression models, the plots of the errors against the estimated values of each estimated value of a variable are particularly helpful in deciding what is affecting the results of the calculations. When three or more random variables $(X_1, X_2, X_3, ...)$ are correlated, a more complicated form of the coefficient of correlation (the coefficient of partial correlation) accounts for the contribution of several variables held constant at some predetermined value on the correlation between two selected variables. In other words, the correlation between X_1 and X_3 is calculated with X_2 held constant at some value. The next example illustrates this point, using the output from a computer software program, rather than showing manual calculations.

Example 10.2. Consider the following paired data:

X = 104, 122, 141, 168, 195, 215, 238, 246, 284, 307, 327, 423.
Y = 32, 41, 28, 56, 63, 49, 78, 91, 71, 60, 95, 110.
Z = 235, 438, 340, 297, 329, 487, 267, 204, 402, 471, 270, 343.

The pair wise correlations are estimated to be:

	Y	**X**	**Z**
Y	1.00	0.87	-0.29
X	0.87	1.00	0.084
Z	-0.29	0.084	1.00

The coefficients of correlation form a matrix that has *1.00* in its main diagonal: X is perfectly correlated with X, Y with Y and so on. The off-diagonal entries are the pair-wise coefficients of correlation. The *partial* correlations (relative to all other variables) are:

	Y	**X**	**Z**
Y	None	0.94	-0.74
X	0.94	None	0.720
Z	-0.74	0.72	None

10.2. QUALITATIVE ASPECTS OF CAUSAL ASSOCIATIONS

Correlations can be useful in many practical analyses but can also be misleading. Once a correlation is estimated, and appears to be reasonable, it can take a meaning well beyond its limitations and become persuasive. The danger is that, without careful reasoning about the rationale for correlating two or more variables, the correlation can be convincing but spurious. A careful review of the literature and theoretical knowledge about the variables that are being studied can guide the risk assessor to decide what variables should be correlated and why. However, a dilemma soon becomes apparent. Without a theory: How can it be asserted that X causes Y? And we may also ask: How *is* theory developed? We can answer some of these questions as follows. The risk assessor often wants to develop a *causal association* between two or more variables. For instance, scientific conjectures can be tested and refined through correlation methods in which a dependent variable is (linearly, in the simplest instance) correlated to one or more independent variables. Naturally, there are correlation measures for the populations and the sample: the mechanisms of hypothesis testing, estimation and inference are available to help with inference from sample data. In other words, if we assume that there is a reason for associating variables X and Y, a priori of the data analysis, we are stating that the relationship between X and Y is that $X \rightarrow Y$: think of it as X causes Y This is not the same as $Y \rightarrow X$ (meaning that the direction of the arrow cannot be reversed, under our initial assertion). Thus, *empirical* correlations are valuable in studying the initial aspects of a relation that can eventually be (still empirically) causal, while maintaining the integrity of probabilistic reasoning based on sample analysis and inference.

A measure of correlation can be used to generate a hypothesis that can then be tested by more complete models, assuming that knowledge and suitable data are available. In other words, a statistical correlation, with plausible biological or physical basis, suggests an initial working hypothesis. Such approach does not result in a *causal relation*, but can yield a plausible *causal association* if mechanistic (the physics of the system) and statistical criteria are met. At least for practical applications in risk assessment, many correlations are *linear*. The (a priori) hypothesis about the *magnitude* of the contribution of the relationship (estimated by the size of the association) and its *direction* (measured by the positive or negative sign associated with the correlation itself) are judged by the results of statistical analysis.

10.2.1. Building a Causal Association

In the event that populations are used there is no inference. However, the magnitude and direction of the correlation or correlations (as well as their variability) are still important to understanding potential causal associations. Some of the key points confronting an analyst or the user of correlations for estimating a measure of causal association are as follows.

Measuring an association. First, the data should be of acceptable accuracy, the units of analysis must correspond to the theoretical basis of the causal relationship, and be justified by reasonable causal association or associations. Second, the samples should be representative of the objectives of the research, statistical design and the sources of the data.

Selecting a measure of association. The selection involves taking account, at the minimum, of the scales over which the data are measured. As discussed in Chapter 2, these scales are nominal, ordinal, interval or ratio. The scale of measurement determines the coefficient of correlation. It is also important to determine whether the causal association has already been theoretically established or been investigated by others and whether the structure of the causal association can be summarized by a linear coefficient of correlation.

Accounting for Probabilistic Causation. A typical causal assertion in practical risk assessment is that:

Agent E probably causes adverse response O in a biological unit at risk that has some *specific characteristics T*, given a time series of *exposure C* and the corresponding *dose rates D*. This statement can be refined and made more precise depending on the availability of information. Cause and effect are represented by the change that the *cause* provokes in the *probability* of the *effect*. The change in the (conditional) probability of response between two individuals of different sex exposed to the same substance is one form of probabilistic causation. The other is the temporal change in the probability; for example, as exposure increases temporally so does the probability of the adverse effect. Given that the context of risk assessment is risk management, the change in the probability should relate to a *physical* linkage between cause and effect.

Accounting for scientific conjectures. Conjectures arise when causal models are unknown, or when results are contradictory.

Among others, such as Koch and Evans, Hill (1965) provides a well-known paradigm for epidemiological causation, which reflects statistical and epistemic aspects, through several criteria. Those criteria were developed in the context of occupational epidemiology to determine when *an association between two variables is perfectly clear-cut and beyond what we would care to attribute to chance.* Specifically, Hill's criteria were formulated to answer the question: *What aspects of that association should we especially consider before deciding that the most likely interpretation is causation?*

10.2.2. Hill's Criteria

In descending order of importance, as assigned to them by Hill, the criteria that can answer his question are as follows.

The first is the *strength of the association*, namely, the magnitude of the relative risk. The example is death from scrotal cancers among chimney-sweepers (*sweeps*) that were determined, by Percival Pott, to have a relative risk of more than *200*! In the second example he gives death rates from lung cancer in smokers, relative to non-smokers, which is about *10*, and that of heavy smokers, where the ratio is between twenty and thirty times higher than in non-smokers. A relevant point is Hill's statement that we *must not be too ready to dismiss cause-and-effect hypothesis merely on the grounds that the observed associations appear slight.*

The second criterion is the *consistency of the observed association*. The issue is whether an empirical association occurs elsewhere. In particular:

"... on the customary tests of significance will appear unlikely to be due to chance. ... Nevertheless whether chance is the explanation or whether the true hazard has been revealed may be sometimes answered only by a repetition of the circumstances and the observations." As Hill continues, *there will be occasions where repetition is absent or impossible and yet we should not hesitate to draw conclusions* about the plausibility of the association.

The third criterion is the *specificity of the association*. The magnitude of the empirical association is specific to a group at risk and to its exposure and health outcome. Even if specificity of association is unavailable, this may be due to a disease that has multiple etiologic factors. Hence, lack of specificity is not always detrimental to empirical causation. The fourth criterion is the *temporal relationship of the association*: cause must either precede or be contemporaneous with an effect. The fifth criterion is *biological gradient*: the increase in the response, given changes in dose, provides more information than just the high rate of response at high doses.

Biological *plausibility* is Hill's sixth criterion. It relates to the biological basis of the association constructed (or hypothesized to exist) between response and exposure. As he states, the *association we observe may be one new to science or medicine and we must not dismiss it too light-heartedly as just too odd*. Nevertheless, the seventh criterion is that there must be *coherence* between the imputed causality and the general state of knowledge about the disease. Finally, preventive measures (the *experiment* criterion) may reduce the probability of response; and the criterion of *analogy* can be useful in determining causal associations. One of Hill's conclusions is notable, *[n]one of my nine viewpoints can bring indisputable evidence for or against the cause-and-effect hypothesis and none can be required as sine qua non*.

As to statistical significance, it is not the sole criterion for judging the validity of a study. Hill remarks that:

"Such tests can, and should, remind us of the effects that the play of chance can create, and they will instruct us in the likely magnitude of those effects. Beyond that they contribute nothing to the 'proof' of our hypothesis."

Hill puts emphasis on the interpretations of results without statistical principles. He asks *whether the pendulum has not swung too far, not only with the attentive pupils but with the statisticians themselves*. He concludes that *like fire, the χ^2 test is an excellent servant and a bad master*:

"Fortunately, I believe we have not yet gone so far as our friends in the USA where, I am told, some editors of journals will return an article because tests of significance have not been applied.... Yet there are innumerable situations in which they are totally unnecessary because the difference is grotesquely obvious, because it is negligible, or because, whether it be formally significant or not, it is too small to be of any practical importance."

Hill's statement that *the clear dose-response* curve admits a *simple explanation and obviously puts the case in a clearer ligh*t must be understood in the context of a simple bivariate relationship between *A* and *B*. Such model, the *simple explanation* of a cancer's process, can be a necessary but not sufficient summary representation of a *causal* process. Cox and Ricci (2002) have reviewed Hill's criteria. They conclude that it is possible that these criteria may not in fact be causal, and that, paradoxically, they actually can support a non-causal association. The reasons for this paradox is that the criteria are ad hoc, they are not quantitative and do not account for heterogeneity in those at risk. For the purpose of this chapter, Hill's criteria and correlations provide an initial insight into testable hypotheses. We begin by introducing one of the best-known coefficients of correlation, Pearson's.

10.3. PEARSON'S COEFFICIENT OF CORRELATION

Perhaps the most common statistical measure of statistical association (that is, correlation) between two (or more) random variables is Pearson's coefficient of correlation between X and Y. It is used with data measured on either an interval or ratio scales. The data is a paired sample from two populations. The statistical measure of correlation is normalized, as the formula below shows, to fall in the interval *(-1, 1)*. The population coefficient of correlation of two random variables X and Y is:

$$\rho = \frac{cov(X,Y)}{var(Y)var(Y)}.$$

When the coefficient of correlation equals *-1*, there is perfect negative correlation between the data; when the value is *1*, there is perfect positive correlation. When the value is *0*, there is no correlation. For *continuous* data, the product-moment (Pearson's) association is the coefficient of correlation, and applies when two variables are normally distributed. The estimator of ρ is symbolized by r: it has a bivariate distribution on a three-dimensional surface like a three-dimensional bell when $\rho = 0$, but becomes elongated when ρ approaches positive or negative *1.0*. The null hypothesis is: H_0: $\rho = \rho_0$; its alternative is: H_1: $\rho \neq \rho_0$. For a sample of size n, Pearson's coefficient of correlation between the random variables X and Y, x_i and y_i being the pairwise sample observations, is estimated by:

$$r = [\sum_{i=1}^{n}(x_i - \bar{x})^2 (y_i - \bar{y})^2]/[\sqrt{\sum_{i=1}^{n}(x_i - \bar{x})^2 \sum_{i=1}^{n}(y_i - \bar{y})^2 }].$$

In this equation, $\bar{x}$ and $\bar{y}$ are the sample means for the observations. Specifically, r is the estimator of ρ.

Example 10.3. Consider the random sample of *10* paired cases:

Age (Y)	1	1	2	2	2	4	5	6	8	10
Exposure (X)	5	5	20	5	8	8	10	20	20	25

Suppose we want to determine: 1) if Y is linearly associated with increased X and, 2) the magnitude and direction of the correlation. The calculations are:

Case #	(Y)	(X)	X*Y	X^2	Y^2
1	1	5	5	25	1
2	1	5	5	25	1
3	2	20	40	400	4
4	2	5	10	25	4
5	2	8	16	64	4
6	4	8	32	64	16

7	5	10	50	100	25
8	6	20	120	400	36
9	8	20	160	400	64
10	10	25	250	625	100
Total	41	126	688	2128	255

Pearson's correlation coefficient has the computational formula:

$$r = \frac{n\sum_i x_i y_i - \sum_i x_i \sum_i y_i}{\{[n\sum_i x_i^2 - (\sum_i x_i)^2][n\sum_i y_i^2 - (\sum_i y_i)^2]\}^{1/2}}.$$

The calculation is as follows:

$$r = [10(688)-(126)(41)]/[10(2128)-(126^2)(10(255)-(41)^2)]^{1/2} = 1714/2167 = 0.79.$$

This calculation yields a positive estimated coefficient of correlation. The direction of the linear relationship between age and exposure is positive and these two variables are correlated ($r = 0.79$). Although intuitively this coefficient of correlation is significant, we can test significance more formally. We let the level of significance be 0.05. Under the null hypothesis (H_0: $\rho = 0$), the two-tailed test uses a transformation formula:

$$\tanh^{-1}(r) = 0.5\ln\frac{(1+r)}{(1-r)},$$

in which *tanh* is the hyperbolic tangent of the value of r; $\tanh^{-1} \sim N(0.5\ln[(1-r)/(1+r)], (n-3)^{-0.5})$. The reason for the transformation is that we now use the standardized normal distribution, rather than the interval $(-1, 1)$. We can obtain a modified z-value, z^*, as follows, for $r = 0.79$:

$$z^* = 0.5\ln\frac{(1+0.79)}{(1-0.79)} = 0.5\ln(8.524) = 1.286$$

which is normally distributed, $N(0, 0.32)$; 0.32 is an approximation calculated from $n^{-1/2} = 10^{-0.5} = 0.316$. This allows us to form the z-ratio: $[(1.286-0)/0.316] = 4.07$. The result is statistically significant because the theoretical z-value, for the two-tailed test and $\alpha = 0.05$, is $|1.96|$. Therefore, the null hypothesis of no linear relationship is rejected. The reader may want to use the t-distribution and determine whether the calculations using the z-distribution still result in the rejection of the null hypothesis, for the same level of significance. Kanji (1999) provides a statistic to test the significance of the results from a study of pair-wise correlations, given the null hypothesis of no difference between the variances of the random variables X and Y. The test uses the F-distribution and the tests statistic:

$$\hat{y} = (\hat{F} - 1)/[(\hat{F} + 1) - 4r^2\hat{F}]^{1/2}.$$

In this equation, $\hat{F} = s_y^2/s_x^2$ and > 1.00; n is the sample size, r^2 is the squared value of Pearson's r. The empirical results are tested against the theoretical value of F, namely: $F_{\alpha,df=n-2}$ to determine whether the null hypothesis of zero correlation should be rejected, or not. The decision rule is that if $\hat{F} < F_{\alpha,df=n-2}$, then the null hypothesis cannot be rejected. Kanji (1999) provides tables for applying the decision rule to different values of α and degrees of freedom. As was the case for the statistics that we have been discussing in past chapters, it is often necessary to find the sample size of a study. The next example shows how we can account for the power of the test.

Example 10.4. A risk assessor is planning a study in which she wishes to test the null hypothesis that the correlation is *0*. The alternative hypothesis is that it is *0.50*. The level of significance equals *0.050* and the test is two-tailed. Using Power and Precision (2001), we calculate that a sample size of *30* will have power of *86.3%*. Suppose that the second objective of this study is to *estimate* the correlation. Based on these parameters and assumptions, the study will enable the risk assessor to report this value with a *95.0%* confidence level of approximately ± *0.28*. That is, an estimated correlation of *0.50* has a *95.0%* confidence interval from *0.17* to *0.73*. The power function for $\alpha = 0.05$, two-tails test, *correlation₁* = *0.500*, *correlation₂* = *0.000* plots as:

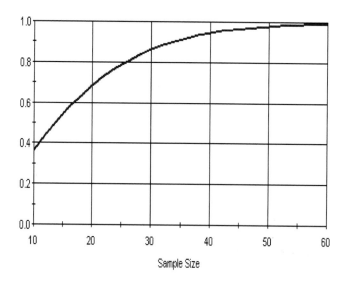

In the example that follows, we develop a meta-analysis of several estimated coefficients of correlation using hypothetical data.

Example 10.5. Suppose that ten hypothetical studies have correlated exposure to ambient air pollution and mortality from cardiovascular diseases. Some studies have resulted in negative correlations, but most have estimated positive correlations, as follows:

Study	Effect	R	Lower 95% CL	Upper 95% CL
arp	death	0.32	0.12	0.54
edo	death	0.4	0.16	0.60
kpr	death	0.31	$-6.12*10^{-3}$	0.57
mdo	death	0.37	0.14	0.56
mjr	death	-0.34	-0.47	-0.19
mjs	death	0.25	$3.49*10^{-02}$	0.44
mrs	death	0.55	0.36	0.70
pdm	death	0.23	$-4.02*10^{-02}$	0.47
pfr	death	-0.56	-0.78	-0.21
rpt	death	0.43	0.25	0.58
	Aggregate	0.19	0.11	0.26

The last row in this example is a meta-analysis of these ten correlations using the software program Comprehensive Meta-Analysis (Borenstein M, Rothstein H, 1999). The meta-analytic result is a coefficient of correlations that equals *0.19 (95%* confidence interval *0.11, 0.26)* that is statistically significant. Suppose that *arp = -0.32, kpr = -0.310* and *pfr = -0.56.* The combined coefficient of correlation would now equal *0.002* with a *p*-value of *0.044.* The positive correlation between exposure and death has almost disappeared.

10.4. ADDITIONAL MEASURES OF ASSOCIATION

There are several statistical measures that estimate the relationship between one variable and another, or between several variables, for *nominal* data. One of them, the contingency coefficient, applies to comparisons among different $r*c$ tables that have different values for r and for c, or to single contingency tables.

10.4.1. Contingency Coefficient φ

If we are considering a single *2*2* contingency table, the contingency coefficient φ falls between *0* and *1*, in which *0* means no association and *1* means perfect association. This coefficient can be used to correlate two random variables measured with nominal data. The null hypothesis is that the association is due to chance. The alternative hypothesis states that the null hypothesis is false. This coefficient also measures the strength and significance of the association in terms of the proportion of the variation of one variable, given the other. The contingency coefficient φ is defined as:

$\varphi = (Pearson's\ chi\text{-}square\ statistic/N)^{1/2}$.

Pearson's *chi*-square statistic is:

$$X_{df}^2 = \sum_{i=1}^{rows} \sum_{j=1}^{cols} \frac{[(x_{ij} - \sum_{j=1}^{cols} y_{ij} \sum_{i=1}^{rows} y_{ij})/N^2]}{(\sum_{j=1}^{cols} y_{ij} \sum_{i=1}^{rows} y_{ij})/N}.$$

In this formula, *cols* symbolizes columns. The interpretation of φ is that the agreement between the two independent variables can be either strong or week, rather than due to chance. This coefficient is identical to the absolute value of Kendall's τ (Borenstein, Rothstein, and Shapiro, 2001).

Example 10.6. Consider the following data set in which the values are associated with three levels of response (modified from ACCSTD.cy3 file, Borenstein, Rothstein, and Shapiro, 2001).

	X	**Y**
Completed	8	8
Declined	2	5
No-response	12	8

The results are:

Coefficient	**Estimate**	**0.95 Confidence Interval**
Phi	0.72	(0.51, 0.94)

10.4.2. Spearman's Coefficient of Correlation

Spearman's bivariate coefficient of correlation is used when the assumption that the distributions of the two variables are normal does not hold. The (one- or two-tailed) null hypothesis is that there is no correlation between the variables. Table 10.1 depicts the structure of the data for Spearman's coefficient of correlation.

Table 10.1. Data and Ranks Used in Estimating Spearman's Coefficient of Correlation

Variable 1	Ranks, Variable 1	Variable 2	Ranks, Variable 1
80	1	45	1
90	2	47	2
100	3	87	5
110	4	65	4
120	5	70	3

This coefficient measures the association between two variables, whose values are at least ordinal. Taking the random variables X and Y, the formula of Spearman's rank correlation coefficient, r_s, when there are no ties in the data, is:

$$r_S = 1 - \frac{[6(\sum_i D_i^2)]}{[N(N^2 - 1)]},$$

in which D_i is the difference between the rank of corresponding values of x_i and y_i, and N is the total number of paired values. The coefficient of correlation lies in the interval $(-1.0 \leq r_s \leq 1.0)$.

Example 10.7. Consider the following hypothetical data:

Social class	High Salary	Medium Salary	Low Salary
A (high)	3	1	0
B (mid)	0	2	2
C (low)	0	1	2

The annual high salaries are those above $45,000, low salaries are below $25,000, and the medium salaries are those between $45,000 and $25,000 per year. The individual data from which we developed the previous table is:

Individual	Social class	Salary
1	A	47,650
2	A	48,000
3	A	51,700
4	A	43,700
5	B	43,940
6	B	42,045
7	B	24,000
8	B	24,500
9	C	26,480
10	C	22,005
11	C	24,555

Specifically, we rank the *11* persons from *1* to *11* (*1* is the highest rank). There are several ties for the social class data. For resolving ties, we calculate the average rank (or the mid-rank). There are four social class-*A* persons, and each individual is therefore ranked as *(1+2+3+4)/4 = 2.50*. For the *B*-class persons the average of 5, 6, 7 and 8 is 6.5. Finally, for the *C*-class persons the average of ranks 9, 10 and 11 is 10.

Next, we calculate the difference in ranks on the two variables for each observation, square those differences and sum them. For example, for the first individual: social class ranks 2.5; annual salary is 3, the difference between these two ranks is 0.5, and the square of the difference is 0.25. The calculations are:

Person	Social class		Salary		D_i	$(D_i)^2$
1	A	(2.5)	47,650	(3)	.5	.25
2	A	(2.5)	48,000	(2)	.5	.25
3	A	(2.5)	51,700	(1)	1.5	2.25

4	A	(2.5)	43,700	(5)	2.5	6.25
5	B	(6.5)	43,940	(4)	2.5	6.25
6	B	(6.5)	42,045	(6)	.5	.25
7	B	(6.5)	24,000	(10)	3.5	12.25
8	B	(6.5)	24,500	(9)	2.5	6.25
9	C	(10.0)	26,480	(7)	3.0	9.00
10	C	(10.0)	22,005	(11)	1.0	1.00
11	C	(10.0)	24,555	(8)	2.0	4.00

We obtain $\sum_i D_i^2 = 48$. Spearman's coefficient of correlation is estimated (with a simplification):

$$r_s = 1 - \frac{[6 \sum_i D_i^2]}{(N^3 - N)} = 0.78.$$

From a table of critical values, the critical value for $N = 11$ and $\chi^2 = 0.01$ is $r_s = 0.723$. Therefore, the correlation is significant, at the 0.01 level of significance.

10.5. CONSISTENCY OF SCIENTIFIC STUDIES

To assess the consistency of a study is to account for its theoretical and empirical basis and to resolve its ambiguities. The use of the results from a consistent study to other situations results in the study's *generality*. *Consistency* is used to assess a study's internal validity, *generality* is used for external validity. Thus, in the presence of competing results, demonstrating that a study has internal consistency and that its results can be generalized increases its acceptability.

10.5.1. Internal Consistency

Internal consistency assesses if the difference in the results obtained by two groups of individuals are due to factors and processes other than those included in the study. Several characteristics of a study can support a finding of internal consistency. These include:

1. *Structural changes.* These relate to the formal structure of the process that is being studied. If the structure of the process modeled has changed, the difference in the empirical results may be due to unaccounted for structural changes, rather than to the factors and relations included in the study.
2. *Selection bias.* If the controls in a case-control study are weaker than the unexposed population at risk, the magnitude of the difference in the responses may be greater than expected.
3. *Censoring.* If censoring occurs in those exposed and unexposed at different rates, and the risk assessor is not aware of censoring, as might be the case in a non-experimental setting, the results can be biased.
4. *Regression to the mean.* This is a phenomenon that results in sets of high numbers, followed by low numbers. For example, lower scores can follow high scores in a test done today if the same test is taken a few days later.
5. *Choice of measuring instrument.* Different analytical instruments have different detection limits. Chemical laboratories' best practices may differ in practice,

performance and accreditation. Answers by respondents in a survey (the document administered to the respondents is an *instrument*) can produce different results if the respondents are familiar with the questions contained in the instrument. Similarly, those who rate an attribute (for example, visibility) may differ in their interpretation of *good visibility*.

6. *Sequential testing of opinions*. If there are two surveys, both covering the same subjects and attributes, the respondent in the second survey may be more open to the interview.

7. *Statistical issues*. A wide range of assumptions is made in building practical causal models based on probabilistic theory and statistical methods. Those assumptions must be tested before a study's results can be acceptable to stakeholders.

8. *Scientific and physical basis of empirical causation*. No empirical model can realistically relate effect to cause unless the physics of the system modeled reflects the state-of-knowledge.

Internal consistency helps the generalization of a study's results to other situations, with measures of concordance (and discordance) providing a quantitative assessment of the strength of raters' opinions.

10.5.2. Generality

The following suggests the range of issues that can be relevant to generalizing from an internally consistent study to risk assessment and management.

1. *Hawthorne effect*. The behavior of those studied, even though those individuals are taken at random from a population, changes to suit the experiment.

2. *Volunteer bias*. It is possible that those that are part of a study are motivated to participate in the study and thus their responses may bias the outcome of such study.

3. *Protocol differences*. Different countries have different protocols for their risk assessment studies. Those protocols may different in significant ways and thus make the generalization of their scientific results problematic.

4. *Budget and expertise*. Experimental and non-experimental studies may be under-funded, not allowing for follow-up or for good statistical and other analytical expertise to be available before and during the study.

5. *Research budgets*. Some countries have larger research budgets directed to risky issues important to that country. It may well be the case that other issues that are equally or even objectively more important to that country are not studied because of political judgments made at the time of allocating the budget. Therefore a bias may be created that may make another country deal less effectively, because of lack of information, with risks.

6. *Legal and regulatory differences between jurisdictions*. It has become common practice in risk assessment to adopt, perhaps with modifications, the risk methods and data of a foreign jurisdiction. Many jurisdictions have developed their risk assessment and management methods based on scientific review, consensus building, legislative processes and legal remedies as an intrinsic part of its overall legal system. Unless the adopting jurisdiction also imports the legal remedies that the originating jurisdiction affords its stakeholders, there can be serious deficiencies that may go unnoticed by the adopting jurisdiction. This is issue is particularly acute when subsequent legal action demonstrates that an agency erred in its risk determinations but this goes unreported in the adopting jurisdiction.

Many causal processes studied in health risk analysis can be described by a common template in which risk management *actions* that change current practices or activities can thereby change the *exposures* to a potentially harmful agent (the "hazard") affecting individuals in a susceptible exposed population; this template, the fundamental causal diagram, is discussed next.

10.6. FUNDAMENTAL CAUSAL DIAGRAM, FCD, IN MODELING CAUSATION[4]

Figure 10.2 depicts the linkages in the discussions.

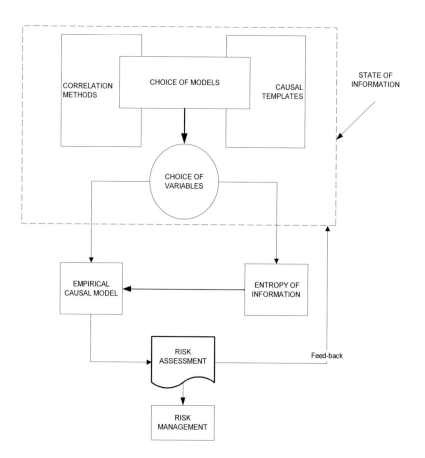

Figure 10.2. Linkages between Methods for Causal Analysis

---

[4] Dr LA Cox jr drafted the material for Section 10.6; additional discussions on these contents can be found in Cox (2001).

Changes in exposures result in changes in expected *illness rates* and hence to changes in adverse *health consequences* (e.g., illness-days or early deaths per capita-year) in those exposed. What we seek is a set of risk management options that decrease adverse health consequences. Figure 10.3 depicts the fundamental causal diagram, FCD, for this work.

$$\Delta u \to \Delta x \to \Delta r \to \Delta c \to \Delta Q$$
$$\qquad \uparrow \qquad \uparrow \qquad \uparrow$$
$$\qquad b \qquad s \qquad m$$

Figure 10.3. The Fundamental Causal Diagram, FCD, for Risk Assessment

The symbols in Figure 10.3 are understood as follows:

Δu = risk management *act* to be evaluated (e.g., a change in _use_ of a product),
Δx = change in *exposure* if act Δu is taken,
Δr = change in *illnesses* (the "_response_" in dose-response models or exposure-response models) caused by Δx,
Δc = change in adverse health _consequences_ caused by Δr,
ΔQ = change in a summary measure of risk (e.g., change in expected quality-adjusted life-years, *QALYs*, lost per capita-year, when this is an appropriate measure) caused by Δc.

This sequence of changes in response to Δu may be modified by other variables, such as:

b = individual _behaviors_ or other factors that affect or modify exposures, given Δu,
s = individual _susceptibility_ and/or other covariates that affect or modify the dose-response relation. This may be a vector of factors (e.g., age, sex, immune status and other health status attributes). Inter-individual variability in dose-response relations can be modeled in terms of differences in s; in this case, s may be a latent variable (such as "frailty") or contain latent or unobserved components, as in finite mixture distribution models.
m = individual _medical_ treatment and/or other factors that affect or modify the illness-consequence relation.

Variables in m, s, and b, may overlap, raising the threat of confounding in interpreting observed statistical associations between Δx (or Δu) and Δc. In many applications, the change in human health consequences, Δc, will consist of a relatively small number (or vector, for multiple health effects) of cases per year occurring in a much larger population, and Δc is well-approximated by a Poisson-distributed random variable (or a vector of them, for multiple health effects). Then, the expected incremental number of cases per year, $E(\Delta c | \Delta u)$, determines the entire Poisson probability distribution for $\Delta c | \Delta u$. This distribution is stochastically increasing in its mean.

The relevance of condition for risk management is that all expected utility-maximizing decision-makers (the class of decision-makers who maximize net expected benefits) who prefer smaller values of Δc to larger ones will also prefer smaller *expected* values of Δc to larger ones (Ross, 1996). In this case, estimating

$E(\Delta c|\Delta u)$ suffices to characterize risk adequately for purposes of risk management decision-making. That is the central task of computational risk assessment.

10.6.1. Computational Aspects

Practical risk assessment has four stages. These stages, hazard identification, exposure assessment, dose-response modeling, and risk characterization for a risk management intervention Δu, correspond roughly to:

1. Instantiating the sequence ($\Delta u \rightarrow \Delta x \rightarrow \Delta r \rightarrow \Delta c$) with specific hazards, exposures, and health effects variables (for hazard identification);
2. Quantifying the input-output relations for the two links $\Delta u \rightarrow \Delta x$ (for exposure modeling) and $\Delta x \rightarrow \Delta r$ (for dose-response modeling); and
3. Quantifying the change in human health consequences Δc caused by a specific exposure change Δx that would either be caused (if positive) or prevented (if negative) by a risk management change Δu.

Mathematically, risk characterization can be performed for a proposed risk management intervention Δu, once the exposure modeling and dose-response modeling steps are complete, by "marginalizing out" the remaining variables, i.e., summing (or integrating, for continuous random variables) over their possible values. For example, this gives the following composition formulas for composing the relations $Pr(\Delta x|\Delta u)$, $Pr(\Delta r|\Delta x)$ and $Pr(\Delta c|\Delta r)$ to compute the probability density and expected value of the human health consequence:

$$Pr(\Delta c|\Delta u) = \Sigma_{\Delta r}Pr(\Delta c|\Delta r)Pr(\Delta r) = \Sigma_{\Delta r}\{Pr(\Delta c|\Delta r)[\Sigma_{\Delta x}Pr(\Delta r|\Delta x)Pr(\Delta x|\Delta u)]\},$$

and:

$$E(\Delta c|\Delta u) = \Sigma_{\Delta r}E(\Delta c|\Delta r)Pr(\Delta r) = \Sigma_{\Delta r}\{Pr(\Delta c|\Delta r)[\Sigma_{\Delta x}Pr(\Delta r|\Delta x)Pr(\Delta x|\Delta u)]\}.$$

These formulas collapse the *entire* causal chain ($\Delta u \rightarrow \Delta x \rightarrow \Delta r \rightarrow \Delta c$) to a single but equivalent risk characterization link ($\Delta u \rightarrow \Delta c$) = $Pr(\Delta c \mid \Delta u)$ relating risk management actions to their probable health consequences. More generally, if the main sequence ($\Delta u \rightarrow \Delta x \rightarrow \Delta r \rightarrow \Delta c$) is embedded in a larger directed acyclic graph (DAG) model with the conditional probability distribution of the value of each node (representing a variable in the model) being determined by the values of the variables that point into it, then the conditional probability distribution for $\Delta c|\Delta u$ can be calculated via computational methods for exact inference in Bayesian networks and causal graphs (Zhang, 1998; Dechter, 1999).

A simpler approximate method, now widely applied in health risk assessment, is Monte Carlo simulation (Cheng and Drudzdel, 2000). Practically, if no Bayesian inference is required, then tools such as @RISK™, CrystalBall™ and Analytica™ can be used to sample values from the probability distributions of input nodes (nodes with only outward-directed arrows) and propagate them forward through the deterministic formulas and conditional probability look-up tables (CPTs) stored at other nodes to create approximate distributions for the values of output nodes (those with only inward-directed arrows). If Bayesian inference is to be used to condition on data while propagating input distributions to obtain output distributions, then software such as the Bayesian Net Toolbox or WinBUGS can be used to perform the more computationally intensive stochastic sampling algorithms (typically, Gibbs Sampling and other Markov Chain Monte Carlo (MCMC) methods) required for accurate but approximate inference using DAGs (Cheng and Drudzdel, 2000; Chang and Tien, 2002).

In summary, a risk assessment model is fully specified using the template in the FCD by specifying its node formulas or look-up probability tables at each node. These determine the value of probability distribution of values for each node conditioned on its inputs (if any). Effective computational inference algorithms and software for quantifying $E(\Delta c|\Delta u)$ and $Pr(\Delta c|\Delta u)$, while conditioning on any relevant data (for individual cases), are available. Therefore, most applied risk assessment efforts can focus on using available data to quantify the component causal relations for the nodes, $Pr(\Delta x|\Delta u, b)$, $Pr(\Delta r|\Delta x, s)$, and $Pr(\Delta c|\Delta r, m)$, that is, the exposure, dose-response, and health consequence models. These components can then be composed (analytically or by Monte-Carlo simulation) via the algorithms and software discussed in this section jointly to compose the causal path from actions to health consequences and to complete the risk assessment by computing $E(\Delta c|\Delta u)$ or $Pr(\Delta c|\Delta u)$.

A simpler approach to risk assessment used before the widespread availability of Monte Carlo risk assessment software applies if all of the conditional relations at the nodes in the FCD are approximately linear and interindividual heterogeneity is ignored. In this case, the following reduced form of a causal structural equation model (SEM) corresponding to the path diagram depicted in the FCD holds:

$E(\Delta c|\Delta u) = (\Delta u)(\Delta x/\Delta u)(\Delta r/\Delta x)(\Delta c/\Delta r).$

Interpretively, the risk due to a change in practices, behavior and so on is represented by the multiplicative sequence:

Risk = (change in use)*(exposure factor)*(unit risk factor)*(consequence factor).

Let action Δu represent a release of a chemical carcinogen into environmental or food pathways. The exposure factor $(\Delta x/\Delta u)$ represents the *additional* mass per person-year that an exposed individual receives per unit of mass released. The incremental "unit risk" factor $(\Delta r/\Delta x)$ is the slope of the dose-response function, measured in units of expected cancers per additional mass-per-year for the exposed individual (as published q_1^* values, *potency* values, in the US EPA usage); and that the consequence factor $(\Delta c/\Delta r)$ gives the expected life-years lost per cancer case caused by this agent. Then the product of these factors gives the increase in individual risk, measured in units of expected life-years lost per unit of release, for this event. To account for interindividual heterogeneity in a population, it suffices in principle to collect the parameters and variables that affect individual risks in response to a change in actions or exposures into a single vector y (e.g., $y = (b, s, m)$ in the FCD) and then discretize y into a finite set of values, interpreted as distinct

"types" or "groups" of individuals, such that the risk characterization relation
$E(\Delta c | \Delta u)$ is (*approximately*) the same for all individuals of the same type.
Classification tree algorithms provide a convenient and computationally practical
approach for automatically constructing such groups from data.

Example 10.8. This example show a classification tree output for case-control data (Friedman et al., 2000) on potential risk factors for the food-borne disease campylobacteriosis.

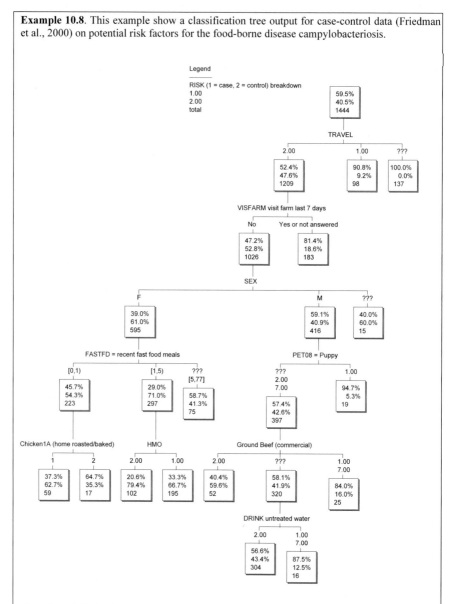

The KnowledgeSeeker™ commercial classification tree program automatically selects
variables (from a list of over 800) and groups or "bins" their values into contiguous ranges

(for *continuous and ordinal categorical variables*) based on how informative they are (*greedily* maximizing expected reduction in classification entropy, discussed later) about whether an individual in this study is likely to be a case (a campylobacteriosis victim) or a control. Those who have recently visited a farm are at high risk (*74.4%* are cases for *VISFARM = 1*, where *1 = yes* and *2 = no*, compared to only *50.1%* among those who have not recently visited a farm). Recent foreign travel (*TRAVEL = 1*) is a very pronounced risk factor. Among domestic non-farm visit cases, *SEX* (*M* or *F*) is most predictive of *RISK* (that is, case or control status), and males are at greater risk than females. Number of meals eaten in a fast food restaurant in the past week (*FASTFD*) is a highly informative predictor of *RISK* for females, and between women with *FASTFD = 1* to *4*, those with *HMO* coverage are more likely to present as cases than those without. For all these variables, the symbol *???* indicates missing data and 7 and 9 and 77 indicate don't know or refused to answer responses. *The leaves of the tree correspond to different population sub-groups* (defined by the combinations of variable-values leading to them) having significantly different risks.

If the groups or individual types are indexed by p (for the relatively homogeneous sub-populations for which risk is to be characterized), then the risk in a heterogeneous population of multiple groups is found by summing over groups:

Population risk from $\Delta u = \Sigma_p E(\Delta c | \Delta u, p) N(p)$,

where the sum runs over all individual types p; $N(p)$ is the size of sub-population p, the number of individuals of this type in the population; and $E(\Delta c | \Delta u, p)$ is the individual risk for members of type p, which is calculated as above. The distribution of $E(\Delta c | \Delta u, p)$ is often approximately log-normal with a relatively small number of distinct value combinations for the variables representing most of the probability density (Druzdzel, 1994). The conceptual units of population risk are the expected additional cases per year (of each health consequence type, if Δc is a vector) caused by Δu.

Population groups indexed by p can be thought of as cells in a multi-factor contingency table (or as *leaves* of a classification tree), each generating an approximately Poisson-distributed number of cases per year having mean and variance:

Expected additional cases/year from group p caused by $\Delta u = E(\Delta c | \Delta u, p) N(p) = k(p) N(p) \Delta u$,

where:

$X(p) = (\Delta x / \Delta u) =$ exposure factor for members of group p. $X(p)$ is in units of exposure per capita-year per unit of Δu,

$R(p) = (\Delta r / \Delta x) =$ unit risk factor (or dose-response factor) for members of group p. Its units are expected additional cases per capita-year per unit of exposure,

$C(p) = E(\Delta c) / \Delta r =$ health consequence factor for group p. Its units are expected adverse health outcomes (e.g., illness-days, fatalities) per case (possibly a vector). If different treatments may be prescribed to members of group p presenting with different symptoms, then $C(p)$ is the sum over all health outcomes weighted by their probabilities, given assumed treatment practices in response to symptoms for members of group p: $C(p) = \Sigma_m E(\Delta c | m) [\Sigma_{sym} Pr(m | sym) Pr(sym | p)]$ where *sym = symptoms*, *m = treatment*.

$k(p) = E(\Delta c)/\Delta u = X(p)R(p)C(p)$, which is the risk ratio for type p individuals (in linear models).

Therefore, we obtain:

$k(p) = (1/\Delta u)\Sigma_{\Delta r}\{E(\Delta c|\Delta r)[\Sigma_{\Delta x}Pr(\Delta r|\Delta x)Pr(\Delta x|\Delta u)]\}$.

If bottom-up measurements or estimates are available at different times or locations for the quantities Δu, Δx, and Δc for individuals in group p, then the ratios:

$X(p) = E(\Delta x|\Delta u)/\Delta u$ (which equals $\Delta x/\Delta u$ in linear models) for group p = *exposure factor*,

and:

$R(p)C(p) = E(\Delta c|\Delta x)/\Delta x$ (which equals $\Delta c/\Delta x$ in linear models) for group p = risk factor,

can be estimated (e.g., by generalized least squares or path analysis software such as SPSS AMOS or SAS CALIS procedures) and then multiplied to estimate $k(p)$. Or, working from the top down instead of bottom-up, if the number of excess cases $\Delta cN(p)$ and their causes Δx or Δu are measured at different times or locations for the population group p, then Poisson regression or other discrete-count regression models can be used to estimate the individual risk ratio for group p:

$k(p) = E(\Delta c|\Delta u)N(p)/\Delta uN(p) = E(\Delta c|\Delta u)/\Delta u$ ratio for members of group p.

Perhaps surprisingly, even if some or all of the types p are unknown or unobserved, the number of types and the coefficients $k(p)$ often can be estimated from data on $\Delta cN(p)$ and Δu via the finite mixture model:

Expected change in cases per year caused by $\Delta u = [\Sigma_p k(p)N(p)]\Delta u$.

If the component risk factors X, R, and C vary over time and if all health effects are aggregated into a summary risk measure such as (perhaps discounted) expected *QALYs* lost, then the total population risk due to a decision or action that generates a change $\Delta u(t)$ in period t is given by summing population risks over all future time periods in the assessment horizon. When all effects in a year are caused by exposures in that year, this sum is:

Total population risk (in QALYs) $= \Sigma_t\Sigma_p X(p, t)R(p, t)C(p, t)Q(p, t)N(p, t)\Delta u(t)$.

Practically, if the dose-response factor and consequence factor are unlikely to vary over time, the double summation simplifies to:

Total population risk $= \Sigma_p R(p)C(p)[\Sigma_t X(p, t)Q(p, t)N(p, t)\Delta u(t)]$.

The QALY-weighting factor $Q(p, t)$ (a vector with one component for each component of Δc if multiple distinct health effects are considered) reflects the relative importance or severity assigned to different health outcomes in different future years. If the required value judgments cannot be made, then varying these relative weights (constrained to be non-negative and normalized to sum to *1*) still allows *undominated* risk management actions (having smallest total risk for *some* choice of the *Q(t)* weights) to be identified. The *QALY* weights often can be assessed as products of single-attribute utility functions (e.g., for duration in a health state and preference for that state). Finally, suppose that a risk management action affects populations via more than one causal path, for instance by changing multiple exposures (possibly to multiple hazardous agents) transmitted via different media such as air, soil, food, and water. Then rational risk management based on assessing the *total* human health consequences of different decisions or actions must sum exposures and their effects over *all* of the causal paths that contribute significantly to human health consequences. For multiple distinct hazards, indexed by *h*, having effects that combine additively, such *cumulative risk assessment* can be expressed by:

Total population risk = $\Sigma_t \Sigma_h \Sigma_p K(p, h, t) \Delta u(t)$ expected QALYs lost, where for simplicity we use the composite factor *K(p, h, t)* to denote the product:

K(p, h, t) = X(p, h, t)R(p, h)C(p, h)Q(p, t)N(p, t) = (exposure of members of group p to hazard h in year t)*(unit risk factor for members of p from exposure to h)*(expected cases per capita-year for members of p exposed to h)*(present value of expected QALYs lost per case for members of group p in year t)*(number of people in group p at time t).

In the important special case where a change *Δu* in practices, processes, or behaviors causes a one-time change in the steady-state levels of exposures and resulting expected illnesses per capita-year in different groups, this linear (sum-of-products) model can be simplified to the following steady-state version:

ΔRisk = Expected change Δc in human health risks caused by Δu = KΔu,

Where:

$K = \Sigma_p k(p)Q(p)N(p),$

in which:

k(p) = E(Δc)/Δu for group p = E(Δc|Δu, p)/Δu = X(p)R(p)C(p), for a linear model,

and:

k(p) = (1/Δu) $\Sigma_{Δr}$ {E(Δc|Δr)[$\Sigma_{Δx}$Pr(Δr|Δx)Pr(Δx | Δu)]}, for a causal graph.

In these equations, the units are expected *QALYs* lost per year in the population due to illnesses or deaths caused by *Δu*. Setting *Q(p) = 1* for all *p* gives the risk in units of cases per year.

In summary, the FCD path analysis (sum-of-products) framework for linear models and the more general Monte Carlo composition-of-components framework (where each component is a probabilistic causal relation) both support effective methods for:

- Estimating the component causal relations from epidemiological (top-down) and/or component (bottom-up) data
- Combining them (via multiplication or composition of conditional probability distributions, respectively) to calculate the change in total population risk caused by a change in exposures (or in the actions that cause the exposures)

Population risks can be expressed in units of additional cases by type (for instance, morbidities of varying severity and mortalities) or additional *QALYs* lost, either per year or over some practical time-horizon. However, before carrying out such risk assessment calculations and interpreting the results as the expected adverse human health effects caused by an actual or proposed change Δu, it is essential to confirm that the variables selected to represent or measure the conceptual quantities Δu, Δx, Δr, and Δc actually do so. That is, they must describe the causal process that they are intended to represent.

10.6.2. Information and Causation: Identifying *Potential* hazards

A key goal of *hazard identification* is to select variables that correctly instantiate the components and overall composition of the conceptual causal model in the FCD. If hazard identification is done correctly, changes in the selected exposure variable will cause changes in illness rates as specified by $Pr(\Delta r|\Delta x, s)$ (or by the unit risk factors $R(p)$ in the linear model); changes in actions will cause changes in exposures as in the selected exposure model $Pr(\Delta x|\Delta u, b)$ (or $X(p)$ in the linear model); and changes in illness rates will produce the changes in adverse health consequences attributed to them by the selected consequence model $Pr(\Delta c|\Delta r, m)$ (or by the consequence factors $C(p)$ in the linear model). In the linear case, the structural equation model:

$\Delta Risk = Expected\ change\ in\ population\ risk = K\Delta u,$

will have a valid causal interpretation as meaning that a change Δu will cause a change in risk that equals $K\Delta u$. This is the key risk assessment information needed to inform choices among different risk management actions corresponding to different choices of Δu. It describes how risk is *predicted* to change when actions are taken.

In general, however, the structural (causal) coefficient K is *not* the same as the "reduced-form equation" coefficient relating measured values of Δu to measured or estimated values of $\Delta Risk$ in aggregate population data. Consider the following simple hypothetical counter-example. The structural equations describing the true causal relations among Age, $\Delta Exposure$ (which equals cumulative exposure received by an individual) and $\Delta Risk$ are specified in this example to be:

$\Delta Risk = Age-\Delta Exposure$ (Structural Equations for Causation),

and:

$\Delta Exposure = (1/3)*Age.$

Thus, *ΔExposure* increases in proportion to age, with proportionality constant of *1/3* in an appropriate system of units; while *ΔRisk* increases in proportion to *Age* and decreases in proportion to *ΔExposure*. All of these effects are causal, so that *ΔExposure* is protective at every age. Then, it is algebraically correct that *Age* = *3*ΔExposure* (although this is not a structural equation: increasing exposure is not the cause of increasing age!). Substituting into the top equation yields the reduced-form model:

ΔRisk = *3*ΔExposure-ΔExposure*,

or:

ΔRisk = *2*ΔExposure* (Reduced-Form Equation for Measured Data).

In other words, the aggregate *statistical* relation is that *ΔRisk* = *2*ΔExposure* in this population, even though the aggregate *causal* relation is that each unit increase in *ΔExposure* causes a unit decrease in *ΔRisk*. The statistical relation would be valid and useful if the goal were to predict the number of illnesses to expect in a community from known or easily measured values of *ΔExposure*. In this case, the reduced-form relation *ΔRisk* = *2*ΔExposure* would be the right one to use. But if a *change* in exposure is contemplated, *Δu* = *50%* increase in *ΔExposure* for every *Age* group, then the resulting change to be expected in illnesses per year is given by the structural model, *ΔRisk* = *-ΔExposure* not the reduced-form model. As a result, the value of *K* in the causal model:

ΔRisk = *KΔu*,

cannot be estimated from previously measured values of *Δu* and *ΔRisk*, even though the corresponding descriptive model can be estimated perfectly. Indeed, if *ΔExposure* = *Δu*, then the causal coefficient (*K* = *-1*) and the inference coefficient (measured ratio of *ΔRisk/Δu* = *+2*) have opposite signs. Past measured values of *Δu* and *ΔRisk* do not contain the information needed to predict how future changes *Δu* will affect future changes in risk. In this example, the correct value of the structural coefficient *K* is given by its defining formula:

$K = \Sigma_p X(p)R(p)C(p)Q(p)N(p)$.

The groups indexed by *p* are age groups. In each age group, $X(p) = 1$, $R(p) = -1$, and $C(p)Q(p) = 1$ if we are just counting cases. Thus, we have:

ΔRisk = *KΔu* = *-NΔu*,

where $N = \Sigma_p N(p)$ = *total number of individuals exposed* and the change in risk per exposed individual is *-Δu*, as the structural equations specify.

10.6.3. Discussion

Our discussions show that an observed strong, consistent, positive statistical association between exposure and risk can contain no information about the causal relationship between them. Suppose that the true relation among three variables is as follows:

$$\Delta X \leftarrow \Delta Y \rightarrow \Delta Z,$$

meaning that changes in Y cause changes in X and changes in Z, but that ΔZ is conditionally independent of ΔX given ΔY (and neither causes the other). If changes in X usually follow changes in Y with a shorter lag than do changes in Z, and if relatively large changes in Y tend to cause relatively large changes in X and in Z, then there can be a *consistent* (found in multiple subjects and settings by multiple reasonably independent investigators using different methods), *strong* (highly statistically significant, clear dose-response or biological gradient between size of changes in X and size of changes in Z), *specific* and *temporal* (that is, changes in Z occur if and only if changes in X occur first) association between measured values of X (or of ΔX) and measured values of Z (or of ΔZ). If X is an exposure variable and Z is an adverse health effect that might plausibly and coherently be hypothesized to be caused by X according to current biological knowledge or speculation (although, in this example, it is not), then the association between changes in X and changes in Z would fulfill all of the most often used epidemiological criteria for judging causation advocated by WHO and other authoritative bodies. Yet, by construction of the example, they in fact provide no evidence at all for causation between ΔX and ΔZ. It follows that the usual epidemiological criteria for judging causality must be enhanced.

This is more than a call to check for the presence of (perhaps unobserved) confounders before declaring an observed exposure-response association to be consistent with the hypothesis of causation. But the broader point is that measured values of ΔX and ΔZ do not contain the key *information* (values of ΔY) needed to specify the correct causal relation between them. Conversely, a strong causal relation may hold between two variables (e.g., a J-shaped or U-shaped dose-response relation for $E(\Delta c | \Delta x)/\Delta x$ as a function of Δx) even if there is *no* statistically significant association between them as measured by linear correlation, ordinal association measures, or other standard measures of association (and hence the usual epidemiological criteria for judging causality would not be met). It is natural, therefore, to consider what information *is* needed to correctly establish causality among action, exposure, response, and health outcome variables for purposes of hazard identification.

10.6.4. Testing for Causation between Measured Variables in a Risk Model

Suppose that measurements on some or all of the variables that are hypothesized to play the roles of Δu, Δx, Δr, Δc, b, s, m in the FCD are available through one or more data sets. Thus, after controlling for b, it is hypothesized that Δx will vary with Δu;

after controlling for *s* (which may overlap with *b*), it is hypothesized that Δr will vary with Δx and so on. The question is: *How can the risk assessor tell whether these instantiated causal hypotheses are consistent with the available data?* And, if many variables have been measured but causality among them is not yet clear, *How can the data themselves be used to identify potential causal relations and to assemble some or all of the FCD* (with its conceptual variables instantiated by corresponding measured variables) *in such a way that the implied causal relations hold?* Hazard identification addresses both tasks: the validation and discovery of causal relations between exposure and response variables. It seeks to identify exposure variables whose past changes are responsible for observed changes in illness rates; and also to identify illness rate and health outcome variables that are likely to be changed by proposed risk management actions (and the changes in exposure they lead to).

In both prospect and retrospect, hazard identification seeks to instantiate the causal link $\Delta x \rightarrow \Delta r$ or to quantify the corresponding node formula $E(\Delta r | \Delta x, p)$ from available data. Although indirect observations instead of an actual experiment can perhaps never prove that a causal model is correct, observational studies can provide data that are more or less consistent with a specified causal theory or model, because the observed data would be relatively probable or improbable, respectively, if the specified model were correct. In the case of the FCD, a repeated task is to discover instances of sub-graphs of the form:

$$A \rightarrow B \leftarrow C,$$

that are implied by or consistent with the data. Ideally, variables in the data set will be found that instantiate this sub-graph template three times: for $(A, B, C) = (\Delta u, \Delta x, b)$, $(\Delta x, \Delta r, s)$, and $(\Delta r, \Delta c, m)$ [or for $(A, B, C) = (\Delta u, \Delta x, p)$, $(\Delta x, \Delta r, p)$, and $(\Delta r, \Delta c, p)$, if *b*, *s*, and *m* are subsumed into a single "type" variable *p* as discussed earlier]. Since *b*, *s*, and *m* may be subsets of variables, brute-force combinatorial search strategies are unlikely to prove useful in data sets with many variables. A more constructive approach is required.

10.6.5. Discussion

The following principles support computationally effective methods for searching among variables to discover and validate instances of the causal model in the FCD. In stating them, we use *A*, *B*, and *C* as generic names for variables. In a typical application, these variables might correspond to the columns in a spreadsheet, with cases (for instance, units of observation, such as individuals in a study population) corresponding to the rows. Each variable has a domain of possible values and a *scale type* (e.g., binary, unordered categorical, ordered categorical, difference scale, ratio scale, proportion, lattice-structured, time series, spatial distribution, etc). We use **D** to denote such a data set, a set of numbers or symbols recording values of variables for different cases or units of observation (perhaps with some variable values unrecorded for some cases) from which hazards are identified. In reality, the evidence typically consists of several heterogeneous data sets with different study populations, variables, and levels and units of observation (e.g., individuals, counties, countries), but this complexity requires no additional fundamental principles.

The following principles help to screen for combinations of variables that might instantiate the FCD, thus supporting automatic hazard identification from data.

Principle 1: Mutual information. For A to be a cause of B, A must be informative about B, meaning that they have *positive mutual information.* As a corollary, for data **D** to provide evidence that *A* is a potential cause of *B*, *A* must provide information about *B*: $I(A ; B) > 0$ in **D**, where $I(A ; B)$ denotes the mutual information between *A* and *B*.

Definitions of entropy and mutual information: If *A* and *B* are discrete random variables, the *entropy* of *A*, in bits, is denoted by $H(A)$. It measures uncertainty about *A*'s value: the information gained by learning its value. It is defined by $H(A) = -\Sigma_a Pr(A = a)log_2 Pr(A = a)$, with the sum ranging over all possible values *a* of *A*. The *mutual information* between *A* and *B*, denoted $I(A; B) = I(B; A)$ is given by:

$$I(A; B) = \Sigma_a \Sigma_b Pr(a, b)log_2[Pr(a, b)/Pr(a)Pr(b)] = H(A)-H(A|B) = H(A)-\Sigma_b Pr(B = b)H(A|B = b).$$

In practice, classification tree software products can automate the calculation of mutual information between variables. Given a dependent variable *B*, such an algorithm searches for variables ("splits") having maximum mutual information with *B*. Thus, Principle 1 can be implemented computationally as: *If A is a cause of B, then A must appear as a split in any classification tree for B* grown by maximizing mutual information. When implemented this way (via classification tree tests for mutual information) for real data sets, the *mutual information* principle holds only generically and for large enough, adequately diverse samples. It does not necessarily hold in sets of measure zero (e.g., if different inputs to a node happen to exactly cancel each other's effects), or if sample sizes are too small to detect the statistical information that *A* provides about *B*. Similar caveats apply to all other principles and tests applied to real, finite data sets.

The concept of mutual information generalizes the traditional principle of *association* between hypothesized causes and effects. One advantage of using mutual information rather than more common statistical measures of association is that it applies even to nonlinear and non-monotonic relations (e.g., $y = x^2$ for the domain $-1 \leq x \leq 1$), for which many measures of association (e.g., correlation) are zero, and thus prone to miss nonlinear causal relations. Conversely, because it is a non-parametric or "model-free" criterion, mutual information is less prone to create false positives than methods that make stronger, possibly incorrect modeling assumptions. For example, if *A* and *B* are two independent random numbers, each uniformly distributed between *0* and *1*, then seeking to instantiate an assumed linear risk-characterization relation $\Delta c = k\Delta x$, by substituting *A* for Δx and *B* for Δc will give a highly statistically significant (and highly misleading) positive correlation of about *0.7* between them, even though there is no true association between them.

The problem is that the assumed parametric model $\Delta c = k\Delta x$ is misspecified for these variables, as examining the residuals of the zero-intercept regression of Δc against Δx immediately reveals. Allowing an intercept would make the linear regression $E(\Delta c|\Delta x) = 0.5+0*\Delta x$, as it should be. Fitting an assumed relation with no intercept introduces a modeling assumption that creates an artificial correlation due solely to model assumption biases, rather than to any true association in the data set itself. By contrast, the mutual information between A and B in this example is zero, and neither appears as a split for the other in classification tree analysis, thus correctly indicating that there is no true association between them in the data. Parametric models and tests for association risk both failing to find true (but nonlinear) associations and finding non-existent (model assumption-driven, rather than data-driven) associations between variables.

Principle 2: Explanation. For A to be directly causally related to B (so that an arrow points from A into B or from B into A in the causal graph), *they must provide information about each other that is not explained away* by (i.e., is not redundant with) information provided by other variables. Technically, A must belong to the *Markov blanket* of B. A *Markov blanket* (in D) of variable B, denoted by $MB(B)$, is defined as a smallest set of variables in D such that B is conditionally independent of all variables not in $MB(B)$, given (i.e., conditioned on the values of) the variables in $MB(B)$. The members of $MB(B)$ are just the parents, children, and spouses of B in the causal graph for the variables (Tsamardinos et al., 2003). Note that if two variables are identical, say $A = C$, then the Markov blanket for B need not be unique, since including either A or C makes the other redundant. In this case, A and C (and any others in their equivalence class) should be collapsed to a single variable before computing $MB(B)$. This extends the issue of multicollinearity in generalized linear models. Interpretively, if A is a direct cause of B and C is a direct cause of A but not of B (as in the diagram $C \rightarrow A \rightarrow B$), then the observed positive mutual information between C and B is *explained away* by A, meaning that B is *conditionally independent* of C given A (i.e., conditional mutual information $I(B; C|A) = 0$, even though their unconditional mutual information satisfies $I(B; C) > 0$. On the other hand, if A is a cause of B, then $I(A; B|C) > 0$ for all possible variables or subsets of variables, C, i.e., the mutual information between A and B is not explained by other variables in D. In terms of classification trees, it is impossible to eliminate A as a split for B by splitting (i.e., conditioning) on one or more other variables before splitting on A. Efficient algorithms exist automatically to compute the Markov blankets of variables even in very large data sets (Aliferis, 2003); but require tests for conditional independence. Classification tree software can be used to perform conditional independence tests for one dependent variable at a time by testing whether conditional mutual information is significantly different from zero (Frey et al., 2003). Alternatively, statistical tests of the residuals in flexible non-parametric ("form-free") regression models (Shipley, 2000) can be used to test conditional independence for one dependent variable at a time. More computationally-intensive commercial software (e.g., BayesiaLab™) will automatically compute Markov blankets for any variable or for entire sets of variables. Effective algorithms to automatically identify the set of a variable's direct parents and children (a subset of its Markov blanket) have recently been developed (Tsamardinos et al., 2003). For A to be directly and causally related to B, it must appear among B's parents and children. The Explanation Principle (or Conditional Independence principle) generalizes the requirement that, to be considered causal, an exposure-response association must *not* be fully explained by confounding (Greenland and Morgenstern, 2001); or by sample selection biases, information biases, or modeling and analysis biases.

Principle 3: Temporal Precedence. If A is a cause of B and both variables are observed over time, then $I[A^-(t), B^+(t)|B^-(t)] > 0$, i.e., the history of A up to and including time t [denoted by $A^-(t)$] is more informative about the future of B after time t [denoted by $B^+(t)$] than is the history of B alone [$B^-(t)$]. This subsumes several more obvious special cases that a change in A must precede resulting changes in B (if the change in A is a one-time *event,* such as a jump, or "change point" in A's level, trend, variance, or dynamic evolution law; or that the start of A must precede the start of B, if A and B are *conditions* that hold over time intervals). We make the following additional points on the Temporal Precedence criterion:

1. They provide an information-theoretic generalization of the criterion of *Granger-Sims causality* for linear (vector autoregressive) time series.
2. The *Markov Blanket* criterion alone cannot distinguish between $A \rightarrow B \rightarrow C$ and $A \leftarrow B \rightarrow C$. The temporal criterion can.
3. *Intervention analysis* (does an intervention such as introduction of a risk management countermeasure or a potential new hazard or exposure at time t affect the subsequent time series?) and *change point detection* (does the time series show that a change happened in its dynamics at about the same time as a known intervention?) are two special cases important in interpretation of epidemiological data and time series.
4. False positives due to *spurious regression* (common in regression-based approaches to multivariate time series analysis of exposures and health effects) can be avoided using the information-theoretic approach, but biases caused by aggregation of time series into discrete reporting intervals can still arise.
5. Software and statistical algorithms are available for automated intervention analysis, change point detection and information-theoretic assessment of whether an exposure or hazard is informative about future health effects.

Although information, explanation, and temporal precedence are the most useful criteria, there are two other principles that refine causal analyses based on the FCD. These two other Principles are:

Principle 4: Compositionality. If $A \rightarrow B \rightarrow C$, then $I(A; C) \leq I(B; C)$ and $Pr(C = c|A = a) = \Sigma_b Pr(C = c|B = b)Pr(B = b|A = a)$. Tests based on compositionality can be used to resolve ambiguities about the direction of causality in Markov-equivalent causal graphs: different causal graphs with identical conditional independence relations among variables. This Principle generalizes the path analysis principle for linear models, that if $y = kx$ is correct, then $var(y) = k^2 var(x)$.

Principle 5: Causal Precedence. Information flows from *determining* (causally antecedent) to *determined* variables. In general, *the entropy of any variable is at least as great as the sum of the entropies of its causal parents (the variables that point into it).* Thus, if A is a cause of B (direct or indirect), then B is not a cause of A (direct or indirect); such transitivity introduces a partial ordering of variables by potential causality. We note that:

- This principle does *not* deny the possibility of feedback among variables (since measurements of the same variable at different dates are treated as different variables in the causal graph and in dynamic Bayesian networks).
- It generalizes Simon causal partial orderings (Pearl, 2000) for deterministic systems of equations.
- Knowledge-based constraints (e.g., health effects may be caused by exposures, but not vice versa) can be imposed through the causal precedence partial ordering.
- Principles 4 and 5 together generalize the traditional *coherence* criterion.

The practical application of the above information-theoretic approach is that A is a potential cause of B in data set D if and only if it satisfies Principles 1-5. Here, A and B can be any two variables, i.e., columns of data set. But in practice, we apply these criteria to the pairs of variables corresponding to ($\Delta u \rightarrow \Delta x$), ($\Delta x \rightarrow \Delta r$), and ($\Delta r \rightarrow \Delta c$) in any proposed instantiation of the FCD, that is:

- Does the proposed Δu (e.g., change in fraction of animals receiving an antibiotic) actually serve as a potential case for the proposed Δx (number of contaminated meals ingested per year)?
- Does Δx potentially cause Δr; i.e., Can historical changes in illness rates be causally linked to changes in exposure or animal drug use via our information-theoretic Principles 1-5?
- Is Δr a potential cause of Δc; i.e., Can changes in treatment failure rates be causally linked to changes in exposure or animal drug use via our information-theoretic Principles 1-5?

These questions can be answered using the algorithms and methods just discussed and exemplified.

10.7. CONCLUSION

In this chapter we developed and exemplified different correlations as well as non-correlation-based methods. In general, correlations and their estimates are *not* causal. The results, even when statistically significant, do not prove causation unless more is added to change the correlation to a causal association. The discussion of correlations covers areas of risk assessment and management that can be useful to formulate initial, often descriptive, assessments of potential relations. The qualitative criteria developed by Hill can obtain a preliminary understanding of potential quantitative relations between exposure and response. Although there are serious questions about the relation between these qualitative criteria and causation, combining them with measures of association can provide initial quantitative guidance to those who have to apply the precautionary principle.

We have also discussed internal consistency and generality. These two concepts are relevant to risk assessment and management because they provide the basis for developing theoretical arguments that are subsumed in the range of the issues developed and exemplified in the past chapters. Finally, this chapter contains an additional discussion of the methods used for causal analysis that were discussed in previous chapters. They have a formal basis in causal diagrams and in information theory through the fundamental causal diagram, FCD. The application of these methods extends causal analyses for risk assessment to the point that risk assessors now have a range of tools that can effectively resolve the vexing issues of confounding and biases in structuring formal causal relations and in statistical estimation.

QUESTIONS

1. Discuss a way in which a correlation may be *causal*. Develop your own example, based on data that is available from your workplace. In the absence of data, refer to a subject you are interested in researching and, after collecting a suitable sample, apply the proper coefficient of correlation. Keep your written material to less than 800 words, including full citations to the data and to the literature that you will use in developing your arguments. Develop both sides of the argument: that is, discuss why your correlation can be considered to be causal associations and why not.
2. How can Hill's criteria be related through simple correlations? Give two examples of your own making and use two hypothetical 2*2 contingency tables using your own data.
3   Consider the following data:

    $X = 3, 10, 7, 6, 13, 9, 12, 15, 9.$
    $Y = 0.01, 0.13, 0.04, 0.16, 0.54, 0.15, 0.32, 0.21, 0.37$

    Assume that the data is ordinal. Select the test that is appropriate to study the correlation of these two random variables and calculate the appropriate correlation. Discuss your results.
4. Discuss the relevance of causal diagrams in the context of developing a conjecture about a mechanism that can lead to a hormetic response in humans; show your development of the appropriate causal diagram, using a BN.

CHAPTER 11.

RISK ASSESSMENT FRAMEWORKS, CALCULATIONS OF RISK AND TOXICOLOGICAL DOSES

Central objective: to provide examples of risk assessment frameworks, calculations of catastrophic risks, probabilistic distributions, and toxicological models such as compartmental and physiologically-based pharmaco-kinetic models (PB-PK).

This chapter begins with a review of some prototypical regulatory frameworks for risk assessment, although their legal implications are not discussed. These consist of a single source for methods, formulae and data developed by organizations such as the American Society for Testing and Materials (ASTM), states (California), and the US Environmental Protection Agency. These frameworks provide practical ways for calculating risks according to regulatory criteria used in managing risks through clean-up or other actions. Specifically, in this context, we review three frameworks and discuss some of their principal formulae for making practical calculation of the concentration of a chemical from ingestion and inhalation.

We also discuss and exemplify methods used to assess events characterized by low probabilities of occurring, but with potentially large consequences, through probabilistic risk analysis, PRA. This aspect of probabilistic analysis is also regulatory. It is often associated with the series of reports, published and known as WASH-1400, which were developed for the US Nuclear Regulatory Commission, in the 1970s, by a team lead by Norman Rasmussen at MIT. The PRA uses *fault-trees* and *event-trees* and generally represents probabilities and consequences through the complement of the cumulative distribution function. Fault-trees describe how multiple and interconnected events can potentially lead to a failure, such as an explosion. Event-trees describe the potential for specific adverse outcomes, such as tens or hundreds prompt deaths from an explosion. As shown throughout this textbook, risk assessments use probabilities and probability distributions to characterize uncertainty and variability. Thus, we also continue our discussions distributions and their theoretical basis to suggest how a risk assessor can choose one distribution over another using mechanistic reasoning.

Specifically, this chapter deals with:

- Three risk assessment frameworks developed by the American Society for Testing and Materials (ASTM), the State of California and the US EPA

- Probability distributions, including binomial, geometric, hypergeometric, Poisson, normal, gamma, *chi*-square, Weibull, exponential, beta and log-normal distributions.
- A summary of compartmental and pharmaco-kinetic toxicological models
- Examples of risk calculations, including the calculations of fatalities/mile and loss of life expectancy
- A simple model of infections
- Fault-tree and event-tree analysis

The connections of the elements of this chapter are:

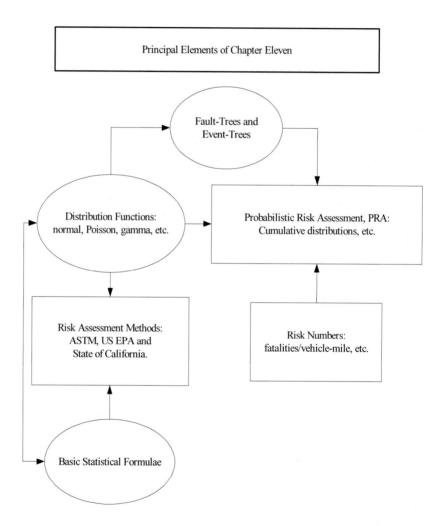

11.1. INTRODUCTION

We introduce the salient aspects of the following risk assessment frameworks:

- The American Society for Testing and Materials, ASTM, *Risk Based Corrective Action*, RBCA
- The State of California *Preliminary Exposure Assessment*, PEA
- The US EPA risk assessment framework
- *Probabilistic Risk Assessment*, PRA, using complements of cumulative distribution functions of accidents, as well as fault trees and event-trees associated with specific accident and cost scenarios

We begin with the American Society for Testing and Materials, ASTM, risk based, corrective actions method, RBCA, which includes aspects of risk assessment developed by the US EPA. A reason for not including more extensive discussions or risk assessment and management frameworks from other jurisdictions, is that these frameworks are available in government documents or the World Wide Web and because their input data change routinely. The interested reader can access her jurisdiction's specific requirements and latest methods and values.

11.2. THE RISK-BASED CORRECTIVE ACTION, RBCA, FRAMEWORK

The American Society for Testing and Materials (ASTM, 2000) has developed RBCA, which consists of a set of methods and data, for conducting environmental risk assessment for cancer and non-cancer endpoints, resulting in Risk-Based Screening Levels, RBSLs. There is an equivalent approach for ecological risks, called *eco*-RBSL, which is not discussed in this book. The RBSLs deal with chemicals likely to contaminate air, soil and water and thus create risks. As inputs for an RBCA, much of the relevant data and cancer slope factors (under the linearizing assumptions made by the US EPA, discussed in chapter 5) found in the US EPA's Integrated Risk Information System, IRIS, database available on line. For non-cancer hazards, media-specific tolerable intakes (or reference doses, RfDs) can also be found in the US EPA databases available on line.

11.2.1. The Risk-Based Corrective Action, RBCA, Framework

In RBCA, there are two basic models to represent *risk* and *hazard* (physical units of the variables are shown within square brackets):

1. *$Risk_{cancer}$ [probability] = (average lifetime compound intake [mg/kg-day])*(slope factor [probability/mg/kg-day]$^{-1}$).*

2. *$Hazard\ Quotient_{non-carcinogens}$ = (average daily intake [mg/kg-day])/(reference dose [mg/kg-day]).*

It follows that risk is measured by a probability; an acceptable level of risk is established by a recognized authority. The hazard quotient does not represent a risk.

The chemical-specific reference dose, RfD, is based on a specific exposure pathway, such as inhalation, and is also established by a recognized authority. The ASTM's definitions for these terms are:

Slope factor: The slope of the dose-response in the low-dose region. When low-dose linearity cannot be assumed, the slope factor is the slope of the straight line from zero dose to the dose associated with a *1%* excess risk. An upper bound on this slope is usually used instead of the slope itself; the upper bound is the *95%* upper confidence limit. The units of the slope factor are usually expressed in [mg/kg-day]$^{-1}$.

Reference Dose: A reference dose is an estimate (with uncertainty typically spanning perhaps an order of magnitude) of a daily exposure [mg/kg-day] to the general human population (including sensitive subgroups) that is likely to be without an appreciable risk of deleterious effects during a lifetime of exposure.

Reference Concentration: A reference concentration is an estimate (with uncertainty spanning perhaps an order of magnitude) of a continuous exposure to the human population (including sensitive subgroups) that is likely to be without appreciable deleterious effects during a lifetime.

In RBCA, the risks associated with mixtures of chemicals are obtained by adding them. Hazards are also added. However, in instances where the carcinogenic compound has several cogeners, such as benzo(a) pyrene (BaP), the cancer potency is relative to their principal cogener, as described in Table 11.1.

Table 11.1. Relative Potency Data for Selected PAHs

Chemical Compound	Relative Cancer Potency (unitless)
Benzo(a)pyrene	1.00
Benzo(a)anthracene	0.10
Benzo(b)flouranthene	0.10
Benzo(k)flouranthene	0.01
Chrysene	0.001
Dibenzo(1,2,3-cd)anthracene	1.0
Indeno(1,2,3-cd)pyrene	0.10

11.2.2. Risk-Based Screening Level, RBSL

We can use the vapor-phase formulae of the RBSL to describe the concepts that are common to the overall RBCA framework. There are two equations used to calculate risk-based screening levels, RBSLs, for inhalation of air pollutants (we are dealing with mass per cubic meter of air, [μg/m^3]). The cancer potency factors and the reference doses, RfDs, are found in the IRIS database maintained by the US EPA.

Non-Carcinogenic Chemicals

$RBSL_{air}$[μg/m^3] = $(THQ*RfC_i*AT_n*365$[days/year]$*1000$[μg/mg]$)/(EF*ED)$.

In this equation:

THQ = Total Hazard Quotient, [dimensionless],
RfC_i = Chronic Reference Concentration for inhalation, [mg/kg-day], chemical specific,
AT_n = Averaging Time for non-carcinogens, [years],
EF = Exposure Frequency, [days/year],
ED = Exposure Duration, [years].

The ASTM provides default values to be used when actual measurements are not available.

Carcinogenic Chemical Compounds

$RBSL_{air}[\mu g/m^3] = (Risk*BW*AT_c*365[\text{days/year}])/(SF_{air}*IR_{air}*EF*ED).$

In this equation:

$Risk$ = policy target excess individual lifetime cancer risk, [dimensionless; a cumulative, lifetime probability generally between $1*10^{-4}$ and $1*10^{-6}$],
AT_c = averaging time for carcinogens, [years],
SF_{air} = carcinogenic slope factor, $[\text{mg/kg-day}]^{-1}$.

The method (ASTM, 1998) for *screening* human health risks uses established (default) physical and chemical data, acceptable dose or risk levels, and formulae (accounting for less than lifetime exposure, when needed) for:

- RBSLs for inhalation of chemicals from the ground water in enclosed spaces
- RBSLs for surface and sub-surface soil and inhalation of vapors and particulate matter
- RBSLs for other pathways

Example 11.1. Non-cancer RBSL for inhalation of toluene by resident near a site:
$RBSL_{resident}[\mu g/m^3 \text{ air}] = (THQ*RfD_i*BW*AT_n*365[\text{days/year}]*1000[\mu g/mg])/(IR_{air}*EF*ED).$

In this equation the variables and the default values are:

THQ = Total Hazard Quotient, [dimensionless] set at 1.00,
RfD_i = Reference Dose for inhalation, [mg/kg-day], chemical specific,
BW = Adult Body Weight, [kg], set at 70 [kg],
AT_n = Averaging Time for non-carcinogens, [years], set at 30 [years],
IR_a = Daily Outdoor Inhalation Rate, [m^3/day], set at 20 [m^3/day],
EF = Exposure Frequency, [days/year], set at 350 [days/year],
ED = Exposure Duration, [years], set at 30 [years].

Using default values and noting the difference in the duration of exposure variable, ED, which is set at thirty years, rather than the $ED = 70$ years for carcinogens, results in the following RBSL:

$RBSL_{non-cancer}[\mu g/m^3] = (1.0*0.2*70*30*365*1000)/(20*350*30) = 730 \ [\mu g/m^3].$

Example 11.2. Cancer RBSL for inhalation of benzene by residents near a site, with the default values, yields:

$RBSL_{cancer}[\mu g/m^3] = (TR*BW*AT_c*365[\text{days/year}]*1000[\mu g/mg])/(SF_{air}*IR_{air}*EF*ED).$

In this equation, the variables and default values are:

TR = Target Excess Individual Lifetime Cancer Risk, [dimensionless], set at 10^{-6} or 10^{-4},
AT_c = Averaging Time for Carcinogens, [years], set at 70 years,

SF_{air} = Inhalation Potency Slope Factor, [mg/kg-day], chemical specific.

Therefore:

$RBSL_{resident}[\mu g/m^3] = (10^{-6}*70*70*365*1000)/(0.029*20*350*30) = 0.294 [\mu g/m^3]$.

The risk framework discussed next, developed by the State of California, is the Preliminary Exposure Assessment.

11.3. STATE OF CALIFORNIA PRELIMINARY EXPOSURE ASSESSMENT, PEA, FRAMEWORK

The PEA contains methods that are relevant to assessing the risk from exposure to various pollutants. The methods are quite similar to those developed by the ASTM and thus will not be developed in detail. We use examples to illustrate how the PEA computes risks.

Example 11.3. The cancer risk model for Volatile Organic Compounds (VOCs) is:

$Risk = (SF_{oral}*C_w*0.0149)+(SF_i*C_w*0.0149)+(SF_o*C_w*0.0325*K_p)$,

in which:

SF_o = Oral Potency Slope, [mg/kg-day],
SF_i = Inhalation Potency Slope, [mg/kg-day],
C_w = Concentration of the Chemical in Ground or Surface Water, [mg/Liter],
RfD_o = Oral Reference Dose, [mg/kg-day],
RfD_i = Inhalation Reference Dose, [mg/kg-day],
K_p = Chemical-specific Dermal Permeability Coefficient for Water Contact.

The risk model for air exposure is:

$Risk = SF_i*C_a*0.149$,

where:

SF_i = Inhalation Cancer Slope Factor, [mg/kg-day],
C_a = Concentration in Air, [mg/m^3].

The calculation of the intake (I) of unspecified airborne chemical is:

$I[mg_{chemical}/kg\text{-}day] = (C_a*IR*EF*ET*ED)/(BW*AT)$,

in which:

C_a = Chemical Concentration in Air, [mg/m^3],
IR = Inhalation Rate, [m^3/hour],
ET = Exposure Time, [hours/day],
EF = Frequency of Exposure, [days/year],
ED = Duration of Exposure, [years],
BW = Body Weight of the Exposed Individual, [kg],

AT = Averaging Time, [days].

Example 11.4. Using a vapor phase chemical such as trichloroethane, TCA, measured at an average of *15* [mg/m^3] of outdoor air, the intake *I* in [mg/kg-day] is:

$I = (1$ [mg/m^3]0.60[m^3/hour]0.50[hours/day]8[hours/day]10[years])/(70[kg]100[days]$) = 0.051$[mg/kg-day].

The 1993 data from IRIS is:

Cancer Slope Factor = (5.7*10^{-2})*[(mg/kg)/day]$^{-1}$
Drinking Water Cancer Unit Risk =1.6*10^{-6}, [μg/liter]
RfD for Water = 4*10^{-3}, [mg/kg/day]

The basis for the cancer risk numbers for *1,1,2*-trichloroethane is an animal study (mouse), using *gavage* as the route for administration. IRIS states that the evidence of carcinogenicity is that of a *possible human carcinogen*.[1] Risks are estimated through the linearized multistage, extra-risk model. For non-cancer effects, the *NOAEL* = 20 [mg/liter] and the *LOAEL* = 200 [mg/liter]. The uncertainty factor for the RfD is *1,000*, which combines such uncertainties as a judgment of *medium* confidence in the study, database and the estimated value of the RfD.

11.4. US EPA RISK ASSESSMENT FRAMEWORK

The US EPA framework for risk assessment is part of a legal-regulatory process for contaminated sites (*Superfund sites*). It consists of four steps (US EPA, 1992):

1. Data collection and evaluation,
2. Exposure assessment,
3. Toxicity assessment, and
4. Risk characterization.

A general equation for calculating an intake of a chemical is:

$I = (C*CR*EFD)/[BW*(1/AT)],$

where (the physical units of these variables were given earlier in this chapter):

I = Intake,
C = Concentration,
CR = Contact Rate,
EFD = Exposure Frequency and Duration,
BW = Body Weight,
AT = Averaging Time.

The units associated with these variables are those used in the previous other frameworks. To account for *variability*, the EPA has stated that:

"because of the uncertainty associated with estimating the true average concentration at a site, the 95 percent upper confidence limit (UCL) of the arithmetic mean should be used for this variable."

---

[1] The US EPA *weight-of-evidence* is C.

The EPA's rationale for choosing this UCL is that it *provides reasonable confidence that the true site average will not be underestimated.* The average concentration is used both to calculate, for a specific site, the pathway-specific reasonable maximum exposure (RME) and an average exposure. What changes in calculating the RME is the concentration, exposure frequency, duration and the other variables selected to give this maximum value.

The UCL used in these calculations is the UCL based on large sample theory and therefore applies to the *population* from which the data is drawn. The US EPA provides a discussion of the difference between lognormal and normally distributed data. The one-sided *95%* UCL, the UCL for *1-α* confidence interval, of the *arithmetic* mean of sample data that are log-normally distributed can be calculated from Gilbert's formula:

$$95\%UCL = exp\left[\bar{x} + 0.5s^2 + s(H/\sqrt{n-1})\right].$$

In this equation, $\bar{x}$ is the arithmetic mean of the sample, s^2 is the sample variance and s is the sample standard deviation. This confidence limit is calculated by taking the natural logarithm, $ln(x_i)$, of each point, x_i, and calculating the sample mean. The formula is:

$$\bar{x} = 1/n \sum_{i=1}^{n} ln(x_i).$$

The formula for the standard deviation of the *log*-transformed data is:

$$sd = 1/(n-1) \sum_{i=1}^{n} [ln(x_i - \bar{x})^2].$$

The value of the *H*-statistic is determined from Gilbert (1987, Table A10). It equals *3.163* for the *95%* confidence limit.

Example 11.5. The data [mg/m³] is:

10, 15, 20, 35, 50, 65, 120, 130, 145, 155, 1,800.

We can calculate the *95%* UCL using the *t*-test:

$$t = \frac{(\bar{x} - \mu)}{s/\sqrt{n-1}},$$

with $\bar{x} = 231.364$, [mg/m³], $df = n-1 = 10$ and $α = 0.025$ (the level of significance is $α = 0.05$, the areas at the two tails are $α/2 = 0.025$); n is the sample size. The *95%* UCL equals *599.908* [mg/m³]:

$231.364 + 2.228(523.087/\sqrt{10}) = 599.908.$

The value 2.228 is found in the standard tables that tabulate the area to the right of the critical values of the t-distribution (with the appropriate number of degrees of freedom). Suppose that the normal distribution were used to obtain the 95% UCL. The 95% UCL for the population mean, using the z-distribution, is:

$$UCL = \bar{x} + z_{0.025}\left(s/\sqrt{n}\right),$$

in which the z-value is 1.96. Therefore, the 95% UCL $= 231.364+1.96(523.087)/(11^{0.5}) = 540.489$ [mg/m^3].

Example 11.6. Consider the same data set and their natural logarithms:

x	ln(x)
10	ln(10) = 2.303
15	ln(15) = 2.708
20	ln(20) = 2.996
35	ln(35) = 3.555
50	ln(50) = 3.912
65	ln(65) = 4.174
120	ln(120) = 4.787
130	ln(130) = 4.868
145	ln(145) = 4.977
155	ln(155) = 5.043
1,800	ln(1,800) = 7.946

We obtain, for the *log-transformed* data: mean $= 4.30$, 95% lower confidence limit $= 3.26$, 95% upper confidence limit $= 3.26$, $sd = 1.50$. We use the formula for the one-tail upper confidence limit and the value for $H_{0.95} = 4.207$ (Table A12 in Gilbert, 1987), to calculate the 95% UCL concentrations:

95% UCL $= \exp[(4.297)+(0.5*1.543^2)+(1.5*4.207/10^{0.5})] = 1,169.18$ [mg/m^3].

Of these two results, 599.908 or $1,083$ [mg/m^3], the US EPA guidelines state that the highest concentration is the most appropriate for risk assessment, because of its precautionary approach. Practical risk assessments must deal with observations that appear outside the overall values of the majority of the sample. This is shown in the two previous examples where the value $1,800$ [mg/m^3] would appear to be inconsistent with the values in the rest of the sample because it is much larger than the next highest value, 155 [mg/m^3]. The influence that some data has on statistical results such as the sample mean and variance has to be studied and formally addressed. When one or more observations are considerably separate from the majority of the data in the sample, these exceptional observations are called *outliers* and must be carefully assessed before they are either rejected from the sample or included in it. We briefly discuss outliers next.

11.4.1 Outliers

Some of the reasons for finding outliers in a data set include:

1. An error of commission (transcribing the wrong number) may have been committed.
2. The outlier is real and belongs to the population. If so, the researcher can investigate the nature of the outlier, explain it on appropriate empirical and theoretical grounds, and then include it in the analysis.
3. The outlier is real but belongs to a different population. In this case, the researcher should omit that observation, but report that omission and its implications.
4. There are several outlying observations. It is probable that two different phenomena are measured in the same sample. Physical and other explanations may be used to describe those outliers.

The analyst may increase the sample size to study if the outliers are, in fact, outliers. Methods such as Monte Carlo simulations or bootstrap can be used to study them. As expected, there are statistical methods to determine the influence of outliers and to assess them. The reader may consult Huber (1981) for details about robust statistical methods, including robust regressions. The effect of outliers, on estimation with the ordinary least squares, increases the size of the variability about the estimated parameters and makes it too large. Estimation becomes inaccurate. Gilbert (1987) gives several practical illustrations of how outliers can be detected using statistical methods. In particular, he also uses *control charts* and provides useful examples of their use. We can use Dixon's and Chochrane's tests (Kanji, 1999) to detect and study outliers.

Example 11.7. Suppose that a sample size of *9* observations is taken at random from a normally distributed population:

4.12, 3.31, 5.98, 2.37, 45.00, 6.10, 3.60, 4.00, 9.10.

The risk assessor is concerned with the values *45.00* and *9.10*, suspecting that both are outliers. She uses Dixon's test (Kanji, 1999) with level of statistical significance, $\alpha = 0.05$. The calculation for ordered values of the sample, in which the largest is first, is based on the ratio:

$$r = (x_2 - x_1)/(x_{n-1} - x_1).$$

Thus, she obtains $r = (9.10\text{-}45.00)/(3.31\text{-}45.00) = 0.86$. Using Table 8 in Kanji (1999) she obtains, for $\alpha = 0.05$ and $n = 9$, the value *0.510*, which is less than *0.86*. Therefore, the observation *45.00* is an outlier, at the level of confidence selected.

The influence of outliers may easily change the direction and significance of the results. For example, outliers affect the evaluation of exceedances of environmental limits, such as environmental standards.

11.5. SELECTED PROBABILITY DISTRIBUTIONS

The empirical frequency distribution is a description of a specific type of pattern of data in a data set. Data can be described graphically (by drawing a frequency curve or histogram) or through numerical measures such as the mean and the standard deviation. Unlike summary measures of central tendency (the mean, for instance) and dispersion (the variance), distributions describe the entirety of the information for the sample or population. This section deals with several distributions of discrete and continuous data generally used in practical risk assessment. The reader should to refer to Applebaum (1996); Dougherty (1990); Apostolakis (1974) and Evans, Hastings and Peacock (2001) for additional details and other distribution functions; their work has influenced the descriptions in this chapter. In general, X indicates the random variable. However, T can serve as well, as can W or Z or some other letter. Typically, $f(X)$ is the distribution function for a univariate distribution; $f(X, Z, ...)$ is the distribution of a multivariate model. For continuous distributions, $f(x)$ generally identifies the *density* function of the random variable X. For discrete distributions, $f(x)$ is the *probability mass* function of the random variable X. For brevity, we cannot discuss all density or mass functions. In the diagrams in this section, the y-axis is the density of the random variable and the x-axis shows the values taken by the random variable.[2] The term *pr(.)* is short-hand for *probability of ...*; *pr* denotes a probability.

Bernoulli outcomes can only have two possible and mutually exclusive states (say, dead or alive); each realization of the random variable must be independent of the other. The two outcomes must be mutually exclusive and exhaust all possibilities. An often-used mathematical representation of these two outcomes is *1* (for the *success*) and *0* for the *non-success* or *failure*. Thus, *dead* could be represented by *1* and *alive* by *0*. Several outcomes can be a binary sequence of ones and zeros: *11100101001010101000*. In this sequence, there are *9 successes* and *11 failures*. The sample size is *20*. If N is the number of trials and the probability remains constant across each trial, we have a series of (identically distributed) Bernoulli outcomes.

11.5.1. Binomial Distribution

One of the common ways to represent Bernoulli (discrete) outcomes is the binomial distribution. This distribution applies when there can only be a failure or a success in the outcome with the following properties:

- There are n repeated experiments that are statistically independent
- Each trial either results in a positive result or in a failure, but not both
- The probability of a positive outcome is constant for each independent experiment

The typical use of this distribution is to determine the probable number of outcomes out of n independent attempts, each attempt having a probability *pr*. The binomial distribution has two parameters, n and *pr*. The statement *pr(s)* is the probability of exactly s *successes* in n trials; the formula for the probability of exactly s outcomes, *pr(s)*, is:

$$pr(s) = \{n!/[s!/(n-s)!]\}(pr^s)(1-pr)^{n-s},$$

with $s = 0, 1, ... , n$. A shorthand notation for the term with the braces, $\{.\}$, is $_nC_s$. The cumulative distribution, the probability of obtaining at most ($\leq$) s successes, is:

---

[2] We have plotted only a few density functions; the references in this chapter provide fuller discussions of these and other distributions as well as the tabulated values used for inference.

$$F(s) = \sum_{s=0}^{n} n! [s!(n-s)!](pr)^s (1-pr)^{n-s} ,$$

with mean $= (n)(pr)$ and variance $= (n)(pr)(1-pr)$. The shorthand representation for a random variable, e.g., X, which is binomially distributed, is $X \sim b(n, pr)$, read as *the random variable X has a binomial distribution with parameters n and pr*. The binomial distribution is approximated by the normal distribution, provided that n is large and that the probability of success and that of failure are not too close to zero.

Example 11.8. Suppose that $X \sim b(10, 0.5)$, calculate $pr(0)$, $pr(1)$, ..., $pr(10)$ (i.e., the probability of 0 successes, the probability of 1 success, up to and including the probability of 10 successes). These probabilities are:

$pr(0) = 10! / [0!(10-0)!](0.5^0)(1-0.5)^{10-0} = 0.00098,$
$pr(1) = 10! / [1!(10-1)!](0.5^1)(1-0.5)^{10-1} = 0.0098,$
and so on, to:
$pr(10) = 10! / [10!(10-10)!](0.5^{10})(1-0.5)^{10-10} = 0.00098.$

If we let N be the size of the population with n subsets. Then, (N, n) is the number of possible subsets of constant size n, in the population of size N. The probability that all n members of a subset respond is pr^n if each of them has an independent probability of responding, pr. Then, $(1-pr)^{N-n}$ is the probability that all of the $(N-n)$ members do not respond, when each of them has an independent probability of not responding equal to $(1-pr)$. These results obtain from $pr(A \ OR \ B) = [pr(A)+pr(B)-pr(A \ and \ B)]$; where $pr(A \ AND \ B) = [pr(A)][pr(B)]$ when A and B are statistically independent.

Example 11.9. Let *four* individuals each of whom has an independent probability of surviving another year equal 0.80. Find the probability that exactly *two* of them will survive. In this example: $N = 4$, $pr = 0.8$ and $x = 2$. The random variable is the *probability of surviving one more year*. Therefore:

$pr(X = 2) = (4, 2) (0.8)^2 (1-0.8)^{4-2} = 0.15.$

Example 11.10. A drug manufacturer develops a new drug used by *10* people. Historically, each user has a 0.01 probability of reacting adversely to the drug and suing the manufacturer. What is the probability that there will be two or more lawsuits? Let the random variable X be the *total number of lawsuits*. Therefore, $pr(X \geq 2) = [1-pr(X < 2)]$, that is: $\{[1-pr(X = 1)]-[pr(X = 0)]\}$. These probabilities are:

$pr(X = 1) = (10, 1)(0.01)^1(0.99)^9 = 0.091,$
and:
$pr(X = 0) = (10, 10) (0.01)^0(0.99)^{10} = 0.904.$
Therefore: $pr \ (X < 2) = (0.091+0.904) = 0.995$. Thus: $pr(X \geq 2) = 1 - 0.995 = 0.005.$

11.5.2 Geometric Distribution

Suppose that a researcher has observed a sequence of Bernoulli outcomes, a series of *0s* and *1s*, but now is interested in calculating the probability of r non-successes that precede the success. The question is: What is the probability of exactly $(r-1)$ failures, before the r-th success? The answer to this question is given by the geometric distribution. As was the case for the binomial distribution, the basis for this distribution is that the outcomes are a sequence of identically distributed Bernoulli random variables. The geometric random variable, Y, is defined as the smallest value of r for which $pr(Y = r) = pr(r)$; its probability mass function is:

$(pr)(1-pr)^n$,

with distribution function:

$1-(1-pr)^{n+1}$.

The formula to calculate the probability $pr(r-1)$ is:

$pr(r-1) = pr(1-pr)^{r-1}$,

where $r = 1, 2, \dots$. The geometric distribution is characterized by a single parameter, r, mean $(1-pr)/(pr)$ and variance $[(1-pr)/(pr^2)]$. In applications, it could represent the waiting time before an event occurs. The term $pr(r)$ is the probability of exactly $(r-1)$ failures preceding the first success, r. The probability of the first success is $pr(r) = (pr)(1-pr)^x$. The cumulative distribution of the failures is:

$$F(x) = \sum_{s=1}^{n} (pr)(1 - pr)^{r-1} .$$

Example 11.11. Let $pr = 0.2$ and $r = 3$. Then:

$pr(r) = (0.2)(0.8)^3 = 0.1024$.

Suppose next that we wish to calculate $pr(Y \le 10)$, with $pr = 0.5$. Then we can calculate:

$F(Y, y = 10) = 1-(0.5)^{10} = 0.990$,

using a simpler formula for the cumulative distribution for the geometric distribution: $F(Y) = 1-(1-pr)^n$.

11.5.3. Hypergeometric Distribution

Suppose that the outcomes of a particular probabilistic process are binary, and thus discrete. Let N be the total number of binary outcomes, and let n be the size of a random sample without replacement from N. Let X be the number of successes: the random variable X counts the number of successes. The probability of observing exactly x successes is:

$$pr(X = x) = \binom{X}{x}\binom{N-X}{n-x} \Big/ \binom{N}{x},$$

The expected value is $n(r/N)$. The variance is:

$$\frac{(nX/N)(1 - X/N)(N - n)}{(n - 1)}.$$

If the number of Bernoulli trials in a population is finite, then there can only be N total trials from which a sample is drawn without replacement. More specifically, the hypergeometric distribution describes the probability of observing exactly r successes, in a sample of size n, drawn from N in which there are k successes in all. If there are k successes, then there must be $N-k$ unsuccessful outcomes. Therefore, $N-k$ and $r \le k$ are the two constraints that must be met in calculating the probability sought. The hypergeometric distribution has three parameters: N, n, and k. The formula for the probability of exactly r successes, for a sample of size n, is:

$$pr(r) = [k!/r!(k-r)!][(N-k)!/(n-r)!][(N-k)-(n-r)]!)/\{N!/[n!(N-n)!]\},$$

with $r = 0, 1, 2, ..., n; r \le k, (n-r) \le (N-k)$. The cumulative distribution is:

$$F(r) = \sum_{r=0}^{x} pr(r).$$

Example 11.12. Suppose that $N = 20$ and that it is known that there are $k = 10$ successes, $n = 5$ and $r = (0, 5)$. A question can be: What is the probability of observing $pr(X = x, x = 0, ..., 5)$? In this problem: $N = 20$, $n = 5$, $k = 10$. The constraints are $(n-r) \le (N-k)$; $(5-r) \le 10$. The answer to the question is given by the following calculations:

$pr(r = 0) = {}_{10}C_0 \, {}_{10}C_5/{}_{20}C_5 = 0.01625,$
$pr(r = 1) = {}_{10}C_1 \, {}_{10}C_4/{}_{20}C_5 = 0.1354,$
$pr(r = 2) = {}_{10}C_2 \, {}_{10}C_3/{}_{20}C_5 = 0.6095,$
$pr(r = 3) = {}_{10}C_3 \, {}_{10}C_2/{}_{20}C_5 = 0.3483,$
$pr(r = 4) = {}_{10}C_4 \, {}_{10}C_1/{}_{20}C_5 = 0.1354,$ and
$pr(r = 5) = {}_{10}C_5 \, {}_{10}C_0/{}_{20}C_5 = 0.01625.$

The binomial distribution approximates the hypergeometric distribution when $k/N = pr$. If n is held at a specific level and N is allowed to tend to infinity ($N \rightarrow \infty$), the variance of the hypergeometric distribution equals the variance of the binomial distribution.

11.5.4. Negative Binomial Distribution (for integer values)

This distribution is used to model events in which the smallest number has a specific characteristic. In other words, the random variable is a specific value, r, for the number of binary successes, for a given sequence of binary outcomes. Consider a sequence of identically distributed Bernoulli-valued random variables and let the probability of failures be $pr(r)$; the complement of this probability is $(1-pr)$. Let k be the number of positive outcomes. The random variable X counts the failures (or non-successes) up to, but not including, the kth success. There are $r+s$ outcomes.

Two parameters, pr and k, characterize the negative binomial distribution. The question that can be answered using this distribution is: How many random failures should occur before observing x number of successes? The formula for the probability of exactly r events is:

$$pr(r) = (r+k-1)!/\{r![(x+k-1)-x]!\}pr^k(1-pr)^x,$$

for $x = 0, 1, 2, ...,$ and $k \geq 0$. The cumulative distribution is

$$F(x) = \sum_{r=0}^{x} pr(r).$$

The mean is $(1-pr)k(pr)^{-1}$ and the variance is $(1-pr)k(pr^{-2})$.

Example 11.13. Suppose that the probability of rejection is known to be *0.01* (the average rejection rate of *one* object per *100*). Then, in a sample of Bernoulli outcomes, what is the probability of finding that the *20th* value is the *3rd* negative value? Let $x = 3$ denote the third physical failure (which is stated as a *success* in the formulae just given). Because the sequence of 20 0s and 1s has been observed, the third physical failure must be associated with 17 physical positive values. Then:

$$pr(X = x, x = 17) = (_{x+k-1}C_{k-1})[pr^k(1-pr)^x] = (_{19}C_2)0.01^3 0.99^{17} = 0.00014.$$

The probability of observing *three* or more successes (that is, the probability of observing *three* or more physical negative results) is:

$$pr(X \geq x, x \geq 3) = [1 - pr(X = 0) - pr(X = 1) - pr(X = 2)].$$

It equals $1 - 0.000057 = 0.9999$, which is obtained from:

$$1 - [_2C_2 pr^3(1-pr)^0 + {_3}C_2 pr^3(1-pr)^1 + {_4}C_2 pr^3(1-pr)^2],$$

with $pr(.) = 0.01$. The expected number of success before the third failure $= (3)(.99)/0.01$, which is approximately *30* with variance $(3)(0.99)/0.01^2 = 297$ The standard deviation is approximately equal to *17* (the square root of *297*).

11.5.5. Poisson Distribution

This distribution applies when the probability of discrete events is very small, and therefore the event is *rare*. Practically, we use this distribution when the average number of events of concern to the risk assessor is constant in a small interval of time or area and the occurrence of these events. The counts are physical: the number of asthma diagnoses in an emergency room, the number of fires of a certain intensity in a geographical area, the number of impurities in a sample and so on. Counts cannot overlap and are independent.

The number of positive events (or outcomes) over that small interval of time or area is characterized by a Poisson random variable that has the following characteristics:

1. The average number of favorable outcomes, per unit time or space, is known (or can be assumed) from past knowledge.
2. The probability of a single positive outcome is proportional to the length of the time

interval or size of the spatial unit and is statistically independent of the positive outcomes that have occurred outside these intervals. These intervals are very small and do not intersect.

3. The probability of more than a single favorable outcome in these intervals is near zero.

The probability, *pr*, of exactly one count in an infinitesimally small interval of time (or in an infinitesimally small area) must approximately equal *(pr)(dt)* or, if an area, *(pr)(ds)*; the probability of two or more counts in either *dt* or *ds* must be approximately equal to zero. If the rate λ at which the counts occur is constant, the Poisson distribution gives the probability of exactly *r* counts evolving over time, *t*, is:

$$pr(r) = [exp(-\lambda t)](\lambda t)^r/r!,$$

with $\lambda > 0, r = 0, 1, ...$.

The cumulative distribution is:

$$F(x) = \sum_{r=0}^{x} pr(r),$$

with $x = 0, 1, 2, ...$ .The mean and the variance are equal to λt. The Poisson distribution approximates the binomial when $N \to \infty$ and $pr \to 0$: $\lambda t = N(pr)$. If time (or area) is nominal (equal *1*), then the formula for the (mass) density simplifies to:

$$pr(r) = [exp(-\lambda)](\lambda)^r/r!.$$

The mean and variance of this distribution are equal: λ. The Poisson distribution changes its shape, depending on the value of λ.

Example 11.14. Suppose that *10* patients arrive at an emergency clinic per *24*-hours and that their distribution is Poisson. Let the random variable X count the number of patients arriving per hour. Then, ($\lambda = 10/24$ or approximately *0.417* patients per hour, which is the required constant rate):

$$pr(X = x) = exp(-0.417)(0.417)^r/r!$$

The probability that *5* patients arrive in an hour is:

$$pr(X = x, x = 5) = exp(-0.417)(0.417^5)/5! = 0.0000692.$$

Example 11.15. Suppose that $\lambda = 2.5$ and that $r = 3$. It follows that:

$$pr(r = 3) = [2.5^3 exp(-2.5)]/3! = 0.214.$$

The mean of this random variable is *2.5* and the variance is *1.581*; if $r = 10$:

$$pr(r = 10) = [2.5^{10} exp(-2.5)]/10! = 2.157*10^{-4}.$$

11.5.6. The Normal Distribution

One of the best-known continuous distributions is the normal (or Gaussian) distribution. It is continuous, unimodal and symmetric about its mean. The mean coincides with the mode and the median. The domain of a random variable that is normally distributed is the entire real axis, from negative infinity $(-\infty)$ to infinity (∞). The shape is that of a (flat) bell.[3]

The mean and variance (the square root of the variance being the standard deviation) characterize the distribution in its entirety; the mean and the variance are sufficient statistics. If two normal distributions have equal mean and standard deviation they are identical.

The density function of the normally distributed random variable X, symbolized by $X \sim N(\mu, \sigma^2)$, has two parameters, μ, the mean, and σ^2, the variance. It has the form:

$$f(x) = [(2\pi)^{1/2} \sigma]^{-1} exp[-(x - \mu)/2\sigma^2] \, ,$$

with $-\infty < x < \infty, -\infty < \mu < \infty, 0 < \sigma < \infty.$

The cumulative distribution of $X \sim N(\mu, \sigma^2)$ is:

$$F(x) = [(2\pi)^{1/2} \sigma]^{-1} \int_{-\infty}^{x} exp\{-1/2[(x - \mu)/\sigma]^2 \}dx \, .$$

Increasing values of μ shift the density function to the left, and vice versa. Changing σ^2, for instance by decreasing it, reduces the width of the density function. The probability that the random variable X takes the value x is:

$$pr(X = x) = [1/(\sigma\sqrt{2\pi})]\{exp[-1/2(x - \mu)^2]\}dx \, .$$

It can be shown that:

$$1/\sqrt{2\pi} \int_{-\infty}^{\infty} exp(-x/2)dx = 1 \, .$$

> **Example 11.16.** The shape of the normal distribution with mean *3* and variance *1*, $X \sim N(3, 1)$, is:

---

[3] The *3-D bell*-shaped distribution is the bivariate Gaussian distribution for two random variables, for example X and Y. The z-axis depicts is the density function of their joint distribution. It is shown later in this chapter, but these results have been used in several earlier chapters.

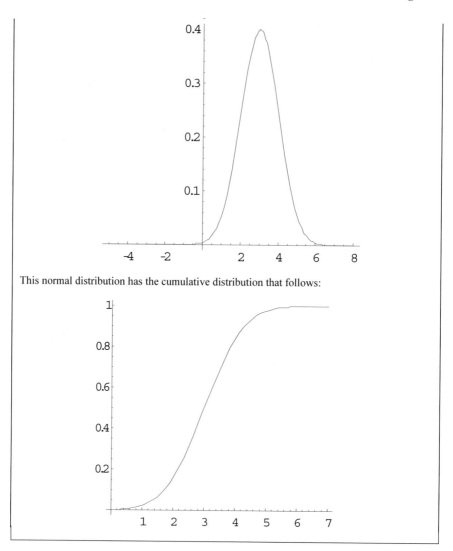

This normal distribution has the cumulative distribution that follows:

We develop the *standardized normal distribution* as follows. Let $y = (x-\mu)/\sigma$, that is, $x = \mu+y\sigma$. For a continuous distribution the *probability* is the product of $f(x)$ and dx; that is, $pr(x) = f(x)dx$. By substitution we obtain:

$pr(x) = 1/(\sigma\sqrt{2\pi})[exp(-1/2y^2)d(\mu+y\sigma)]$.

To find the density function of this random variable, let: $d(\mu+y\sigma) = 0+\sigma dy$. Then, we obtain:

$pr(x) = 1/(\sqrt{2\pi})exp(-1/2y^2)dy$.

The standardized normal distribution of the random variable $Z \sim N(0, 1)$ is:

$$f(z) = (2\pi)^{1/2} \exp(-z^2/2),$$

with $-\infty < z < \infty$. Its cumulative density function is:

$$\Phi(z) = 1/\sqrt{2\pi} \int_{-\infty}^{z} \exp[-(y^2/2)]dy.$$

The probability that the random variable Z is between the limits a and b ($a < b$) is:

$$pr(a < Z < b) = [\Phi(b) - \Phi(a)].$$

What are the quantities $\Phi(a)$ and $\Phi(b)$? They are general statements about probabilities (i.e., areas under the standardized density function). Thus $\Phi(a)$ is an area valued up to some value of $Z = a$, which means that $z = a$. And $\Phi(b)$ is another area, for $z = b$. The probability distribution $N(0, 1)$ has no unknown population parameters. It can be shown that:

$$pr(a < X < b) = pr[(a-\mu)/\sigma] < Z < [(b-\mu)/\sigma] = \Phi[(b-\mu)/\sigma] - \Phi[(a-\mu)/\sigma].$$

These points lead to the question: Is it possible to reduce all normal population distributions to this simpler representation? Doing so is useful because it avoids the otherwise labor intensive burden of having to calculate integrals to get probability values whenever population parameters change. The answer is that it is possible to do so through the transformation $z = (x-\mu)/\sigma$. We show, in Table 11.2, a subset of z-values.

Table 11. 2. Selected Values of the z-Statistic

z-value	Between mean and z	Greater than z	z value	Between mean and z	Beyond z
0.00	.0000	0.5000	1.50	.4332	0.0668
0.01	.0040	0.4960	1.64	.4495	0.0505
0.02	.0080	0.4920	1.70	.4554	0.0446
0.03	.0120	0.4880	1.80	.4641	0.0359
0.04	.0160	0.4840	1.90	.4713	0.0287
0.05	.0199	0.4801	2.00	.4772	0.0228
0.06	.0239	0.4761	2.10	.4821	0.0179
...	...	...	...	...	...

In Table 11.2 there are three separate sections; each comprises three columns. The first column shows values of z, arranged in sequence. The first z value is *0.00*, and runs down to *0.06*. We omit several z-values, start again at the top of the middle section with a z-value of *1.50*, and so on. The z value describes the proportion of the distribution that lies above or below a particular value of *z*. The middle section ends with a *z* value of *2.10*, and the third column starts with *3.40*. In the more extensive tables, such as those found in statistical textbooks there are more *z* values, generally from *0.00* to *4.00*, in small increments.

The sum of two or more normally distributed random variables is also normally distributed, if those random variables are statistically independent. Moreover, if a normally distributed random variable is decomposed into two random variables, these two are also normally distributed (Cramer-Levy theorem). Importantly, when the samples are small, we use the *t*-distribution. This distribution is also unimodal and symmetric about its mean, median and mode. The discussions just made apply to the *t*-distribution, noting that it has a special parameter, the number of degree of freedom, which determines its shape by controlling the way the *t*-distribution tends to the normal distribution. It is often stated that the normal distribution can be used if the sample size is greater than *30*; the reader should be aware that this is a rule of thumb only.

11.5.7. Gamma, Γ, Distribution

This continuous distribution is defined over the positive half of the real axis. The random variable that follows the gamma distribution can involve *r* outcomes that occur at a constant rate, λ. The gamma distribution of outcomes over time models the time it takes for the number of outcomes, *r*, to be observed. This distribution has two parameters, λ and *r*. The gamma density function of the random variable *T*, with realizations *t*, is:

$$f(t) = [\lambda^r / \Gamma(r)] t^{(r-1)} exp(-\lambda t),$$

provided that $t \geq 0$, $r > 0$ and $\lambda > 0$; otherwise $f(t) = 0$. The gamma *function*, $\Gamma(r)$, is $\Gamma(r) = \int_0^\infty x^{r-1} exp(-x) dx$. The cumulative density function of the gamma distribution is:

$$F(t) = \lambda^r / \Gamma(t) \int_0^t \tau^{r-1} exp(-\lambda \tau) d\tau,$$

with mean r/λ and variance $r\lambda^2$. The shape parameter is *r*: its value influences the shape of the density function. If $r \leq 1$, the density function is concave upwards. For *r* > *1* it is shaped as an asymmetric bell, with a maximum at $t = (r-1)/\lambda$. As *r* increases in value, the gamma distribution becomes more symmetric.

The gamma distribution function, $\Gamma(3, 1)$ is depicted in Figure 11.1.

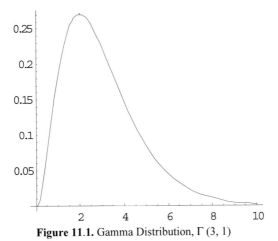

Figure 11.1. Gamma Distribution, Γ (3, 1)

If the shape parameter is an integer, c, then the distribution function simplifies to:

$$F(x) = 1 - [exp(-x/b)][\sum_{i=0}^{c-1} (x/b)^i /i!] ,$$

in which the scale parameter is $b > 0$. The probability density function is:

$$f(x) = \frac{(x/b)^{c-1} exp(-x/b)}{b(c - 1)} .$$

In this case, the mean is bc and the variance b^2c.

11.5.8. Exponential Distribution

The exponential distribution, when $r = 1$, is a special case of the gamma distribution. The interpretation of λ can be changed to account for the number of outcomes that precede a failure. In this interpretation, the $(r-1)$ successes that precede the failure are modeled by the geometric cumulative distribution, with the probability that the kth outcome is a success being equal to $(1-\lambda)^k$. Then, provided that λ is small and k large, the probability that the kth outcome is positive is approximately $exp(-k\lambda)$. The density function of the exponential distribution is: $f(t) = \lambda exp(-\lambda t)$, with $\lambda > 0$ and $t \geq 0$. The cumulative distribution is: $F(t) = 1-exp(-\lambda t)$. The mean and variance are λ^{-1} and λ^{-2}, respectively. When the rate at which adverse outcomes occur is small and t is not too long, then $f(t)$ is approximately equal to λt.

11.5.9. *Chi*-Square Distribution

This continuous distribution has already been discussed and we merely restate it here, while considering its *central* form. The *chi*-square density function ($v = df$) is:

$$f(x) = \frac{x^{(v-2)/2} exp(-x/2)}{2^{v/2} \, \Gamma(v/2)} ,$$

provided that $x \geq 0$. The mean and variance are v and $2v$, respectively. When the number of degrees of freedom is 1, χ^2 ($df = 1$), it is the square of the normal distribution (Dougherty, 1990). The chi-square density function for five degrees of freedom is depicted in Figure 11.2.

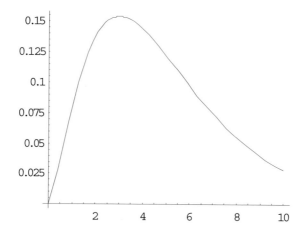

Figure 11.2. Chi-square Distribution, df = 5.

11.5.10. Weibull Distribution

This is a distribution of independent, continuous, minimum values from a parent distribution that has a finite left bound (such as the gamma distribution); we have used in earlier chapters. The Weibull distribution can represent the distribution of lifetimes of physical objects. The *two*-parameter Weibull density function is:

$$f(x) = (\alpha/\beta)(x/\beta)^{\alpha-1} exp[-(x/\beta)^\alpha] ,$$

with $x \geq 0$, $\alpha > 0$, $\beta > 0$; otherwise it is 0. Its cumulative distribution is:

$$F(x) = 1-exp(-x/\beta)^\alpha.$$

The mean equals $\beta\Gamma(1/\alpha+1)$; the variance is $\beta^2\{\Gamma(\alpha/2+1)-\Gamma(1/\alpha+1)]^2\}$. If $\alpha < 1$, the density function is shaped as a reverse J. When $\alpha = 1$, the Weibull density becomes the exponential distribution; when $\alpha > 1$ the density function is *bell*-shaped and symmetric.

The two-parameters Weibull distribution can account for constant, decreasing, and increasing failure rates (Evans, Hastings, and Peacock, 2000). For $\alpha = 1$, $\beta = 3$, the Weibull distribution function plots as depicted in Figure 11.3.

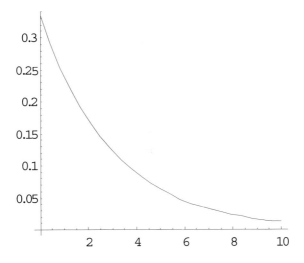

Figure 11.3. Weibull Distribution, $\alpha = 1$, $\beta = 3$

11.5.11. Beta Distribution

The beta distribution takes continuous values within an interval that, for our discussion, is limited to values between *0* and *1*, inclusive of these values. Other intervals can be normalized to *(0, 1)* by using a transformation of *t*, namely: $t' = [t-t(0)]/[t(1)-t(0)]$. The density function of a beta-distributed random variable is:

$$f(t) = \{[\Gamma(\alpha+\beta)]/[\Gamma(\alpha)\,\Gamma(\beta)]\}t^{\alpha-1}(1-t)^{\beta-1},$$

with α and $\beta > 0$ and $0 \leq t \leq 1$; otherwise $f(t) = 0$. The beta function is defined as:

$$\beta(\alpha, \beta) = \int_0^1 t^{\alpha-1}(1-t)^{\beta-1}\,dt,$$

which has been shown to equal $\{[\Gamma(\alpha+\beta)]/[\Gamma(\alpha)\,\Gamma(\beta)]\}$. The cumulative distribution of the beta distribution is:

$$F(t) = \{[\Gamma(\alpha+\beta)]/[\Gamma(\alpha)\,\Gamma(\beta)]\} \int_0^t \tau^{\alpha-1}(1-\tau)^{\beta-1}\,d\tau,$$

with $0 \leq t \leq 1$. The mean is $\alpha/(\alpha+\beta)$ and the variance equals $\alpha\beta/[(\alpha+\beta)^2(\alpha+\beta+1)]$. The shape of the beta distribution can change, depending on the values taken by α and β. Thus, for instance, if both of these parameters are less than *1* the function is U-shaped; if these two parameters are equal, the function is symmetrical; and if α and β equal *1*, the function is a straight horizontal line, $f(t) = 1$. If α and β are both greater than *1*, the function has a single maximum. Depending on the values of α and β the function can look like a *J* or show a right-skew or a left-skew. For shape parameters that equal *4* and *2*, the beta density function plots as depicted in Figure 11.4.

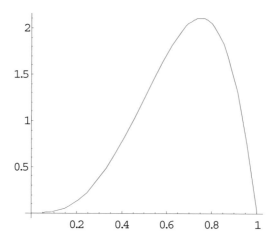

Figure 11.4. Beta Distribution with Shape Parameters 4 and 2

Example 11.17. Suppose that the random variable X yields a number n of observations, and that these observations are ordered from the first observed to the last. Let us define an interval from x_r to x_{n-s+1}. We also define a new random variable, X_1, to describe the events that occur in this interval. Then, $\alpha = (n-r-s+1)$ and $\beta = (r+s)$. A question can be: What is the probability that $F(X_1 > 0.95)$? To answer this question, we use the cumulative distribution of the beta distribution:

$$F(X_1 > 0.95) = 1-F(X_1 \leq 0.95).$$

Letting $n = 10$, $r = 1$, and $s = 1$, with $\alpha = 9$ and $\beta = 2$ we obtain:

$$1-[\Gamma(10)/\Gamma(9)\Gamma(2)] \int_0^{0.95} x^8 (1 - x)dx = 1-[10!/8!1!(x^9/9-x^{10}/10)] = 0.091.$$

Thus, the answer is that there is almost a ten percent probability that *95%* of the values will fall within these two extreme values (Dougherty, 1990).

Following Dougherty (1990), assume that the limits of the interval are the *optimistic* and *pessimistic* values of the time of completing a particular task. The beta distribution can be used to solve for the expected time of completion and to answer questions involving the probability of obtaining a particular result. The generalized beta distribution is:

$$f(x) = \frac{(x - a)^{a-1} (b - x)^{b-1}}{(b - a)^{\alpha+\beta-1} B(\alpha, \beta)},$$

with $a < x < b$, otherwise $f(x) = 0$. The mean and variance are $\{a+[(b-a)\alpha]/(\alpha+\beta)\}$ and $\{a+[(b-a)^2 \alpha\beta]/[(\alpha+\beta)^2(\alpha+\beta+1)]\}$, respectively. Assume that the units of time are given in weeks and that $a = 2$, $b = 7$, $\alpha = 2$ and $\beta = 2$. The expected number of (consecutive weeks) for completion is *five* weeks: $\{2+[(7-2)(3)/(3+2)]\} = 5$, using the expected value. The probability that the random variable X (which measures the time to complete the tasks) is less than 6 is:

$$F(x) = [4!\,(5^4\,)2!1!\,]\int_2^6 (x-2)(7-x)dx = 0.8912.$$

11.5.12. Log-Normal Distribution

A log-normally distributed random variable has continuous outcomes that have been changed from their natural units to their corresponding natural logarithms. This distribution has a scale and shape parameters α and β, respectively. The log-normal density function is:

$$f(x) = [(2\pi)^{1/2}\beta x]^{-1}\{exp[-ln(x-\alpha)^2]/2\beta\},$$

with $-\infty < \alpha < \infty$, $\beta > 0$ and $x \geq 0$. A more familiar version is:

$$f(x) = [x\sigma(2\pi)^{1/2}\,]^{-1}\,exp\{-[ln(x/m)]^2\,/2\sigma^2\,\}\,,$$

in which m is the median (> 0), where $m = exp(\mu)$ and $\mu = ln(m)$. The cumulative distribution does not have a closed-form solution. The mean and variance of a log-normal random variable are $m[exp(\sigma^2/2)]$ and $\{m^2[exp(\sigma^2)](exp(\sigma^2-1)\}$. Using an alternative representation, the mean and variance are $[exp(\alpha+\beta^2/2)]$ and $\{exp(2\alpha+\beta^2)[exp(\beta^2-1)]\}$, respectively. The log-normal distribution is common in risk assessment because it reflects the physical dimensions of a variable that cannot take negative values. We note that the arithmetic mean of the log-normal data, which can be used to estimate the population mean of log-normally distributed data, is biased (Gilbert, 1987). The minimum variance unbiased estimator, $mean_{UMV}$, of the population mean, $\mu_{log-normal}$, of a log-normal distribution, is:

$Mean_{UMV} = [exp(arithmetic\ sample\ mean\ of\ ln(data))]\{\psi[n,\ variance\ of\ the\ ln(data)/2]$

That is (Gilbert, 1987):

$$\bar{x}_{UMV} = [exp(\bar{x}_{ln(x)}\,)]\Psi_n(s_{ln(x)}^2/2)\,.$$

The formula for the variance$_{UMV}$ is given in Gilbert (1987). The function $\psi_n(.)$ has been tabulated by Gilbert (1987, Table A9).

Example 11.18. Consider the following data, x_i and their natural logarithms, $ln(x_i)$.

$x_i = 4.12, 3.31, 5.98, 2.37, 45.00, 6.10, 3.60, 4.00, 9.10, 2.49.$
$ln(x_i) = 1.416, 1.197, 1.788, 0.863, 3.807, 1.808, 1.281, 1.386, 2.208, 0.912.$

The *geometric* mean of this data is $exp(1.667) = 5.296$. The lower and upper 95% confidence limits for the log-transformed data are: 1.052, and 2.281; the geometric *sd* equals 0.859. Using Table A9 (Gilbert, 1987), $= \psi(10, 0.369) \approx 1.389$:

$Mean_{UMV} = (5.926)(1.38) = 8.18,$

and:

$Var_{UMV} = exp[(2)(1.667)][(1.38)^2 - 1.746] = (28.05)(1.929 - 1.746) = 5.13.$

The se_{UMV} of the $mean_{UMV} = 8.18$ is calculated from $5.13^{1/2} = 2.26.$

The upper and lower confidence limits for the population mean from a log normal distribution are:

$$LCL_\alpha = exp(\bar{x}_{ln(x)} + 0.5s^2_{ln(x)} + s_{ln(x)}H_\alpha/\sqrt{n-1}),$$

and:

$$UCL_{1-\alpha} = exp(\bar{x}_{ln(x)} + 0.5s^2_{ln(x)} + s_{ln(x)}H_{1-\alpha}/\sqrt{n-1}).$$

In these expressions, and the expressions that follow, the subscript $ln(x)$ means that the estimator uses the natural log of the data. For example, the *one*-sided UCL, using the data in the example and Table A12 in Gilbert (1987) for determining H, we obtain:

$UCL_{.95} = exp(1.667 + 0.5*0.737 + 0.858*2.90/\sqrt{9}) = 17.57.$

Note that $\sqrt{s^2_{ln(x)}} = \sqrt{0.737} = 0.858$. The sample geometric mean, GM, is:

$$G.M. = \prod_1^n x_i^{1/n}.$$

It is a biased estimator of the population mean of a *log*-normally distributed population. The correction factor for this bias is $exp\{[-(n-1)/2n]\sigma_{ln(x)}^2\}.$

The median of a *log*-normal population has the following approximate confidence interval:

$$exp(\bar{x}_{ln(x)})exp(-t_{\alpha/2,n-1}s_{ln(x)}) \le exp(\mu_{ln(x)}) \le exp(\bar{x}_{ln(x)})exp(t_{1-\alpha/2,n-1}s_{ln(x)}).$$

Tables for the *t*-statistic can be found in most statistical textbooks; for example, $t_{0.95, 10} = 1.812$; $t_{0.95, 20} = 1.725$ and $t_{0.95, 1,000} = 1.685.$

We conclude this section by showing a bivariate joint distribution. In figure 11.5, we depict a joint normal distribution. Consider the joint distribution of two normally distributed random variables, X and Y. Let the means be *0* and *2*, and the variance-covariance (*1, 1.2*) and (*1.2, 4*). The joint distribution, pdf_{XY}, plots as depicted in Figure 11. 5.

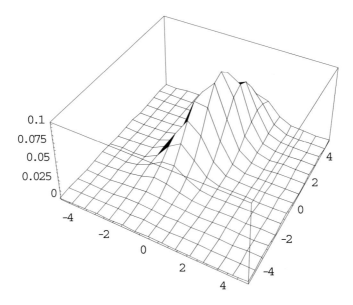

Figure 11.5. Joint Distribution Function

A scatter plot generated by the pdf_{XY} is ($n = 200$), in which the x-axis is horizontal and the y-axis is vertical, is depicted in Figure 11.6.

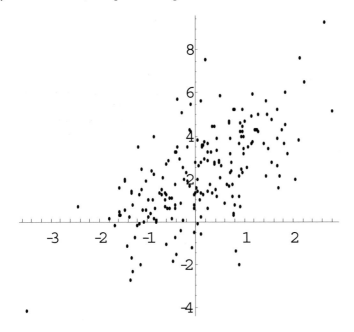

Figure 11.6. Scatter Diagram of the Joint Distribution Function

Example 11.19. Using JMP4 and a subset of data file DRUG, we describe the basic analysis inherent to developing *joint* confidence levels. The sample size for X is thirty, and the sample size for Y is also thirty. The z-axis is the probability density axis for the random variables X and Y, it is not shown but is perpendicular to the page. The two ellipses are slices parallel to the X-Y plane, which is shown.

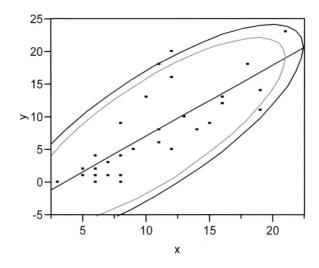

In this graph, the inner ellipse corresponds to $\alpha = 0.10$, $(1-\alpha = 0.90)$ and the outer ellipse to $\alpha = 0.05$, $(1-\alpha = 0.95)$.

11.6. THE POISSON PROCESS IN RISK ASSESSMENT

Several probabilistic processes are discussed in most textbooks on probability theory, in particular Kalbfleisch and Prentice (1980), Ross (1983), Grimmet and Stirzaker (1982), and Cox (2002) whom we follow. We have used the Markov process in Chapter 5. In this section we develop a few new concepts. We begin by defining the constant, discrete time, hazard function:

$h(t) = pr(T = t | T \geq t)$.

It has expected value $E(T) = 1/pr$, which means that $S[E(T)] = S(1/pr) = (1-pr)^{E(T)}$ where $E(T)$ is the expected number of discrete time periods until the failure arrives. If the hazard rate is constant it can be shown that $S[E(T)] = 1/e$, where $exp = e = 0.2718....$ Therefore, $pr[T > E(T)] \approx 0.37$.

In continuous time:

$S(t) = exp(-ht)$,

where $h(t) = 1/E(T)$ is the constant hazard rate. The interpretation is that $h(t)dt$ is approximately equal to the *pr(next failure occurs in the interval of time t+dt, given that the failure has not occurred by time t)*. By definition, $F(t) = pr(T \leq t)$ is the cumulative distribution function. If the hazard function is continuous and increasing, $h(t) \leq 1/E(T)$ and $pr[T > E(T)] \geq (1/e)$ and, more generally, $S(t) \geq (1/e)^{t/E(T)}$ for $0 \leq t \leq E(T)$. If the hazard function is continuous and decreasing, $pr(T > t) = (1/e)^{m(t)}$ where $m(t)$ is the expected number of arrivals in the interval of time $(0, t)$, provided that the probabilistic process has a renewal. In other words, the process starts again after the failure has occurred. If $m(t) = t/[E(T)]$, then the process is said to be a Poisson process (Ross, 1983).

The following results, also discussed in previous chapters, are relevant:

1) $h(t) = f(t)/S(t)$,

2) $f(t) = -\{d[S(t)]/dt\}$,

3) $S(t) = exp[-H(t)]$,

where $H(t)$ is the cumulative hazard function. Thus we obtain $h(t) = -d[logS(t)]/dt$, from which $H(t) = -logS(t)$.

11.6.1. The Poisson Process

The (homogeneous) Poisson process is characterized by a constant and continuous hazard function $h(t) = \lambda$. That is, for a number of accidents (N) arriving in the time interval $(0, t)$, the probability density function of the random variable $N(t)$ is:

$$pr[N(t) = n] = [(\lambda t)^n exp(-\lambda t)]/(n!),$$

for $n = 0, 1, 2, \ldots$. Note that $S(t) = pr[N(t) = 0] = exp(-\lambda t)$. The variance of $N(t) = \lambda t$.

Example 11.20. Following Cox (personal communication), let the number of health claims arrive at a health insurance office be *two* claims per week. What are the probabilities that:

1) No claims arrive during the first two weeks in June?
2) Exactly two claims arrive during the last two weeks in June.

Assume, for ease of calculation, that June has exactly four weeks. The hazard rate is constant $\lambda = 2$ claims per week. Therefore, by direct application of the formulae of the Poisson process with constant hazard rate:

1) $pr(no\ claims\ for\ two\ weeks) = S(2) = exp(-2\lambda) = exp(-4) = (1/2.718)^{-4} = 0.0183$.
2) $pr(3\ claims\ in\ two\ weeks) = pr[N(2) = 3] = [(\lambda)(2)^3(exp(-2\lambda)/(3!)] = 0.195$.

If the cumulative density function for the time to first arrival is $pr[T < t] = F(t)$, then the probability density function of the random variable T is:

$$f(t) = d[F(t)]/dt,$$

where $d[.]/dt$ is the first derivative of the function $F(t)$ with respect to t. The example that follows should crystallize the use of these functions. A question can be: what is the probability density function of the random variable T, when $h(t) = \lambda$? In other words, how do we get $f(t)$? We can answer this question as follows. We let $S(t) = exp(-\lambda t)$, from which:

$$F(t) = 1 - S(t) = [1 - (exp - \lambda t)],$$

and

$$f(t) = d[f(t)]/dt = [0 - (exp - \lambda t)(-\lambda)] = [(\lambda)(exp - \lambda t)].$$

This is the exponential probability density function with parameter λ. Generally, individuals exposed to a dangerous physical or chemical agent are characterized by different probabilities of response. If the individual probabilities of response are fairly small (say, less than 0.1 per unit of time), and the number of individuals is large, then the total number of cases is approximated by a Poisson distribution. More specifically, let the individual probabilities be pr_i, and let $\Sigma_i pr_i = \lambda$, with $n \to \infty$, with the largest value of pr_i approaching 0 so that the summation equals λ. Then, the total number of cases will have a probability distribution that will approximate the Poisson distribution, with parameter λ.

Example 11.21. Suppose that the risk assessor is interested in modeling the probability of forest fires, perhaps because she wishes to relate those to injuries and deaths from burns and smoke inhalation. For simplicity's sake, she is defining the causes of fire as those from lightning strikes and those from camping activities. The assumptions are that those two causes are rare, and the causal events are statistically independent. Thus, modeling using the Poisson process is appropriate. Let λ_1 and λ_2 be the intensities of lighting fires (per year) and camping fires (also per year). Take the following two questions (assume no other information): 1) What is the probability that a year elapses without a fire; and 2) What is the probability that a fire has started and that fire was caused by a camper? The answer to question *1* is calculated as follows. Let: *pr(no fires) = pr(no fire from source 1 AND from source 2)*. The *survival* function is *[pr(no fires from source 1)][pr(no fire from source 2)] = [exp-$\lambda_1 t$][pr-$\lambda_2 t$]*. The answer to question *2* is calculated with $[\lambda_1/(\lambda_1 + \lambda_2)]$.

To conclude, the arrival process generated by all n possible sources, taken together, is also Poisson with parameter $\Sigma_i \lambda_i$, provided that the assumptions of a Poisson process hold. In other words, the Poisson random variable X can be added to the Poisson random variable Y such that $X + Y$ are also Poisson-distributed with intensity $(\lambda_1 + \lambda_2)$. Moreover, the probability the next event will arrive from the ith source is $[\lambda_i/\Sigma_i \lambda_i]$, the summation being taken from i to n.

The compound Poisson process is characterized by Poisson-distributed arrivals that are independent from each other. The mean and variance of the cumulative damage caused by those arrivals is proportional to the elapsed time before arrival. It turns out that the compound Poisson process is approximately normally distributed. Note these two results (Cox, 2002):

1. Let the number of cases of disease in a sufficiently large population follow the Poisson distribution with mean N. Then, there is an approximately 0.95 probability that the expected number of cases is in the interval $N \pm 1.96\sqrt{N}$, provided that N is sufficiently large.
2. Let S_N be the cumulative damage from the first N arrivals characterized by a compound Poisson process. S_N will be approximately normal with expected value $[(N)E(X)]$ and variance $[(N)var(X)]$. The cumulative damage done in the interval $(0, t)$ is approximately normal with mean $\{\lambda t[E(X)]\}$ and variance $\{\lambda t[E(X^2)]\}$, provided that N is sufficiently large.

These results can be used to deal with risk questions such as: How much cumulative (that is, total) damage can occur from several specific events, within a specific interval of time, $(0, t)$? Let us assume that each event independently causes a randomly occurring amount of damage $X(k)$, from the kth accident, and that the amount of damage done by each different accident has identical probability distributions, with mean $E(X)$ and variance $var(X)$. The cumulative damage from N randomly occurring events arriving at period t is given by:

$$S_N = X(1) + X(2) +, ..., + X(N).$$

This compound Poisson random variable has mean:

$$E(S_N) = [E(N)E(X)],$$

and variance:

$$var\ (S_N) = \{E(N)var(X) + var(N)[E(X)]^2\}.$$

From the properties of the Poisson distribution, $E(N) = var(N) = \lambda t$. We obtain:

$$E(S_N) = (\lambda t)[E(X)],$$

with:

$$var(S_N) = (\lambda t)[E(X^2)].$$

11.7. PROBABILISTIC RISK ASSESSMENT (PRA)

The acronym PRA stands, particularly in reliability or failure analysis of technological systems, for *probabilistic risk analysis*. An event, such as an explosion at a petrochemical facility, can cause prompt deaths, injuries as well as delayed deaths and injuries. More precisely, that event (e.g., an explosion equivalent to *1* ton of TNT) can cause a range of deaths. Suppose that the context is occupational. The explosion may cause from zero prompt deaths, if nobody is around to, let us say, *100* prompt deaths if the explosion occurs when employees, visitors and workers are on-site. Note that we are dealing with a random variable, the number of prompt deaths.

A summary statistic for this random variable is the expected value of the distribution of those deaths, accounting for each probability of the discrete number of possible deaths from a single explosion of certain yield. We find that, in the PRA literature, risk is defined as an expected value:

$$Risk = \sum_{i=1}^{n}(pr_i)(magnitude_i).$$

The magnitude of the consequences must be homogeneous: prompt deaths, delayed deaths, injuries requiring hospitalization, property damage and so on. This definition of risk also applies to consequences measured in dollars, as shown in Table 11.3.

Table 11.3. Cumulative Frequency Calculations for Discrete Values

Magnitude (Cost, $)	Number of events (count), n_i	Cumulative number of events (count), Σn_i	Cumulative frequency, F_i
10	20	20	0.476
20	10	30	0.714
30	6	36	0.857
40	3	39	0.929
50	2	41	0.976
60	1	42	1.000

Thus, we obtain $R = (0.476*10+0.714*20+0.857*30+0.929*40+0.976*50+1.00*60) = \$20.48$, which is an expected value. A more informative representation uses the distribution. If the events are discrete, their empirical cumulative frequency is: $F_i = n_i/(\Sigma n_i+1)$. Table 11.4 depicts the results.

Table 11. 4. Cumulative Frequency Calculations for Binned Data

Magnitude (min - max)	Number of events	Cumulative number of events	Cumulative frequency, F_i, $F_i = n_i/(\Sigma n_i+1)$
10 to 100	15	15	0.533
101 to 1,000	10	25	0.833
1,001 to 10,000	3	28	0.933
10,001 to 50,000	1	29	0.967

The cumulative frequency is interpreted as the probability of an event being either less than or equal to a specific value or other quantity. Thus, the cumulative frequency of the events characterized by a magnitude between *1,001* and *10,000* is less than or equal to *0.933*. On this rationale, it is also possible to calculate cumulative frequencies greater than a specific magnitude. Such frequency is calculated as $(1-F_i)$, that is $(1-0.933) = 0.067$, which is the complement of *0.933*. The complement of the cumulative distribution is often called the *risk curve* in technological risk assessments, Figure 11.7.

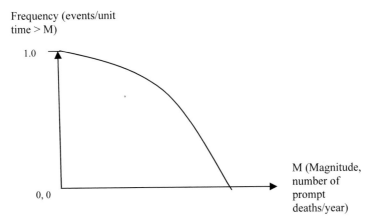

Figure 11.7. Complement of the Cumulative Distribution

This curve is based on the density function of the random variable X, $f(x)$, which measures the number of prompt deaths from a single explosion of a certain yield. Its cumulative distribution is:

$$F(X) = \int_0^X f(x)dx \ .$$

By definition, the complement of $F(X)$, preferentially used in a PRA, is:

$$1 - F(X) = 1 - \int_x^I f(x)dx \ .$$

Often, the data to fit the complement of the distribution are available near the high probabilities of events of small magnitude. The parameters of this curve are estimated using the MLE.

To summarize these concepts, consider Systems A and B as alternatives to reach a specific objective. The random variable X measures the magnitude of the adverse outcomes, as counts of prompt deaths. As depicted in Figure 11.8, the MLE curve of System A is uniformly riskier than System B because the risk curve for A dominates the risk curve for B everywhere. However, the 95% confidence intervals do not show that dominance because the confidence bounds about each curve overlap. The variability about the risk curve B (the straight line crossing it) overlaps with the uncertainty about the risk curve A (straight line crossing it). Developing these curves is a statistical estimation problem, in which the complement of the cumulative distribution appropriate to the events is fit to the data; their use in risk management deals with the dominance of one curve over another, for given choices of technological options.

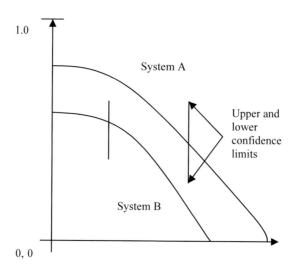

Figure 11.8. Comparison of Complements of Cumulative Distributions

Cox (2002) describes several aspects of deterministic and stochastic *dominance* that are relevant to resolving some of the difficulties induced by overlapping confidence bounds. Risk curves can be used to assess the performance of different people, technological systems and other generators of the adverse events, if the comparison is based on the same outcome. An example is given by consider two systems, *A* and, as shown in Table 11.5.

Table 11.5. Comparisons of Two Discrete Cumulative Distributions

Net Cost System A (min-max)	Net Cost System B (min-max)	System A, # of events	System B, # of events	System A, cumulative # of events	System B, cumulative # of events	F_{iA}	F_{iB}
10-100	0-50	20	15	20	15	0.54	0.24
101-1,000	51-100	12	20	32	35	0.86	0.56
1,001-10,000	101-500	3	25	35	60	0.94	0.96
10,001-50,000	501-5,000	1	1	36	61	0.97	0.98

11.7.1. Fault-Trees and Event-Trees

Assessing risky situations requires formulating the way in which a causal event can take place and then studying the potential paths that the consequences take, given that initiating event. This initiating event in an event tree, the *top event* in the fault-tree, is described in terms of more elementary events that can cause (or, at least, lead to) it. Consider the situation in which event *A* is *switch fails to open (at time t)*, event *B* is *back-up fails (at time t)*, event *C* is *stand-by generator is out of operation (at time t)*. These events lead to event *D*, *electricity is unavailable (at time t)*. The top event, *E*, is *failure to operate (at time t)*. Using logical operators, we obtain a causal diagram as follows: *A OR B* cause *D*; *D AND C* cause *E*. The logical structure of the fault-tree is described in Figure 11.9.

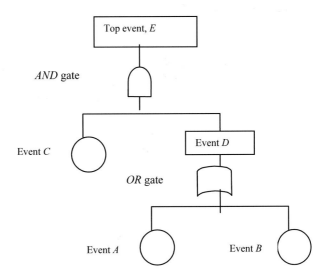

Figure 11.9. Elementary Fault-Tree

The *OR* gate is *union* of two or more events. In other words, the resulting event is true if any or all of the events entering that gate are true. The *AND* gate is the *intersection* of all the events entering that *AND* gate. Operations, such as complementation and multiplication, provide the algebra for these logical gates. Success or failure can be modeled using these gates. There are theorems that can be used to show the duality of these alternative representations using the rules of Boolean algebra (Bedford and Cooke, 2001).

In general, the analysis is Boolean, although new methods now available in the literature extend these representations by accounting for fuzziness. The logical representations of the events, their logical combinations and the inclusion of the relevant events leading to the top event form the fault-tree. The inclusion and exclusion of events suggests using boundaries that explicitly depend on the representation. These boundaries are temporal and system-specific. Even though the representations are consistent with physical events and are quantitative, the qualitative and subjective elements of the analysis cannot be overlooked because the model of the system under assessment requires expert engineering or other judgment.

The fault-tree can be set up as a probabilistic representation of the (otherwise deterministic) network leading to the top event, stated as either the failure or the success of a particular component of a physical system. Simply put, a random variable takes binary values, such as either *1* or *0*; *1* indicates *true* and *0* indicates *false*. The probability of the *AND* gate is calculated, for independent events entering that gate, as the product of the probabilities. For the *OR* gate, the probability is obtained as the sum of the probability of the events minus the probability of their intersection, to avoid double counting. This is the direct result of elementary probability theory that deals with the union (OR) and the intersection (AND) of independent events. Specifically, the basic calculations are as follows. Let:

pr(null event) = 0,

and *pr(event A OR event B) = pr(A OR B) = pr(A)+pr(B)-pr(A AND B).* A simple calculation, based on the *rare events* assumption, reduces the last statement to: *pr(top event) ≈ ∑pr_i(sub-events),* under the assumptions discussed by Bedford and Cooke (2001). An important aspect of the PRA is that fault-trees can be combined with event-trees as depicted in Figure 11.10 to characterize the initiating events, intermediate event, the top event and the set of exposures and consequences that result from that top event (Ricci, Sagan and Whipple, 1984). Fault-tree and event-tree can provide a plausible and sound causal linkage between physical events.

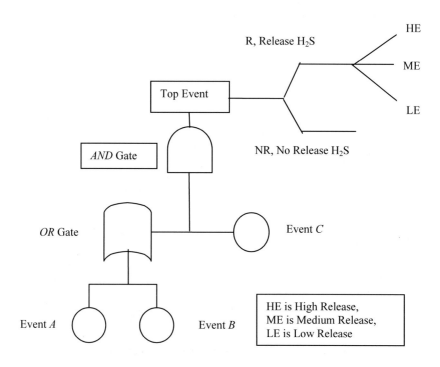

Figure 11.10. Combined Fault-Tree and Event-Tree

Example 11.22. Suppose that the probabilities of the *independent* events *A*, *B* and *C* are rare, say *pr = 0.01.* Use the rare event approximation: *pr(D) ≈ pr(A AND B)+pr(A AND C) =*

$pr(A)pr(B)+pr(A)pr(C)$. That is: $pr(D) \approx (0.01)(0.01)+(0.01)(0.01) = 0.0002$. Suppose that gate B is a *NOT* gate. What is the probability at that gate? Define the probability of success as pr_s and the probability of failure is $(1-pr_s)$. For an *AND* gate leading to a success and several sub-events the probability is $\Pi_i(pr_i)$; letting B be the complement of A, the probability of the *NOT* gate is $[1-\Pi_i(pr_i)]$.

The event-tree depicts the sequences of events leading to failure. For instance, the top event D leads to release: [*failure occurs AND release of H$_2$S occurs AND high exposure occurs*]. For example, the sequence [*failure occurs AND release of H$_2$S occurs AND high exposure occurs*] becomes [$pr(D)pr(R)pr(HE|R)$], in which the conditioning accounts for possible dependencies between events. The event-tree depicts the sequence of immediately preceding and necessary events leading to a distribution of outcomes.

The construction of both of these trees requires understanding the physical processes being modeled and assigning probabilities. These can be either subjective or frequentistic. For many fault-trees, frequencies may be available from testing the performance of components to failure. The discussion of fault trees and event trees has glossed over the practical difficulties inherent to developing complex trees. Fortunately, techniques exist to reduce the complexity and calculations required in the sense of eliminating chains of events that do not materially influence outcomes or are physically impossible (Bradford and Cooke, 2001).

The results from combining fault-trees and event-trees are used to generate cumulative (or, as it is more usual, complements of the cumulative distribution) of the consequences. Table 11.6 summarizes these concepts.

Table 11.6. Discrete Values for Probability and Magnitude of the Consequence

	Probability of exceeding M (prompt fatalities)/year			
Magnitude, M	10	250	1,200	3,700
Probability	10^{-1}	10^{-2}	10^{-4}	10^{-6}

Cohen and Davies (in Ricci, Sagan and Whipple, 1984) provide some indications of the potential risk associated with major industrial accidents, show in Table 11.7.

Table 11.7. Risk Comparisons for Technological Systems

	Fatalities with weeks of accident	**Delayed fatalities (expected between 20 and 40 years after the accident)**
Oil refinery	1,500–18,500, (includes all casualties)	NA
Power plant chlorine	100	NA

A tolerable risk curve consists of the relations between magnitude, M, and probability, pr. This curve could be characterized as (Borel, 1992) $1/pr = [1+R_{tolerable}(1/M-1/M_{tolerable})]$. Figure 11.11 depicts the curve.

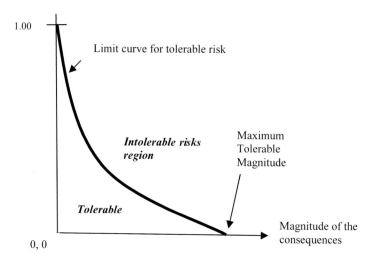

Figure 11.11 Limit Curve for Tolerable Risks, Complement of the Cumulative Distribution of the Magnitude of the Consequences

The curve depicted in Figure 11.11 identifies those combinations of probability and magnitude that either fall on the tolerable risk curve, above it or beneath it. Any combination that falls above the curve of tolerable risk is, by definition, not tolerable, although any specific combination of probability and magnitude may be tolerated on other grounds. Unlike the complements of the cumulative distribution discussed in the preceding paragraphs, the curve on Figure 11.1 is constrained to a specific shape.

The usefulness of this type of analysis, from a risk management standpoint, is that the complements of the cumulative distributions can also be compared, if the random variable, for example prompt illnesses, is used. Thus, accidents in petrochemical facilities compare with accidents in fertilizer plants, if the same consequences are common to both. Furthermore, these curves can be plotted against a curve that depicts the *limit of tolerability* of accidents that can result in catastrophic losses.

11.8. RISK CALCULATIONS IN ACCIDENT ANALYSIS

In this section we exemplify simple accident analysis based on travel. The calculations give an insight on how physical units are used to obtain simple representations of the effect of an adverse event, such as an accident.

1. Suppose that there are *5,000* prompt accidental deaths, per year due to car crashes in a country. Let the yearly total number of kilometers driven be $1.5*10^6$ in that country. Then:

    $5,000$[deaths/year]$/1.5*10^6$[km/year] $= 3.33*10^{-3}$[deaths/kilometer traveled].

2. Suppose that the average number of kilometers driven per day is *10* km and that the number of driving days per year is *300*. Then:

   $3.33*10^{-3}$[deaths/kilometer traveled]*10[km/day]*300[days/year] = *9.99* [deaths/year].

3. Suppose that the monetary loss associated with average death is *$100,000*. Then:

   9.99[deaths/year]*$100,000$[$/death] = *999,000* [$/year].

4. Suppose that there are *5,000* deaths per year due to car crashes and that the resident population is *20* million people. Then:

   5,000[deaths/year]/*20*[million residents] = $2.5*10^{-4}$[deaths/resident-year].

   If this number is multiplied by the expected remaining life, say *60* years from the *16th* birthday (the assumed legal age for driving), then:

   $2.5*10^{-4}$[deaths/resident-year]*80[years of expected lifetime at birth] = *0.020*[death/resident].

5. Calculate the reduction in life expectancy due to prompt death from a car crash in the country. Assuming that an age- and sex-specific life expectancy is *30* years, then:

   0.020[death/resident]*30[years] = *0.60*[death/resident-year].

6. Assume that there are *10,000* yearly deaths due to car accidents in country *XYZ*. Assume a general population *N* of *50,000,000* persons in that year. Assume that the average driver drives *10,000* km per year and that there is *1.00* [person/car-year]. A question is: What is the number of deaths per kilometer, in the general population of country *XYZ*, for that year? The answer is:

   10,000[cause-specific deaths/year]*$1/50,000,000$[1/N]*$1/1$[person/car-year]*$1/10,000$[car-year/kilometer driven] = $2*10^{-8}$[cause-specific deaths/kilometer-year driven].

7. Suppose that a city has a population *N* of *100,000* persons in a year and that a proposed activity can raise the individual background yearly cancer risk by $6.5*10^{-6}$ in that city. Then:

   Expected annual cases$_{city}$ = $(6.5*10^{-6})$*$100,000$[N/year/activity] = 0.65[N/year/activity].

8. Suppose that a district in that city has a population of *1,000* [N], that the individual increased yearly risk is $1.0*10^{-5}$ and that those persons are exposed for *15* years to the hazard. Then:

   Expected cases$_{district}$ = $1.0*10^{-5}$*1.000[N/year/activity]*15[years] = *0.15*[N/activity].

11.9. SIMPLE DYNAMIC MODEL OF INFECTION

Not all risk assessments are about cancer or toxic effects. Many risk assessors are concerned with predicting the spread of contagious diseases that are caused by bacteriological agents. This section deals with the way a risk assessor might practically model the spread of an infection in a population at risk. In this model, we use incidence rather than prevalence.

Suppose that the incidence rate for new cases of the disease in a period of time is $I(t)$, and that we want to calculate the time that it takes for the incidence to double, t_2. Let $I(t = 0)$ be the initial incidence and assume that the process of increase is exponential, at a rate r [new cases/year], then we can describe this process as:

$I(t) = I(t=0)exp(rt)$.

By substitution:

$2*I(t=0) = I(t=0)exp(rt_2)$.

Thus, taking the natural logs of both sides of the expression and simplifying, we obtain:

$t_{1/2} = ln(2)/r = 0.693/r$.

Let $r = 0.05$ [cases/year]. Therefore, doubling time is $0.693/0.05 = 13.86$ [years].

Suppose that the initial distribution of new cases of a contagious disease concerns the risk assessor. Let $I(t)$ be the *incident* cases in a population of size N. Let k [incident cases/year] be the rate at which the number of cases changes, assume that this rate is independent of individual behavior or other factors. We also assume that transmission is direct. Let k be the relative growth rate, $[dI(t)/dt]/I(t)$. In a period of time dt, $I(t)$ will cause $kI(t)$ new cases to occur. The fraction $I(t)/N$ of $kI(t)$ is already infected. Therefore the new cases in the period dt are:

$\{kI(t)-kI(t)[I(t)/N]\}$.

It follows that:

$d[I(t)]/dt = kI(t)-kI(t)[I(t)/N]$.

In the initial phase of the diffusion of the disease, $1-(1/N) \approx 1.00$: N is much larger than $I(t)$. Therefore:

$d[I(t)]/dt = kI(t)$.

This differential equation has the solution: $I(t) = I(t = 0)exp(kt)$. If we let $k = 2$ cases/year and the initial incidence be *five* cases we obtain:

$I(t = 1.5) = 5exp(2*1.5) = 100.4$ [cases],

at the end of *1.5* years. Note that this is a situation of continuous compounding. If k is not constant but varies inversely with time, $k = m/t$, the model becomes:

$dI(t)/dt = m[I(t)/t]$.

The solution of this differential equation is: $I(t) = I(t = 0)t^m$. The doubling time equals $\sqrt[m]{(2-1)}t$; the value of m depends on the disease and is determined empirically.

11.10. TOXICOLOGICAL MODELING FOR RISK ASSESSMENT: Key Organs, Physico-Chemical Reactions, Compartments and PB-PK Models

In this section, we discuss the role of two key organs, the liver and the kidneys, which are implicated in building mechanistic models of dose-response and in developing uptake formulae discussed earlier in this chapter. Important physiological and biochemical processes take place in these organs and therefore matter in determining the mass of a toxic agent reaching the target organ. Clearly, other organs, such as the lungs, are also important to understanding the mechanism of disease and death. The role of those organs is not discussed for the sake of brevity. We hope that, understanding the physiological processes described, gives a sense of the importance of mechanistic (biologically plausible) models in risk assessment.

11.10.1. The Liver

The liver is an important organ because it is a chemical filter for blood. The relationship between total flow and cleared flow containing the untransformed chemical through an organ can quantify the efficiency of chemical removal from the liver. Flow (replacing plasma for blood in some of the discussion) in the liver system consists of:

Plasma (portal vein) → *Liver* → *Fraction not metabolized or extracted entering systemic blood flow = (1-Fraction extracted and metabolized).*

Some of the key concepts governing the functions of the liver are *extraction* and *clearance* rates. The extraction rate, using concentrations in that organ, is:

Extraction Ratio = (Concentration out of the Liver)/(Concentration into the Liver).

Specifically, the liver extraction ratio is:

[(Unbound Fraction of the Toxic)(Liver Clearance)]/[(Blood Flow Rate)(Unbound Fraction)(Clearance)].

Clearance rate is:

Clearance Rate = (Blood Flow into the Liver)(Extraction Rate).

This expression applies if the toxic agent does not impede metabolic processes in the liver and does not bind to proteins. Theoretical (*intrinsic*) liver clearance is calculated as:

$$Clearance_{theoretical} = (Maximum\ Enzyme\ Reaction\ Rate)/(Disassociation\ Constant) = V_{max}/k_{max}.$$

The coefficient k_{max} is low when the chemical toxin is tightly bound to a protein. The extraction ratio falls in the interval (*0, 1*). *1* means extremely high clearance and *0* means extremely low clearance ratio. Clearance values range from *1,500* [milliliters/minute] to *0*: the higher numbers indicate high clearance rates. The clearance process for the liver consists of the inputs and outputs depicted in Figure 11.12.

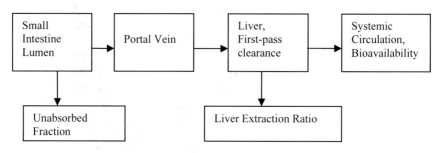

Figure 11.12. Liver Clearance Process

11.10.2. The Kidneys

The process of kidney clearance can be simplified to three steps, after the arrival of bound proteins or soluble chemicals in the blood to this organ:

Filtration (unbound chemical in plasma) → *Active secretion mechanisms (chemical-dependent)* → *Passive fraction of filtrated plasma that is reabsorbed (through kidney tubules).*

Urine (an excretion pathway) is formed in these steps. Kidney clearance is:

Kidney Clearance = (Fraction of Chemical Unbound in Plasma)[(Glomerular Filtration Rate)(Secretion Clearance)][1-(Fraction of Chemical Reabsorbed through Kidney's Tubules)].

The kidneys excrete several diverse substances, such as polar toxicants as well as hydrophilic metabolic byproducts. The basic processes controlling clearance by the kidney, which occur in the nephron, are depicted in Figure 11.13.

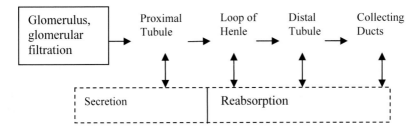

Figure 11.13. Kidney Clearance Processes

Filtration, a molecular size-limited process, is a passive process characterized by a rate of about *180* liters of plasma/day, for the average non-diseased human (Hodgson and Levi, 1987). Reabsorption by the proximal tubules is about *75%* of the filtrated material, in which the soluble chemical is necessary for the proper functioning of clearance. Tubules can secrete passively and actively; secretion is generally a function of urine *pH*. Polar chemicals concentrate from the plasma into the tubules, are not reabsorbed and are thus rapidly secreted. Toxicant-protein complexes that yield free toxicants in plasma are secreted by the tubules and become disassociated to keep the free and bound toxicants in equilibrium. This process is governed by solubility, *pH* and by the disassociation constant, *pK$_a$* (the negative of the logarithm of the disassociation constant). The loop of Henle regulates osmolal reabsorption, while distal tubules reabsorb water and ions.

Liver and kidney are physiological compartments that, coupled with blood vessels, lungs, soft tissue and other important organs, can be modeled to predict the fate and chemical changes that a chemical can undergo, after ingestion, inhalation or other. We turn to compartmental modeling next.

11.10.3. Compartmental Models

The *compartment* model is a toxicological model that makes no assumptions about a specific physiological model of an animal or human being. Such a model deals with the reactions that occur in idealized physiological systems in which compartments are chemically homogeneous units. Specifically, a compartment model is a physicochemical lumped parameter system, without the details of blood flow and diffusion. A simple compartment model, in which a single dose is input through intravenous injection, is depicted in Figure 11.14.

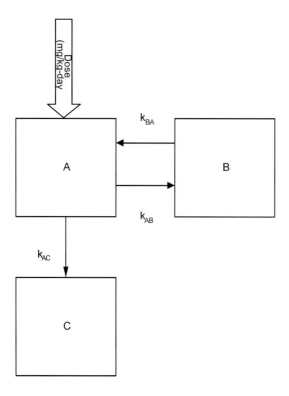

Figure 11.14. Two-compartment Model with First Order Elimination Rate

The basis of compartment models is:

(free chemical)+(chemical receptor)↔(chemical-receptor)→(adverse response).

Compartment modeling (Boroujerdi, 2002) uses the following assumptions. Only the free chemical interacts with a receptor to lead to an adverse response; the concentration of the free chemical in circulation in the blood is proportional to the free chemical at the receptor. Accordingly, the first order chemical reaction of concentration, C, is modeled as a differential equation (the symbol ∝ means *proportional*):

$dC/dt \propto C,$

so that:

$dC/dt = \pm k_1 C.$

The zero order reaction is modeled as:

$dC/dt = k_0.$

Boroujerdi (2002) gives the details of these models. The first order reaction model has solution:

$C(t) = C(t=0)exp(-k_1 t)$,

in which $C(t = 0)$ is the initial concentration of the chemical. The fraction of the concentration of the chemical in circulation in the body at some time t is:

$C(t)_{body}/C(t=0) = exp(-k_1 t)$.

Figure 11.15 depicts the relationships between concentration, C, and time, t, for first and zero order reactions.

Concentration

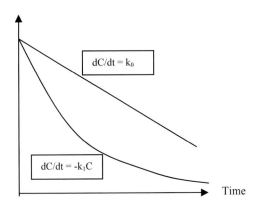

Figure 11.15. Solutions to the First and Zero-order Reactions Differential Equations

The elimination of chemical XYZ from the blood compartment via an excretion and metabolic routes is modeled by the linear differential equation $dM/dt = -k_{tot}M$, in which M is mass, $k_{tot} = k_1 + k_2$. Then, the half-time, $t_{1/2}$, for total elimination is calculated from $M(t) = M(t = 0)exp(-kt)$, which results in $t_{1/2} = 0.693/k_{tot}$.

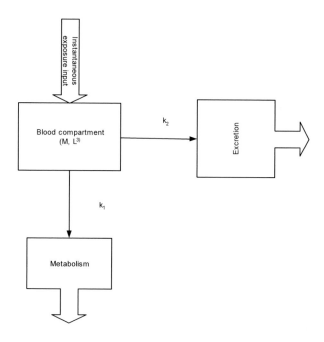

Figure 11.16. Compartment Model, Instantaneous Exposure

Example 11.23. Suppose that a single exposure to chemical XYZ occurs at *1200* hours and that at *1800* hours there are *10* μg_{XYX}/milliliter of blood [M/L^3]. Suppose also that XYZ has a half-life of *8* hours. Then:

$k_{total} = 0.693/8 = 0.087$[hours^{-1}].

The concentration in blood at time *0* is:

$(10$[μg_{XYX}/milliliter of blood]$)*1/exp[(-0.087(8)] = 20.06$ [μg_{XYX}/milliliter of blood].

At 2400 hours the concentration in blood is:

$C(t = 2400-1200) = 20.06*exp[-0.087(12)] = 5.865*10^{-4}$ [μg_{XYX}/milliliter of blood].

The fraction of chemical XYZ after five hours from initial exposure is $exp[-0.087(5)] = 0.65$.

11.10.4. Physiologically-Based Pharmaco-Kinetic Models (PB-PK)

In risk assessments that use toxicological data one often finds pharmacokinetic-physiologically based models (PB-PK). PB-PK models describe the distribution of a chemical in the animal or human body. PB-PK models are important to risk assessment because they simulate the changes of a chemical toxicant and predict the ultimate mass of the chemical reaching a particular organ. In these models, the exchange of chemicals is accounted by such model parameters as blood flow rates, organ volumes and blood flows. The values of the PB-PK parameters are modeled theoretically and fit to experimental data (Spear and Bois, 1994; Cox and Ricci,

1992). Figure 11.17 depicts some of the relations that can be described using PB-PK models (adapted from Spear et al., 1991 and from Spear and Boys, 1994).

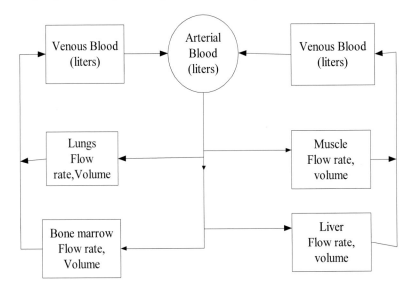

Figure 11.17. Physiologically-Based Pharmaco-Kinetic Model

An example of chemical mass balance for a PB-PK model is:

[Δ(chemical in blood)/Δt] = [Δ(input quantity from organs or tissue into blood)/Δt]–[Δ(output quantity from organs or tissue)/Δt]+dose(t).

The symbol Δ represents a (discrete) change. For continuous change, Δt is replaced by dt, and thus the mathematical model, a differential equation, for a physiological compartments that is flow-limited, is:

$Volume_{blood}[dC_{blood}/dt] = [Q_{liver}C_{liver}/R_{liver}+....]-[Q_{liver}+....]C_{blood}+d(t)$.

In this equation, R is dimensionless. The units for the terms in the last equation are: $[L^3/M/L^3]*[1/T] = [M/T] = [L^3/T]*[M/L^3]$. Different aspects of the physiological model require additional mass balances and differential equations. For example, the amount of chemical contaminant in an organ or tissue is:

$\Delta[(chemical\ in\ organ\ or\ tissue)/\Delta t] = [\Delta(input\ quantity\ into\ organ\ or\ tissue)/\Delta t]-[\Delta(output\ quantity\ from\ organs\ or\ tissue)/\Delta t]$.

The continuous form of this model can be written as:

$Volume_{organ}[dC_{organ}/dt] = [Q_{organ}-C_{organ}/R_{organ}]$.

The details can be found in Boroujerdi (2002). In developing PB-PK models it should be noted that some organs are richly perfused by blood and other are poorly perfused. Similarly, fat is an important compartment for storage of lipophilic toxicants.

An example of the values of coefficients for two species and their dimensions used in modeling PB-PK are show in Table (Andersen et al., 1986):

Table 11.8. Example of Coefficients' Values Used in PB-PK Models

	Average, Rat	Average, Human
Body Weight (kg)	0.25	70
Liver (% Body Weight)	4.00	3.14
Fat	7.00	23.10
Cardiac output (L/hr)	5.10	348.00
Liver/Blood partition coefficient	0.732	1.46
Lung/Blood	6.19	12.40

As stated, compartmental and PB-PK toxicological models are deterministic. Spear et al. (1991) and Cox and Ricci (1992) discuss the variability of the parameters (the coefficients) of the deterministic differential equations that mathematically describe PB-PK models. The methods include Monte Carlo simulations and CART, discussed in Chapters 4 and 6 of this book.

11.11. CONCLUSION

The ASTM and PEA methods are essentially deterministic frameworks for risk assessment. Data on water, food, and other intakes of contaminants for children and others are variable and that variability can, however, be formally accounted for using Monte Carlo simulations or other methods. These frameworks, including the US EPA's, can and do include that possibility. The US EPA has provided statistical guidance for risk assessments, as discussed throughout this textbook. The reason for using probabilistic methods is that they add confidence in the output of risk assessment and thus lead to more confident decision-making. The results of risk assessments based on the type of frameworks discussed in this chapter develop relative comparisons of risk, but not necessarily absolute comparisons. They can provide bounds on the variability of the results but not on uncertainty.

This chapter also includes some commonly encountered formulae and models used for practical risk assessment and completes the discussion of toxicological models. It should be clear that the input data into those formulae are determined with the methods discussed in Chapters 4-11. The methods for PRA are important and can be used not only in the analysis of catastrophic events but also for routine ones. Fault-tree methods describe the physical chain of events and model the probabilities leading to the *top event*. This is the event from which the adverse consequences to humans or the environment emanate. The event-tree represents the ramifications of the consequences and their magnitudes. Their combination is important for making comparisons of risky activities as well as in understanding how risky outcomes can happen. The combination of probabilistic analysis and mechanistic portrayal of the linkages leading to an event likely to cause damage and the consequences of such is a necessary step for sound risk management.

Pharmaco-kinetic models caps the discussions of fundamental mechanisms that are relevant to causal reasoning in health risk assessment. In particular, we attempt to clarify the difference between *exposure*, as used in the frameworks and formulae, and *dose* to a particular organ or tissue. The importance of these models is that they provide a much more accurate and defensible input into a damage function, than exposure alone. A difficulty is that the data used in these models may not be available for the chemical of interest to the risk assessor. Finally, we include a simple dynamic and deterministic model of infections as a further example of models that can be brought to bear on health risk assessment.

QUESTIONS

1. How would you calculate the RBSL for inhalation by residents near a site emitting benzene, using acceptable lifetime individual risk levels of 1/10,000 and 1/100,000? What is the implication of these calculations for risk management? How would you choose the appropriate dose-response model? Keep your answer to less than 400 words.

2. How would you develop a simple probabilistic model rather than using pre-established values, to describe the variability in the RBSL's calculated using the ASTM formulae?

3. The following table that depicts the number of accidents associated with Systems A and B:

Time (year)	# of accidents, System A	# of accidents, System B
1995	100	40
1996	200	90
1997	100	60
1998	80	10

Develop the cumulative frequency distributions for *A* and *B*. Plot these distributions. What qualitative conclusions can you make based on the plots? Would you be willing to extrapolate the number of accidents for these two systems to year 2005. Give a simple idea of how you would characterize the *uncertainty* in the extrapolation. How would you characterize the variability about your forecast to the year 2005?

4. Draw two risk curves (cumulative complements of the distributions for *X* and *Y*) in which System X is only *partially* less risky than System Y. How would you practically show the *uncertainty* about each risk curve? How would you make a choice of either *X* or *Y*, when these two systems are only partially superior to one another? Explain using graphs.

5. What is the key implication of the exponential model of the diffusion of the disease? Should there be a term in this model that accounts for saturation

and for decline of the incidence numbers? If so, how would you include it in the model?

6. Suppose that r = 1. Calculate the negative binomial probability of the random variable X for pr = 0.5 and n = 10.

7. Calculate the sample mean, median, mode and the sample CV (CV = mean/variance) for the data (in ppm): 10, 15, 20, 35, 50, 65, 120, 130, 145, 155, 1,235, and 1,800. Suppose that the value 1,800 is omitted because it is due to a transcription error and should be replaced with 180. Calculate the sample mean, standard deviation and the coefficient of variation. What conclusion do you draw from comparing the CV of these two samples (the sample that includes the value 1,800 and the sample that includes the value 180)? Suppose that the value 1,800 is not an error. How would you justify omitting it? Discuss your reasons.

CHAPTER 12.

PRACTICAL ANALYSIS OF DECISIONS FOR RISK MANAGEMENT

Central objective: to describe and exemplify how public and private decisions can be evaluated using decision trees, utility and game theory

This chapter provides a review and examples of practical methods used for modeling the outcomes of risky choices, given a set of potential choices and decision criteria, while accounting for the role of chance. Those methods apply to making public or private decisions that involve either one or more stakeholders. The central point made in this chapter is that formal justifications for risky decisions are generally preferable to informal ones because we can use them for predictions, are defensible in administrative or other proceedings, and provide a complete and transparent accounting of the assumptions made in selecting a choice from the set of choices. More specifically, environmental and health decisions involve studying the outcomes from alternatives and the selection of one choice from many, based on complex facts, uncertainty, variability and causal arguments. Assessing all of these choices, and selecting the preferred (or optimal) one, requires analytical approaches that should be theoretically sound, transparent and accessible. Nonetheless, formal justifications do not stand alone in making decisions. They enhance decision-making by supplementing, but not substituting, policy, political and legal factors. Within this context, this chapter focuses on:

- Decision-trees and tables applied to making risky decisions, from the perspective of the single decision maker
- Aspects of single and multi-attribute utility theory
- Aspects of game theory for two or more decision-makers who confront risky decisions
- Criteria for making justifiable decisions, including the expected value of information, which assist in the decision of when to stop gathering costly information

The private decision-maker's objective is to maximize, at least in the long-run, her expected net profits.[1] In this sense, her objective is relatively unambiguous. On the other hand, a public manager can have vague objectives, such as safeguarding health and welfare of the public, which can arise under precautionary precepts. Making decisions about risky choices involves either a single decision-maker or several decision-makers who may have different degrees of beliefs in the outcomes. Their

---

[1] In the short-term, she may operate at a loss to stay in business.

choices should therefore account for the role of *chance* (the *states-of-nature*) that can foil a decision-maker's attempts to reaching her objectives.[2] The actions that a decision-maker considers are, at least in principle, under her control; however, the *states-of-nature* are not. Therefore, a structured probabilistic approach becomes necessary for representing the outcomes of each choice and in justifying the selection and implementation of one of them. The essential elements of this chapter link as follows.

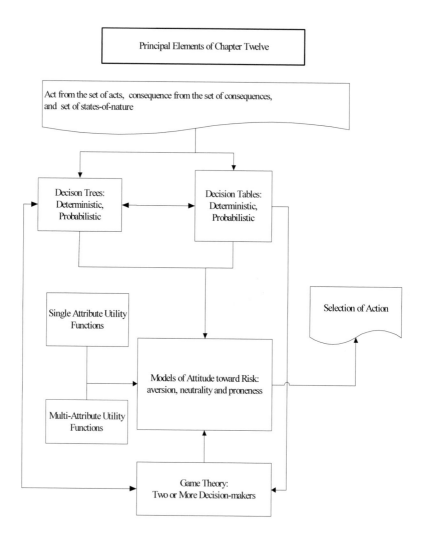

---

[2] In this chapter, we only deal with the role of chance (nature) and random events. Naming *nature* for chance follows tradition.

The methods of decision analysis formulate the ways through which the alternatives open to the decision makers are linked to the pay-offs likely to result from each choice, accounting for uncertainty. Each pay-off is either an expected value or some other value such as expected utility. The final managerial choice is justified through a decision rule. For example, we can use Bayesian methods to calculate the expected value of losses and gains under a set of options, accounting for the *loss* associated with making the incorrect choice. Additionally, *value of information* formulae help to determine when to gather costly information and when to stop gathering that information. We can also ignore uncertainty, but we do so at some cost.

12.1. INTRODUCTION

Our decisional framework is one in which the decision-maker can choose from several courses of action (her set of acts) that result in potential outcomes or consequences, the *pay-offs*. However, *chance* can interfere with this determinism: the states-of-nature represent how uncertainty can foil her plans. The reasoning is as follows. If there were no uncertainty, action a_1 would surely lead to outcome o_1. However, chance interferes by allowing the possibility that a_1 can either lead to o_1 with pr_1 or to o_2, with pr_2 but not both. The basic relations affecting a risky choice are:

Act, from the set of acts $\rightarrow$ *Consequence, from the set of consequences* $\leftarrow$ *States-of-nature.*

Practically, the first step in decision-making consists of developing the complete, fully exhaustive, enumeration of the potential actions associated with each choice. The second step is the selection of methods for analysis and the criteria to make a choice. The third is the causal step. The fourth step consists of accounting for uncertainty and variability. The fifth and final step is the assessment of the results and the discussion of the issue that remain unresolved. These steps require understanding the scientific basis for the causal links between hazards, pollution emissions, exposure and response. As Anand (2002) states, the *simple, general message is that decision-making without scientific evidence, or well-defined probabilities, is far from unusual.* Yet, it does not have to be so. To see why, let us develop some of the determinants of a process for structuring risky decisions, which leads to either reducing or eliminating hazardous conditions. The determinants include considering:

- Who evaluates the potential outcomes of the risk management plan?
- Who makes decisions?
- Who is accountable throughout the life of the plan and its implementation?
- What are the objectives of the plan?
- Who is at risk, who benefits, and who pays?
- How is the choice of actions implemented, by whom, and how will its outcomes be monitored?
- Who will enforce an implementation or punish the lack of implementation?
- What system of feedbacks will be put in place to inform the stakeholders?

More specifically, the process for making risky decisions consists of three components, linked as shown in Figure 12.1; these are:

1. *Acts*, from the set of acts, as discrete and separable actions or options available to the decision-maker,
2. *Pay-offs* or other *consequences*, from the set of pay-offs or the set of consequences, modeled using damage, utility and value functions,
3. *States-of-nature*, from the discrete and fully enumerated set of states-of-nature, representing the effect of chance.

In Figure 12.1, a decision rule may be a choice of the act with the largest positive expected value of the net benefits, or the *minimax* value, or the *maximin* value or an act established by some other formal decision rule. We discuss these decision rules in the next subsection.

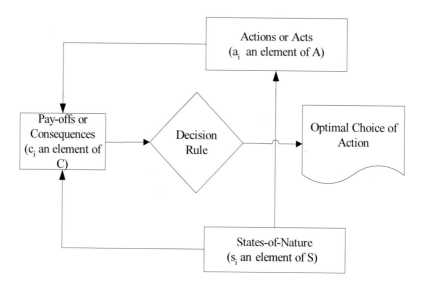

Figure 12.1. Essential Components of Risky Decision-Making

This diagram is useful in the contexts of the sequence of the following steps:

1. Define the objective of the initial decision,
2. Define the mutually exclusive and fully exhaustive alternatives that satisfy that objective,
3. Define the consequences associated with each alternative,
4. Use a decision-theoretic method to select the optimal solution, given the optimization rules inherent to the method adopted.

Stakeholders can deal with changes or violations of a socially desirable risk minimization plan through:

- Litigation under constitutional (e.g., taking without just compensation), tort (e.g., negligence law), environmental (e.g., appropriate section of the Clean Water Act), administrative (e.g., appropriate section of the Administrative Procedures Act) law
- The enactment of new or amended regulation (e.g., standards, guidelines, economic instruments such as emission banking)
- Self-policing and self-insuring
- Private, public or mixed forms of prospective and retrospective compensation to those at risk or affected
- Extra-judicial dispute resolution methods and contracts

These possibilities may not be mutually exclusive and can be used as a portfolio of alternatives. Recollecting Anand's comment, principled decision-making that recognizes and accounts for uncertainty and variability leads to more informed and accurate decision-making. Practically, it is useful to consider the following aspects of decision-making:

- Decisions made with deterministic numbers and causal linkages between the inputs and the outputs (complete certainty)
- Decisions made with variability in some or all of the inputs and linkages
- Decisions made with uncertainty

12.2. REPRESENTING RISKY CHOICES

A useful way to represent the relationships of interest to the decision-maker (called the *normal form*) consists of a payoff matrix in which $c_i \in C$, (portraying the net gains or losses associated with each action, including the *do-nothing* action) for the fully exhaustive and mutually exclusive collection of available actions, $a_i \in A$, and states-of-nature, $s_i \in S$.[3] Table 12.1 depicts the central idea of this construct deterministically, with losses associated with three actions being dependent on the states-of-nature.

Table 12.1. Deterministic Decision Table in Normal Form

	States-of-Nature (losses in dollars):			
Actions:	s_1	s_2	s_3	**Minimum Loss**
No action, a_1	$c_{11} = 0$	$c_{12} = 2$	$c_{13} = 5$	0
Low Intervention, a_2	$c_{21} = 10$	$c_{22} = 4$	$c_{21} = 3$	3
High Intervention, a_3	$c_{31} = 20$	$c_{32} = 40$	$c_{33} = 60$	20

The decision-maker must also have a criterion for choosing between the pay-offs. A criterion for choice is to minimize the maximum loss, the *minimax* criterion. Applying this criterion to the data in Table 12.1 indicates that action a_1 is minimax. Therefore, this is the action that should be taken, barring considerations external to this analysis. We add uncertainty or variability (as is the case) using probabilities as follows. Suppose that the states-of-nature are as described in Table 12.2. In this

---

[3] The formal discussion of the differences and methods for analysis can be found in Cox (2001) and were introduced in earlier chapters of this textbook.

Table, the intersection of x_1 and s_1 is the probability of obtaining observation x_1 and state-of-nature s_1; that is, $pr(x_1 \text{ AND } s_1) = 0.25$.

Table 12.2. Probabilistic Decision Table

	pr(X = xᵢ *AND* S = sᵢ)		
	States-of-Nature		
Observations	s_1	s_2	s_3
x_1	0.25	0.41	0.50
x_2	0.50	0.42	0.30
x_3	0.25	0.17	0.20
Probability	1.00	1.00	1.00

From the data in Table 12.1 and 12.2, the expected value of the losses is calculated and shown in Table 12.3.

Table 12.3. Decision Table and Expected Value of the Losses, Given the States-of-Nature

	E(Loss), l(sᵢ, aᵢ), for three states-of-nature		
Actions	s_1	s_2	s_3
a_1	0.25*0 = 0	0.41*2 = 0.82	0.50*5 = 2.5
a_2	0.50*10 = 5.0	0.42*4 = 1.68	0.30*3 = 0.90
a_3	0.25*20 = 5.00	0.17*40 = 6.8	0.20*60 = 12.0

Other tables can yield further insights into making decisions. For example, the *regret* table depicts the *regret* of *not having guessed the right state-of-nature*. The *regret* values in the cells of a regret matrix are calculated (in dollars) as [*loss (aᵢ, sⱼ)- (minimum loss value of sⱼ)*]. Table 12.4 depicts the regrets.

Table 12.4. Decision Table and Regrets

	S₁	S₂	S₃	Regret calculation for the states-of-nature: [*loss(aᵢ, sⱼ)-(minimum loss value of sⱼ)*]
No intervention, a₁	0	2	2.5	$s_1 \rightarrow 0\text{-}0 = 0.0$
Low Intervention, a₂	5	1.68	0.9	$s_2 \rightarrow 2.1\text{-}0 = 2.1$
High Intervention, a₃	5	68	12	$s_3 \rightarrow 5\text{-}0 = 5.0$

If expected values are used in the loss and pay-off tables, *risk* is defined as (Chernoff and Moses, 1986) $E[loss\ (a_i,\ s_j)]\text{-}E(minimum\ loss\ value\ of\ s_j)$, where $E(.)$ symbolizes the expected value. Subjective probabilities can be used to calculate the *Bayesian Expected Loss*, BEL, for a given state-of-nature, as $BEL(s_j) = \sum (pr_j)[L(s_j, a_i)]$.

12.2.1. Discussion

On these considerations, we can now increase the elements of a decision in risk management to include (Granger-Morgan and Henrion, 1990; Kleindorfer, Kunreuther and Schoemaker, 1993) the:

1. Set of states-of-nature, $\{s \in S\}$,
2. Set of (non-overlapping, complete and exhaustive) potential decisions, $\{d \in D\}$, or acts, $\{a \in A\}$,
3. Set of prior information, $\{x \in X\}$ and the prior distribution function $f(x)$ when using Bayesian methods,
4. Loss function, $L(d, x)$,
5. Bayesian decision rule, using the Bayesian Expected Loss, which consists of the minimization of the expected loss, $Min\{E[L(d, x)]\}$, for all $\{d \in D\}$. The optimal Bayesian choice minimizes the Bayesian expected loss,
6. Representation of the problem in a table (the normal form) or as a decision tree (the extended form), and
7. Calculations and conclusions.

$E[L(d, x)]$ is the expectation of the loss function for the random variable X with continuous density function $f(x)$: $E[L(d, x)] = \int [L(d, x)]f(x)dx$. If the random variable has a discrete distribution, then:

$E[L(d, x)] = \sum_I pr_i(x_i)[L(d, x)]$.

Many practical applications of these concepts use discrete distributions. The analyst must make sure that the approximations made using discrete representations are consistent with the goals and objectives of the decision-maker. Anand (2002) has developed the theory of *decision-making under ambiguity*. The new concept is that individuals are sensitive to ambiguity. Because of ambiguity, according to Anand, individual behavior does not conform to the axioms of subjective probability. He proposes the following non-probabilistic representation, in which there are three actions and three states, as shown in Table 12.5.

Table 12.5. Decision-Making under Ambiguity

States	Actions		
	Ban	*Restrict*	*Do nothing*
High Risk	0.9	0.3	0
Low Risk	0.6	0.7	0.4
Safe	0	0.3	1.0
Maximum	0.9	0.7	1.0
Minimum	0	0.3	0

In Table 12.5, *1* is the best outcome and *0* is the worst. The maximum and minimum determine the *maximin* (restrict, *0.7*) and *maximax* (do nothing, *1.0*).

12.3. THE PRACTICE OF MAKING RISKY DECISIONS

A preferable act, given the set of acts and the selected criterion for choice, is not only optimal but also *dominant*. Forms of dominance include deterministic dominance, single outcome with either strong stochastic dominance or weak stochastic dominance, multiple outcomes with temporal dominance, multiple outcomes with strong stochastic dominance, weak stochastic dominance, and cardinal dominance. The methods for reducing the size of the decision analysis problem include conjoint and disjoint decision rules, elimination by *aspects*, and linear (additive) difference rules (Kleindorfer, Kunreuther, and Schoemaker, 1993; Cox, 2002).

Example 12.1. Following Wilson and Crouch (1987) consider an agricultural community of 100 individuals who have to decide whether or not to use a known cancer-causing pesticide that is also known to double their agricultural yields, per year over 20 years. Suppose that the use of the pesticide can cause 1.00 expected case of cancer per 100 exposed. Thus, the balancing of the risk with the benefits is one cancer for a doubling of the yield. Production yields one-time benefits of *$5,000,000* and a *1%* risk of cancer. Each cancer is valued at *$1* million. The expected number of cancers is *20*:

A	Benefits	Risks	Time Periods	Action
1	Status quo	Status quo	Status quo	Do nothing
2	Single (one year) increased net revenues by $5,000,000 through increased production	1% risk of 2000 excess cancers, over 30 years = 20 excess cancers	30 years	All output is consumed in the first year, but the risks are distributed over thirty years
3	$5,000,000 is invested at 10%, yielding at the end of the 30 years $87 millions	At year 30, 87- (0.01*2000) = 67 expected lives saved (at $1 million/life)	30 years	Invest $5,000,000 (therefore do not produce at all)

Over *20* years, the total number exposed is *20*[years]***100*[individuals/year] = *2,000* [individuals]. Using the present discounted value, actions *1* and *2*, with the expected value of the number of cancers equal to *(0.01)(2000)* = *20*, with the convention that each cancer is valued at one million dollars, the net present value, *PDV*, discounted over *30* years, is *$1.15* million. Because the value of the benefits is *$5.0* millions (in year one), the net benefits are much larger than the costs. It follows that action *2* should be undertaken.

Figure 12.2 depicts two risk levels (e. g., zero risk and controlled risk at $1.0*10^{-6}$ additional lifetime cancer risk).

Probability density, f(X)

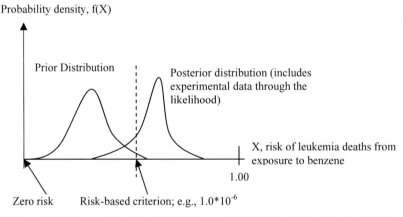

Figure 12.2. Prior and Posterior Distributions with Risk Threshold of One in a Million

Risk-based cut-off criteria, such as *zero risk* or the *trivial (de minimis)* risk, can be used in environmental decision-making and policy to control hazards at a given level of individual (or aggregate) risk, such as $1.0*10^{-6}$ per expected lifetime individual cancer risk. Other risk-based methods use probability distributions to decide on a tolerable risk cut-off number. In Figure 12.2, the location of the 10^{-6} risk value depends on experimental information. Using Bayes' theorem, we show a situation in which the posterior distribution shifts to the right of the prior distribution.

12.3.1. Single Attribute Utility Theory

In our discussion, we assume that the actions of the stakeholders are rational (meet von Neumann--Morgenstern utility axioms) and as the individual members of society prefer one outcome relative to another, so does society. Table 12.6 depicts the function $u(x)$ that describes increasing or decreasing *utility* for a single attribute, x, which is the (single) attribute relevant to the decision.

Table 12.6. Utility Functions for Risk Neutrality, Aversion and Proneness (Risk Seeking Behavior) for an Attribute that Is *Increasing* in Utility

Risk Attitude	Definition	Utility Function (increasing utility)	Comments
Aversion	The expected value of a non-degenerate lottery is preferred to that lottery	$u(x) = a_1 + b_1[-\exp(-c_1 x)]$, b_1 and $c_1 > 0$.	1) At least two consequences per lottery. 2) No division by 0. 3) exp = e. 4) If the coefficient d equals -1, the utility function is decreasing, if it is equal to 1, then it is increasing.
Neutrality	The expected value of a non-degenerate lottery is indifferent to that lottery	$u(x) = a_2 + b_2(c_2 x)$.	
Proneness (risk seeking)	The expected value of a non-degenerate lottery is less preferred to that lottery	$u(x) = a_3 + b_3[\exp(c_3 x)]$, where b_3 and $c_3 > 0$.	

Figure 12.3 depicts these three single-attribute utility functions.

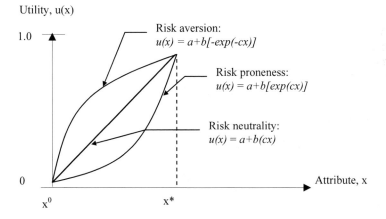

Utility, u(x)

1.0

Risk aversion:
$u(x) = a + b[-\exp(-cx)]$

Risk proneness:
$u(x) = a + b[\exp(cx)]$

Risk neutrality:
$u(x) = a + b(cx)$

0   Attribute, x

x^0 x^*

Figure 12.3. Shape of Single Attribute Utility Functions for Risk Neutrality, Aversion and Proneness for *Increasing* Utility

For example, the attribute *cost* shows decreasing utility because a large cost is less preferred to a small cost. The coefficients a_i and b_i are scaling coefficients so that

$u(x)$ is bounded between 0 and 1. For the attitude towards risk described, the values of x are ordered, from low to high (meaning that x_0 is of lesser magnitude than x_1 and so on). If the attribute has *decreasing* utility, the functions begin at *1.0* and decrease as the magnitude of the attribute increases. Utility is bounded between *0* and *1*; that is $(0 < u(x) < 1)$ and the range of x is between x^0 and x^*. The linkages between attitude toward risk and a lottery are described in Table 12.7.

Table 12.7. Risk Attitudes, Decision Rules and Utility Functions (with decreasing utility)

Attitude	Lottery	Expected Value	Decision	Utility Function
Risk Averter	p, C_1, $1-p$, C_2 $= L_1$	$E(L_1)$ $= pr(C_1)+(1-pr)(C_2)$	Expected value is preferred to the lottery	$u(x)$
Risk Neutral	p, C_1, $1-p$, C_2 $= L_2$	$E(L_2)$ $= pr(C_1)+(1-pr)(C_2)$	Indifference between the lottery and the expected value	$u(x)$
Risk Prone (or seeker)	p, C_1, $1-p$, C_2 $=L_3$	$E(L_3)$ $= pr(C_1)+(1-pr)(C_2)$	Lottery is preferred to the expected value	$u(x)$

Risk *neutrality* assumes that a decision-maker is indifferent between solutions that have the same *expected* value. The solution is formulated by considering the expected value. For example, if the expected values are the same and if a decision-maker is risk neutral, he should be indifferent between two actions that have the same expected value. Alternatives attitudes to risk neutrality are *risk aversion* and *risk seeking behavior*, described in Tables 12.6 and 12.7. Cox (2002) discusses and exemplifies many of the issues with this approach concluding that:

"[d]espite (its) limitations, social utility theory makes a useful starting point for analyzing societal risk management decisions, just as (expected utility) theory makes a useful starting point and baseline for discussions of individual decision-making."

Example 12.2. Suppose that two alternatives have identical cost, but one produces *$100* in benefits while the other either generates *$0* or *$200*, with equal probabilities. The attribute is measured in dollars.

The expected value of the lottery, under the second alternative, is: $E(benefit) = (0.50)(0)+(0.050)(200) = \$100$. If the decision-maker prefers the certain outcome to the lottery, he may be willing to accept a lesser benefit than *$100 for sure*, rather than having to face the lottery at all.

Example 12.3. Consider the utility function: $u(x) = 1-exp(-x/k)$, in which $u(x)$, $x >> 0$. Let $k = 0.5$, $u(x) = 1-exp(-x/0.50)$; the results are:

u(X)	X
0.18	0.01
0.86	0.10
0.99	0.50
0.9999	0.80

The assessment of a single decision-maker's attitude towards risk is exemplified through choices consisting of (non-degenerate) lotteries and the certainty equivalents, as follows. For the single attribute x, $(x_0 < x < x_2)$, let (x_0, x_2) be the minimum and the maximum values of that attribute. The *equi-probable* lottery $[0.5(x_0)+0.5(x_2)]$ means that the decision-maker faces either a 50% chance of getting either her best outcome or a *50%* chance of getting the least preferred outcome, but not both. The certain value is calculated as x_1, which is the point where the decision-maker states that she is *indifferent* between the value of the equiprobable lottery and that certain value, namely x_1. This point is called the *certainty equivalent* (CI) for that lottery. For example if the certain value is *$4,000* and the lottery has the expected value *$6,000*, then *$4,000* is the certainty equivalent number, if that is what the decision-maker asserts. The process is repeated several times (and is idealized to a continuous function labeled *risk aversion* in Figure 12.2). If the single decision-maker preferred the certain outcome to the value of the lottery (which is measured by the expected value of the lottery) she would be *risk averse* (Keeney, 1980; Keeney and Smith, 1982). Figure 12.4 depicts a single attribute utility function for income in which CE is the certainty equivalent.

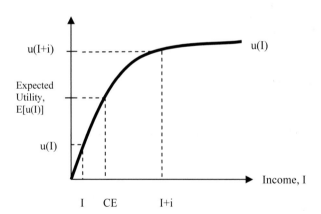

Figure 12.4. Single Attribute Utility Function for Income, U(I).

The remaining steps include:

- An assessment of the rankings obtained by the utility functions
- A discussion of the rationale for substantial differences

- An assessment of the estimated attitude for risk discovered
- A resolution or reconciliation of the differences

Example 12.4. The function $u(I)$ shows the diminishing marginal utility of income, I: as income increases it provides less utility. Although some can disagree, this seemingly paradoxical suggestion has good theoretical grounding. The expected utility of the annual income, $E[u(I)]$, is $pr[u(I)]+\{1-pr[u(I+i)]\}$. The expected income is $[pr(I)+(1-pr)(I+i)]$. If $I = 0$ or $I = i$, and each of these two outcomes has equal probability, then $[(0.50)(0)+(0.50)(i)] = i/2$.

In theory, what matters is the utility of income, *not* income itself. The certainty equivalent of income, for a given utility function, is the income that corresponds to the expected utility of income. The function is nonlinear and concave from below. Therefore, the certainty equivalent will be less than the utility associated with the average (or expected) income.

When the function is linear, certainty equivalent and expected income would be the same. Expected utility, $E[u(a_r)]$, of an action is calculated as:

$$E[u(a_{r=i})] = \Sigma_i[pr(a_{r=i})u(x_{r=i,c})].$$

The notation $r = i$ means that the *i*th row remains constant, but the summation is over all of the consequences, subscript c, corresponding to the *i*th action. The decision rule is to choose the action that yields the maximum expected value of the net benefits.

Example 12.5. The mechanism for developing the data to which a utility function is fit is as follows. Assume that the decision-maker is facing a loss. Define the extrema of the utility function: stated as percentages, the minimum and maximum of the attribute's *loss in percentage units* are *0%* loss and *100%* loss.

Utility is measured on the axis $(0 < u(x) < 1)$. It naturally follows that the $u(x = 100\%) = 0$ and that $u(x = 0\%) = 1$. Let the sure loss equal *80%*, and let the lottery to which the decision-maker be indifferent be $[0.5*u(x)+(1 -0.5)*u(x)]$. That is:

$$u(80\%) = 0.5(0)+0.5(1) = 0.50.$$

This calculation determines the utility of the loss that equals *80%*. Next, using this result:

$$u(50\%) = 0.5u(80\%)+0.5u(0\%) = 0.5(0.5)+0.5(1) = 0.75.$$

Similar reasoning leads to the next calculation:

$$u(90\%) = 0.5u(80\%)+0.5u(100\%) = 0.5(0.5)+0.5(0) = 0.25.$$

The three data points are $u(80\%) = 0.50$; $u(50\%) = 0.75$; $u(90\%) = 0.25$. We also know two additional points. These are: $u(100\%) = 0.00$ and $u(0\%) = 1.00$, as discussed. With these five points we can fit a utility function:

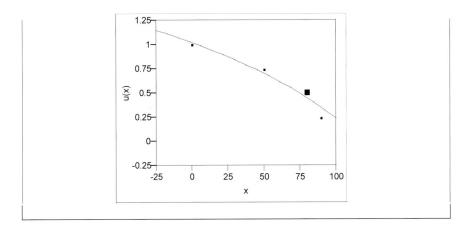

In summary, the essential steps for the practical use of utility functions require:

1. Selecting an objective,
2. Determining its attributes,
3. Developing the appropriate utility functions through elicitation and fitting to the data the utility function,
4. Assessing the forms of independence (discussed later in this chapter in the section dealing with multi-attribute utility analysis) to specify the form of utility functions.

12.3.2. Value of Information Criteria

Suppose the question now is: What if a risk manager directed the assessor to ignore uncertainty and variability? Ignoring uncertainty and variability suggests taking a value from a set, considering it to be deterministic, and proceeding with the calculations. But: What is the *best* value? Here we run into some problems with a deterministic choice because it is difficult to justify choosing a value without statistical guidance. If the set of numbers is a sample, we are on surer grounds because statistical theory can help. For example, if we want a number that is the most likely to occur, in some cases the best choice is the mean of the population, μ_x. It is demonstrably best by the maximum likelihood principle, provided that the population distribution is symmetrical about the mean. In general, more information leads to greater expected utility relative to decisions made with less information (Cox, 2002). This result is a restatement of the decision rule (Cox, 2002):

Abandon an action that has a probability, pr, of an adverse effect if and only if E(pr|Relevant Information) > the tolerable risk (probability), pr.

12.3.2.1. Expected Values of Information

As discussed in this chapter, it can be quite important to be able to decide when to gather additional information and when to stop such data collection effort. A way to deal with this problem is to develop criteria based on the information and its value to the decision-maker. Suppose that someone proposes to develop a study that can resolve the uncertainty in a particular situation. The question is: How does the manager decide if it is worth it to spend money and how much should be spent? The risk assessor can deal with this question as follows. He defines the *expected value with perfect information, EVwPI,* calculated as the expected value of the best possible pay-offs, *BP-O,* over the states-of-nature, for the actions that can be undertaken (Render, Stair and Balakrishnan, 2003):

$$EVwPI = \sum_{j=1}^{J} (pr_j)(BP - O)_j,$$

in which the probabilities are those associated with each state-of-nature, and there are *J* states-of nature. Note that some actions may not contribute to this calculation because only the best pay-offs are relevant, given the state-of-nature.

Example 12.6. Suppose that the set of best pay-offs for the alternatives open to the decision-maker, given only two states-of-nature, are *200* and *0* units, respectively. Suppose also that each of the two states-of nature is equally likely (each state-of nature has *0.50* probability). Then: *EVwPI = (0.5)(200)+(0.5)(0) = 100.* Suppose the largest expected monetary value of the decisions, *EV$_{max}$,* has already been estimated to be *45* units and that the study that would resolve uncertainty costs *70* units. *EV$_{max}$* is determined from selecting, from all actions and their expected values, the maximum expected value (Render, Stair and Balakrishnan, 2003). That is, the decision-maker considers the following information:

	States-of-Nature		
Actions	*Positive (0.5)*	*Negative (0.5)*	*Expected Values (resources)*
A	200	-180	(0.5)(200)+(0.5)(-180) = 10
B	100	-10	(0.5)(100)+(0.5)(-10) = 45
C	0	0	(0.5)(0)+(0.5)(0) = 0

The *expected value of perfect information, EVPI,* is: *EVPI = (EVwPI)-(EV$_{max}$) = 100-45 = 55.* The expected value of sample information, *EVSI,* is:

EVSI = E(best outcome, using the sample obtained a zero cost)-E(best outcome, without sample information).

Specifically, if the best expected value *without* the cost of taking the sample is *50*, and the best expected cost without any sample information is *40*, then *EVSI = 50-40 = 10* units. Supposing the proposed study would cost *70* units. Then, because EVPI is less that *70*, the decision-maker should not spend *70* units of resources on the proposed study.

Change the representation of the analysis to that of a decision-tree, the calculation of the expected values at each chance node (a circle) of the tree consists of folding back the calculations from the branches to the right of each chance node, as shown on Figure 12.5. This method is fully discussed and exemplified in section 12.5.

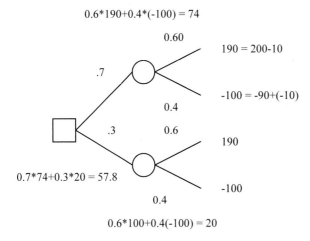

0.6*190+0.4*(-100) = 74

0.60

190 = 200-10

.7

-100 = -90+(-10)

0.4

.3   0.6

190

0.7*74+0.3*20 = 57.8

-100

0.4

0.6*100+0.4(-100) = 20

Figure 12.5. Decision Tree with Expected Values Calculated by Folding Back

The cost of sampling is *10* units, therefore the pay-offs are reduced by *10* units wherever they occur. Thus, if the top branch of the tree involves sampling, instead of *200* and *-100* which do not include the cost of sampling, we now have two pay-offs: *190* and *-110*. The top branch has an expected value of *74* units. This expected value is then related to the initial decision, the square box. Before doing so, we have to calculate the expected value of the bottom branch. The bottom branch also involves pay-offs with the cost of sampling, but emanates from the decision with a probability that is calculated as the complement of *0.7*, namely *0.3*. The average cost associated with this decision is *57.80* units. For example, if the best expected value of the branch without the sample option (not shown) were *30*, the EVSI is *EVSI = 67.8 -30 = 37.8* units. The reason for the value *67.8* is that the definition of the *EVSI* is based on the best (meaning the largest) expected value *with* the sample information but *without* the cost, thus we add *10* to *57.8*. We then subtract the expected values of the optimal decision *without* the *information* provided by the sample, which is *30*.

If a tree had several decision nodes and chance nodes, we can calculate several expected values, including the expected value of the "do not act" alternative. Granger-Morgan and Henrion (1990) define the *expected value of including uncertainty* (EVIU) for the random variable X as:

$$EVIU = \int [L(d_{iu}, x)-L(d_B, \mu_x)]f(x)dx = E[L(d_{iu}, x)]-E[L(d_B, \mu_x)].$$

In these expressions, d_{iu} is the decision based on ignoring uncertainty, d_B is the Bayesian decision, $f(x)$ is the prior distribution, $L(.)$ is the *loss* function and $E[.]$ is the expected value of the quantity within the square bracket. The loss function is used as follows. If an estimate does not correspond to the true value of the population parameter, it is biased. Correcting for bias involves relating the distance between the true value and the biased estimate through a loss function. Suppose we are estimating the mean value of the random variable X and that its distribution is known. Let, for this description, the bias, a, be positive. We can form the new random variable, $Y = X+a$. When $a = 0$, $Y = X$. Suppose that the loss of utility due to the bias increases as a function of the square of Y, Y^2, $L(Y) = 5.0Y^2$. As Chernoff and Moses (1986) demonstrate that $L(a) = k[E(X+a)^2]$. $L(a)$ reaches its minimum, $k[E(X+a)^2]$ is a parabola, where $a = -\mu_X$, Thus, that minimum expected loss equals $kE(X-\mu_X)^2$. The expected value of the Bayesian decision can be either smaller than or equal to the expected value of the decision taken with ignoring uncertainty. Therefore, the Bayesian decision (d_B), the decision that minimizes the expected loss for a given prior distribution function, is either superior to or equal to a decision based on a deterministic number. In other words, it always pays to account for variability (represented by the prior probability distribution). Granger-Morgan and Henrion (1990) provide several examples that the reader may want to study in detail. We summarize one that shows that, in the case of linear loss function, using the expected value of the prior distribution *equals* the result that is obtained using the Bayesian decision rule. This result applies to quadratic loss functions as well, but not to asymmetric loss functions.

Example 12.7. Consider the linear loss function $L(d_i, x) = a_i+b_ix$ (for each d_i there is a loss function). Then, $E[L(d, x)] = L[d, E(x)]$ (a result that follows from the algebra of expectations). Then $d_B = Min\{E[L(d, x)]\}$, because $E[L(d, \mu_x)]$ equals $Min\{E[L(d, x)]\}$. When the loss function is linear, using either the expected value or the Bayesian decision rule yields the same optimal result. However, if the loss function is asymmetric about the true expected value, the Bayesian decision rule will yield a different result and be superior to the decision based on the expected value.

A metric that measures the value of information is the *difference*, Δ, between the expected value of the decision taken under decision rules such as either to: 1) minimize the expected value of the maximum damage, or 2) maximize the expected value of the minimum benefit. Because gathering information is costly, we should account for the change in cost associated with the change in the information. The expected marginal (meaning change in cost due to a unit change in the amount of information) cost of gaining new information is:

$[\Delta(EV|\Delta Information)]/[\Delta(cost\ of\ the\ additional\ unit\ of\ new\ information)]$,

in which the term $\Delta(cost)$ is the change in cost associated with gaining the new information.

Example 12.8. Consider the following two situations in which there is information on exposure, E, and response, R, at time 1, and information on E and R, at time 2: (E_1, R_1) and (E_2, R_2) for the two periods. What changes is the information in these two periods on E and R. This example involves probabilities that are intervals rather than single numbers. The way to deal with them in calculating the various expectations is shown next, for two situations in which the information is $\{40, 70, 60, 20\}$ at time period 1 and $\{55, 70, 60, 20\}$, at time period 2. In period 1:

	E_1	E_2	Expectations based on: ($0.8 \le pr \le 0.9$)
R_1	40	70	$[(0.8)(40)+(0.2)(70)] = 46$; $[(0.9)(40)+(0.1)(70)] = 42$.
R_2	60	20	$[(0.8)(60)+(0.2)(20)] = 52$; $[(0.9)(60)+(0.1)(20)] = 56$

In period *2*:

	E_1	E_2	Expectations based on: $(0.8 \leq pr \leq 0.9)$
R_1	55	70	$[(0.9)(55)+(0.1)(70)] = 56.5$; $[(0.9)(40)+(0.1)(70)] = 42$
R_2	60	20	$[(0.8)(60)+(0.2)(20)] = 52$; $[(0.9)(60)+(0.1)(20)] = 56$

The *change* in the expected value of information yields *56.5-52 = 4.5* units. Forming the ratio *4.5/Δ(cost)* yields the marginal value of the additional information.

The next section outlines some of the technical aspects of utility functions to give a sense of the basis for the theory and to indicate some of the reasons for its advantages and limitations. Fundamentally, although utility theory is important in making risky choices, applications of the theory to practical situation requires some understanding of its foundations and assumptions. We follow Luce and Raiffa (1957), Chernoff and Moses (1986), Savage (1972) and Pratt, Raiffa and Schleifer (1995) to develop the essential aspects of utility theory and the implications for practical decision-making, including dealing with ignorance.

12.4. FOUNDATIONS OF UTILITY THEORY

The rationale for utility theory is axiomatic. Following the practice adopted in the earlier sections of this chapter, we have used acts, states-of-nature, consequences and a utility function. Continuing, we can take two actions, the set $\{A = a_1, a_2\}$, to develop the axioms necessary to deal with *partial* ignorance.

These principal axioms include:

I. The set of actions is not empty (a solution can always be found).
II. The preferred action is independent of the scale of measurement of the utilities.
III. The preferred action is invariant to the description of the acts used to represent them.
IV. If $a_1 \in A$ and a_2 is either preferred to a_1 or a_2 is indifferent to a_1, then $a_2 \in A$.
V. If $a_1 \in A$, then a_1 is admissible.
VI. If the class of actions includes actions that are weakly dominated or that are equivalent to actions in the initial set, the optimality of the early actions is unaffected.
VII. If an action is less then optimal, the addition of another will not change its sub-optimality.
VIII. If two decision makers have the same set of actions and states, and one of them has a payoff that is independent of the action, then the optimal choice will be the same as that of the other decision maker.
IX. If the decision maker considers a_1 and a_2 to be optimal; their mixture will also be optimal

If axiom IX does not hold, then the decision maker faces what Luce and Raiffa call *complete ignorance*. To account for *complete ignorance*, the added axioms are:

X. The choice of the optimal set is independent of the states.
XI. The deletion of a state with the same payoff as another does not alter the optimal choice.

XII.　The deletion of a state equivalent to a probability mixture found in other states does not alter the optimal choice

Savage (1972) developed a method for dealing with incomplete information using a decision-maker's subjective probability distribution (mass or density functions, depending on the data) over the states relevant to make a rational decision. This method has the effect of *reducing the decision problem from one of uncertainty to one of risk*. Savage assumed that the decision-maker is *consistent* with respect to axioms I, III, IV, V VII, VIII and IX. He also developed axioms of *weak ordering*, (among other things) that:

1.　Preferences are well defined,
2.　Two events can be compared through their probabilities,
3.　Within the set of actions, there is at least one pair of actions that is not indifferent to the decision maker.

Savage derived two important theorems that justify probabilistic decision-making using utility theory. The first theorem results in *personal* probabilities; the second establishes the relationship between probabilities and utilities. Using a *personal* probability means that elicitations can yield the relationship between expected utility and the decision maker's preferences. Practically, four conditions provide a coherent method for evaluation through utility theory (Chernoff and Moses, 1986), Table 12.

Table 12.8. Practical Conditions for Evaluation Using Utility Theory

Condition	Description	Comment
Outcome's Valuation	The measure of value is the utility	Utilities are measured numerically
Alternative's Valuation	The single measure for evaluation is the utility of the outcomes	This measure is the only measure used
Knowledge of states	Probabilistic measures describe knowledge about the states	The subjective belief of the decision-maker is a probability
Stochastic independence	The probability of the state does not depend on the action	Must be determined to hold

The decision rules are axiomatic, are theoretically sound, explicit, and have been tested (Von Winterfeldt and Edwards, 1986; Keeney, 1980; Pratt, Raiffa and Schlaifer, 1995). However, there are some. Allais (1953) developed a classic paradox, but many other paradoxes exist. Consider Table 12.9, the values are measured in dollars Allais (1953).

Table 12.9. Data for Allais' Paradox, Pay-offs in Dollars

	States-of-Nature		
	S_1	s_2	s_3
Action 1	4,000	0	0
Action 2	3,000	3,000	3,000
Action 3	4,000	0	3,000
Action 4	3,000	3,000	3,000

According to this table, the single decision-maker who prefers action *1* to action *2* must also prefer action *3* to action *4*. The reason is that, if the decision-maker preferred action *1* to action *2*, then, by the *sure thing* principle, she should also choose action *3* to action *4*. A *sure thing*, informally, is a choice that a decision maker would make regardless of whether other contingent events occur or not. For example, the purchase of a good may be contingent on two opposite events. If the purchase is made regardless of the probability of either event, then this choice is called a *sure* choice. What is common to actions *1* and 2 is the fact that state *3* has zero pay-off; what is common to actions *3* and *4* is that states *3* and *4* also have in common the same pay-off *($3,000)*.

In part to avoid those paradoxes, Jeffrey (1983) has suggested that the rational decision-maker should maximize the conditional expected utility of an option relative to the set of available and considered options. The basis of Jeffrey's theory, rather than focusing on outcomes, uses *propositions, desirability* measures, and a calculus based on conjunction, disjunction and other logical operators, rather than lotteries. Although paradoxes are not fatal to decision theory, the reader may refer to Pratt, Raiffa, and Schlaifer (1995) for additional information.

12.4.1. Multi-Attribute Utility

In most risk assessments, a single attribute can generally be insufficient to account for the factors that characterize utility. Methods from multi-attribute utility analysis are used in those situations. Decision theory can be used to describe the changes in the set of attributes, $x_1, x_2, ..., x_n$, characterizing a specific choice, y. Elicitation, by questionnaire, can yield estimates of utility, $u(\Delta x_i)_j$, where Δx_i is the change in the ith-attribute; the subscript j applies to the jth individual respondent (Keeny and Raiffa, 1967; Keeney, 1980). For practical reasons, attributes must be *preferentially independent*, meaning that the preferences between two sets of attributes depend on the difference among the attributes in the two sets, and not in differences among the attributes within a set. The second requirement is that any single attribute is *utility independent* of the others. Ordering by preference is independent of the preferences for any of the other attributes. The third is *difference independence*: the difference in the preferences between two attributes depends only on the level of an attribute, x_i, not on the level of other attributes. The last is *additive independence*: the order of the preference for lotteries depends only on their marginal distribution functions (Keeny and Raiffa, 1967).

The general form of a multi-attribute utility function is (Keeney, 1974; Keeney 1977; Fishburn, 1967; Keeney, 1980):

$$u(x_1, ..., x_n) = f[u_1(x_1), ..., u_i(x_i), ..., u_n(x_n); k_1, ..., k_i, ..., k_n].$$

Utility functionally depends on *n* attributes, x_i and scale constants k_i, for *n* attributes. Three types of independence provide the conditions that determine the form of the utility functions. In the formulae below (Keeney, 1980), utility is measured on the scale *0* and *1*, including these two points and $(0 \leq k_i \leq 1)$.[4]

Utility independence. This form of independence relates to the uncertainty for preferences. Attribute *X* is utility independent of all other attributes if the lottery for two values of *X*, say x_1 and x_2, is preferred to the lottery in which the values of *X* are x_3 and x_4, for all levels of any other attribute. If attributes are utility independent (or at least one of them is) and preferentially independent, the utility function is multiplicative:

$$u(x) = \Pi_i k_i u_i(x_i),$$

or:

$$1+ku(x) = \Pi_i[1+Kk_i u_i(x_i)],$$

in which $K > -1$.

Additive Independence. This form of independence exists when the order of preferences for lotteries does not depend on their joint distribution, but does depend on their marginal distributions. The multi-attribute utility function for additively independent attributes is also additive ($i = 1, ..., n; n \geq 2$):

$$u(x) = \Sigma_i k_i u_i(x_i).$$

Preferential Independence. This form of independence occurs when, in a subset of two attributes of the consequence, the *order* of the preferences for the consequences is independent of the levels taken by other attributes. This is a deterministic form of independence. The general case of the multi-attribute utility function for preferential independence is somewhat complicated; the reader can refer to Keeny (1980) for details. If the sum of the scaling coefficients equals *1.00*, $k = 0$, the multi-attribute utility function is:

$$u(x) = \Sigma_i k_i u_i(x_i).$$

If the scaling constant is less than zero, attributes are *substitutes* and the respondent is risk averse over the attributes. Otherwise, the attributes are *complements* and the respondent prefers risks. For $k = 0$, the respondent is risk neutral. If $k \neq 0$, the multi-attribute utility function takes the multiplicative form (Keeney, 1980):

$$1+ku(x) = \Pi_i[1+Kk_i u_i(x_i)].$$

Example 12.9. Assuming utility independence, the utility function for two attributes, *x* and *y*, is:

$$u(x, y) = [k_x u_x(x)]+[k_y u_y(y)]+\{(1-k_x -k_y)[u_x(x)u_y(y)]\}.$$

---

[4] In this discussion upper case letters are for attributes, lower case for values of an attribute: If X_1 is an attribute; one of its values is symbolized by x_i.

The development of the function itself requires establishing intervals of values on suitable scales of measurement for x and for y, say $(0, 10$ measured in arbitrary units, where 1 is the best value and 10 the worst) and with the utilities being valued between 0 and 1, such that $u(10,10)$ has the least utility. That is: $u(10,10) = 0$, and: $u(0, y) = u(x, 0) = 1$. Similarly, for each of the two individual attributes: $u_x(10) = 0$ and $u_x(0) = 1$ as well as $u_y(10) = 0$ and $u_y(0) = 1$. Using this information at the point $(0, 10)$:

$$1 = k_x u_x(x) + [k_y u_y(y)] + \{(1 - k_x - k_y)[u_x(x)u_y(y)]\},$$

that is:

$$1 = k_x[u(x) = 1] + k_y[u(x) = 0] + \{(1 - k_x - k_y)[u(x)u(y)]\}.$$

Therefore, $k_x = 1$, because $\{(1 - k_x - k_y)[u(x)u(y)]\} = 0$, $u(y) = 0$ and $k_y = 1$.

The assessor can determine whether the initial specification of the utility function is consistent with the findings from individual elicitations or questionnaires. The shape of the function, monotonicity and scale coefficients allow the assessment of the appropriateness of the initial specification of the utility functions and can be used to modify it. The construction of utility function must account for the assumptions made about the forms of independence (Keeny and Raiffa, 1967). If some assumptions are not met, mathematical transformations are available to develop a practical and theoretically sound utility function. The techniques available to obtain $u(x)$ or value functions, $v(x)$, include conjoint analysis (Hair, Anderson and Black, 1995). These methods provide the means to assess qualitative and quantitative attributes; they can include hypothetical factors and their attributes, through the rating (as a preference) of the *part-importance* of factors given by those elicited. However, attributes such as the number of lives saved per year cannot be mapped to a utility number because the utility function is *state-dependent* and thus is not unique. Individual single or multi-attribute utility functions can be aggregated over all of the individuals from whom the utility functions have been estimated. For example, in decisions involving risks to human and environmental health, the benefits are measured on a cardinal scale. Thus, given a set of individual $u_i(x_i)$, what is sought is an *aggregate* utility function or a utility function for the individual utilities, $U(u_1, ..., u_n)$ with the attributes being implicit. Most public decisions deal with measuring the change in total social welfare inherent to a change from a baseline (or status quo level) to that resulting from the imposition of a guideline or standard. It follows that the change from a utility level $u_0(.)$ to $u_1(.)$ determines the (positive or negative) change in aggregate utility, under certain assumptions (Cox, 2002). The principles that justify aggregating utility function are *coherence* and *efficiency*. For example, coherence informally means that the axioms, discussed earlier in this chapter, must apply at the aggregate level. The resulting aggregate utility function is a composition of weighted individual utility functions (Kirkwood, 1979). There also are ethical justifications based on two concepts (Keeney and Kirkwood, 1975). The first is that the aggregate utility should depend on the individual utilities: if an individual utility function is applicable to all, then that individual function would be the aggregate utility function. The second is that if all but one is indifferent between two options, the societal preference between those two options should be that of the outstanding individual. For assessing the risks, costs and benefits of policy, a linear *aggregate* utility function can be an appropriate first approximation.

12.4.2. Discussion

The analysis of risky decisions should include qualitative and quantitative discussions of the implications and effects of (temporally and spatially) changing attitudes and values. Although there is no guarantee that all stakeholders will be satisfied with a risk analysis, nevertheless such inclusion enhances its acceptability. In particular, the search for an optimal decision becomes clearer when alternative value systems are characterized with accuracy. Attributes range over several domains of knowledge relevant to making risky decisions. Table 12.10 includes a summary of some generic issues that can complicate an empirical analysis based on utility theory.

Table12.10. Classes of Attributes for an Environmental Decision

Attribute Class	Issue	Example	Comment
Environmental	Measurements. Incomplete or lack of knowledge forces. Missing data, improper unit of analysis	Intangible values, indirect effects	Uniqueness, option values, residual risks
Health and safety	As above	Intangible values	Risk trade-offs
Technological	As above	Choice of level of technology to mitigate hazards	Risk trade-offs
Individual and societal	As above; aggregation artifacts	Welfare functions are difficult to specify	As above. Interpersonal valuations are difficult

The content of this table suggests that using multi-attribute utility theory can be a relatively complex and costly effort and thus may be used in dealing with complex problems where the value of the pay-off is high. As discussed, the framework for making decisions is based on acts, states-of-nature and adverse outcomes or consequences. The theory, first developed by John von Neumann and Oscar Morgenstern (vN-M), consists of choices between acts dictated by choices over the probabilities of (possibly adverse) outcomes. The value of an act is measured by an ordinal function of values associated with each act. Ordinal means that a value of an action is preferred to the value of another if the value of the latter is greater than the former. More recent theories deal with the limitations of vN-M work in decision making under uncertainty. For example, the state-dependent utility theory (Karni, 1985), the risk measure is *pr(consequence|state-of-nature)*. Kahneman and Tversky (1979) have developed a prospect theory in which gain, loss or the status quo characterize the consequences. Machina (1982) has dealt with decision theory in terms of consequences measured on a monetary scale, unlike vN-M's utility scale, accounting for the severity of an outcome (not just the magnitude of the consequences). Nonetheless, the basic concepts discussed in this chapter, such as *certainty equivalent* and choices based on lotteries still apply. The difficulties inherent to scaling *severity*, not just *magnitude*, have been resolved by Dyer and Sarin (1982). Using vN-M utility theory, they have developed a scale of values in which equal intervals correspond to equal changes in severity.

Expected utility can be calculated in such a way that different decision-makers' attitude and strength of belief towards risk can be shown on the same scale of measured values. This assessment is captured by the coefficient of subjective attitude against risk; Bell (1985) used regret theory to demonstrate that, in practice, decision under uncertainty cannot be limited to considering magnitude alone. He stated that the magnitude of the consequence and the magnitude of the lost opportunity are two necessary attributes. In Bell's work, risk is the bivariate cumulative distribution over these two attributes. The concept of *temporal risk aversion* extends the traditional discussions about static risk aversion (Machina, 1984). To give but a brief description of dynamically evolving risks, we follow Cox's discussion of *uncertain risks* (Cox, 2002). An uncertain risk exists when the hazard function in which time is the argument is not known. In this situation, there is a set of probable trajectories for risk, one for each state of nature, each of which has a *pr(s)*. Until there is no adverse effect the decision maker experiences the expected value of the hazard function, namely *E[h(t); T ≥ t, pr(s)]*. When the adverse event occurs, at time *t+1*, this information updates *pr(s)* to *pr(s, t+1)*. In general, for a decreasing hazard function, the arrival of the adverse event will cause a jump upwards in the trajectory. This shift can measure the *disappointment* in the outcome.

12.5. DECISION TREES

We begin with the description of an enumeration, without any probabilities. Let $S_1 =$ {*a, b*} and $S_2 =$ {*1, 2, 3*}; their enumeration is depicted in Figure 12.6.

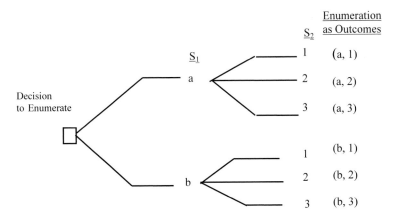

Figure 12.6. Deterministic Tree, $S_1 =$ {a, b}, $S_2 =$ {1, 2, 3} and Enumerations

A decision tree is a graph that, given an initial decision, depicts subsequent decisions and probabilities, to provide potential outcomes that emanate from the single, initial choice. Decision trees are particularly useful in representing decisions involving different levels of knowledge about the probabilistic outcomes, as the example that follows shows.

Example 12.10. Suppose that three different instruments, I_1, I_2, and I_3 yield *60%, 20%* and *20%* of the measurements taken at a site. These instruments incorrectly report measurements with probability *0.01, 0.005* and *0.001*, respectively. What is the probability that a measurement is faulty, assuming that the measurements are examined at random?

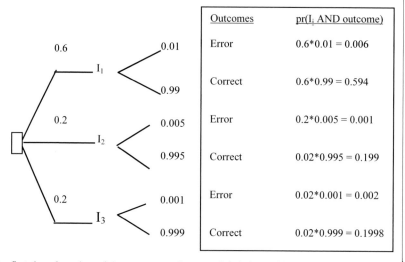

	Outcomes	pr(I_i AND outcome)
0.01	Error	0.6*0.01 = 0.006
0.99	Correct	0.6*0.99 = 0.594
0.005	Error	0.2*0.005 = 0.001
0.995	Correct	0.02*0.995 = 0.199
0.001	Error	0.02*0.001 = 0.002
0.999	Correct	0.02*0.999 = 0.1998

The first three branches of the tree set up the potential choices of instruments. For example, the probability of a measurement from instrument I_1 is *0.60*. Instrument I_1 can give an erroneous reading with probability *0.01*. Otherwise the reading is correct with probability *[1-pr(incorrect reading)] = 1-0.01 = 0.99*. The second set of branches sets out the probabilities of either an erroneous or a correct reading, depending on the instrument. The erroneous measurement is symbolized by *y* (the random variable is *Y*).

From the theorem of total probability we can obtain the formula for determining the probability of an erroneous reading, $pr(Y = y)$, that is $pr(y)$, which is:

$pr(y) = pr(I_1)pr(y|I_1)+pr(I_2)pr(y|I_2)+pr(I_3)pr(y|I_3)$
$= (0.6)(0.01)+(0.2)(0.005)+(0.2)(0.001)$
$= 0.0072.$

Bayes' theorem yields the probability that that measurement is read by instrument I_1:

$pr(I_1|y) = [pr\ (I_1)pr(y|I_1)]/\{[pr(I_1)pr(y|I_1)]+[pr(I_2)pr(y|I_2)]+[pr(I_3)pr(y|I_3)]\}$
$= [(0.6)(0.01)]/\{[(0.6)(0.01)]+[(0.2)(0.005)]+[(0.2)(0.001)]\} = 0.08.$

Recollect that the expected value of a pay-off is:

$$EV = \sum_{i=1}^{j}(pr_j)(Pay - off_j).$$

Specifically, for the decision table immediately below, the expected values are calculated in the last column, using arbitrary probabilities:

Actions:	s_1 (pr = 0.50)	s_2 (pr = 0.50)	Expected Values
No action, a_1	$c_{11} = 200$	$c_{12} = -180$	(0.5)(200)+(0.5)(-180) = 10
Low Intervention, a_2	$c_{21} = 100$	$c_{22} = -20$	(0.5)(100)+(0.5)(-20) = 40
High Intervention, a_3	$c_{31} = 0$	$c_{32} = 0$	(0.5)(0)+(0.5)(0) = 0

A similar calculation can be made for regrets, R_j, obtaining the expected value of the regrets, EVR:

$$EVR = \sum_{i=1}^{j} (pr_j)(R_j).$$

For the table of regrets immediately below, the EVRs are given in the last column, using arbitrary probabilities:

Actions:	s_1 (pr = 0.50)	s_2 (pr = 0.50)	Expected Values
No action, a_1	200-200 = 0	0-(-180) = 180	(0.5)(0)+(0.5)(180) = 90
Low Intervention, a_2	200-100	0-(-20) = 20	(0.5)(100)+(0.5)(20) = 60
High Intervention, a_3	200-0	0-(-0) = 0	(0.5)(200)+(0.5)(0) = 100

In the following example, low probability outcomes are shown as inequalities to reflect the difficulty of assigning *crisp* probabilities. The diagram (under the expected pay-offs, *E(pay-off)* column) depicts the extremes of the expected values, calculated by accounting for the inequalities. Other expected pay-offs can be calculated to satisfy these inequalities.

Example 12.11. The decision tree in this example consists of a discrete distribution (with three probability masses).

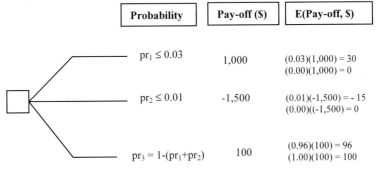

Probability	Pay-off ($)	E(Pay-off, $)
$pr_1 \leq 0.03$	1,000	(0.03)(1,000) = 30 (0.00)(1,000) = 0
$pr_2 \leq 0.01$	-1,500	(0.01)(-1,500) = - 15 (0.00)((-1,500) = 0
$pr_3 = 1-(pr_1+pr_2)$	100	(0.96)(100) = 96 (1.00)(100) = 100

The method for calculating the expected pay-offs consists of eliminating the inequality sign and calculating the pay-off for each point in the interval. Thus, $pr = 0.03$ and $pr = 0.00$ bracket the inequality $pr_1 \leq 0.03$. The maximum expected value is $96.00, which is obtained from $(100\$)[1-(0.03+0.01)] = \$96$; all other pay-offs are calculated accordingly.

Example 12.12. The decision maker wishes to choose between two alternatives, *1* and *2*; each of which has a choice from the set *{a, b, c, d}*. Each alternative is characterized by a probability mass distribution with three points:

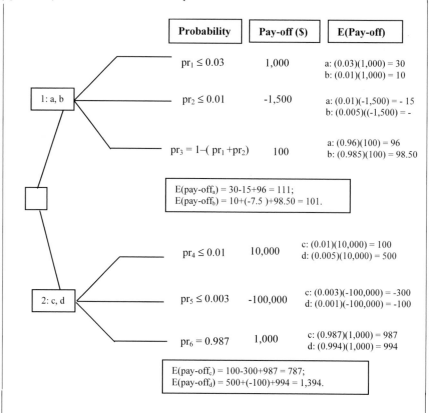

Probability	Pay-off ($)	E(Pay-off)
$pr_1 \leq 0.03$	1,000	a: (0.03)(1,000) = 30 b: (0.01)(1,000) = 10
$pr_2 \leq 0.01$	-1,500	a: (0.01)(-1,500) = - 15 b: (0.005)((-1,500) = -
$pr_3 = 1-(pr_1 + pr_2)$	100	a: (0.96)(100) = 96 b: (0.985)(100) = 98.50

1: a, b

E(pay-off$_a$) = 30-15+96 = 111;
E(pay-off$_b$) = 10+(-7.5)+98.50 = 101.

$pr_4 \leq 0.01$	10,000	c: (0.01)(10,000) = 100 d: (0.005)(10,000) = 500
$pr_5 \leq 0.003$	-100,000	c: (0.003)(-100,000) = -300 d: (0.001)(-100,000) = -100
$pr_6 = 0.987$	1,000	c: (0.987)(1,000) = 987 d: (0.994)(1,000) = 994

2: c, d

E(pay-off$_c$) = 100-300+987 = 787;
E(pay-off$_d$) = 500+(-100)+994 = 1,394.

The decision-maker only has partial information on the probabilities. Therefore, the tree shows lower and upper bounds on each probability statement (the pay-offs are in arbitrary units). The decision rule is to minimize the expected maximum loss. Thus, between the alternative choices, choice *c* with an expected value of *987.00*, minimizes the maximum expected value.

Example 12.13. A risk assessor has used two data sets and two models to estimate the tolerable exposures, measured in parts per million to a toxic agent found in soil. He has used $1*10^{-5}$ as the tolerable risk level, but is unable to determine which of these results is most credible. The risk manager has hired an independent consultant who, based on her experience with the substance, its biological effects and knowledge of exposure-response, has developed the decision tree that follows.

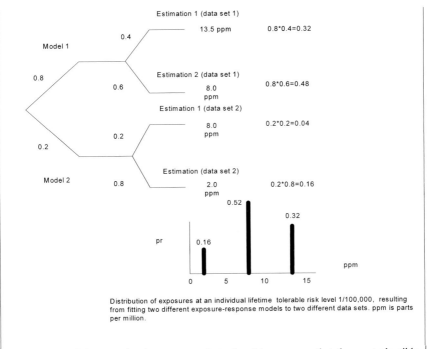

Distribution of exposures at an individual lifetime tolerable risk level 1/100,000, resulting from fitting two different exposure-response models to two different data sets. ppm is parts per million.

On the basis of these results, he recommends to the risk managers that the most plausible exposure level is *8.00* ppm.

Although our examples are contrived, the usefulness of decision trees can be summarized as follows. If the decision-maker is still not satisfied with these analyses, he can ask for some sensitivity analysis or even hire other experts to corroborate the probabilities assigned in the first set of calculations. Importantly, both assumptions and results are shown in a clear and consistent manner. The decision-maker is fully informed, in a way that is replicable, using other degrees of belief or frequencies. By exposing the models and data sets in a transparent way that accounts for the beliefs in these, an assessor's work can be evaluated for omissions, biases and to study the effect of defaults on tolerable exposure levels.

12.6. GAME THEORY

This chapter has so far dealt with the single decision maker who faces a choice that can have multiple outcomes. Suppose that two decision-makers face one another in the context of addressing a risky decision. Each attempts to gain from the strategies taken by the other. The theory and methods of game theory are useful in assessing this situation. An illustration of game theory follows, using a static game (time is not considered a factor affecting the results of the game). More than two decision-makers can be modeled. In these examples, we follow Straffin's (1993).

Example 12.14. Suppose that the decision-makers are Row and Column and that they face the following pay-offs measured on a scale from *-10* to *+10*, (dimensions omitted) with deterministic pay-offs:

Row's		Column's Strategies	
		Act	*Do not act*
Strategies	*Act*	+10, +10	-10, +10
	Do not act	+10, -10	-10, -10

If Row chooses to act, she may gain *+10*, provided that Column takes the same course of action. But if Column chooses not to act, while Row acts, then Row loses *10* (that is, *-10*) and Column wins *+10*. If, on the other hand, Row does not act when Column chooses to act, the gains are reversed: Row wins *+10* but Column loses *10*. If both decision-makers do not act both of them lose. The ordering of the options for Column is thus:

Row prevails over Column (*+10, -10*),
Second best solution: both gain (*+10, +10*),
Third best solution: both lose (*-10, -10*), and
Fourth best strategy: Row looses and Column wins (*-10, +10*).

In summary, a game requires at least two decision-makers and a set of numerical pay-offs that measure the value of the outcomes of the game to the decision makers. Each decision maker has a finite, exhaustive set of strategies that are mutually exclusive and separable, and the strategies determine the outcomes.

Example 12.15. Consider the following static game between Row and Column:

Row's		Column's Strategies	
		Act	*Do not act*
Strategies	*Act*	(2, -2)	(-3, 3)
	Observe	(0, 0)	(2, -2)
	Do not act	(-5,5)	(10, -10)

This is a *zero sum* game because the pay-offs are symmetric in the sense that if Column wins 2, then Row looses 2. The selection of a strategy suggests a number of rules. First a strategy *dominates* another if all of the outcomes in that strategy, *A*, are at least as preferable as those of another strategy, *B*, and at least one outcome of *A* is greater than the corresponding outcome in *B*. A rational decision-maker will invariably opt for the dominant strategy. Second, a *saddle point* occurs when an outcome is both less than or equal to a pay-off in a row and greater then or equal to a payoff in the intersecting column of the matrix of pay-offs.

Example 12.16. Consider the following zero sum game, omitting Column pay-offs, which are symmetric in magnitude to Row's, but with opposite sign). The pay-off for Row and Column, noting that Row has four strategies to Column's three, is set up in italics:

5	*4*	*3*	*6*	*3*	*maximin*
-15	*4*	*4*	*-5*	*-15*	
10	*8*	*3*	*6*	*3*	
0	*11*	*-10*	*-15*	*-15*	*maximin*
10	**11**	*3*	**6**		
		minimax			

> The pay-off that is the smallest for Row and the largest for Column is the deterministic *maximin* payoff or payoffs associated with one or more strategies. It has value *3*. The minimum of the maxima, the *minimax*, also has value *3*. This game has two *saddle points*, meaning that a minimum for one stakeholder is a maximum for the other, or vice versa.

If the minimax and the maximin do not occur in the same column of the pay-off matrix, then there is no saddle point. Straffin (1993) gives complete examples and additional elaborations, as well as theorems related to this discussion. Finally, each stakeholder can develop probabilistic strategies and use expected values to assess the outcomes of a strategy.

12.7. CONCLUSION

In this chapter we have developed some of the additional aspects of formal decision-making for risky decisions either by a single decision maker or by two or more decision makers. Decision trees, decision tables and game theory are formal means to assess these decisions. The methods discussed in this chapter link to cost-benefit analysis, discussed in Chapter 3, and complement its usefulness. Monetary measures of value such as the dollar, euro or other can now be used with utility functions and avoid some of the issues associated with valuations based on money alone. Pay-offs and consequences are determined from exposure or dose-response and economic models. Each tree or table is an explicit statement of causation, more or less simplified, depending on the resolution of the tree or a table. Representation of the decisions through either decision-trees or tables is a way to capture information in a pleasing way that succinctly shows all of the critical information to the decision-maker or stakeholders. These can modify the contents and thus study alternatives in a formal and consistent way. Decision trees and tables are practical and tried methods for assessing the outcomes of a set of options that account for time in the sense that the tree implicitly accounts for it. Clearly, these representations can only be as good as the state-of-the-art (data, methods and so on) allow. They also depend on the assessors' familiarity with the problem being addressed, as well as time and budget. Decision trees can capture the process of net risk, cost benefit analysis, probabilistically and dynamically because they can use subjective probability estimates, can be used to deal with sequential updating and experimentation, as well as with the propagation of the uncertainty and variability in the components of the process itself. There is commercially available computer software that allows the analysis of risky decisions through decision trees and multi-attribute utility analysis in diverse contexts, ranging from health to financial risks. The software includes *Analytica 2.0* and *Decision Right*.

QUESTIONS

1. Making a practical decision can involve many variables and relations. Some suggest that making decisions using scenarios (for example, selecting one or more outlying values, calculating the output and then relating that output to a base case for comparison) is particularly helpful in risk management. Consider a simple calculation involving 10 independent variables, each with a distribution consisting of the minimum, most plausible and maximum values only. The combinatorial explosion results in 3^{10} possible scenarios

(Granger-Morgan and Henrion, 1990). That is, a manager has to deal with approximately 59,000 scenarios that would have to be assessed and some analyzed. Although many can be excluded from an analysis, the representations of even 100 scenarios can be overwhelming and force ad hoc choices. Can a decision tree help? Limit your answer to 800 words or less.

2. Develop a two-branches decision tree, starting with the single decision of having to travel home from a distance of 50 km. Use at least two chance nodes and two branches after these nodes. Express your values in money, rather than utilities. Calculate, by assigning your own judgmental probabilities, the expected values associated with each choice and select the best. Do you think that using money rather than utility would change your conclusion? Why and why not?

3. Why is money not the best measure of value when dealing with risky events? Limit your answer to 400 words or less and use an example of your own making.

4. Could you develop a probabilistic version of a two-person game? Where would you place the probability numbers? Develop a suitable table and show your work; use your own numbers.

5. Provide a critique of utility theory in the context of catastrophic outcomes associated with the release of large amounts of air pollutants in a small area. Limit your answer to 400 words or less.

6. How can one be sure that all of the attributes of a choice are known? What can be done to obtain a full enumeration of the attributes? Assuming that you can enumerate the attributes relevant to a decision: What can be done when a subset of these is not measured and measuring them is too costly? Limit your answer to less than 600 words.

7. Anand (2002) proposes a non-probabilistic representation of a decision problem when probabilities are absent. He uses Laplace's Principle of Insufficient Reason that *assigns equal likelihoods to each state and calculates the expected outcome for each action on that basis.* Please discuss briefly.

REFERENCES

Abbey D.E., Huang B.L., Burchette, R.J., Vencuren T., Mills P.K., (1995) Estimated Long-term Ambient Concentrations of PM_{10} and Development of Respiratory Symptoms in Nonsmoking Population. *Archives of Environmental Health*, 50:139-152.

Abbey D.E., Huang B.L., Burchette, R.J., Vencuren T., Mills P., (1993) Long-term Ambient Concentrations of Total Suspended Particulates, Ozone, and Sulfur Dioxide and Respiratory Symptoms in a Nonsmoking Population. *Archives of Environmental Health*, 48:33-45.

Abdel-Rahman M.S., Kandry A.M., (1995) Studies on the Use of Uncertainty Factors in Deriving RFDs. *Human and Ecological Risk Assessment*, 1:614 -623.

Abelson P.H., (1992) Exaggerated Carcinogenicity of Chemicals. *Science*, 256:1609.

Abelson P.H., (1993) Pathological Growth of Regulations. *Science*, 260:1859.

Adler M.A., Posner E.A., (Eds), (2001) *Cost-Benefit Analysis: Legal, economic, and philosophical perspectives*. University of Chicago Press, Chicago.

Agency for Toxic Substances and Disease Registry, ATSDR, (2001) *Guidance Manual for the Assessment of Joint Toxic Action of Chemical Mixtures*. US Dept. of Health and Human Services, Washington, DC.

Allais M., (1953) Le Comportement de l'Homme Rationnel Devant le Risque: Critique des postulats et axioms de l'ecole americaine. *Econometrica*, 21:503-524.

Aldberg L., Slob M., (1993) Hazardous Concentrations. *Ecotoxicology and Environmental Safety*, 25:48-63.

Allen A., Raper S., Mitchell J., (2001) Uncertainty in the IPCC's Third Assessment Report. *Science*, 293: 430-433.

Ambrosone C.B., Kadlubar F.F., (1997) Toward an Integrated Approach to Molecular Epidemiology, *American Journal of Epidemiology*, 146:912.

American Society for Testing and Materials (ASTM) (2000), *Standard Guide for Risk-Based Corrective Action*. Conshohocken.

Ames B.N., Gold L.S., (1990) Too Many Rodent Carcinogens: Mitogenesis increases mutagenesis. *Science*, 9249:70-74.

Anand P., (2002) Decision-Making when Science Is Ambiguous. *Science*, 295:1839.

Andersen M.E., Fabrikant, J., (1986) Dose Route Extrapolations Using Inhalation Toxicity Data to Set Drinking Water Standards and the Risk Assessment Process for Methylene Chloride. *Toxicology and Applied Pharmacology*, 3:12-31.

Anderson D.R., (2000) *Model Selection and Inference: A practical information-theoretic approach*. Springer, New York.

Andrews J.D., Moss T.R., (1993) *Reliability and Risk Assessment*, Longman, Harlow.

Apostolakis G.E., (1974) *Mathematical Methods of Probabilistic Safety Analysis*. UCLA-ENG-7464.

Applebaum D., (1996) *Probability and Information: An integrated approach*. Cambridge University Press, Cambridge.

Applegate J.S., (2000) The Precautionary Preference: An American Perspective on the Precautionary Principle. *Human and Ecological Risk Assessment*, 6: 413-423.

Arrow J.K., Fisher A.C., (1974) Preservation, Uncertainty and Reversibility. *Quarterly Journal of Economics*, 87:312-320.

Arrow J.K., Lind R.C., (1970) Uncertainty and Evaluation in Public Investment Decisions. *American Economic Review*, 60:364-375.

Ashby J., Tennant R.W., (1991) Definitive Relationship Among Chemical Structure, Carcinogenicity and Mutagenicity for 301 Chemicals Tested by the U.S. NTP. *Mutation Research*, 257:229-238.

Ashford N., Ryan C.W., Caldart C.C., (1983) Law and Science Policy in Federal Regulation of Formaldehyde. *Science*, 222:894-398.

Ballester F., (1996) Air Pollution and Mortality in Valencia, Spain: A Study Using the APHEA Methodology. *Journal of Epidemiology and Community Health*. 50:527-540.

Barlow R.E., Proschan F, (1975) *Statistical Theory of Reliability and Life Testing: Probabilistic models*. Holt, Rinehart and Winston, New York.

Barnes D.G., Dourson M., (1988) Reference Dose (RfD): Description and use in risk assessment. *Regulatory Toxicology and Pharmacology*, 8:471-488.

Bates G., Lipman Z., (1988) *Corporate Liability for Pollution*. LBC Information Services, Pyrmont.

Baumol W.J., (1977) *Economic Theory and Operations Analysis*. Prentice-Hall, Englewood Cliffs.

Baumol W.J., (1968) On the Social Rate of Discount. *American Economic Review*, 36:788-799.

Bedford T., Cooke R., (2001) *Probabilistic Risk Analysis: Foundations and methods*. Cambridge University Press, Cambridge.

Bell D.E., (1985) Disappointment in Decision Making Under Uncertainty. *Operations Research*, 33:1 – 27.

Bernabini M., (1987) *Deconvolution and Inversion*. Blackwell Scientific, Oxford.

Birkett D., (2002) *Pharmacokinetics Made Easy*. McGraw-Hill, Sydney.

Bishop R.C., Heberlein T.A., (1979) Measuring Values of Extra-Market Goods: Are indirect measures biased? *American Journal of Agricultural Economics*, 61:926-948.

Bishop R.C., Heberlein T.A., Kealy M.J., (1983) Contingent Valuation of Environmental Assets: Comparisons with a simulated market. *Natural Resources Journal*, 23:619-630.

Blair A., Spirtas R., (1971) Use of Occupational Cohort Studies in Risk Assessment, CR Richmond, PJ Wesch, ED Copenhaver (Eds), *Health Risk Analysis*. Franklin Institute Press, Philadelphia.

Bockstael N.E., Bishop M., (2000) On Measuring Economic Values for Nature. *Environmental Science & Technology*, 34:1384-1388.

Bogen K.T., (1994) A Note on Compounded Conservatism. *Risk Analysis*, 14:471-475.

Borenstein M., Rothstein H., (1999) *Comprehensive Meta-Analysis* Englewood, NJ.

Borenstein M., Rothstein H., Cohen J, 2001, *Power and Precision*. Englewood.

Boroujerdi M., (2002) *Pharmacokinetics, Principles and Applications*. McGraw-Hill, NY.

Boyd R., (1991) Confirmation, Semantics, and the Interpretation of Scientific Theories, in R Boyd, P Gasper, JD Trout (Eds) *The Philosophy of Science*. MIT Press, Cambridge, MA.

Box G.E.P., Jenkins G.M., (1976) *Time Series Analysis Forecasting and Control*. 2nd Ed., Holden-Day, San Francisco.

Bradford D.F., (1970) Benefit-Cost Analysis and Demand Curves for Public Goods. *Kyklos*, 23:775-786.

Braun-Fahrlander C., Ackermann-Liebrich U., Schwartz J., Gnehm H.P., Rutishouser M., Wanner U., (1992) Air pollution and Respiratory Symptoms in Preschool Children. *American Review of Respiratory Disease*, 145:42-53.

Breiman L., Friedman J., Olshen R., Stone C., (1984) *Classification and Regression Trees*. Chapman and Hall, NY.

Breslow N.E., Day N., Days W., (1980) *Statistical Methods in Cancer Research, Vol 1, The Analysis of Case-Control Studies*. IARC Scientific Publications No. 32, Lyon.

Brookshire D., (1982) Valuing Public Goods: A comparison of survey and hedonic approaches. *American Economic Review*, 72:165-179.

Brown J., Rosen H., (1982) On Estimation of Structural Hedonic Price Models. *Econometrica*, 50:765-771.

Bryner G., (1999) *New Tools for Improving Government Regulation: an assessment of emission trading and other market-based regulatory tools*. Price Waterhouse Coopers Endowment for the Business of Government, http/www.endowment.pcwglobal.com

Buck S.F., (1960) A Method of Estimation of Missing Values in Multivariate Data Suitable for Use with an Electronic Computer. *Journal of the American Statistical Association*, B22:302-309.

Bukowski J., (1995) Correlated Inputs in Quantitative Risk Assessment: The effect of distributional shapes. *Risk Analysis*. 15:215-219.

Bull S.B., (1993) Sample Size and Power Determination for a Binary Outcome and an Ordinal Exposure when Logistic Regression Analysis Is Planned. *American Journal of Epidemiology*, 137:676-683.

Burden R.L., Faires J.D., (1989) *Numerical Analysis*. 4th Ed., PWS-Kent, Boston.Burnham K.P.,

Calabrese E.J., Baldwin L.A., (2001) Special Issue on the Scientific Foundations of Hormesis. *Toxicology*, 31: 351-695.

Calabrese R., McCarthy P., Kenyon J., (1987) The Occurrence of Chemically Induced Hormesis. *Health Physics*, 52:531-545.

Carson R.T., (2000) Contingent Valuation: A user 's guide. *Environmental Science & Technology*, 34: 1413-1418.

Chatfield C., (1975) *The Analysis of Time Series: Theory and practice*. Chapman & Hall, London.

Chen C.J., Chuang Y.C., Lin T.M., Wu H.Y., (1985) Malignant Neoplasms among Residents of a Blackfoot Disease Endemic Area in Taiwan: High arsenic artesian well water and cancers. *Cancer Research*, 45:5895-5899.

Chernoff H., Moses L.E., (1986) *Elementary Decision Theory*. Dover, Mineola.

Chihara C.S., (1990) *Constructibility and Mathematical Existence*. Oxford University Press, NY.

Chiou H.Y., et al., (1995) Incidence of Internal Cancers and Ingested Inorganic Arsenic: A seven-year follow-up study in Taiwan. *Cancer Research*, 5:1296-1305.

Clayton D., Mills M, *Statistical Models in Epidemiology*. Oxford Science, Oxford, (1994).

Coase R., (1960) The Problem of Social Cost. *Journal of Law and Economics*, 3:1-12.

Cohen J., (1995) *How Many People Can the Earth Support?* W.W. Northon, Philadelphia.

Cohen J., (1977) *Statistical Power Analysis in the Behavioral Sciences*. Academic Press, New York.

Cohen P.R., (1985) *Heuristic Reasoning About Uncertainty: An artificial intelligence approach*. Morgan Kaufmann, San Mateo, CA.

Colford J.M., Tager, I., Ricci P.F., Hubbard A., Horner W., (1999) Methods for Assessing the Public Health Impact of Outflows from Combined Sewer Systems. *Journal of the Air & Waste Management Association*, 49:454 -469.

Collett D., (1991) *Modelling Binary Data*. Chapman & Hall, London.

Committee on the Biological Effects of Ionizing Radiation, BEIR, (1980) National Academy of Sciences. *Health Effects of Ionizing Radiation*, Washington, DC.

Coombs R., Dawes J.P., Tversky A., *Mathematical Psychology*. Prentice-Hall, (1970).

Cooper H.M., (1979) Statistically Combining Independent Studies: A meta-analysis of sex differences in conformity research. *Journal of Personality and Social Psychology*, 37:131-139.

Cooter R., (1982) The Cost of Coase. *Journal of Legal Studies*, 11:1- 36.

Cox D.R., Oakes W., (1984) *Analysis of Survival Data*. Chapman and Hall, New York.

Cox, D.R., (1965) Regression Models and Life Tables. *Journal of the Royal Statistical Society*, Ser. B 34:187-201.

Cox L.A. jr., (1986) (Ch. 5) in *Benefit Assessment: The state of the art*. Bentkover J.D. (Ed.), D. Reidel, Dordrecht.

Cox L.A., jr., Ricci P.F., (2002) Empirical Causation and Biases in Epidemiology: Issue and solutions. *Technology*, 9:23-39.

Cox L.A., jr., (1989) *Cancer Risk Assessment for Chemical Carcinogens: From statistical to biological models*, Unpublished Manuscript, Denver.

Cox L.A. jr., Ricci P.F., (1989) Risks, Uncertainty, and Causation: Quantifying human health risks, in *The Risk Assessment of Environmental and Human Health Hazards*. Paustenbach D.J. (Ed.), Wiley, NY.

Cox L.A. jr., Ricci P.F., (1992) Reassessing Benzene Cancer Risks Using Internal Doses. *Risk Analysis*, 12:401-412.

Cox, L.A. jr, (2002) *Risk Analysis: Foundations, models and methods*. Kluwer, Boston.

Crump K.S., (1984) A New Method for Determining Allowable Daily Intakes. *Fundamental and Applied Toxicology*, 4:854-867.

Crump K.S., (1984) Issues Related to Carcinogenic Risk Assessment from Animal Bioassay Data, in P.F. Ricci, L.A. Sagan, C.G. Whipple (Eds), *Technological Risk Assessment*. NATO ASI Series, No. 81, M. Nijhoff Publishers, The Hague.

Currim I.S., Sarin R.K., (1984) A Comparative Evaluation of Multiattribute Consumer Preference Models, *Management Science*, 30:543-561.

Cuzick J., Sassieni P., Evans S., (1992) Ingested Arsenic, Keratoses, and Bladder Cancer. *American Journal of Epidemiology*, 136:417-426.

Cytel, (2001) *StaExact5 User Manual 1 and 2*. Cytel Software Corp., Cambridge.

Daly H.E., (1992) Allocation, Distribution and Scale: Towards an economy that is efficient, just and sustainable. *Ecological Economics*, 6:185-198.

Dasgupta A.K., Pearce D.W., (1978) *Cost-Benefit Analysis: Theory and Practice*. Macmillan, NY.

Davis H., Svendsgaard R., (1990) U-Shaped Dose-Response Curves: Their occurrence and implications for risk assessment. *Journal of Toxicology & Environmental Health*, 30:71-87.

Dawson-Saunders B., Rapp R.G., (1994) *Basic and Clinical Biostatistics*. Prentice-Hall, London (1994).

Day N.E., (1985) Statistical Considerations, Wald N.J, Doll R.(Eds), *Interpreting Negative Epidemiologic Evidence for Carcinogenicity*. IARC Scientific Pub. No. 65, Lyon.

De Finetti B., (1970) *Theory of Probability*. Vol. 1. Wiley, Chichester.

Dempster A.P., Rubin D., Layard M., (1977) Maximum Likelihood from Incomplete Data via the EM Algorithm. *Journal of the Royal Statistical Society*, B39:1-17.

Dixon J.A., Sherman P.B., (1986) *Economics of Protected Areas: A new look at benefits and costs*. Earthscan, London.

Doll R., (1985) Purpose of the Symposium, in Wald N.J., Doll R., (Eds), *Interpreting Negative Epidemiologic Evidence for Carcinogenicity*. IARC Scientific Pub. No. 65, Lyon.

Doll R., Peto R., (1981) The Causes of Cancer: Quantitative Estimates of Avoidable Risk of Cancer in the United States Today, *Journal of the National Cancer Institute*, 66:1195- 1284.

Donahue P., (1989) Diverting the Coaesian River: Incentive schemes to reduce unemployment spells. *Yale Law Journal*, 99:549-574.

Dougherty E.D., (1990) *Probability and Statistic for the Engineering, Computing and Physical Sciences*. Prentice Hall, Englewood Cliffs.

Dyer J.S., Sarin R.K., (1979) Group Preference Aggregation Rules Based on Strength of Preference, *Management Science*.

Dzeroski S., (1996) Inductive Logic Programming and Knowledge Discovery in Databases, in Fayyad U.M, Piatetsky-Shapiro G., Smyth P., Uthusuramy R., (Eds) *Advances in Knowledge Discovery and Data Mining*. MIT Press, Cambridge.

Ecologically Sustainable Development Steering Committee, (1992) *Draft National Strategy for Ecologically Sustainable Development*, AGPS, Canberra.

Eells E., (1991) *Probabilistic Causality*. Cambridge University Press, Cambridge.

Efron B., Tibshirani R.J., (1993) *An Introduction to the Bootstrap*. Chapman and Hall, (1993).

Efron B., Tibshirani R.J., (1991) Statistical Data Analysis in the Computer Age. *Science*, 253:390-393.

Ehling U.H., et al., (1983) Review of the Evidence for the Presence or Absence of Thresholds in the Induction of Genetic Effects by Toxic Chemicals. *Mutation Research*, 123:281 (1983).

Ellesberg D., (1961) Risk, Ambiguity, and the Savage Axioms. *Quarterly Journal of Economics*, 75:528-538.

Epstein R.L., Carnielli P., (1989) *Computability Functions, Logic, and the Foundations of Mathematics*. Wadsworth and Brooks/Cole, Belmont.

Farrel D., (1987) Information and the Coase Theorem. *Journal of Economic Perspectives*, 1:113-123.

Faustman E.M., (1994) Dose Response Assessment for Developmental Toxicity: characterization of the database and determination of NOAELs. *Fundamental and Applied Toxicology*, 23:478-485.

Federal Register, 29 May 1992, 57(104): 22888.

Feinstein A.R., (2002) *Principles of Medical Statistics*. Chapman & Hall/CRC, Boca Raton.

Ferejhon J., Page T., (1978) On the Foundations of Intertemporal Choice. *American Journal of Agricultural Economics*, 23:60-81.

Ferson S., Long T.F., (1997) Deconvolution Can Reduce Uncertainty in Risk Analysis in Newman M, Strojan C., (Eds) *Logic and Measurement*. Ann Arbor Press, Ann Harbor.

Ferson S., (1996) Judgment Under Uncertainty: Evolution May not Favor a Probabilistic Calculus. *Behavioral and Brain Science*, 19:24-28.

Fishburn P.C., (1967) Methods for Estimating Additive Utilities. *Management Science*, 13:435-443.

Fisher R.A., (1948) Combining Independent Tests of Significance. *American Statistician*, 2:30-37.

Fleiss J.L., (1981) *Statistical Methods for Rates and Proportions*. Wiley, NY.

Freedman R., (1983) Note on Screening Regression Equations. *American Statistician*, 37:152-158.

Freeman M., (1979) *The Benefits of Environmental Improvements*. Johns Hopkins University Press, Baltimore.

Friedman J.H., (1995) An Overview of Predictive Learning and Function Approximation, in *From Statistics to Neural Networks: theory and pattern recognition applications*, Cherkassky V., JH Friedman J.H., Wechsler H., (Eds). Springer, Berlin.

Galles D., Pearl J., (1997) Axioms of Causal Relevance. *Artificial Intelligence*, 97:9-16.

Gardenfors P., Sahlin N.E. (Eds), (1988) *Decision, Probability, and Utility: Selected readings*. Cambridge Univ. Press, Cambridge.

Gardner A., (1987) *An Artificial Intelligence Approach to Legal Reasoning*, The MIT Press, Cambridge.

Gelman A., Carlin J.B., Stern H.S., Rubin D.B., (1995) *Bayesian Data Analysis*. Chapman & Hall, London.

Gilbert R.O., (1987) *Statistical Methods for Environmental Pollution Monitoring*. Van Nostrand Reinhold, NY.

Gilks W.R., (1996) Full Conditional Distributions, in Gilks W.R., Richardson S., Spiegelhalter D.J., (Eds) *Markov Chain Monte Carlo in Practice*. Chapman & Hall/CRC, Boca Raton.

Gill J., et al., (1989) The Rat as an Experimental Animal. *Science* 245: 269-272.

Gill R.D., (1984) Understanding Cox's Regression Model: A martingale approach. *Journal American Statistical Association*, 17:386-391.

Glass G.V., (1997) Integrating Findings: The meta-analysis of research. *Review of Research in Education* 5:351-363.

Glass G.V., (1983) Synthesizing Empirical Research: Meta-Analysis, in Ward and Reed (Eds) *Knowledge Structure and Use: Implication for synthesis and interpretation*, Temple Univiversity Press, Philadelphia.

Granger C.W.J., Newbold P., (1986) *Forecasting Economic Time Series*. 2nd Ed., Academic Press, Orlando.

Granger-Morgan M., Henrion M., (1990) *Uncertainty: A guide to dealing with uncertainty in quantitative risks and policy analysis*. Cambridge University Press, Cambridge.

Gratt L.B., (1996), *Air Toxic Risk Assessment and Management: Public health risks from routine operations*. Van Nostrand Reinhold, NY.

Green W.H., (1997) *Econometric Analysis*. 3rd Ed., Prentice-Hall, Upper Saddle River.

Greenland G., (1994) Can Meta-analysis Be Salvaged? *American Journal of Epidemiology*, 140:783-785.

Greenland S., (2003) Quantifying biases in causal models: classical confounding vs collider-stratification bias. *Epidemiology*, 14:300-306.

Greenland S., Morgenstern H., (2001) Confounding in health research. *Annual Review of Public Health*, 22:189-212.

Grimmett G.R., Stirzaker D.R., (1982) *Probability and Random Processes*. Oxford University Press, Oxford.

Guerrero P., (1997) *Overview and Issues on Emission Allowances Trading Programs*. US General Accounting Office, http/www.gao.gov/index/reports/1997/reptoc.htm

Guess H.A., (1989) Behavior of the Exposure Odds Ratio in Case-Control Study When the Hazard Function Is not Constant over Time. *Journal of Clinical Epidemiology*, 42:1179-1184.

Guttmann Y.M., (1999) *The Concept of Probability in Statistical Physics*. Cambridge University Press, Cambridge.

Hall P., (1987) On the Bootstrap and Likelihood-based Confidence Regions. *Biometrika*, 74:481-488.

Hallenbeck W.H., Cunningham K.N., (1988) *Quantitative Risk Assessment for Environmental and Occupational Health*. Lewis, Ann Harbor.

Hammersley J.M., Handscomb D.C., *(1965) Monte Carlo Methods*. Methuen, London.

Hampel F, (2001) An Outline of Unifying Statistical Theory. *II Int. Symposium on Imprecise Probabilities and Their Applications*, Ithaca, NY.

Hanemann W.M., (1994), Valuing the Environment through Contingent Valuation. *Journal of Economic Perspectives*, 8:19-43.

Harris C.C., Hollstein M., (1993) Clinical Implications of the p53 Tumor Suppressor Gene. *New England Journal of Medicine*, 329:1318-1322.

Harrison D., Rubinfeld D.L., (1978) Hedonic Prices and the Demand for Clean Air. *Journal of Environmental Economics and Management*, 5:81-92.

Hong Y., Apostolakis G., (1993) Conditional Influence Diagrams in Risk Management. *Risk Analysis*, 13:623-627.

Heckerman D., (1996) Bayesian Networks for Knowledge Discovery, in Fayyad U.M., Piatetsky-Shapiro G., Smyth P., Uthurusamy R., (Eds) *Advances in Knowledge Discovery and Data Mining*. MIT Press, Cambridge.

Hedges L.V., Olkin I., (1985) *Statistical Methods for Meta-Analysis*. Academic Press, Orlando.

Helsel D.R., (1992) Less than Obvious. *Environmental Science and Technology*, 23:261-265.

Higgins B.G., et al., (1995) Effect of Air Pollution on Symptoms and Peak Expiratory Flow Measurements in Subjects with Obstructive Airways Disease. *Thorax*, 50:149-157.

Hildebrandt B., (1987) Overdose Toxicity Studies versus Thresholds: Elements of biology must be incorporated into risk assessment. *Archives of Toxicology*, 61:217-221.

Hill A.B., (1965) The Environment and Disease: Association or Causation? *Proceedings Royal Society of Medicine, Section on Occupational. Medicine*, 58:295-303.

Hjorth J.S.U., (1994) *Computer Intensive Statistical Methods: Validation, model selection and bootstrap*. Chapman & Hall, London.

Hodgson E., Levy P.E., (1987) *A Textbook of Modern Toxicology*. Appleton and Lange, Norwalk.

Hoek G., Brunekreef B., (1993) Acute Effects of Winter Air Pollution Episode on Pulmonary Function and Respiratory Symptoms of Children. *Archives of Environmental Health*, 48:328-336.

Hoek G., Brunekreef B., (1994) Effects of Low-Level Winter Air Pollution Concentrations on Respiratory Health of Dutch Children. *Environmental Research*, 64:136-143.

Hoogenven R.T., et al., (1999) An Alternative Exact Solution of the Two-Stage Clonal Growth Model of Cancer. *Risk Analysis*, 19:9-15.

Hopenyan-Rich C., Biggs M.L., Smith A.H., (1998) Lung and Kidney Cancer Mortality Associated with Arsenic in Drinking Water in Cordoba, Argentina. *International Journal of Epidemiology*, 27:561-268.

Hopenyan-Rich C., Biggs M.L., Smith A.H., (1996) Bladder Cancer Mortality Associated with Arsenic in Drinking Water in Argentina. *Epidemiology*, 7:117-125.

Hori H., (1975) Revealed Preference for Public Goods. *American Economic Review*, 65:197-210.

Howard R.A., (1988) Uncertainty about Probability: A decision analysis perspective. *Risk Analysis*, 8:91-96.

Howell D.C., (1995) *Fundamental Statistics for the Behavioral Sciences*. Duxbury Press, New Haven.

Howson C., Urbach P., (1993) *Scientific Reasoning: The Bayesian approach*. Open Court, Chicago.

Hueting R., Bosch P., in Kuik O., Verbruggen H., (1996) (Eds) *In Search of Indicators of Sustainable Development*. Springer-Verlag, Berlin.

Huber P.J., (1981) *Robust Statistics*. Wiley, NY.

Hurwicz L., (1957) *Optimality Criterion for Decision Making under Ignorance*, Cowles Commission Discussion Paper, no. 370, (1951), in Luce R.D., Raiffa H., (Eds) *Games and Decisions*. Wiley, NY.

Ichida J.M., et al., (1993) Evaluation of Protocol Change in Burn-care Management Using the Cox Proportional Hazards Model with Time-dependent Covariates. *Statistics in Medicine*, 12:301-313.

Ingelfinger J.A., Mosteller F., Thibodeau L.A., Ware J.H., (1987) *Biostatistics in Clinical Medicine*. 2nd Ed., MacMillan, NY.

International Commission on Radiological Protection, ICRP, (1996) *Recommendations of the International Commission on Radiological Protection.* ICRP 9,Geneva.

International Union for the Conservation of Nature, (1980) *World Conservation Strategy.* Gland.

Jansson P.A., (Ed), (1984) *Deconvolution and Applications in Spetroscopy.* Academic Press, Orlando.

Jeffrey R.C., (1983) *The Logic of Decision.* University of Chicago Press, Chicago.

Jensen F.V., (1996) *An Introduction to Bayesian Networks.* Springer-Verlag, NY.

Johnsen T.H., Donaldson J.B., (1985) The Structure of Intertemporal Preferences and Time Consistent Plans. *Econometrica*, 53: 1451 - 1458.

Johnston J., (1991) *Econometric Methods, 3rd Ed.*, McGraw-Hill, Auckland.

Judge G.G., Griffiths W.E., Hill R.C., Lutkepohl H., Lee T.-C., (1980) *The Theory and Practice of Econometrics.* 2nd Ed, Wiley, NY.

Justinian, (1979) *The Digest of Roman Law*, Penguin, London.

Kahn H.A., Sempos C.T., (1989) *Statistical Methods in Epidemiology.* Oxford University Press, Oxford (1989).

Kalos M.H., Whitlock P.A., (1986) *Monte Carlo Methods.* Vol. 1, Wiley, NY.

Kalbfleisch J.D., Prentice R.L., (1980) *The Statistical Analysis of Failure Time Data.* Wiley, NY.

Kanji G.K., (1999) *100 Statistical Tests.* Sage, London (1999).

Katsouyanni K., Schwarz J., Spix C., Toulumi G., Zmirou D., Zanobetti A., Wojtyniak B., Vonk J.M., Tobias A., Ponka A., Medina A., Bacharova L., Anderson H.R., (1996) Short Term Effects of Air Pollution on Health: A European approach using epidemiological time series data, the APHEA protocol. *Journal of Epidemiology & Community Health*, 50:8-19.

Katz M.L., Rosen H.S., (1998) *Microeconomics.* Irwin McGraw-Hill, Boston.

Keeny, R.L., (1974) Multiplicative Utility Functions. *Operations Research,* 22:22-31.

Keeney R.L., (1977) The Art of Assessing Multiattribute Utility Functions. *Organizational Behavior and Human Performance*, 19:267-276.

Keeney R.L., (1980) *Siting Energy Facilities.* Academic Press, NY.

Keeney R.L., Kirkwood C., (1975) Group Decisionmaking using Cardinal Social Welfare Functions. *Management Science*, 22:430-437.

Keeney R.L., Smith G.R., (1982) Structuring Objectives for Evaluating Possible Nuclear Material Control and Accounting Regulations. *IEEE Transactions on Systems, Man and Cybernetics*, MC-12:743-747.

Keeny R.L., Raiffa H., (1967) *Decisions with Multiple Objectives: Preferences and value tradeoffs.* Wiley, NY.

Keeney R.L., Lathrop J.F., Sicherman A, (1986) An Analysis of Baltimore Gas and Electric Company's Technological Choice. *Operations Research*, 34: 28-35.

Kennedy P.A., (1992) *A Guide to Econometrics.* 3rd Ed., MIT Press, Cambridge.

Kennedy P.A., (1998), *Guide to Econometrics.* 4th Ed., MIT Press, Cambridge.

Khoury M.J., D.K. Wagener, (1995) Epidemiological Evaluation of the Use of Genetics to Improve the Predictive Value of Disease Risk Factors. *American Journal of Human Genetics*, 56:935-942.

Kirkwood C., (1979) Pareto Optimality and Equity in Social Decision Analysis. SMC-9 *IEEE Transactions: System, Man & Cybernetics*, 89:101-108.

Klassen C.D., (2001) *Cassarett and Doull's Toxicology - The Basic Science of Poisons*. 6th Ed., McGraw-Hill, NY.

Kleinbaum D.G., Morgenstern B, (1982) *Epidemiological Research: Principles and quantitative methods*. Lifetime Learning, Belmont.

Kleindorfer, P.R., Kunreuther H.C., Schoemaker P.H., (1993) *Decision Sciences: An integrated perspective*. Cambridge University Press, Cambridge.

Klir G.J., Folger T.A., (1988) *Fuzzy Sets, Uncertainty, and Information*. Prentice Hall, Englewood Cliffs.

Kneese A.V., Sweeney J., (1985) *Handbook of Environmental Resources Economics*. North-Holland, Amsterdam.

Kodell, R.L., Gaylor D.W., Chen J.J., (1987) Using Average Lifetime Dose Rate for Intermittent Exposure to Carcinogens. *Risk Analysis*, 7:339-344.

Kofler E., Meges G., (1976) *Decision under Partial Information*, Springer, Berlin.

Krantz R., Luce R.D., Suppes P., Tversky A., (1971) *Foundations of Measurement*, Vol 1. Academic Press, NY (1971).

Kreps, D.M., (1988) Notes on the Theory of Choice. Westview Press, Boulder.

Kreps D.M., Porteous E.L., (1978) Temporal Resolution of Uncertainty and Dynamic Choice Theory. *Econometrica*, 46:185-1991.

Krewski D., Zhu Y., (1995) A Simple Data Transformation for Estimating Benchmark Dose in Developmental Toxicity Experiments. *Risk Analysis*, 15:29-34.

Liang K.Y., McCullagh P., (1993) Case Studies in Binary Dispersion. *Biometrics*, 49:623-632.

Light R.J., Pillemer B.B., (1984) *Summing Up: The Science of Reviewing Research*. Harvard University Press, Cambridge.

Linhart H., Zucchini W., (1986) *Model Selection*. Wiley, NY.

Lindley D.V., (1984) *Bayesian Statistics: A review*. SIAM, Philadelphia.

Lippmann, M., Ito K., (1995) Separating the Effects of Temperature and Season on Daily Mortality from those of Air Pollution in London: 1965-1972. *Inhalation Toxicology* 7:85-91.

Little R.J.A., Rubin D.B., (1987) *Statistical Analysis with Missing Data*. Wiley, NY.

Little R.J.A., Rubin D.B., (1990) The Analysis of Social Science Data with Missing Values, in Fox J., Long J.S. (Eds) *Modern Methods of Data Analysis*. Sage, Newbury Park.

Ljung G.M., (1986) Diagnostic Testing of Univariate Time Series Models. *Biometrika*, 73:725-732.

Luce R.D., Raiffa H., (1957) *Games and Decisions*. Wiley, NY.

Luce R.D., Narens L., (1990) Measurement, in Eatwell J., Milgate M., Newmann P. (Eds), *The New Palgrave: Time series and statistics*. WW Norton, NY.

Luce R.D., Narens L., Measurement Scales on the Continuum, *Science* 236:1527 (1987).

Luce R.D., Where Does Subjective Utility Theory Fail Prescriptively? *Journal of Risk and Uncertainty*, 5:5 (1992).

Lumina Decision Systems, (1999) *Analytica 2*. Los Gatos, CA.

Machina M., (1984) Temporal Risk Aversion and the Nature of Induced Preferences. *Journal of Economic Theory*, 45:199-204.

Madigan D., Raftery A., (1994) Model Selection and Accounting for Uncertainty in Graphical Models Using Occam's Window. *Journal of the American Statistical Association*, 89:1535-1542.

Makridakis S., Wheelwright S.C., McGee V., (1983) *Forecasting: Methods and Applications*. Wiley, NY.

Maler K.G., (1974) *Environmental Economics*. Johns Hopkins University Press, Baltimore MD.

Mantel N., Haenszel W., (1959) Statistical Aspects of the Analysis of Data From Retrospective Studies of Disease. *Journal of the National Cancer Institute*, 22:719-726.

Marcantonio R.J., (1992) *A Monte Carlo Simulation of Eight Methods for Handling Missing Data under Varying Conditions of Multivariate Normality*. Unpublished Doctoral Dissertation, University of Illinois, Chicago.

Marglin S., (1963) The Social Rate of Discount and the Optimal Rate of Investment. *Quarterly Journal of Economics*, 95:23-29.

Maritz M., (1981) *Distribution-Free Statistical Methods*. Chapman and Hall, London.

Maugh R., (1978) Chemical Carcinogens: How Dangerous Are Low Doses? *Science*, 202:37-39.

May R., (1976) Simple Mathematical Models with Very Complicated Dynamics. *Nature*, 261: 459-467.

McCann J., Horn L., Kaldor J., (1984) An Evaluation of the Salmonella (Ames) Test Data in the Published Literature: Application of statistical procedures and analysis of mutagenic potency. *Mutation Research*, 134:1-12.

McMahon T., (1973) Size and Shape in Biology. *Science*, 179:1201-1203.

Mehta C., Patel N., (1999) *StatXact5, Statistical software for exact nonparametric inference*. User Manual, Cytel, Cambridge.

Mendelshon R., Brown G.M., (1983) Revealed Preference Approach for Valuing Outdoor Recreation. *Natural Resources Journal*, 23:607-619.

Milne M.J., (1996) On Sustainability: The environment and management accounting. *Management and Accounting Research*, 7:135-148.

Mitchell R.C., Carson R.T., (1989) *Using Surveys to Value Public Goods: The contingent valuation method*. Resources for the Future, Washington DC.

Moolgavkar S.H., Luebeck E.G., Hall T.A., Anderson E.L., (1995) Air Pollution and Daily Mortality in Philadelphia. *Epidemiology* 6:476-484.

Moolgavkar S.H., Dewanji A., Venzon D.J., A (1988) Stochastic Two-stage model for Cancer Risk Assessment, I: The hazard function and the probability of tumor. *Risk Analysis*, 8: 383-390.

Moore R.E., (1979) Methods and Applications of Interval Analysis, *SIAM Studies on Applied Mathematics*, Philadelphia.

Morales K.H., et al., (2000) Risk of Internal cancers from Arsenic in the Drinking Water. *Environmental Health Perspectives* 103:684-694.

Myerson R., Sattertwhite M., (1983) Efficient Mechanisms for Bilateral Trading. *Journal of Economic Theory*, 29:265-281.

Nachman D.C., (1975) Risk Aversion, Impatience, and Optimal Timing Decisions. *Journal Economic Theory*, 11:196-203.

Neapolitan R.E., (1992) A Survey of Uncertain and Approximate Inference, in *Fuzzy Logic for the Management of Uncertainty*. Zadeh L., Kacprzyk J., (Eds), Wiley, NY.

Norris J.R., (1997) *Markov Chains.* Cambridge University Press, Cambridge.

Office of Technology Assessment, OTA, (1988) *Assessment of Technologies for Determining Cancer Risks in the Environment.* Washington, DC.

Oreskes N., Shrader-Frechette K., Belitz K., (1994) Verification, Validation and Confirmation of Numerical Models in the Earth Sciences. *Science,* 263:641-644.

Ostro B.D., et al., (1995) Air Pollution and Asthma Exacerbations among African-American Children in Los Angeles. *Inhalation Toxicology* 7:711-719.

Orwin R.G., (1983) A Fail-Safe N for Effect Size. *Journal of Educational Statistics,* 8:157-164.

Office of Technology Assessment, OTA, (1986) *Technologies for Detecting Heritable Mutations in Human Beings.* OTA-H-298, Government Printing Office, Washington DC.

Packel E.W., Traub J.F., (1987) Information-Based Complexity. *Nature* 328:29-32.

Palmer, A.R., Mooz W.E., Quinn T.H., Wolf K.A., (1980) *Economic Implications of Regulating Chlorofluorocarbon Emissions from Nonaerosol Applications.* R-2524-EPA, RAND, Santa Monica.

Pearl J., (Ed.), (1988) *Probabilistic Reasoning in Intelligent Systems: Networks of plausible inference.* Morgan Kaufmann, San Mateo.

Pearl J., (2000) *Causality: Models, reasoning and inference.* Cambridge University Press, NY.

Pearl J., (1986) Fusion, Propagation, and Structuring in Belief Networks. *Artificial Intelligence,* 29:241-250.

Pearson K., (1933) On a Method of Determining Whether a Sample of Size n Supposed to Have Been Drawn from a Parent Distribution Having a Known Probability Integral Has Been Drawn at Random. *Biometrika,* 25:379-385.

Pearson E.S., Hartley H.O., (1996) *Biometrika Tables for Statisticians.* 3rd Ed., Cambridge University Press, Cambridge.

Pearson L., (1996) Incorporating ESD Principles in Land-Use Decision-Making: Some issues after *Teoh. Environmental and Public Law Journal,* 13: 47-75.

Perera F.P., (1996) Uncovering New Clues to Cancer Risks. *Scientific American,* 94: 40-46.

Perera F.P., Weinstein W., (1982) Molecular Epidemiology and Carcinogen-DNA Adduct Detection: New approaches to studies of human cancer causation. *Journal of Chronic Diseases,* 35:12-21.

Petitti D., (1994) *Meta-analysis, Decision Analysis, and Cost-effectiveness Analysis: Methods for quantitative synthesis in medicine.* Oxford University Press, NY.

Pisciotto A., Graziano G., (1980) Induction of Mucosal Glutathione Synthesis by Arsenic. *Biochemical and Biophysical Acta,* 628:241-249.

Pope III, C.A., Kanner R.E., (1993) Acute Effects of PM_{10} Pollution on Pulmonary Function of Smokers with Mild to Moderate Chronic Obstructive Pulmonary Disease. *American Review of Respiratory Disease* 147:1336-1342.

Pope III C.A., et al., (1992) Daily Mortality and PM_{10} Pollution in Utah Valley. *Archives of Environmental Health* 47: 211-218.

Polya G., (1954) *Patterns of Plausible Inference.* Princeton University Press, Princeton.

Popper K.R., (1959) *The Logic of Scientific Discovery,* Basic Books, NY.

Porter P.S., et al., (1988) The Detection Limit, *Environmental Science and Technology.* 22:856-260.

Pratt J.W., Raiffa H., Schlaifer R., (1995) *Introduction to Statistical Decision Theory*. MIT Press, Cambridge.

Raaschou-Nielsen O., Nielsen M.L., Gehl J., (1995) Traffic-related Air Pollution: exposure and health effects in Copenhagen street cleaners and cemetery workers. *Archives of Environmental Health*, 50:207-212.

Raiffa H., (1970) *Decision Analysis: Introductory Lectures on Choices under Uncertainty*. Addison-Wesley, Boston.

Rao J.N.K., Wu C.F.J., (1988) Resampling Inference with Complex Survey Data. *Journal of the American Statistical Association*, 83:231-238.

Ramsey F.P., (1990) *Philosophical Papers*. Mellor D.H. (Ed.), Cambridge University Press, Cambridge.

Reilly .J, Stone P.H., Forest C.E., Webster M.D., Jacoby H.D., Prinn R.G., (2001) Uncertainty and Climate Change Assessments. *Science*, 293: 430-433.

Render B., R.M. Stair, N. Balakrishnan, (2003) *Managerial Decision Modeling with Spreadsheets*. Prentice-Hall, Upper Saddle River.

Reports of the Working Groups on Ecologically Sustainable Development, AGPS, (1991) Canberra, Australia.

Resampling Stats, Inc., (1997) stats@resample.com.

Research Education Association, (1978) *Problem Solver: Statistics*. Research and Education Association, Piscataway.

Reynolds S.H., et al., (1987) Activated Oncogenes in B6C3F1 Mouse Liver Tumors: Implications for risk assessment. *Science*, 237:1309-1313.

Rice D.P., 91966) *Estimating the Cost of Illness*. US Dept. of Health, Education and Welfare, Washington, DC.

Ricci P.F., (1995) Non-Glamorous Natural Resources: Legal and Scientific Aspects of Their Evaluation under Uncertainty. *Australasian Journal of Natural Resources Law and Policy*, 2:59-283.

Ricci P.F., (1979) A Comparison of Air Quality Indices for 26 Canadian Cities. *Journal of the Air Pollution Control Association*, 29: 1242.

Ricci P.F., (1985) *Principles of Health Risk Assessment*. Prentice-Hall, Englewood Cliffs.

Ricci P.F., Cox L.A. Jr., (1987) De Minimis Considerations in Health Risk Assessment. *Journal of Hazardous Materials*, 15: 77-80.

Ricci P.F., Gray N.J., (1988) Toxic Torts and Causation: Towards an equitable solution in Australian law (I). *University of New South Wales Law Journal*, 21:787-806. (1998).

Ricci P.F., Gray N.J., (1999) Toxic Torts and Causation: Towards an equitable solution in Australian law (II). *University of New South Wales Law Journal*, 22: 1-20.

Ricci P.F., Glaser E.R., Laessig R.E., (1979) A Model To generate Rent Surfaces: Applications and results for a public project. *Water Resources Research*,15:539-545.

Ricci P.F., Hughes D.M., (2000) Environmental Policy and the Necessity for Risk Assessment: a contribution from PB-PK models and Bayesian Networks, in Beer T., (Ed.), *Air Pollution and Health Risks*. CSIRO, Melbourne.

Ricci P.F., Molton L.S., (1981) Risk and Benefit in Environmental Law. *Science* 214:1096-2001.

Ricci P.F., Perron L.E., Emmett B.A., (1997) Environmental Quality, Expenditures Patterns and Urban Location: Effects on income groups. *Socio-Econ. Planning Sciences*, 11: 249-257.

Ricci P.F., Sagan L.A., Whipple C., (1984) *Technological Risk Assessment*. NATO Advanced Study Institute No. 41, Suijthoff, The Netherlands.

Ricci P.F., Wyzga R., (1983) An Overview of Cross-sectional Studies of Mortality and Air Pollution and Related Statistical Issues. *Environment International*, 9:177-190.

Robins J., Wasserman L., (1988) On the Impossibility of Inferring Causation from Associations without Background Knowledge, in Glymour C., Cooper G., (Eds), *Computation and Causality*. AAAI/MIT Press, Cambridge.

Rodgers W., (1979) A Hard Look at *Vermont Yankee*: Environmental choices under close scrutiny. *Georgetown law Journal*, 67:699-725.

Rodgers W., (1994) *Environmental Law*, West Publishing Company, St. Paul.

Roe E., (1997) *Taking Complexity Seriously: Policy analysis, triangulation and sustainable development*. Kluwer, Boston.

Roemer W., Hoek Brunekreef G., (1993) Effect of Ambient Winter Pollution on Respiratory Health of Children with Chronic Respiratory Symptoms. *American Review of Respiratory Disease*, 147:118-24-31.

Romp G., (1997) *Game Theory, Introduction and Applications*. Oxford University Press, NY.

Ronan J., (1973) Effects of Some Probability Displays of Choice. *Organizational Behavior*, 9:1-6.

Rosen H.S., (1992) *Public Finance*. 3rd Ed., Irwin, Homewood.

Rosenthal R., (1987) Comparing and Combining Judgment Studies, In: *Design, analysis, and meta-analysis*. Cambridge University Press, NY.

Rosenthal R., Rubin D.B., (1982) Comparing Effect Sizes of Independent Studies. *Psycology Bulletin*, 92:500-507.

Rosner B., (1990) *Fundamentals of Biostatistics*. 3rd Ed., Duxbury Press, Belmont.

Ross S.M., (1983) *Stochastic Processes*. Wiley, NY.

Roth R., (Ed.), (1985) *Game-Theoretic Models of Bargaining*. Cambridge University Press, Cambridge.

Roth S.H., Skrajny B., Reiffenstein R.J., (1995) Alternation of the Morphology and Neurochemistry of the Developing Mammalian Nervous System by Hydrogen Sulfide. *Clinical and Experimental Pharmacology and Physiology*, 22:379-380.

Royall R., (1999) *Statistical Evidence: A likelihood paradigm*. Chapman & Hall, Boca Raton.

Saldiva P.H.N., Schwartz, J, (1994) Association between Air Pollution and Mortality due to Respiratory Diseases in Children in Sao Paulo, Brazil: A preliminary report. *Environmental Research* 65:218-226.

Samuelson W., (1984) Bargaining under Asymmetric Information. *Econometrica*, 52:995-1005.

Savage L.J., (1972) *The Foundations of Statistics*. 2nd Rev. Ed., Dover, NY.

Scarlett J.F., Griffiths J.M., Stracjan D.P., Anderson H.R., (1995) Effect of Ambient Levels of Smoke and Sulfur Dioxide on the Health of a National Sample of 23 Year Old Subjects in 1981. *Thorax*. 50:764-768.

Schork B., et al., (1998) The Future of Genetic Epidemiology. *Trends in Genetics*, 14:266-273.

Schwartz J., (1993) Air Pollution and Daily Mortality in Birmingham, Alabama. *American Journal of Epidemiology*, 137:1136-1141.

Schwartz J., (1994) Air Pollution and Hospital Admission for the Elderly in Birmingham, Alabama. *American Journal of Epidemiology*, 139:589-594.

Schwartz J., PM_{10}, (1994) Ozone, and Hospital Admissions for the Elderly in Minneapolis-St. Paul, Minnesota. *Archives of Environmental Health*, 49:366-370.

Schwartz J., (1995a) Air pollution and Hospital Admissions for Respiratory Disease. *Epidemiology*, 7:20-725.

Schwartz J., (1995b) Short term fluctuations in air pollution and hospital admissions of the elderly for respiratory disease. *Thorax*, 50:531-534.

Schwartz J., et al., (1993) Particulate Air Pollution and Hospital Emergency Room Visits for Asthma in Seattle. *American Review of Respiratory Disease*, 147:826-834.

Scott M.J., Bilyard G.R., Link S.O., Ulibarri C.A., Westerdhal H.E., Ricci P.F., Seely H.E., (1998), Valuation of Ecological Resources and Function, *Environmental Management*, 22:49-68.

Selvin S., (2001) *Epidemiologic Analysis, A case-oriented approach.* Oxford University Press, Oxford.

Shafer G., (1996) *The Art of Causal Conjecture*, MIT Press, Cambridge.

Shapiro S., (1994) Point/Counterpoint: meta-analysis of observational studies. *American Journal of Epidemiology*, 140:771-775.

Shastri L., (1998) *Semantic Networks: An Evidential Formalization and its Connectionist Realization*, Morgan Kaufmann, San Mateo.

Shaw I.C., Chadwick J., (1998) *Principles of Environmental Toxicology.* Taylor & Francis, London.

Shenoy P.P., (1992) Valuation-Based System for Bayesian Decision Analysis, *Operation Research*, 40:463-470.

Sielken R.L. jr., (1985), Some Issues in the Quantitative Modeling Portion of Cancer Risk Assessment. *Regulatory Toxicology and Pharmacology*, 5:175-181.

Sielken R.L., Valdez-Flores C., (1999) Probabilistic Risk Assessment's Use of Trees and Distributions to Reflect Uncertainty and Variability and to Overcome the Limitations of Default Assumptions. *Environment International*, 25:755-764.

Siegel S., (1956) *Non-Parametric Statistics for the Behavioral Sciences.* McGraw-Hill, NY, (1956).

Sivak, A.M., Goyer M., Ricci P.F., (1987) Nongenotoxic Carcinogens: Prologue, nongenotoxic mechanisms in carcinogenesis. in *Bambury Report* No. 25, Cold Spring Harbor.

Smith A.H., (1992) Cancer Risks from Arsenic in Drinking Water. *Environmental Health Perspectives*, 97:259-264.

Smith A.H., (1998) Marked Increase in Bladder and Lung Cancer Mortality in a Region of Northern Chile Due to Arsenic in Drinking Water. *American Journal of Epidemiology*, 147:660-667.

Smith V.K., (1979) Indirect Revelation of the Demand for Public Goods: An overview and critique. *Scottish Journal of Political Economy*, 26:183-187.

Solow R., (1974) The Economics of Resources or the Resources of Economics. *American Economic Review*, 64:1-8.

Sombé L.A., (1990) *Reasoning Under Incomplete Information in Artificial Intelligence.* Wiley, NY.

Spear R.C., Boys F.Y., (1994) Parameter Variability and the Interpretation of PB-PK Modeling Results. *Environmental Health Perspectives*, 102:61-70.

Spear R.C., Bois F.Y., Modeling Benzene Pharmacokinetics across Three Sets of Animal Data: parametric sensitivity and risk implications. *Risk Analysis*, 11: 641 (1991).

Spiegelhalter D.J., et al., (1996) Hepatitis B: a case study in MCMC methods, in Gilks W.R., Richardson S., Spiegelhalter D,J., (Eds) *Markov Chain Monte Carlo in Practice*. Chapman & Hall/CRC, Boca Raton.

Spitzer W.O., (1991) Meta-Analysis: A quantitative approach to research integration. *Journal of Clinical Epidemiology*, 44:103-110.

Sprent P., (1989) *Applied Nonparametric Statistical Methods*. Chapman and Hall, London.

Straffin P.D., (1993) *Game Theory and Strategy*. Mathematical Association of America, No. 46.

Sunyer J., Saez M., Murillo C., Castellsague J., Martinez F., Anto J.M., (1993) Air Pollution and Emergency Room Admissions for Chronic Obstructive Pulmonary Diseases: a 5-year study. *American Journal of Epidemiology*, 137:701-705.

Sunstein C.R., (2002) *Risk and Reason: Safety, law and the environment*. Cambridge University Press.

Suppes F., (1999) The Positive Model of Scientific Theories, in *Scientific Inquiry, Readings in the Philosophy of Science*. Klee R. (Ed.), Oxford University Press, NY.

Tietenberg T., (2000) *Environmental and Natural Resources Economics*. 5th Ed., Addison-Wesley.

Taubes G., (1993) EMF - Cancer Links: Yes, no, and maybe. *Science*. 262:649-652.

Traub J.F., (1998) *Information-Based Complexity*. Academic Press, NY.

Tribe L., (1971) Trial by Mathematics: Precision and ritual in the legal process. *Harvard Law Review*, 84:1329-1340.

Tsai S.-M., Wang T.-N., Ko Y.-C., (1998) Cancer Mortality Trends in a Blackfoot Disease Endemic Community of Taiwan Following Water Source Replacement. *Journal of Toxicology and Environmental Health*, Part A, 55(6):389-395.

Tsuda T., et al., (1995) Ingested Arsenic and Internal Cancers: A historical cohort study followed for 33 years. *American Journal of Epidemiology*, 14:198-203.

US Nuclear Regulatory Commission, (1981) *Fault Tree Handbook*. NUREG/CR-0942, Washington, DC.

US EPA (1973) 600/5-73-013, *The State of the System Model* (SOS). Washington, DC.

US EPA, (1994) *Report on the Workshop on Cancer Risk Assessment Guidelines and Issues*, EPA/630/R-94/005a, Washington, DC.

US EPA, (1989) Risk Assessment Guidance for Superfund, Vol. 1, *Human Health Evaluation Manual*. EPA/540/1-89/002, Washington, DC.

US EPA, (1988) *Superfund Exposure Assessment Manual*. EPA/540/1-88/001, US EPA ORD, Washington, DC.

US EPA, (1995) *The Use of Benchmark Dose in Health Risk Assessment*. EPA/630/R-94/007, Washington, DC.

US EPA, (1996) *Proposed Guidelines for Cancer Risk Assessment*. EPA/600/P-92/003C, Washington, DC.

US EPA, (2002) *Guidelines for Cancer Risk Assessment*. EPA, Washington, DC.

US EPA, (2003) *Supplemental Guidance for Assessing Cancer Susceptibility from Early Life Exposure to Carcinogens.* EAP/630/R-03/003, Washington, DC.

US EPA, (2000) *Science Policy Council Handbook: Risk characterization.* EPA Science Policy Council, Washington, DC. EPA/100/B-00/002, Washington, DC.

US EPA, (1995) *Exposure Factor Handbook.* ORD/OHEA, EPA/600/P-95/002Ba, Washington, DC.

US EPA, *Guidelines for Exposure Assessment.* ORD/OHEA, EPA/600/2-92/001, Washington DC (1992).

US Geological Survey, USGS, (2000) *A Retrospective Analysis of the Occurrence of Arsenic in Ground-Water Resources of the United States and Limitations in Drinking-Water-Supply Characterization.* WRIR Report 99-4279, Reston.

US National Research Council, (2001) *Arsenic in Drinking Water 2001 Update.* National Academy Press, Washington, DC.

US National Research Council, (1999) *Arsenic in Drinking Water.* National Academy Press, Washington, DC.

US National Research Council, (1986) *Drinking Water and Health.* Vol. 6, National Academy Press, Washington, DC.

US National Research Council, (1994) *Science and Judgment in Risk Assessment,* National Academy Press, Washington, DC.

van Fassen B., (1976) To Save the Phenomena. *Philosophy* 73:623-634.

Viscusi W.K., (1996) Economic Foundations of the Current Regulatory Reform Efforts. *Journal of Economic Perspectives,* 10:120-129.

von Plato J., (1994) *Creating Modern Probability.* Cambridge University Press, Cambridge.

Von Winterfeldt D., Edwards W., (1986) *Decision Analysis and Behavioral Research.* Cambridge University Press, Cambridge.

Wachter K.W., (1988) Disturbed by Meta-Analysis? *Science,* 241:1047-1049.

Wachter K.W., Straf M.L., (1993) *The Future of Meta-Analysis.* Sage, Newbury Park.

Wagner W.E., (2000) The Precautionary Principle and Chemical Regulation in the US. *Human and Ecological Risk Assessment,* 6: 459-465.

Walpole R.E., (1974) *Introduction to Statistics.* 2nd Ed., Macmillan, NY.

Walters S., Griffiths R.K., Ayres J.G., (1994) Temporal Association between Hospital Admissions for Asthma in Birmingham and Ambient Levels of Sulfur Dioxide and Smoke. *Thorax,* 49:133-140.

Waters M.D., Stack H.F., Jackson M.A., (1999) *Short Term tests for Defining Mutagenic Carcinogens.* IARC, Lyon.

Watson J., et al., (1987) *Molecular Biology of the Gene.* 4th ed, Benjamin/Cummings, Menlo Park.

Wei W.W.S., (1990) *Time Series Analysis: Univariate and multivariate methods.* Addison-Wesley, Redwood City.

Weterings R.A.P.M., Opschoor J.B., (1992) *The Ecocapacity as a Challenge to Technological Development.* Advisory Council for Research on Nature and Environment, Rijswijk.

Wigley T.M., Raper S.C.B., (2001) Interpretation of High Projections for Global-Mean Warming. *Science,* 293: 451-453.

Wiggering H., Rennings K., (1997) *Environmental Geography,* 32:1-10.

Wilson R., Crouch E.A.C., (1987) Risk Assessment and Comparisons: An introduction. *Science,* 267:236-240.

Wittgenstein L., (1922) *Tractatus Logico-Philosophicus*. Routledge and Kegan Paul, London.

Wolf F.M., (1986) *Meta-Analysis: Quantitative Methods for Research Synthesis* Sage, Newbury Park.

Wonnacott J.R., Wonnacott T.H., (1970) *Econometrics*. Wiley, NY.

Wu M.M., Kuo T.L., Hwang Y.H., Chen C.J., (1989) Dose-response Relation between Arsenic Concentrations in Well water and Mortality from Cancers and Vascular Diseases. *American Journal of Epidemiology*, 130:1123-1130.

Xu X., Dockery D.W., Christiani D.C., Li B., Huang H., (1995) Acute Effects of Total Suspended Particles and Sulphur Dioxide on Preterm Delivery: A community-based cohort study. *Archives of Environmental Health*, 50:407-412.

Xu X., Dockery D.W., Christiani D.C., Li B., Huang H., (1995) Association of Air Pollution with Hospital Outpatient Visits in Beijing. *Archives of Environmental Health*, 50:214-218.

Yang S.C., Yang S.P., (1994) Respiratory Function Changes from Inhalation from Polluted Air. *Archives of Environmental Health*. 49:182-188.

Zadeh L., (1978) Fuzzy Sets as a Basis for Possibility Theory. *Fuzzy Sets and Systems*, 1:3-12.

Zeckhauser R.J., Viscusi W.K., (1990) Risk within Reason. *Science* 248:559-562.

GLOSSARY

This Glossary describes the principal terms of art used in risk assessment and management. It is developed from definitions developed by the US EPA, WHO and other agencies. The reader should consult the worldwide web pages of these organizations for further definitions and details.

1. **Absorption** – The chemical uptake of liquids, solids or gases by solids or liquids. An absorbed dose is the mass of a chemical that can become toxic after metabolic or other processes have taken place.

2. **Acceptable Daily Intake (ADI)** – The intake of a chemical measure as a concentration over a lifetime of exposure that creates no significant adverse health effects in the exposed. The standard unit for the ADI is the mg/kg-day.

3. **Adduct** – A chemical moiety that is bound to the DNA or a protein.

4. **Acute Exposure** – Exposure to an agent (e.g., a chemical) for 14 days or less.

5. **Adsorption** – The adhesion of a layer of molecules (of gases, solutes, or liquids) to the surface of solids or liquids with which they come into contact.

6. **Background Exposure** – The quantity of a chemical that occurs naturally in an environmental medium such as water.

7. **Benchmark Dose (BMD)** – The lower confidence limit on the dose that produces a specified change in an adverse response. For example, a $BMD_{0.10}$ would be the dose at the 95% lower confidence limit on the 10% response. The benchmark response (BMR) is 10%. The BMD is determined by modeling the dose-response, given observable data to which the model is fit.

8. **Benchmark Dose Model** – A probabilistic dose-response model used with either experimental toxicological or epidemiological data to calculate a benchmark dose or concentration if an exposure-response model is used.

9. **Biomarkers** – Indicators signaling specific events in biological systems. They can mark for exposure, sensitivity or other biological characteristic.

10. **Carcinogen** – An agent capable of causing cancer. (A carcinogen can be a promoter, initiator or have other effects on the progression of cancer.

11. **Case-control study** – An experimental epidemiological study of the relationship between an adverse outcome and one or more causal agents. A group of individuals with specific exposure and outcome is related to another group of similar individuals who are not exposed. Exposure is not explicitly controlled, but it is approximated or reconstructed. The epidemiological measure is prevalence data.

12. **Ceiling Value** – A concentration of an agent that should not be exceeded, even momentarily.

13. **Chronic Exposure** – Exposure to an agent for 365 days or more.

14. **Cohort study** – An experimental epidemiological study of a specific group of individuals who are initially free from the disease and are followed for a specified period of time and some develop a response. Exposure is explicitly controlled in this study. The epidemiological measure used is incidence.

15. **Cross-sectional study** – An observational epidemiological study in which the exposure-response of those in the study is assumed to be unaffected by time.

16. **Default value, coefficient or parameter** – An arbitrary number generally set by informed consensus to be used in the absence of other relevant and appropriate information.

17. **Dose** – The amount of a chemical or other agent administered to an individual, stated in mg/kg-day or other appropriate unit. Different types of doses are used in risk assessment. For example, administered dose is the dose to which an individual is first exposed. The absorbed dose is the dose reaching the target organ.

18. **Dose-response relationship** – The quantitative relationship between the amount of dose to an agent and the incidence of the adverse response.

19. **Epidemiology** – The study of temporal and spatial patterns of diseases in humans.

20. **Exposure** – The initial mass or concentration reaching an individual from an external source without penetrating the organism or being transformed at the point of contact. Exposure can be transient, continuous, discontinuous, steady state or other.

21. **False negative** – An empirical result that is found to be negative, while it is not.

22. **False positive** – An empirical result that is found to be positive, while it is not.

23. **Genotoxicity** – An adverse effect on the genome of living cells that can be expressed as a mutagenic, clastogenic or carcinogenic event from an unrepaired mutation.

24. **Half-life** – The time at which half of the initial mass is halved.

25. **Hormesis** – An empirical result in which low dose or exposure is associated with a beneficial result while at higher dose or exposure the effect is adverse.

26. **Immediately Dangerous to Life or Health (IDLH)** – The maximum concentration of an agent from which a person can escape within 30 minutes, without any impairing symptoms or irreversible health effects.

27. **Incidence** – The number of new diseased individuals, in a suitable period of time.

28. **Intermediate Exposure** – Exposure for a period from 15 to 364 days.

29. **In vitro** – Occurring in a *test* tube or other container such as a Petri dish.

30. **In vivo** – Occurring in a live organism.

31. **Lethal Concentration 50** (LC_{50})– The concentration at which 50% of those exposed die.

32. **Lethal Time 50** (LT_{50}) – The time at which 50% of those exposed die.

33. **Lowest Observed Adverse Effect Level** (LOAEL) – The lowest exposure to a toxic agent in a study that produces a statistical or biological significant increase in the frequency of adverse health effects in the exposed.

34. **Minimal Risk Level** (MRL) – An estimated daily exposure to a toxic agent that is unlikely to result in a significant risk of non-cancerous health outcome, for a specific route and duration of exposure.

35. **Mode of Action** - A sequence of key events and processes, starting with interaction of an agent with a cell, proceeding through operational and anatomical changes, and resulting in *cancer*. Examples of modes of carcinogenic action include mutagenicity, mitogenesis, inhibition of cell death, and immune suppression.

36. **Modifying Factor** (MF) – A constant (> 0; default = 1.00) used in developing a minimal risk level (MRL) to reflect an uncertainty in data that is not accounted for by uncertainty factors (UF).

37. **Morbidity** – Being diseased. Morbidity rate can be measured by the incidence or prevalence of a disease in a group of individuals. It is generally stated as number of diseased individuals, per 100,000 individuals, per year or other denominator number of individuals. Cause-specific morbidity means that a particular disease – e.g., lung cancer – is the disease. Morbidity rates can be cause-specific and age-standardized.

38. **Mortality** – Death. Mortality rate is measured by the incidence or prevalence of death in a group of individuals. It is generally stated as number of dead individuals, per 100000, per year or other number of individuals. Mortality can be cause-specific and age-standardized.

39. **Mutagen** – An agent that causes a change to the DNA.

40. **No Observed Adverse Effect Level** (NOAEL) – The level of exposure or dose of a toxic agent that results in no observable or statistically significant increase in the frequency of the adverse health effect. Often stated as the highest experimental dose (mg/kg-day) or exposure at which no adverse effect is observed; there is no statistically significant difference in the response of the controls and the exposed.

41. **No Observed Effect Level** (NOEL) – The highest dose that does not affect the morphology, functional capacity or other aspect of the animal on test.

42. **Odds Ratio** (OR) – A probabilistic measure of association between exposure and response in a group exposed, relative to a group not exposed. An odds ratio greater than one states that exposure is more damaging than no exposure

43. **Permissible Exposure Limit** (PEL) – An Occupational Health and Safety Administration (US OSHA) allowable exposure level in workplace air, averaged over 8-hours for 40 hours per week.

44. **Pharmacokinetics** – The study of the fate of a chemical substance in a live organism.

45. **Pharmacokinetic Model** – It consists of equations that describe the temporal course of a chemical or its metabolites in a biological system. These models can be compartmental or physiological.

46. **Physiologically Based Pharmaco-kinetic Model** (PB-PK) – It consists of one or more physiological organs, e.g., the lung, the venous and arterial system, with volumes and flow rates and other factor, such as membrane permeability, alveolar ventilation rates and so on. This model also uses biochemical data such as blood/air partition coefficient, metabolic parameters and so on.

47. **Point of Departure** (POD) – It represents the point at which the extrapolation to lower doses begins. It is measured by the estimated dose (expressed in human-equivalent terms) near the lowest observations.

48. **Prevalence** – The total number of cases of a disease in a population at a specific point in time.

49. **Prospective study** – A cohort study where the information about exposure and response are obtained after the study has begun.

50. **Reference Concentration** (RfC) – An estimate of the continuous inhalation exposure for humans (including sensitive sub-groups) that can be without appreciable risk of non-cancer health during the lifetime of the exposed.

51. **Reference Dose** (RfD) – An estimate of the daily dose for humans that is likely to be without risk of adverse effect during a lifetime. The RfD is derived from the NOAEL by applying uncertainty factor that reflect the uncertainty used to estimated the RfD, and a modifying factor, based on the professional judgment on the entire database for the agent. In the US, it is not applied to cancer.

52. **Retrospective Study** - The cohort or other epidemiological study in which those exposed in the past are studied and the exposure is reconstructed from the beginning of the study to the discovery of the adverse effect.

53. **Risk** - The probability that an adverse effect will be the result of a particular exposure.

54. **Risk Factor** – A variable in a causal model that is related to adverse response. That variable (or factor) may act with other factors additively or in some other way (e.g., multiplicatively).

55. **Risk-Ratio** – The ratio formed by the responses in which those with the risk factor are related to those without that risk facto. If the ratio is greater than 1.00, the risk factor increased the risk to those exposed to it.

56. **Short-Term Exposure Limit** (STEL) -- The American Conference of Industrial Hygienists (ACGIH) maximum concentration for workers consists of exposure duration of no more than 15 minutes continuously. No more than 4 exceedances per day can occur and there must be at least 60 minutes without exceedances. The daily threshold limit value, time weighted average (TLV-TW) cannot be exceeded.

57. **Target Organ Toxicity** – The organ (e.g., lung) specific adverse effect that is toxic.

58. **Teratogen** – A chemical that causes a structural defect that changes the morphology of the organism.

59. **Threshold Limit Value** (TLV) – An American Conference of Industrial Hygienists concentration of a substance to which most workers can be exposed without adverse effect. The TLV can be expressed as a Time weighted Average (TWA) or as a short-term exposure limit (STEL).

60. **Time Weighted Average** (TWA) – An allowable exposure averaged over a normal 8-hours workday for a 40-hours work limit.

61. **Tolerable Daily Intake (TDI)** – The amount of a toxic chemical that can be taken daily for a duration that equals the expected lifetime of the individual at risk, without significantly adding to the background risk.

62. **Toxic Dose** (TD_{50}) – The estimated dose of a chemical which is expected to cause a specific toxic effect 50% in a defined population.

63. **Toxicokinetics** – A study of the adsorption, distribution and elimination of toxic compounds in the living organism.

64. **Uncertainty Factor** (UF) – A whole number used in calculating the Minimal Risk Level (MRL), the Reference Dose (RfD) or the Reference Concentration (RfC) from toxicological data, but not epidemiological results where standard statistical measures are used instead. The UF accounts for variation in the sensitivity of humans, uncertainty in the conversions from animal to humans, extrapolations or inference from less-than lifetime studies, and uncertainty in using the LOAEL value rather than the NOAEL. The default number for each UF is 10. For certainty UF = 1. A UF = 3 can be used on a case-by-case basis; 3 is the (approximate) logarithmic average of 1 and 10. Each area that produces uncertainty is given a UF number, which is multiplied to obtain the total UF value.

LIST OF SYMBOLS

Symbol	Definition
$\pm$	Plus or minus
$\mid$	Conditioning; conditioned on
$*$	Multiplication
$\sim$	Is approximately equal to
$\equiv$	Identically equal to
$\neq$	Not equal
$\leq$	Less than or equal to
$\geq$	Greater than or equal to
Δ	Discrete change, an increment or a decrement
$\partial(.)/\partial x$	Partial derivate of the quantity in parentheses with respect to x
∞	Infinity
$\int f(x)dx$	Indefinite integral
$\sqrt{}$	Square root
$\prod$	Multiplication
$\sum$	Summation
$>>$	Much greater than
$\#$	Number, count
$\lvert . \rvert$	Absolute value
$e(.)$	Exponential function; $e = 2.71828\ldots$
$\exp(.)$	Exponential function; $\exp = 2.71828\ldots$
s.d., sd	Standard deviation of the sample
$Var(.)$	Variance for sample or population
μ	Population mean
σ^2	Population variance
λ	Rate
$E(.)$	Expected value
$<.>$	Expected value
$\log_2(.)$	Logarithm of (.) to the base 2
x^n	x raised to the nth power
$\sqrt[n]{x}$	nth root of x
pr	Probability
$F(.)$	Cumulative distribution function
$f(.)$	Density function
$H(.)$	Cumulative hazard function
$h(.)$	Hazard function
OR, O.R.	Odds ratio
RR, R.R.	Relative risk
$\int_a^b f(x)dx$	Definite integral
$\otimes$	Convolution
$pr(Y\mid X)$	Probability of event Y, given X; conditional probability
$pr(Y\ AND\ X)$	The joint probability of event X and event Y

Symbol	Definition
pr(Y OR X)	The probability of either event X or event Y or both
nCr	The number of combinations of n objects , r taken at the time
d(.)/dx	The first derivative of the quantity between parentheses, with respect to x
ln$x$	Logarithm to the base e of the number x
n!	Factorial of the number n
$a \in A$	A is an element of the set A
f^{-1}(.)	Inverse of f(.)
log$x$	Logarithm to the base 10 of the number x
X → Y	X is an input to Y
$\forall$	For all ...
i.i.d.	Independent and identically distributed
dx	Infinitesimal change
$\cup$	Union
L, M, T	Length, Mass and Time
X~N(.); U~T(.), Z~LN(.) ...	The random variable is normally distributed, the random variable U has a triangular distribution, the random variable Z is log-normally distributed ...
$\bar{x}, \bar{y}$...	The sample means for observations taken on X, Y and so on
$\propto$	Proportional to

This List of Symbols defines the symbols used in this textbook. Some of the symbols described in this List of Symbols may sometime have more than one use. For example, f(x) can mean the value of the function, f, for the value x. However, f(X) can also symbolized the density or mass function of the random variable X.

INDEX

A

Acceptable daily intake (ADI), 203
Actions (acts), 422, 423
Adequate,
-margin of safety, 7, 8, 51
Additional risk, 184
Agency for Toxic Substances and Disease Registry,
ATSDR, 221
Affinity,
- binding, 215
Agreement,
- test, 324, 329
Air Pollution and Health, European Approach
(APHEA), 123, 124
Akaike information criterion (AIC), 159, 162, 163,
195
Allais paradox, 434
Allometric,
- formulae, 205
Ambiguity,
- decision-making, 419, 424
- risk, 419, 421
American Petroleum Institute, 10
American Society for Testing and Materials (ASTM),
369, 371
- risk based corrective action (RBCA), 371
- risk based screening level (RBSL), 372
Ample margin of safety, 9, 12
Amplitude, 159, 160
Analysis of variance (ANOVA), 301, 302, 304
AND (logical), 58, 400, 401, 402
Arbitrary and capricious, 27, 29
Area under the curve (AUC), 215, 218
Arsenic, 22, 215
Asbestos, 9
Association, causal 341
Auto-Regressive, Integrated, Autocorrelated, ARIMA,
Process, 114, 155, 158
Attribute, (see utility)

B

Banking,
- emissions, 105
Bartlett test, 300
Bayesian
- analysis, 5, 40, 58, 64
- expected loss, 423
- network, 166, 264, 267, 268

- factors, 35
- theorem, 32, 59, 442
Best available control
technology, 105
Biological Effects of Ionizing
Radiations (BEIR), 190, 200,
201
Beliefs, 31, 33, 61, 62
Benchmark,
- concentration, 192,
- dose-response model, 192
193, 196, 197, 211
Benzene, 8, 12, 21, 185, 186
Birth-death process,
- multistage, 188
- MVK, 189
Blanket,
- Markov, 363
Bootstrap, 240, 241
Bubble, 105
Burden of proof, 7, 24

C

Calibration, 47
California preliminary
exposure assessment (PEA),
372
Carcinogens, 176
Causal path, 259
Causation,
- deterministic, 259
- fundamental diagram, FCD,
351, 353, 354
- legal, 26
- probabilistic, 268, 270, 339
- scientific, 26, 41, 275, 366
CERCLA, 9
Chi-square, 147
-test, 193, 289, 316, 317, 330,
341, 346
- distribution, 145, 316
Classification and Regression
Tree, CART, 166, 270, 271,
272
Clearance, 215, 407, 408
Clean Air Act, 8, 9, 10, 11,
28, 49, 79, 80, 105, 106
Clean Water Act, 8
Cochran's Q test, 158, 307
Compartment,
- model, 409, 410, 412
- reactions, 410, 411
Concordance,

ENVIRONMENTAL POLLUTION

1. J. Fenger, O. Hertel and F. Palmgren (eds.): *Urban Air Pollution – European Aspects.* 1998   ISBN 0-7923-5502-4
2. D. Cormack: *Response to Marine Oil Pollution – Review and Assessment.* 1999   ISBN 0-7923-5674-8
3. S.J. Langan (ed.): *The Impact of Nitrogen Deposition on Natural and Semi-Natural Ecosystems.* 1999   ISBN 0-412-81040-9
4. C. Kennes and M.C. Veiga (eds.): *Bioreactors for Waste Gas Treatment.* 2001   ISBN 0-7923-7190-9
5. P.L. Younger, S.A. Banwart and R.S. Hedin: *Mine Water.* Hydrology, Pollution, Remediation. 2002   ISBN 1-4020-0137-1; Pb 1-4020-0138-X
6. K. Asante-Duah: *Public Health Risk Assessment for Human Exposure to Chemicals.* 2002   ISBN 1-4020-0920-8; Pb 1-4020-0921-6
7. R. Tykva and D. Berg (eds.): *Man-Made and Natural Radioactivity in Environmental Pollution and Radiochronology.* 2004   ISBN 1-4020-1860-6
8. L. Landner and R. Reuther (eds.): *Metals in Society and in the Environment.* A Critical Review of Current Knowledge on Fluxes, Speciation, Bioavailability and Risk for Adverse Effects of Copper, Chromium, Nickel and Zinc. 2004
                                                   ISBN 1-4020-2740-0; Pb 1-4020-2741-9
9. P.F. Ricci: *Environmental and Health Risk Assessment and Management.* Principles and Practices. 2006   ISBN 1-4020-3775-9